PREFACE

This book had its nucleus in some lectures given by one of us (J.O'M.B.) in a course on electrochemistry to students of energy conversion at the University of Pennsylvania. It was there that he met a number of people trained in chemistry, physics, biology, metallurgy, and materials science, all of whom wanted to know something about electrochemistry. The concept of writing a book about electrochemistry which could be understood by people with very varied backgrounds was thereby engendered. The lectures were recorded and written up by Dr. Klaus Muller as a 293-page manuscript. At a later stage, A.K.N.R. joined the effort; it was decided to make a fresh start and to write a much more comprehensive text.

Of methods for direct energy conversion, the electrochemical one is the most advanced and seems the most likely to become of considerable practical importance. Thus, conversion to electrochemically powered transportation systems appears to be an important step by means of which the difficulties of air pollution and the effects of an increasing concentration in the atmosphere of carbon dioxide may be met. Corrosion is recognized as having an electrochemical basis. The synthesis of nylon now contains an important electrochemical stage. Some central biological mechanisms have been shown to take place by means of electrochemical reactions. A number of American organizations have recently recommended greatly increased activity in training and research in electrochemistry at universities in the United States. Three new international journals of fundamental electrochemical research were established between 1955 and 1965.

In contrast to this, physical chemists in U.S. universities seem—perhaps partly because of the absence of a modern textbook in English—out of touch with the revolution in fundamental interfacial electrochemistry which has occurred since 1950. The fragments of electrochemistry which are taught in many U.S. universities belong not to the space age of electrochemically powered vehicles, but to the age of thermodynamics and the horseless carriage; they often consist of Nernst's theory of galvanic cells (1891) together with the theory of Debye and Hückel (1923).

Electrochemistry at present needs several kinds of books. For example, it needs a textbook in which the whole field is discussed at a strong theoretical level. The most pressing need, however, is for a book which outlines the field at a level which can be understood by people entering it from different disciplines who have no previous background in the field but who wish to use modern electrochemical concepts and ideas as a basis for their own work. It is this need which the authors have tried to meet.

The book's aims determine its priorities. In order, these are:

1. Lucidity. The authors have found students who understand advanced courses in quantum mechanics but find difficulty in comprehending a field at whose center lies the quantum mechanics of electron transitions across interfaces. The difficulty is associated, perhaps, with the interdisciplinary character of the material: a background knowledge of physical chemistry is not enough. Material has therefore sometimes been presented in several ways and occasionally the same explanations are repeated in different parts of the book. The language has been made informal and highly explanatory. It retains, sometimes, the lecture style. In this respect, the authors have been influenced by *The Feynmann Lectures on Physics*.

2. Honesty. The authors have suffered much themselves from books in which proofs and presentations are not complete. An attempt has been made to include most of the necessary material. Appendices have been often used for the presentation of mathematical derivations which would obtrude too much in the text.

3. Modernity. There developed during the 1950's a great change in emphasis in electrochemistry away from a subject which dealt largely with solutions to one in which the treatment at a molecular level of charge transfer across interfaces dominates. This is the "new electrochemistry," the essentials of which, at an elementary level, the authors have tried to present.

4. Sharp variation is standard. The objective of the authors has been to begin each chapter at a very simple level and to increase the level to one which allows a connecting up to the standard of the specialized monograph. The standard at which subjects are presented has been intentionally

---ANNOUNCING---

PLENUM/ROSETTA EDITIONS

Plenum, one of the world's leading scientific and technical publishers, is pleased to announce its new paperback line—*plenum/rosetta editions*. Those advanced-level, scientific works in chemistry, biology, physics, etc., written or edited by leading authorities in each field which are most suitable for school use will be published under the new imprint.

VOLUME 1
MODERN ELECTROCHEMISTRY

An Introduction to an Interdisciplinary Area

John O'M. Bockris
Professor of Electrochemistry
University of Pennsylvania, Philadelphia, Pennsylvania

and
Amulya K. N. Reddy
Professor of Electrochemistry
Indian Institute of Science, Bangalore, India

A Plenum/Rosetta Edition

Library of Congress Cataloging in Publication Data

Bockris, John O'M.
   Modern electrochemistry.

   "A Plenum/Rosetta edition."
   1. Electrochemistry. I. Reddy, Amulya K. N., joint author. II. Title.
QD553.B63   1973   541'.37   73-13712
ISBN 0-306-25001-2 (v. 1)

A Plenum/Rosetta Edition
Published by Plenum Publishing Corporation
227 West 17th Street, New York, N.Y. 10011

First paperback printing 1973

© 1970 Plenum Press, New York
A Division of Plenum Publishing Corporation

United Kingdom edition published by Plenum Press, London
A Division of Plenum Publishing Company, Ltd.
Davis House (4th Floor), 8 Scrubs Lane, Harlesden, London, NW10 6SE, England

All rights reserved

No part of this publication may be reproduced in any form
without written permission from the publisher

Printed in the United States of America

variable, depending particularly on the degree to which knowledge of the material appears to be widespread.

5. One theory per phenomenon. The authors intend a *teaching book*, which acts as an introduction to graduate studies. They have tried to present, with due admission of the existing imperfections, a simple version of that model which seemed to them at the time of writing to reproduce the facts most consistently. They have for the most part refrained from presenting the detailed pros and cons of competing models in areas in which the theory is still quite mobile.

In respect to references and further reading: no detailed references to the literature have been presented, in view of the elementary character of the book's contents, and the corresponding fact that it is an introductory book, largely for beginners. In the "further reading" lists, the policy is to cite papers which are classics in the development of the subject, together with papers of particular interest concerning recent developments, and in particular, reviews of the last few years.

It is hoped that this book will not only be useful to those who wish to work with modern electrochemical ideas in chemistry, physics, biology, materials science, etc., but also to those who wish to begin research on electron transfer at interfaces and associated topics.

The book was written mainly at the Electrochemistry Laboratory in the University of Pennsylvania, and partly at the Indian Institute of Science in Bangalore. Students in the Electrochemistry Laboratory at the University of Pennsylvania were kind enough to give guidance frequently on how they reacted to the clarity of sections written in various experimental styles and approaches. For the last four years, the evolving versions of sections of the book have been used as a partial basis for undergraduate, and some graduate, lectures in electrochemistry in the Chemistry Department of the University.

The authors' acknowledgment and thanks must go first to Mr. Ernst Cohn of the National Aeronautics and Space Administration. Without his frequent stimulation, including very frank expressions of criticism, the book might well never have emerged from the Electrochemistry Laboratory.

Thereafter, thanks must go to Professor B. E. Conway, University of Ottawa, who gave several weeks of this time to making a detailed review of the material. Plentiful help in editing chapters and effecting revisions designed by the authors was given by the following: Chapters IV and V, Dr. H. Wroblowa (Pennsylvania); Chapter VI, Dr. C. Solomons (Pennsylvania) and Dr. T. Emi (Hokkaido); Chapter VII, Dr. E. Gileadi (Tel-

Aviv); Chapters VIII and IX, Prof. A. Despic (Belgrade), Dr. H. Wroblowa, and Mr. J. Diggle (Pennsylvania); Chapter X, Mr. J. Diggle; Chapter XI, Dr. D. Cipris (Pennsylvania). Dr. H. Wroblowa has to be particularly thanked for essential contributions to the composition of the Appendix on the measurement of Volta potential differences.

Constructive reactions to the text were given by Messers. G. Razumney, B. Rubin, and G. Stoner of the Electrochemistry Laboratory. Advice was often sought and accepted from Dr. B. Chandrasekaran (Pennsylvania), Dr. S. Srinivasan (New York), and Mr. R. Rangarajan (Bangalore).

Comments on late drafts of chapters were made by a number of the authors' colleagues, particularly Dr. W. McCoy (Office of Saline Water), Chapter II; Prof. R. M. Fuoss (Yale), Chapter III; Prof. R. Stokes (Armidale), Chapter IV; Dr. R. Parsons (Bristol), Chapter VII; Prof. A. N. Frumkin (Moscow), Chapter VIII; Dr. H. Wroblowa, Chapter X; Prof. R. Staehle (Ohio State), Chapter XI. One of the authors (A.K.N.R.) wishes to acknowledge his gratitude to the authorities of the Council of Scientific and Industrial Research, India, and the Indian Institute of Science, Bangalore, India, for various facilities, not the least of which were extended leaves of absence. He wishes also to thank his wife and children for sacrificing many precious hours which rightfully belonged to them.

CONTENTS

VOLUME 1

CHAPTER 1
Electrochemistry

1.1	Introduction	1
1.2	Electrons at and across Interfaces	3
1.2.1	Many Properties of Materials Depend upon Events Occurring at Their Surfaces	3
1.2.2	Almost All Interfaces Are Electrified	3
1.2.3	The Continuous Flow of Electrons across an Interface: Electrochemical Reactions	7
1.2.4	Electrochemical and Chemical Reactions	8
1.3	Basic Electrochemistry	12
1.3.1	Electrochemistry before 1950	12
1.3.2	The Treatment of Interfacial Electron Transfer as a Rate Process: The 1950's	17
1.3.3	Quantum Electrochemistry: The 1960's	19
1.3.4	Ions in Solution, as well as Electron Transfer across Interfaces	22
1.4	The Relation of Electrochemistry to Other Sciences	26
1.4.1	Some Diagrammatic Presentations	26
1.4.2	Some Examples of the Involvement of Electrochemistry in Other Sciences	28
1.4.3	Electrochemistry as an Interdisciplinary Field, Apart from Chemistry?	29
1.5	Electrodics and Electronics	31
1.6	Transients	32

1.7	Electrodes are Catalysts	34
1.8	The Electromagnetic Theory of Light and the Examination of Electrode Surfaces	35
1.9	Science, Technology, Electrochemistry, and Time	38
1.9.1	Do Interfacial Charge-Transfer Reactions Have a Wider Significance Than Has Hitherto Been Realized?	38
1.9.2	The Relation between Three Major Advances in Science, and the Place of Electrochemistry in the Developing World	39

CHAPTER 2

Ion–Solvent Interactions

2.1	Introduction	45
2.2	The Nonstructural Treatment of Ion–Solvent Interactions	48
2.2.1	A Quantitative Measure of Ion–Solvent Interactions	48
2.2.2	The Born Model: A Charged Sphere in a Continuum	49
2.2.3	The Electrostatic Potential at the Surface of a Charged Sphere	52
2.2.4	On the Electrostatics of Charging (or Discharging) Spheres	54
2.2.5	The Born Expression for the Free Energy of Ion–Solvent Interactions	56
2.2.6	The Enthalpy and Entropy of Ion–Solvent Interactions	59
2.2.7	Can One Experimentally Study the Interactions of a Single Ionic Species with the Solvent?	61
2.2.8	The Experimental Evaluation of the Heat of Interaction of a Salt and Solvent	64
2.2.9	How Good Is the Born Theory?	68
Further Reading		72
2.3	Structural Treatment of the Ion–Solvent Interactions	72
2.3.1	The Structure of the Most Common Solvent, Water	72
2.3.2	The Structure of Water near an Ion	76
2.3.3	The Ion–Dipole Model of Ion–Solvent Interactions	80
2.3.4	Evaluation of the Terms in the Ion–Dipole Approach to the Heat of Solvation	88
2.3.5	How Good Is the Ion–Dipole Theory of Solvation?	93
2.3.6	The Relative Heats of Solvation of Ions on the Hydrogen Scale	95
2.3.7	Do Oppositely Charged Ions of Equal Radii Have Equal Heats of Solvation?	96
2.3.8	The Water Molecule Can Be Viewed as an Electrical Quadrupole	98
2.3.9	The Ion–Quadrupole Model of Ion–Solvent Interactions	99
2.3.10	Ion-Induced-Dipole Interactions in the Primary Solvation Sheath	102
2.3.11	How Good Is the Ion–Quadrupole Theory of Solvation?	103
2.3.12	The Special Case of Interactions of the Transition-Metal Ions with Water	108
2.3.13	Some Summarizing Remarks on the Energetics of Ion–Solvent Interactions	113
Further Reading		116
2.4	The Solvation Number	117
2.4.1	How Many Water Molecules Are Involved in the Solvation of an Ion?	117

2.4.2	Static and Dynamic Pictures of the Ion–Solvent Molecule Interaction	120
2.4.3	The Meaning of Hydration Numbers	123
2.4.4	Why Is the Concept of Solvation Numbers Useful?	124
2.4.5	On the Determination of Solvation Numbers	125
Further Reading		132
2.5	The Dielectric Constant of Water and Ionic Solutions	132
2.5.1	An Externally Applied Electric Field Is Opposed by Counterfields Developed within the Medium	132
2.5.2	The Relation between the Dielectric Constant and Internal Counterfields	136
2.5.3	The Average Dipole Moment of a Gas-Phase Dipole Subject to Electrical and Thermal Forces	139
2.5.4	The Debye Equation for the Dielectric Constant of a Gas of Dipoles	142
2.5.5	How the Short-Range Interactions between Dipoles Affect the Average Effective Moment of the Polar Entity Which Responds to an External Field	145
2.5.6	The Local Electric Field in a Condensed Polar Dielectric	147
2.5.7	The Dielectric Constant of Liquids Containing Associated Dipoles	152
2.5.8	The Influence of Ionic Solvation on the Dielectric Constant of Solutions	155
Further Reading		158
2.6	Ion–Solvent–Nonelectrolyte Interactions	158
2.6.1	The Problem	158
2.6.2	The Change in Solubility of a Nonelectrolyte Due to Primary Solvation	159
2.6.3	The Change in Solubility Due to Secondary Solvation	160
2.6.4	The Net Effect on Solubility of Influences from Primary and Secondary Solvation	163
2.6.5	The Case of Anomalous Salting in	164
Further Reading		168
Appendix 2.1	Free Energy Change and Work	168
Appendix 2.2	The Interaction between an Ion and a Dipole	169
Appendix 2.3	The Interaction between an Ion and a Water Quadrupole	171

CHAPTER 3

Ion–Ion Interactions

3.1	Introduction	175
3.2	True and Potential Electrolytes	176
3.2.1	Ionic Crystals Are True Electrolytes	176
3.2.2	Potential Electrolytes: Nonionic Substances Which React with the Solvent to Yield Ions	176
3.2.3	An Obsolete Classification: Strong and Weak Electrolytes	177
3.2.4	The Nature of the Electrolyte and the Relevance of Ion–Ion Interactions	180
Further Reading		180
3.3	The Debye–Hückel (or Ion-Cloud) Theory of Ion–Ion Interactions	180
3.3.1	A Strategy for a Quantitative Understanding of Ion–Ion Interactions	180

3.3.2	A Prelude to the Ionic-Cloud Theory	183
3.3.3	How the Charge Density near the Central Ion Is Determined by Electrostatics: Poisson's Equation	186
3.3.4	How the Excess Charge Density near the Central Ion Is Given by a Classical Law for the Distribution of Point Charges in a Coulombic Field	187
3.3.5	A Vital Step in the Debye–Hückel Theory of the Charge Distribution around Ions: Linearization of the Boltzmann Equation	189
3.3.6	The Linearized Poisson–Boltzmann Equation	190
3.3.7	The Solution of the Linearized P–B Equation	191
3.3.8	The Ionic Cloud around a Central Ion	193
3.3.9	How Much Does the Ionic Cloud Contribute to the Electrostatic Potential ψ_r at a Distance r from the Central Ion?	199
3.3.10	The Ionic Cloud and the Chemical-Potential Change Arising from Ion–Ion Interactions	201
Further Reading		202
3.4	**Activity Coefficients and Ion–Ion Interactions**	**202**
3.4.1	The Evolution of the Concept of Activity Coefficient	202
3.4.2	The Physical Significance of Activity Coefficients	204
3.4.3	The Activity Coefficient of a Single Ionic Species Cannot Be Measured	206
3.4.4	The Mean Ionic Activity Coefficient	207
3.4.5	The Conversion of Theoretical Activity-Coefficient Expressions into a Testable Form	209
Further Reading		212
3.5	**The Triumphs and Limitations of the Debye–Hückel Theory of Activity Coefficients**	**212**
3.5.1	How Well Does the Debye–Hückel Theoretical Expression for Activity Coefficients Predict Experimental Values?	212
3.5.2	Ions Are of Finite Size, Not Point Charges	219
3.5.3	The Theoretical Mean Ionic-Activity Coefficient in the Case of Ionic Clouds with Finite-Sized Ions	222
3.5.4	The Ion-Size Parameter a	224
3.5.5	Comparison of the Finite-Ion-Size Model with Experiment	227
3.5.6	The Debye–Hückel Theory of Ionic Solutions: An Assessment	230
3.5.7	On the Parentage of the Theory of Ion–Ion Interactions	237
Further Reading		238
3.6	**Ion–Solvent Interactions and the Activity Coefficient**	**238**
3.6.1	The Effect of Water Bound to Ions on the Theory of Deviations from Ideality	238
3.6.2	Quantitative Theory of the Activity of an Electrolyte as a Function of the Hydration Number	240
3.6.3	The Water-Removal Theory of Activity Coefficients and Its Apparent Consistency with Experiment at High Electrolytic Concentrations	243
Further Reading		246
3.7	**The So-Called "Rigorous" Solutions of the Poisson–Boltzmann Equation**	**246**
Further Reading		250
3.8	**Temporary Ion Association in an Electrolytic Solution: Formation of Pairs, Triplets, etc.**	**251**

3.8.1	Positive and Negative Ions Can Stick Together: Ion-Pair Formation	251
3.8.2	The Probability of Finding Oppositely Charged Ions near Each Other	251
3.8.3	The Fraction of Ion Pairs, According to Bjerrum	253
3.8.4	The Ion-Association Constant K_A of Bjerrum	257
3.8.5	Activity Coefficients, Bjerrum's Ion Pairs, and Debye's Free Ions	260
3.8.6	The Fuoss Approach to Ion-Pair Formation	261
3.8.7	From Ion Pairs to Triple Ions to Clusters of Ions	265
Further Reading		266
3.9	The Quasi-Lattice Approach to Concentrated Electrolytic Solutions	267
3.9.1	At What Concentration Does the Ionic-Cloud Model Break Down?	267
3.9.2	The Case for a Cube-Root Law for the Dependence of the Activity Coefficient on Electrolyte Concentration	269
3.9.3	The Beginnings of a Quasi-Lattice Theory for Concentrated Electrolytic Solutions	271
Further Reading		272
3.10	The Study of the Constitution of Electrolytic Solutions	273
3.10.1	The Temporary and Permanent Association of Ions	273
3.10.2	Electromagnetic Radiation, a Tool for the Study of Electrolytic Solutions	274
3.10.3	Visible and Ultraviolet Absorption Spectroscopy	275
3.10.4	Raman Spectroscopy	276
3.10.5	Infrared Spectroscopy	278
3.10.6	Nuclear Magnetic Resonance Spectroscopy	278
Further Reading		279
3.11	A Perspective View on the Theory of Ion–Ion Interactions	279
Appendix 3.1	Poisson's Equation for Spherically Symmetrical Charge Distribution	282
Appendix 3.2	Evaluation of the Integral $\int_{r=0}^{r\to\infty} e^{-(\varkappa r)}(\varkappa r)\, d(\varkappa r)$	283
Appendix 3.3	Derivation of the Result $f_+ = (f_+^{\nu_+} + f_-^{\nu_-})^{1/\nu}$	284
Appendix 3.4	To Show That the Minimum in the P_r versus r Curve Occurs at $r = \lambda/2$	284
Appendix 3.5	Transformation from the Variable r to the Variable $y = \lambda/r$	285
Appendix 3.6	Relation Between Calculated and Observed Activity Coefficients	285

CHAPTER 4

Ion Transport in Solutions

4.1	Introduction	287
4.2	Ionic Drift under a Chemical-Potential Gradient: Diffusion	289
4.2.1	The Driving Force for Diffusion	291
4.2.2	The "Deduction" of an Empirical Law: Fick's First Law of Steady-State Diffusion	293

4.2.3	On the Diffusion Coefficient D	296
4.2.4	Ionic Movements: A Case of the Random Walk	299
4.2.5	The Mean Square Distance Traveled in a Time t by a Random-Walking Particle	301
4.2.6	Random-Walking Ions and Diffusion: The Einstein–Smoluchowski Equation	304
4.2.7	The Gross View of Non-Steady-State Diffusion	307
4.2.8	An Often Used Device for Solving Electrochemical Diffusion Problems: The Laplace Transformation	309
4.2.9	Laplace Transformation Converts the Partial Differential Equation Which Is Fick's Second Law into a Total Differential Equation	312
4.2.10	The Initial and Boundary Conditions for the Diffusion Process Stimulated by a Constant Current (or Flux)	313
4.2.11	The Concentration Response to a Constant Flux Switched on at $t=0$	317
4.2.12	How the Solution of the Constant-Flux Diffusion Problem Leads On to the Solution of Other Problems	323
4.2.13	Diffusion Resulting from an Instantaneous Current Pulse	328
4.2.14	What Fraction of Ions Travels the Mean Square Distance $\langle x^2 \rangle$ in the Einstein–Smoluchowski Equation?	332
4.2.15	How Can the Diffusion Coefficient Be Related to Molecular Quantities?	338
4.2.16	The Mean Jump Distance l, a Structural Question	339
4.2.17	The Jump Frequency, a Rate-Process Question	340
4.2.18	The Rate-Process Expression for the Diffusion Coefficient	342
4.2.19	Diffusion: An Overall View	342
Further Reading		345
4.3	**Ionic Drift under an Electric Field: Conduction**	**345**
4.3.1	The Creation of an Electric Field in an Electrolyte	345
4.3.2	How Do Ions Respond to the Electric Field?	349
4.3.3	The Tendency for a Conflict between Electroneutrality and Conduction	351
4.3.4	The Resolution of the Electroneutrality-versus-Conduction Dilemma: Electron-Transfer Reactions	351
4.3.5	The Quantitative Link between Electron Flow in the Electrodes and Ion Flow in the Electrolyte: Faraday's Law	353
4.3.6	The Proportionality Constant Relating the Electric Field and the Current Density: The Specific Conductivity	354
4.3.7	Molar Conductivity and Equivalent Conductivity	357
4.3.8	The Equivalent Conductivity Varies with Concentration	360
4.3.9	How the Equivalent Conductivity Changes with Concentration: Kohlrausch's Law	363
4.3.10	The Vectorial Character of Current: Kohlrausch's Law of the Independent Migration of Ions	364
Further Reading		367
4.4	**The Simple Atomistic Picture of Ionic Migration**	**367**
4.4.1	Ionic Movements under the Influence of an Applied Electric Field	367
4.4.2	What Is the Average Value of the Drift Velocity?	368
4.4.3	The Mobility of Ions	369
4.4.4	The Current Density Associated with the Directed Movement of Ions in Solution, in Terms of the Ionic Drift Velocities	371
4.4.5	The Specific and Equivalent Conductivities in Terms of the Ionic Mobilities	373
4.4.6	The Einstein Relation between the Absolute Mobility and the Diffusion Coefficient	374

4.4.7	What Is the Drag (or Viscous) Force Acting on an Ion in Solution?	377
4.4.8	The Stokes–Einstein Relation	379
4.4.9	The Nernst–Einstein Equation	381
4.4.10	Some Limitations of the Nernst–Einstein Relation	382
4.4.11	A Very Approximate Relation between Equivalent Conductivity and Viscosity: Walden's Rule	385
4.4.12	The Rate-Process Approach to Ionic Migration	387
4.4.13	The Rate-Process Expression for Equivalent Conductivity	391
4.4.14	The Total Driving Force for Ionic Transport: The Gradient of the Electrochemical Potential	394
Further Reading		399
4.5	**The Interdependence of Ionic Drifts**	**399**
4.5.1	The Drift of One Ionic Species May Influence the Drift of Another	399
4.5.2	A Consequence of the Unequal Mobilities of Cations and Anions, the Transport Numbers	400
4.5.3	The Significance of a Transport Number of Zero	402
4.5.4	The Diffusion Potential, Another Consequence of the Unequal Mobilities of Ions	406
4.5.5	Electroneutrality Coupling between the Drifts of Different Ionic Species	410
4.5.6	How Does One Represent the Interaction between Ionic Fluxes? The Onsager Phenomenological Equations	411
4.5.7	An Expression for the Diffusion Potential	413
4.5.8	The Integration of the Differential Equation for Diffusion Potentials: The Planck–Henderson Equation	417
Further Reading		420
4.6	**The Influence of Ionic Atmospheres on Ionic Migration**	**420**
4.6.1	The Concentration Dependence of the Mobility of Ions	420
4.6.2	Ionic Clouds Attempt to Catch Up with Moving Ions	422
4.6.3	An Egg-Shaped Ionic Cloud and the "Portable" Field on the Central Ion	423
4.6.4	A Second Braking Effect of the Ionic Cloud on the Central Ion: The Electrophoretic Effect	424
4.6.5	The Net Drift Velocity of an Ion Interacting with Its Atmosphere	425
4.6.6	The Electrophoretic Component of the Drift Velocity	427
4.6.7	The Procedure for Calculating the Relaxation Component of the Drift Velocity	427
4.6.8	How Long Does an Ion Atmosphere Take to Decay?	428
4.6.9	The Quantitative Measure of the Asymmetry of the Ionic Cloud Around a Moving Ion	429
4.6.10	The Magnitude of the Relaxation Force and the Relaxation Component of the Drift Velocity	430
4.6.11	The Net Drift Velocity and Mobility of an Ion Subject to Ion–Ion Interactions	432
4.6.12	The Debye–Hückel–Onsager Equation	434
4.6.13	The Theoretical Predictions of the Debye–Hückel–Onsager Equation versus the Observed Conductance Curves	435
4.6.14	A Theoretical Basis for Some Modifications of the Debye–Hückel–Onsager Equation	438
Further Reading		439
4.7	**Nonaqueous Solutions: A New Frontier in Ionics?**	**440**
4.7.1	Water Is the Most Plentiful Solvent	440

xvi CONTENTS

4.7.2	Water Is Often Not an Ideal Solvent	441
4.7.3	The Debye–Hückel–Onsager Theory for Nonaqueous Solutions	442
4.7.4	The Solvent Effect on the Mobility at Infinite Dilution	443
4.7.5	The Slope of the Λ versus $c^{\frac{1}{2}}$ Curve as a Function of the Solvent	445
4.7.6	The Effect of the Solvent on the Concentration of Free Ions: Ion Association	447
4.7.7	The Effect of Ion Association upon Conductivity	448
4.7.8	Even Triple Ions Can Be Formed in Nonaqueous Solutions	450
4.7.9	Some Conclusions about the Conductance of Nonaqueous Solutions of True Electrolytes	452
Further Reading		452

Appendix 4.1	The Mean Square Distance Traveled by a Random-Walking Particle	453
Appendix 4.2	The Laplace Transform of a Constant	454
Appendix 4.3	A Few Elementary Ideas on the Theory of Rate Processes	455
Appendix 4.4	The Derivation of Equations (4.257) and (4.258)	458
Appendix 4.5	The Derivation of Equation (4.318)	460

CHAPTER 5

Protons in Solution

5.1	The Case of the Nonconforming Ion: The Proton	461
5.2	Proton Solvation	462
5.2.1	What Is the Condition of the Proton in Solution?	462
5.2.2	Proton Affinity	466
5.2.3	The Overall Heat of Hydration of a Proton	467
5.2.4	The Coordination Number of a Proton	468
Further Reading		470

5.3	Proton Transport	470
5.3.1	The Abnormal Mobility of a Proton	470
5.3.2	Protons Conduct by a Chain Mechanism	474
5.3.3	Classical Proton Jumps and Proton Mobility	476
5.3.4	Do Proton Jumps Obey Classical Laws?	478
5.3.5	Quantum-Mechanical Proton Jumps and Proton Mobility	480
5.3.6	Water Reorientation, a Prerequisite for Proton Jumps	481
5.3.7	The Rate of Water Reorientation and Proton Mobility	482
5.3.8	A Picture of Proton Mobility in Aqueous Solutions	484
5.3.9	The Rate-Determining Water-Rotation Model of Proton Mobility and the Other Anomalous Facts	485
5.3.10	Proton Mobility in Ice	486
5.3.11	The Existence of the Hydronium Ion from the Point of View of Proton Mobility	487
5.3.12	Why Is the Mechanism of Proton Mobility So Important?	487
Further Reading		488

5.4	Homogeneous Proton-Transfer Reactions and Potential Electrolytes	488
5.4.1	Acids Produce Hydrogen Ions and Bases Produce Hydroxyl Ions: The Initial View	488
5.4.2	Acids Are Proton Donors, and Bases Are Proton Acceptors: The Brönsted View	489
5.4.3	The Dissolution of Potential Electrolytes and Other Types of Proton-Transfer Reactions	491
5.4.4	An Important Consequence of the Brönsted View: Conjugate Acid–Base Pairs	493
5.4.5	The Absolute Strength of an Acid or a Base	494
5.4.6	The Relative Strengths of Acids and Bases	495
5.4.7	Proton Free-Energy Levels	500
5.4.8	The Primary Effect of the Solvent upon the Relative Strength of an Acid	504
5.4.9	A Secondary (Electrostatic) Effect of the Solvent on the Relative Strength of Acids	507
Further Reading		511

CHAPTER 6

Ionic Liquids

6.1	Introduction	513
6.1.1	The Limiting Case of Zero Solvent: Pure Liquid Electrolytes	513
6.1.2	The Thermal Dismantling of an Ionic Lattice	514
6.1.3	Some Features of Ionic Liquids (Pure Liquid Electrolytes)	515
6.1.4	Liquid Electrolytes Are Ionic Liquids	517
6.1.5	The Fundamental Problems in Pure Liquid Electrolytes	518
Further Reading		522
6.2	Models of Simple Ionic Liquids	522
6.2.1	The Origin of Liquid Electrolyte Models	522
6.2.2	Lattice-Oriented Models	523
	6.2.2a The Experimental Basis for Model Building	523
	6.2.2b The Need to Pour Empty Space into a Fused Salt	523
	6.2.2c The Vacancy Model: A Fused Salt Is an Ionic Lattice with Numerous Vacancies	526
	6.2.2d The Hole Model: A Fused Salt Is Full of Holes like Swiss Cheese	527
6.2.3	Gas-Oriented Models for Liquid Electrolytes	529
	6.2.3a The Cell-Theory Approach	529
	6.2.3b The Free Volume Belongs to the Liquid and Not to the Particles: The Liquid Free-Volume Model	530
6.2.4	A Summary of the Models for Liquid Electrolytes	532
Further Reading		533
6.3	Quantification of the Hole Model for Liquid Electrolytes	533
6.3.1	An Expression for the Probability That a Hole Has a Radius between r and $r + dr$	533
6.3.2	The Fürth Approach to the Work of Hole Formation	536
6.3.3	The Distribution Function for the Size of the Holes in a Liquid Electrolyte	537

6.3.4	What Is the Average Size of a Hole?	539
Further Reading		541
6.4	**Transport Phenomena in Liquid Electrolytes**	**541**
6.4.1	Some Simplifying Features of Transport in Fused Salts	541
6.4.2	Diffusion in Fused Salts	542
	6.4.2a Self-Diffusion in Pure Liquid Electrolytes: It May Be Revealed by Introducing Isotopes	542
	6.4.2b Results of Self-Diffusion Experiments	544
6.4.3	The Viscosity of Molten Salts	547
6.4.4	What Is the Validity of the Stokes–Einstein Relation in Ionic Liquids?	550
6.4.5	The Conductivity of Pure Liquid Electrolytes	553
6.4.6	The Nernst–Einstein Relation in Ionic Liquids	555
	6.4.6a The Nernst–Einstein Relation: Its Degree of Applicability	555
	6.4.6b The Gross View of Deviations from the Nernst–Einstein Equation	557
	6.4.6c Possible Molecular Mechanisms for Nernst–Einstein Deviations	560
6.4.7	Transport Numbers in Pure Liquid Electrolytes	564
	6.4.7a Some Ideas about Transport Numbers in Fused Salts	564
	6.4.7b The Measurement of Transport Numbers in Liquid Electrolytes	566
	6.4.7c A Radiotracer Method of Calculating Transport Numbers in Molten Salts	571
	6.4.7d A Stokes' Law Approach to a Rough Estimate of Transport Numbers	572
Further Reading		573
6.5	**The Atomistic View of Transport Processes in Simple Ionic Liquids**	**574**
6.5.1	Holes and Transport Processes	574
6.5.2	What Is the Mean Lifetime of Holes in Fused Salts?	576
6.5.3	Expression for Viscosity in Terms of Holes	577
6.5.4	The Diffusion Coefficient from the Hole Model	577
6.5.5	A Critical Test of a Model for Ionic Liquids Is a Rationalization of the Heat of Activation of $3.7RT_m$ for Transport Processes	580
6.5.6	An Attempt to Rationalize $E_D = E_\eta = 3.7RT_m$	581
6.5.7	The Hole Model, the Most Consistent Present Model for Liquid Electrolytes	584
Further Reading		587
6.6	**Mixture of Simple Ionic Liquids—Complex Formation**	**587**
6.6.1	Mixtures of Simple Ionic Liquids May Not Behave Ideally	587
6.6.2	Interactions Lead to Nonideal Behavior	588
6.6.3	Can One Meaningfully Refer to Complex Ions in Fused Salts?	589
6.6.4	Raman Spectra, and Other Means of Detecting Complex Ions	590
Further Reading		593
6.7	**Mixtures of Liquid Oxide Electrolytes**	**594**
6.7.1	The Liquid Oxides	594
6.7.2	Pure Fused Nonmetallic Oxides Form Network Structures Like Liquid Water	594
6.7.3	Why Does Fused Silica Have a Much Higher Viscosity Than Do Liquid Water and the Fused Salts?	597
6.7.4	The Solvent Properties of Fused Nonmetallic Oxides	601

6.7.5	Ionic Additions to the Liquid-Silica Network: Glasses	603
6.7.6	The Extent of Structure Breaking of Three-Dimensional Network Lattices and Its Dependence on the Concentration of Metal Ions	604
6.7.7	The Molecular and Network Models of Liquid Silicate Structure	606
6.7.8	Liquid Silicates Contain Large Discrete Polyanions	610
6.7.9	The "Iceberg" Model	615
6.7.10	Fused-Oxide Systems in Metallurgy: Slags	616
Further Reading		618
Appendix 6.1	The Effective Mass of a Hole	619
Appendix 6.2	Some Properties of the Gamma Function	620
Appendix 6.3	The Kinetic Theory Expression for the Viscosity of a Fluid	621
Index		xxxiii

CONTENTS

VOLUME 2

CHAPTER 7
The Electrified Interface

7.1	Electrification of an Interface	623
7.1.1	The Electrode–Electrolyte Interface: The Basis of Electrodics	623
7.1.2	New Forces at the Boundary of an Electrolyte	623
7.1.3	The Interphase Region Has New Properties and New Structures	626
7.1.4	An Electrode Is Like a Giant Central Ion	626
7.1.5	The Consequences of Compromise Arrangements: The Electrolyte Side of the Boundary Acquires a Charge	627
7.1.6	Both Sides of the Interface Become Electrified: The So-Called "Electrical Double Layer"	629
7.1.7	Double Layers Are Characteristic of All Phase Boundaries	630
7.1.8	A Look into an Electrified Interface	632
Further Reading		639
7.2	Some Problems in Understanding an Electrified Interface	640
7.2.1	What Knowledge Is Required before an Electrified Interface Can Be Regarded as Understood?	640
7.2.2	Predicting the Interphase Properties from the Bulk Properties of the Phases	641
7.2.3	Why Bother about Electrified Interfaces?	642
7.2.4	The Need to Clarify Some Concepts	643
7.2.5	The Potential Difference across Electrified Interfaces	644
	7.2.5a What Happens when One Tries to Measure the Absolute Potential Difference across a Single Electrode–Electrolyte Interface	644

xxii CONTENTS

	7.2.5b	The Absolute Potential Difference across a Single Electrified Interface Cannot Be Measured	648
	7.2.5c	Can One Measure Changes in the Metal–Solution Potential Difference?	650
	7.2.5d	The Extreme Cases of Ideally Nonpolarizable and Polarizable Interfaces	653
	7.2.5e	The Development of a Scale of Relative Potential Differences	655
	7.2.5f	Can One Meaningfully Analyze an Electrode–Electrolyte Potential Difference?	659
	7.2.5g	A Thought Experiment Involving a Charged Electrode in Vacuum	660
	7.2.5h	The Test Charge Must Avoid Image Interactions with the Charged Electrode	660
	7.2.5i	The Outer Potential ψ of a Material Phase in Vacuum	663
	7.2.5j	What is the Relevance of the Outer Potential to Double-Layer Studies?	665
	7.2.5k	Another Thought Experiment Involving an Uncharged, Dipole-Covered Phase	667
	7.2.5l	The Dipole Potential Difference $^M\Delta^S\chi$ across an Electrode–Electrolyte Interface	670
	7.2.5m	The Sum of the Potential Differences Due to Charges and Dipoles: The Absolute Electrode–Electrolyte (or Galvani) Potential Difference	670
	7.2.5n	The Outer, Surface, and Inner Potential Differences	673
	7.2.5o	An Apparent Contradiction: The Sum of the $\Delta\phi$'s across a System of Interfaces Can and the $\Delta\phi$ across One Interface Cannot Be Measured	674
	7.2.5p	What Deeper Understanding Has Been Hitherto Gained Regarding the Absolute Potential Difference Across an Electrified Interface?	677
7.2.6		The Accumulation and Depletion of Substances at an Interface	679
	7.2.6a	What Would Represent Complete Structural Information Regarding an Electrified Interface?	679
	7.2.6b	The Concept of Surface Excess	680
	7.2.6c	Does Knowledge of the Surface Excess Contribute to Knowledge of the Distribution of Species in the Interphase Region?	683
	7.2.6d	Is the Surface Excess Equivalent to the Amount Adsorbed?	684
	7.2.6e	Is the Surface Excess Measurable?	685
	7.2.6f	The Special Position of Mercury in Double-Layer Studies	687
Further Reading			687
7.3		The Thermodynamics of Electrified Interfaces	688
7.3.1		The Measurement of Interfacial Tension as a Function of the Potential Difference across the Interface	688
7.3.2		Some Basic Facts about Electrocapillary Curves	690
7.3.3		A Digression on the Electrochemical Potential	693
	7.3.3a	Definition of Electrochemical Potential	693
	7.3.3b	Can the Chemical and Electrical Work Be Determined Separately?	695
	7.3.3c	A Criterion of Thermodynamic Equilibrium between Two Phases: Equality of Electrochemical Potentials	696
	7.3.3d	Nonpolarizable Interfaces and Thermodynamic Equilibrium	697
7.3.4		Some Thermodynamic Thoughts on Electrified Interfaces	698

7.3.5	Interfacial Tension Varies with Applied Potential: Determination of the Charge Density on the Electrode	701
7.3.6	Electrode Charge Varies with Applied Potential: Determination of the Electrical Capacitance of the Interface	703
7.3.7	The Potential at Which an Electrode Has a Zero Charge	706
7.3.8	Surface Tension Varies with Solution Composition: Determination of the Surface Excess	707
7.3.9	Reflections on Electrocapillary Thermodynamics	714
7.3.10	Retrospect and Prospect in the Study of Electrified Interfaces	715
Further Reading		717
7.4	**The Structure of Electrified Interfaces**	**718**
7.4.1	The Parallel-Plate Condenser Model: The Helmholtz–Perrin Theory	718
7.4.2	The Double Layer in Trouble: Neither Perfect Parabolas nor Constant Capacities	719
7.4.3	The Ionic Cloud: The Gouy–Chapman Diffuse-Charge Model of the Double Layer	722
7.4.4	Ions under Thermal and Electric Forces near an Electrode	724
7.4.5	A Picture of the Potential Drop in the Diffuse Layer	728
7.4.6	An Experimental Test of the Gouy–Chapman Model: Potential Dependence of the Capacitance, but at What Cost?	732
7.4.7	Some Ions Stuck to the Electrode, Others Scattered in Thermal Disarray: The Stern Model	733
7.4.8	A Consequence of the Stern Picture: Two Potential Drops across an Electrified Interface	734
7.4.9	Another Consequence of the Stern Model: An Electrified Interface Is Equivalent to Two Capacitors in Series	735
7.4.10	The Relative Contributions of the Helmholtz–Perrin and Gouy–Chapman Capacities	737
7.4.11	Some Questions Regarding the Sticking of Ions to the Electrode	738
7.4.12	An Electrode Is Largely Covered with Adsorbed Water Molecules	739
7.4.13	Metal–Water Interactions	740
7.4.14	The Orientation of Water Molecules on Charged Electrodes	741
7.4.15	How Close Can Hydrated Ions Come to a Hydrated Electrode?	741
7.4.16	Is It Only Desolvated Ions which Contact-Adsorb on the Electrode?	742
7.4.17	The Free-Energy Change for Contact Adsorption	742
7.4.18	What Determines the Degree of Contact Adsorption?	743
7.4.19	How Is Contact Adsorption Measured?	745
7.4.20	Contact Adsorption, Specific Adsorption, or Superequivalent Adsorption	748
7.4.21	Contact Adsorption: Its Influence of the Capacity of the Interface	749
7.4.22	Looking Back to Look Forward	752
7.4.23	The Complete Capacity–Potential Curve	753
7.4.24	The Constant-Capacity Region	753
	7.4.24a The So-Called "Double Layer" Is a Double Layer	753
	7.4.24b The Dielectric Constant of the Water between the Metal and the Outer Helmholtz Plane	756
	7.4.24c The Position of the Outer Helmholtz Plane and an Interpretation of the Constant Capacity	757
7.4.25	The Capacitance Hump	761
7.4.26	How Does the Population of Contact-Adsorbed Ions Change with Electrode Charge?	762
7.4.27	The Test of the Population Law for Contact-Adsorbed Ions	769
7.4.28	The Lateral-Repulsion Model for Contact Adsorption	776

7.4.29	Flip-Flop Water on Electrodes	779
7.4.30	Calculation of the Potential Difference Due to Water Dipoles	781
7.4.31	The Excess of Flipped Water Dipoles over Flopped Water Dipoles	782
7.4.32	The Contribution of Adsorbed Water Dipoles to the Capacity of the Interface	788
Further Reading		790
7.5	The Competition between Water and Organic Molecules at the Electrified Interfaces	791
7.5.1	The Relevance of Organic Adsorption	791
7.5.2	The Forces Involved in Organic Adsorption	792
7.5.3	Does Organic Adsorption Depend on Electrode Charge?	793
7.5.4	The Examination of the Water Flip-Flop Model for Simple Cases of Organic Adsorption	797
7.5.5	At What Potential Does Maximum Organic Adsorption Occur?	798
Further Reading		801
7.6	Electrified Interfaces at Metals Other than Mercury	801
Further Reading		803
7.7	The Structure of the Semiconductor–Electrolyte Interface	803
7.7.1	How Is the Charge Distributed inside a Solid Electrode?	803
7.7.2	The Band Theory of Crystalline Solids	804
7.7.3	Conductors, Insulators, and Semiconductors	806
7.7.4	Some Analogies between Semiconductors and Electrolytic Solutions	811
7.7.5	The Diffuse-Charge Region inside an Intrinsic Semiconductor: The Garrett–Brattain Space Charge	813
7.7.6	The Differential Capacity Due to the Space Charge	816
7.7.7	Impurity Semiconductors, n Type and p Type	818
7.7.8	Surface States: The Semiconductor Analogue of Contact Adsorption	821
7.7.9	Semiconductor Electrochemistry: The Beginnings of the Electrochemistry of Nonmetallic Materials	823
Further Reading		823
7.8	A Bird's-Eye View of the Structure of Charged Interfaces	824
7.9	Double Layers between Phases Moving Relative to Each Other	826
7.9.1	The Phenomenology of Mobile Electrified Interfaces: Electrokinetic Properties	826
7.9.2	The Relative Motion of One of the Phases Constituting an Electrified Interface Produces a Streaming Current	829
7.9.3	A Potential Difference Applied Parallel to an Electrified Interface Produces an Electro-osmotic Motion of One of the Phases Relative to the Other	831
7.9.4	Electrophoresis: Moving Solid Particles in a Stationary Electrolyte	832
Further Reading		835
7.10	Colloid Chemistry	835
7.10.1	Colloids: The Thickness of the Double Layer and the Bulk Dimensions Are of the Same Order	835
7.10.2	The Interaction of Double Layers and the Stability of Colloids	836
7.10.3	Sols and Gels	839
Further Reading		841
Appendix 7.1	Measurement of the Electrode–Solution Volta Potential Difference	841

CHAPTER 8
Electrodics

8.1	Introduction	845
8.1.1	The Situation Thus Far	845
8.1.2	Charge Transfer: Its Chemical and Electrical Implications	846
8.1.3	Can an Isolated Electrode–Solution Interface Be Used as a Device?	849
8.1.4	Electrochemical Systems Can Be Used as Devices	851
8.1.5	An Electrochemical Device: The Substance Producer	851
8.1.6	Another Electrochemical Device: The Energy Producer	855
8.1.7	The Electrochemical Undevice: The Substance Destroyer and Energy Waster	859
8.1.8	Some Basic Questions	861
8.2	The Basic Electrodic Equation: The Butler–Volmer Equation	862
8.2.1	The Instant of Immersion of a Metal in an Electrolytic Solution	862
8.2.2	The Rate of Charge-Transfer Reactions under Zero Field: The Chemical Rate Constant	865
8.2.3	Some Consequences of Electron Transfer at an Interface	868
8.2.4	What Is the Rate of an Electron-Transfer Reaction under the Influence of an Electric Field?	869
8.2.5	The Two-Way Electron Traffic across the Interface	873
8.2.6	The Interface at Equilibrium: The Equilibrium Exchange-Current Density i_0	876
8.2.7	The Interface Departs from Equilibrium: The Nonequilibrium Drift-Current Density i	879
8.2.8	The Current-Producing (or Current-Produced) Potential Difference: The Overpotential η	880
8.2.9	The Basic Electrodic (Butler–Volmer) Equation: Some General and Special Cases	883
8.2.10	The High-Field Approximation: The Exponential i versus η Law	888
8.2.11	The Low-Field Approximation: The Linear i versus η Law	892
8.2.12	Nonpolarizable and Polarizable Interfaces	894
8.2.13	Zero Net Current and the Classical Law of Nernst	897
8.2.14	The Nernst Equation	901
8.2.15	The Nernst Equation: Its Sphere of Relevance	906
8.2.16	Looking Back	908
Further Reading		909
8.3	The Butler–Volmer Equation: Further Details	910
8.3.1	The Need for a Careful Look at Some Quantities in the Butler–Volmer Equation	910
8.3.2	The Relation between Structure at the Electrified Interface and the Rate of Charge-transfer Reactions	911
8.3.3	The Interfacial Concentrations May Depend on Ionic Transport in the Electrolyte	916
8.3.4	What Is the Physical Meaning of the Symmetry factor β?	917
8.3.4a	The Factor β Is at the Center of Electrode Kinetics	917
8.3.4b	A Preliminary to a Second Theory of β: Potential-Energy-Distance Relations of Particles Undergoing Charge Transfer	918
8.3.4c	A Simple Picture of the Symmetry Factor	922

	8.3.4d	Is the β in the Butler–Volmer Equation Independent of Overpotential?	926
8.3.5		Summing-up of Further Details on the Butler–Volmer Equation	928
Further Reading			929
8.4		The Current–Potential Laws at Other Types of Charged Interfaces	930
8.4.1		Semiconductor n–p Junctions	930
8.4.2		The Current across Biological Membranes	937
8.4.3		The Hot Emission of Electrons from a Metal into Vacuum	942
8.4.4		The Cold Emission of Electrons from a Metal into Vacuum	944
Further Reading			946
8.5		The Quantum Aspects of Charge-Transfer Reactions at Electrode–Solution Interfaces	947
8.5.1		A Few Words on the Mechanics of Electrons	947
8.5.2		The Penetration of Electrons into Classically Forbidden Regions	950
8.5.3		The Probability of Electron Tunneling through Barriers	953
8.5.4		The Distribution of Electrons among the Energy Levels in a Metal	956
8.5.5		Under What Conditions Do Electrons Tunnel between the Electrode and Ions in Solution?	959
8.5.6		The Tunneling Condition and the Proton-Transfer Curve	966
8.5.7		Electron Tunneling and the De-electronation Reaction	973
8.5.8		A Perspective View of Charge-Transfer Reactions at an Electrode	974
8.5.9		The Symmetry Factor β: A Better View	975
8.5.10		Quantifying the Charge-Transfer Picture	977
8.5.11		Some Desirable Refinements and Generalizations	981
8.5.12		Surveying the Progress	983
Further Reading			986
8.6		Electrodic Reactions and Chemical Reactions	986
Further Reading			989
Appendix 8.1		The Number of Electrons Having Energy E_F Striking the Surface of a Metal from the Inside	990

CHAPTER 9

Electrodics: More Fundamentals

9.1	Multistep Reactions	991
9.1.1	The Question of Multistep Reactions	991
9.1.2	Some Ideas on Queues, or Waiting Lines	992
9.1.3	The Overpotential η Is Related to the Electron Queue at an Interface	993
9.1.4	A Near-Equilibrium Relation between the Current Density and Overpotential for a Multistep Reaction	994
9.1.5	The Concept of a Rate-Determining Step	997
9.1.6	Rate-Determining Steps and Energy Barriers for Multistep Reactions	1002
9.1.7	How Many Times Must the Rate-Determining Step Take Place for the Overall Reaction to Occur Once? The Stoichiometric Number ν	1004
9.1.8	The Order of an Electrodic Reaction	1008

9.1.9	Blockage of the Electrode Surface during Charge Transfer: The Surface-Coverage Factor	1014
Further Reading		1017
9.2	The Transient Behavior of Interfaces	1017
9.2.1	The Interface under Equilibrium, Transient, and Steady-State Conditions	1017
9.2.2	How an Interface Is Stimulated to Show Time Variations	1019
9.2.3	Some Ideas on the Understanding of Transients	1020
9.2.4	Intermediates in Electrodic Reactions and Their Effects on Potential–Time Transients	1026
9.2.5	Experimental Methods for the Determination of Partial Coverage, with Adsorbed Entities, of the Surface of Electrocatalysts	1029
	9.2.5a Radiotracer Method	1030
	9.2.5b Galvanostatic Transient Method	1030
	9.2.5c Potentiostatic Transients	1032
	9.2.5d The Potential-Sweep, or Potentiodynamic, Method	1033
Further Reading		1035
9.3	Transport in the Electrolyte Effects Charge Transfer at the Interface	1036
9.3.1	Ionics Looks after the Material Needs of the Interface	1036
9.3.2	How the Transport Flux Is Linked to the Charge-Transfer Flux: The Flux-Equality Condition	1038
9.3.3	Appropriations from the Theory of Heat Transfer	1040
9.3.4	A Qualitative Study of How Diffusion Affects the Response of an Interface to a Constant Current	1041
9.3.5	A Quantitative Treatment of How Diffusion to an Electrode Affects the Response with Time of an Interface to a Constant Current	1044
9.3.6	The Concept of Transition Time	1047
9.3.7	Convection Can Maintain Steady Interfacial Concentrations	1050
9.3.8	The Origin of Concentration Overpotential	1052
9.3.9	The Diffusion Layer	1055
9.3.10	The Limiting Current Density and Its Practical Importance	1059
	9.3.10a Polarography: The Dropping-Mercury Electrode	1060
	9.3.10b The Rotating-Disc Electrode	1070
9.3.11	The Steady-State Current–Potential Relation under Conditions of Transport Control	1072
9.3.12	Transport-Controlled De-electronation Reactions	1074
9.3.13	What Is the Effect of Electrical Migration on the Limiting Diffusion-Current Density?	1075
9.3.14	Some Summarizing Remarks on the Transport Aspects of Electrodics	1076
Further Reading		1079
9.4	Determining the Stepwise Mechanism of an Electrodic Reaction	1080
9.4.1	How One Tries to Determine the Reaction Mechanism	1080
9.4.2	Which Is the Rate-Determining Step in the Iron Deposition and Dissolution Reaction?	1083
9.4.3	The Transfer Coefficient α and Reaction Mechanisms	1089
9.4.4	Summarizing Remarks Concerning Mechanistic Studies	1090
Further Reading		1093
9.5	More on Mechanism Determination	1093
9.5.1	Why Review Mechanism Determination?	1093

xxviii CONTENTS

9.5.2	What Is Mechanism Determination?	1094
9.5.3	Stages in the Elucidation of a Reaction Mechanism	1095
9.5.4	The Elucidation of the Overall Reaction, the Entities in Solution, and the Surface Coverage	1096
9.5.5	Some Techniques for Mechanism Determination	1099
	9.5.5a The Determination of Reaction Order	1099
	9.5.5b The Determination of the Transfer Coefficients	1100
	9.5.5c The Determination of the Stoichiometric Number	1101
	9.5.5d Auxiliary Methods	1104
9.5.6	Mechanism Determination for Saturated Hydrocarbons	1107
Further Reading		1110

9.6	Current–Potential Laws for Electrochemical Systems	1111
9.6.1	The Potential Difference across an Electrochemical System	1111
9.6.2	The Equilibrium Potential Difference across an Electrochemical Cell	1114
9.6.3	The Problem with Tables of Standard Electrode Potentials	1115
9.6.4	The pH–Potential Diagrams: A General Representation of Equilibrium Potential Differences across Cells	1121
9.6.5	Are Equilibrium Cell Potential Differences Useful?	1124
9.6.6	Electrochemical Cells: A Qualitative Discussion of the Variation of Cell Potential with Current	1129
9.6.7	Electrochemical Cells in Action: Some Quantitative Relations between Cell Current and Cell Potential	1132
Further Reading		1137

| 9.7 | The Grand Divide | 1138 |

CHAPTER 10

Electrodic Reactions of Special Interest

10.1	Electrocatalysis	1141
10.1.1	A Chemical Catalyst and an Electrocatalyst	1141
10.1.2	At What Potential Should Electrocatalysts Be Compared?	1143
10.1.3	Electrocatalysis in Simple Redox Reactions	1146
	10.1.3a How Does the Electrocatalytic Rate Depend upon the Substrate Work Function at the Reversible Potential?	1146
	10.1.3b Can the Exchange-Current Density Depend upon the Work Function?	1149
10.1.4	Electrocatalysis in Reactions Involving Adsorbed Species	1153
	10.1.4a Electrocatalysis in the Hydrogen-Evolution and - Dissolution Reaction	1153
	10.1.4b The Electrocatalytic De-electronation of Hydrocarbons	1156
	10.1.4c The Dependence of the Rate upon Substrate for the Oxidation of Ethylene	1161
	10.1.4d The Special Position of Platinum as an Electrocatalyst	1166
10.1.5	Special Features of Electrocatalysis	1168
	10.1.5a The Effect of the Electric Field	1168
	10.1.5b Reactivity at Low Temperatures	1169
	10.1.5c The Activation of an Electrocatalyst	1170
	10.1.5d Increasing the Power Output by Changing the Reaction Path	1171
	10.1.5e The Use of Porous Electrodes	1171
Further Reading		1172

10.2	The Electrogrowth of Metals on Electrodes	1173
10.2.1	The Two Aspects of Electrogrowth	1173
10.2.2	The Reaction Pathway for Electrodeposition	1175
10.2.3	Stepwise Dehydration of an Ion; the Surface Diffusion of Adions	1177
10.2.4	Mechanism Determination on Surfaces Which Change with Time	1182
10.2.5	The Time Variation of the Average Adion Concentration in Response to the Switching on of a Constant Current	1185
10.2.6	The Contributions of Double-Layer Charging and Faradaic Reaction to the Total Deposition-Current Density	1190
10.2.7	The Time Variation of the Overpotential and the Rate-Determining Step in Electrodeposition	1192
10.2.8	The Contribution of Charge Transfer and Surface Diffusion to the Total Overpotential for Electrodeposition at Steady State	1199
10.2.9	From Deposition to Crystallization	1202
10.2.10	Some Devices for Building Lattices from Adions: Screw Dislocations and Spiral Growths	1203
10.2.11	Microsteps and Macrosteps	1207
10.2.12	How Steps from a Pair of Screw Dislocations Interact	1210
10.2.13	Crystal Facets Form	1212
10.2.14	Deposition on Single-Crystal and Polycrystal Substrates	1218
10.2.15	How the Diffusion of Ions in Solution May Affect Electrogrowth	1218
10.2.16	Organic Additives and Electrodeposits	1221
10.2.17	The Simultaneous Deposition of More Than One Metal: Alloy Deposition	1223
10.2.18	The Sometimes Unavoidable Complication: Hydrogen Codeposition	1227
Further Reading		1230
10.3	The Hydrogen-Evolution Reaction	1231
10.3.1	A Reaction with a Special History	1231
10.3.2	What Are the Possible Paths for the Hydrogen-Evolution Reaction?	1233
10.3.3	What Mechanisms Are Possible in Hydrogen Evolution?	1235
10.3.4	How One Determines the Path and Rate-Determining Step of the Hydrogen-Evolution Reaction	1237
	10.3.4a The Determination of the Exchange-Current Density	1238
	10.3.4b The Determination of the Transfer Coefficient	1238
	10.3.4c The Determination of Reaction Order with Respect to Hydrogen Ions in Solution	1243
	10.3.4d The Stoichiometric Number v	1244
	10.3.4e The Determination of Hydrogen Coverage	1245
	10.3.4f The Heat of Adsorption of Atomic Hydrogen on the Electrode	1247
	10.3.4g Isotopic Separation Factors	1248
	10.3.4h What Are the Probable Mechanisms for Hydrogen Evolution?	1250
Further Reading		1250
10.4	The Electronation of Oxygen	1251
10.4.1	The Importance of the Oxygen-Electronation Reaction	1251
10.4.2	The Evaluation of One of the Mechanisms of Oxygen Electronation	1253
10.4.3	Catalysis and the Oxygen Reaction	1256
10.4.4	Some Special Difficulties with Electrodic Reactions Having Small Exchange-Current Densities	1259
10.4.5	An Electrodic Method of Purifying Solutions	1261
10.4.6	Observing Very Slow Reactions near Equilibrium	1263
Further Reading		1263

CHAPTER 11

Some Electrochemical Systems of Technological Interest

11.1	Technological Aspects of Electrochemistry	1265
11.2	Corrosion and the Stability of Metals	1267
11.2.1	Civilization and Surfaces	1267
11.2.2	Charge-Transfer Reactions Are the Origin of the Instability of a Surface	1268
11.2.3	A Corroding Metal is Analogous to a Short-Circuited Energy-Producing Cell	1269
11.2.4	The Mechanism of the Corrosion of Ultrapure Metals	1273
11.2.5	What Is the Electronation Reaction in Corrosion?	1275
11.2.6	Thermodynamics and the Stability of Metals	1277
11.2.7	Potential–pH (or Pourbaix) Diagrams: Uses and Abuses	1281
11.2.8	The Corrosion Current and the Corrosion Potential	1285
11.2.9	The Basic Electrodics of Corrosion in the Absence of Oxide Films	1287
11.2.10	An Understanding of Corrosion in Terms of Evans Diagrams	1291
11.2.11	Which Step in the Corrosion Process Controls the Corrosion Current?	1296
11.2.12	Metals, pH, and Air	1297
11.2.13	Some Common Examples of Corrosion	1301
11.2.14	Electrodic Approaches to Increasing the Stability of Metals	1306
	11.2.14a Corrosion Inhibition by the Addition of Substances to the Electrolytic Environment of a Corroding Metal	1306
	11.2.14b Corrosion Prevention by Charging the Corroding Metal with Electrons from an External Source	1309
11.2.15	Passivation: The Transformation from a Corroding and Unstable Surface to a Passive and Stable Surface	1315
11.2.16	The Mechanism of Passivation	1318
11.2.17	The Dissolution–Precipitation Model for Film Formation	1321
11.2.18	Spontaneous Passivation: Nature's Method of Stabilizing Surfaces	1323
11.2.19	A Competition in Models for Passivation?	1324
11.2.20	The Thermodynamics of Passivation	1326
11.2.21	Hydrogen Diffusion into a Metal	1328
11.2.22	The Preferential Diffusion of Absorbed Hydrogen to Regions of Stress in a Metal	1330
11.2.23	Interstitial Hydrogen Can Crack Open a Metal Surface	1333
11.2.24	Surface Instability and the Internal Decay of Metals: Stress-Corrosion Cracking	1335
11.2.25	Surface Instability and Internal Decay of Metals: Hydrogen Embrittlement	1338
11.2.26	Charge Transfer and the Stability of Metals	1345
11.2.27	The Cost of Corrosion	1346
11.2.28	A Bird's-Eye View of Corrosion	1347
Further Reading		1349
11.3	Electrochemical Energy Conversion	1350
11.3.1	The Present Situation in Energy Consumption	1350
11.3.2	How Are the Hydrocarbon Fuels Used at Present?	1352
11.3.3	The Pollution of the Atmosphere with Products from Internal-Combustion Reactions and Its Possible Effect on World Temperature and Sea Levels	1353
	11.3.3a Products of Combustion Other than Carbon Dioxide	1353
	11.3.3b Carbon Dioxide	1355

CONTENTS xxxi

11.3.3c	Uncertainties in Predicting the Future Pollution of the Atmosphere	1357
11.3.4	Thermal-Combustion Engines Waste the Chemical Energy Available from Burning Hydrocarbons in Air	1357
11.3.5	Direct Energy Conversion	1358
11.3.6	Direct Energy Conversion by Electrochemical Means	1361
11.3.7	The Maximum Intrinsic Efficiency in Electrochemical Conversion of the Energy of a Chemical Reaction to Electric Energy	1361
11.3.8	The Actual Efficiency of an Electrochemical Energy Converter	1366
11.3.9	The Physical Interpretation of the Absence of the Carnot Efficiency Factor in Electrochemical Energy Conversion	1366
11.3.10	Cold Combustion	1369
11.3.11	Making V near V_e Is the Central Problem of Electrochemical Energy Conversion	1369
11.3.12	The Electrochemical Quantities Which Must Be Optimized for Good Energy Conversion	1374
11.3.13	The Power Output of an Electrochemical Energy Converter	1376
11.3.14	The Electrochemical Engine	1378
11.3.15	Was the Wrong Path Taken in the Development of Power Sources at the End of the Nineteenth Century?	1379
11.3.16	Electrodes Burning Oxygen from Air	1382
11.3.17	The Special Configurations of Electrodes in Electrochemical Reactors	1382
11.3.18	Electrochemical Electricity Producers: The Two Basic Types	1385
11.3.19	Examples of Electrochemical Generators	1386
	11.3.19a The Hydrogen–Oxygen Cell	1388
	11.3.19b Reformer-Supplied Hydrogen–Air Cells	1389
	11.3.19c Hydrocarbon–Air Cells	1391
	11.3.19d Dissolved-Fuel Fuel Cells	1393
	11.3.19e Natural Gas and CO–Air Cells	1393
11.3.20	The Relations between Electrochemical Energy Conversion and the Future Dominance of Atomic Energy as the Source of Power	1395
	11.3.20a Will Atomic Power Sources Compete for Any of the Uses Foreseen for Electrochemical Power Sources?	1395
	11.3.20b Will Electrochemical Means Be Used to Convert Nuclear Power to Electricity?	1396
	11.3.20c What Is the Relation between Electricity Storage and Atomic Energy?	1397
11.3.21	A Summary of the Direct Conversion of Chemical Energy to Electricity	1398
Further Reading		1400
11.4	**Electricity Storage**	**1401**
11.4.1	Conventional and Descriptive Terminology in Energy Conversion and Storage	1403
11.4.2	The Important Quantities in Electricity Storage	1404
	11.4.2a Electricity Storage Density	1404
	11.4.2b Energy Density	1407
	11.4.2c Power	1410
	11.4.2d Desirable Trends	1412
11.4.3	Classical Electricity Storers	1413
	11.4.3a The Lead–Acid Storage Battery	1413
	11.4.3b A Dry Cell	1415
	11.4.3c Two Relatively New Electricity Storers	1416

11.4.4	The Large Gap between the Maximum Feasible and the Present Actual Energy Densities of Electricity Storers	1418
11.4.5	Outlines of Some Possible Future Electricity Storers	1420
	11.4.5a Electricity Storage in Hydrogen	1420
	11.4.5b Storage by Using Alkali Metals	1422
	11.4.5c Storers Involving Nonaqueous Solutions	1424
	11.4.5d Storers with Zinc in Combination with an Air Electrode	1427
11.4.6	The Respective Realms of Applicability of Electrochemical Energy Converters and Electricity Storers	1428
11.4.7	Electrochemical Electricity Storage in a Nutshell	1430
Further Reading		1432
Index		**xxix**

VOLUME 1

MODERN ELECTROCHEMISTRY

CHAPTER 1
ELECTROCHEMISTRY

1.1. INTRODUCTION

It is hoped by the writers of this book that many of its readers will become electrochemists, i.e., full-time researchers in the field which deals with chemical transformations produced by electric currents and with the production of such currents from the transformation of chemical substances. If that hope is to be realized, then the reader has much to learn for electrochemistry is a field which is *interdisciplinary*. Electrochemistry may be considered to be born out of Chemistry by Electricity, but it involves far more than a knowledge of the chemistry of substances and the physics of electric fields.

The reader and future electrochemist must also realize that, even if he digests all that is in this book, he will still be walking on a set of shifting logs, for modern electrochemistry is a rapidly developing branch of knowledge, mostly awakened to life in the 1950's and only beginning to call out lustily in the 1960's.

This book attempts to present a perspective view of the contemporary scene in the field. Much of the theoretical material presented here is bound to be improved, and some revised, during the next few decades. This is especially true because the book intentionally contains a dearth of solidly sure, but unstimulating, thermodynamic treatments and an excess of conceptual, model-oriented approaches which, though associated with a temporary and uncertain character, undoubtedly stimulate criticisms and new developments.

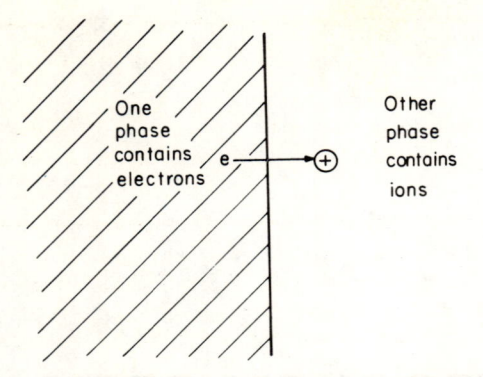

Fig. 1.1. The fundamental act in electrochemistry. (Often, the electron-containing phase is a metal; and the ion-containing phase, an aqueous solution. But germanium in contact with a molten salt would also involve electron- and ion- containing systems.)

What should one choose to give as introductory material in a field as wide as electrochemistry? Should one start by considering the quantum mechanics of the exchange of electrons between a metal and ions in the solution with which the metal is in contact (Fig. 1.1)? Solve the quantum mechanical equations in the first 10 pages, and obtain the interfacial reaction rate as a function of the potential difference across the interphase?[†] This would not be a good way, we think, to teach electrochemistry at present, and for two reasons. Firstly, the quantum mechanical formulation of the reaction rate across the interface—the heart of electrochemistry—has grown slowly, and a lot of scaffolding is still up. Secondly, because the viewpoint taken here may be expressed by the statement "Electrochemistry is the study of phenomena at electrified interfaces," is an unconventional one for Anglo-American readers (though in accord with a viewpoint long accepted in Europe), we do need to have a fairly detailed understanding of the surroundings of the solid–solution interphase before we attempt to find out about electric charges crossing it. In presenting the preliminary

---

[†] It is essential in electrochemistry to distinguish between the terms *interface* and *interphase*. An interface formed by two phases is the apparent two-dimensional surface of contact of these two phases. It is an apparent surface because, in reality, when two phases come together, there is a *region* in which there is a continuous transition from the properties of one phase to the properties of the other. If one aims to refer specifically to this three-dimensional transition region, then it is more appropriate to use the term interphase.

IONICS	ELECTRODICS
Concerns ions in solution and in the liquids arising from melting solids composed of ions.	Concerns the region between an electronic and an ionic conductor and the transfer of electric charges across it.

Fig. 1.2. A way to divide the two quite different aspects of the field of electrochemistry. In this book, the point of view is presented that the electrodic area should be the realm associated with electrochemistry. Ionics is a necessary adjunct field (just as is the theory of electrons in metals and semiconductors, which is adequately dealt with in books on the solid state).

(electrolyte-oriented) material in the first part of our book, we are presenting material which was formerly understood to constitute electrochemistry itself. (For, as explained in Section 1.3.1, there was a hiatus in the development of interfacial electrochemistry from about 1910 to 1950, and researchers became so interested in the study of the ionic solution surrounding the interface that they identified electrochemistry with this study and turned away from the events at the interface.)

So, we have deemed it better to go stepwise and to retain a certain aura of conventionality in arrangement, first treating the preliminary subject of the behavior of ions in solutions—the field which we suggest should be called "ionics"—then considering the main part of electrochemistry—"electrodics"—, the theory of charged interfaces and the conditions governing the transfer of charges across them (see Fig. 1.2).

In this chapter, an attempt is made to explain a little of what present-day fundamental electrochemistry is about.

1.2. ELECTRONS AT AND ACROSS INTERFACES

1.2.1. Many Properties of Materials Depend upon Events Occurring at Their Surfaces

The great interdisciplinary area of materials science treats the properties of materials. This composite field of study is made up of metallurgy along with the physics and chemistry of the solid state. It seeks to interpret, for example, their mechanical properties predominantly in terms of the interactions of the atoms and molecules which constitute the bulk of the material. This approach has respectable precedents because the beginning of our understanding of the constitution of materials came with the X-ray work of Laue in 1912 and the quantum theory of specific heats proposed by Einstein in 1907. Such a bulk-oriented approach to solid materials has been the origin of most of the progress which has been hitherto made toward understanding their properties. For example, it is possible to derive theoretically the heat contents of solids, or their electrical conductivity. Further, much of the work on solids has contributed to the growing appreciation of the theory of the bonds between atoms (the essence of chemistry) and the energy states of electrons in solids.

Such studies of the bulk properties can give, however, only a very partial interpretation of the observable behavior of materials. Some phenomena which escape interpretation concern important practical aspects of materials. For example, metals sometimes undergo unexpected breakage when strained, and such sudden failures are not always explicable in terms of the bulk properties of the metal. Such phenomena may be explicable in terms of certain events associated with the surface, and these events cannot be predicted from a knowledge of the bulk properties. It may be found that an atomic-scale defect at the surface, caused by a surface reaction, is the source of a spreading crack which, assisted by an applied or residual stress, may work its way spontaneously through several millimeters of solid in a few hours.

Further, there is a general fact (one of the more important from the viewpoint of our limited resources) which we have to interpret, namely, the decay of materials. Metals, with the exception of those that we, therefore, call *noble*, have a limited life when they are in contact with moist air and tend to revert to the oxides from which many of them were extracted. Little of this general decay of materials (which limits the practical life of a ship to some 30 years or perhaps causes the undercarriage of an aircraft to break even when subjected to normal loads) is due to *erosion*, i.e., the wearing away by rubbing, as of rocks by rain. Most of the decay is due to much more subtle electrochemical surface phenomena, knowledge of the basic principles of which did not begin to spread until the late 1940's.

Lastly, it is important to recall that very many of the molecules which govern the mechanics of biological processes are colloidal. This means that their size is between about 100 and 10,000 Å and that their stability as separate entities depends predominantly upon the forces which exist between their surfaces and the surrounding ionic solutions, for it is the repulsive interaction between the electrified interfaces of these particles which prevents them from coalescing when they approach each other. Correspondingly, these repulsive interactions govern the behavior at the many interfaces within biological systems. Proteins and other biological macromolecules

MATERIALS SCIENCE	SURFACE SCIENCE
Has dealt hitherto largely with the **bulk** properties of solids and with the constitution of molecules. This approach is the origin of most of our present knowledge of the properties of materials.	Many properties of materials—for example, their stability or instability in contact with other materials—cannot be explained adequately in terms of bulk properties. Much of this needed, future knowledge of materials is in the realm of **surface** science.

Fig. 1.3. Many properties of materials depend primarily upon events occurring on their surfaces.

such as salts of deoxyribonucleic acid (DNA) are substances of this kind, in which the *surface situation* dominates the other influences in determining stability.

Thus, some important properties of materials and the behavior of entities which determine the properties of living systems depend on surface phenomena and surface properties, i.e., *surface science* (Fig. 1.3).

1.2.2. Almost All Interfaces Are Electrified

Consider the bulk of a solid material, e.g., a metal. It may be pictured, in an elementary way, as consisting of charged particles, positive ions and electrons, with the total negative charge balancing the total positive charge so that there is no *net* electric charge in the bulk of the material.

Imagine, now, a thought experiment in which a limitingly thin knife —perhaps a narrow laser beam—cuts through a piece of metal kept in vacuum. The cutting process bares two surfaces. What happens at each freshly created metal–vacuum interface? There is a small degree of spillover of the electrons from the metal into vacuum (*cf.* Section 7.2) resulting in a disturbance to the former balance of electrical charges in the surface region of the metal. Having lost some electronic charge, the metal surface becomes positively charged; and having gained some electronic charge, the vacuum side of the interface becomes negatively charged. Thus, the metal–vacuum interface has becomes electrified (Fig. 1.4).

If air[†] is allowed to come into contact with the metal, further possibilities arise. What one is dealing with is, in fact, the interface between a

---

[†] In practice, the term *air* often means moist air.

metal and a moisture film; and, on the moisture side of this interface, there are water molecules, dissolved oxygen molecules, and perhaps hydrogen ions (H^+). Hence, the interface under consideration may be that between an electronic conductor and an ionic conductor. The water molecules, which are electrical dipoles (*cf.* Section 2.3.1), may form an oriented layer on the metal surface. Such an oriented dipole layer is equivalent to two sheets of charge (Fig. 1.4). Another possibility is the exchange of electrons between the metal and particles in the moisture film. For example, the electron transfer may proceed thus,

$$2H^+ + 2e \to H_2$$

and thereafter in the opposite direction, thus,

$$H_2 \to 2H^+ + 2e$$

These electron flows in the two directions—from metal to hydrogen ions and from hydrogen molecules to metal—only become equal at equilibrium, by which stage there would have been a *net* electron flow in one of the directions. If, for example, this net electron flow is from metal to solution, then the metal has suffered a certain depletion of its negative charge and, thus, its surface acquires a net positive charge. Correspondingly, the solution side of the interface acquires a net negative charge.

It is seen therefore that, irrespective of whether one is considering a spillover of electrons from a metal surface into vacuum, a net orientation of dipolar neutral molecules, or the net flow of electrons in electron transfer processes, there are two results. Firstly, there is the charging of the two sides of the interface between a metal and another phase (Fig. 1.4), and, secondly, there is the development of a potential difference across the inter-

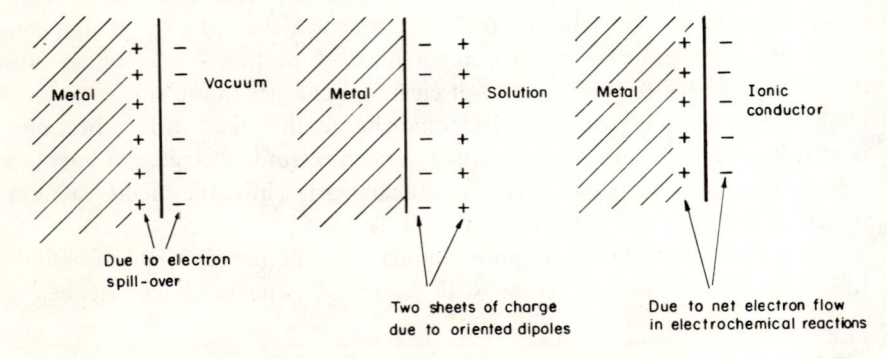

Fig. 1.4. Electrified interfaces.

phase. The charging proceeds until the charges on the two phases are equal in magnitude but opposite in sign.

The argument about a net charge of one sign arising on each surface (of the two phases meeting at an interface) has been couched in terms of a metal and moist air. But the argument can be generalized. Almost all interfaces, therefore, are electrified, and the surfaces of material carry excess charges. It is these excess charges which affect the surface properties of materials and often make them deviate from those of the bulk.

There are widespread consequences of the fact that the surfaces of nearly all phases carry a net electric charge and form one side of an electric double layer.† Among them is the fact that the electrical aspect of the surface forces is an important factor in the determination of properties of the interphase. Correspondingly, of course, those properties of solids which depend on happenings at the surface depend on the electric charge separation and electric field strength across the interphase. Very many aspects of the behavior of surfaces, in practice, involve electrochemical considerations.

1.2.3. The Continuous Flow of Electrons across an Interface: Electrochemical Reactions

It has been argued in the above section that almost all surfaces carry an excess electric charge and that interfaces are electrified. However, the argument was made by considering an isolated piece of material unconnected to a source or sink of electrons.

Suppose now that the metal, an electronic conductor, is connected to a power supply,‡ i.e., to a source of electrons so large in capacity that the 10^{19} to 10^{20} electrons* drawn from the source leave it unaffected in any significant way. To make the discussion specific, consider that the electronic conductor is a platinum plate and the ionically conducting phase is an aqueous solution of HI.

Then, by connecting the power source to the metal plate, it becomes possible for electrons to flow from the source to the surface of the platinum plate. Before this was done, the electrified platinum–solution interface was

---

† Some of these consequences concern semiconductor devices, now the central aspect of solid-state electronics.
‡ Actually, a power supply has two terminals, and one must also consider how the metal–electrolyte interface is connected to the other terminal, but this consideration is postponed to the next section.
* An Avogadro number ($\sim 10^{23}$) of electrons produce 1 gram equivalent of substance in an electrochemical reaction; hence 10^{19} to 10^{20} electrons produce 10^{-4} to 10^{-3} g-eq.

in equilibrium. Under these equilibrium conditions, the platinum plate had a certain net surface charge, and the ionically conducting solution, an equal and opposite excess charge; and, further, the electron flows across the interface, associated with the electron-transfer reactions, were occurring to an equal extent in the two directions. What happens when the disequilibrating shower of extra charges from the power source arrives at the surface of the platinum? The *details* of what happens—the mechanism—is a long story, told partly in Chapters 8 and 10. However, the essence of it is that the new electrons overflow, as it were, the electron capacity of the metal plate and *cross the surface* to strike and neutralize the ions, e.g., hydrogen ions, in the solution phase. This process can proceed continuously because the power source supplying the electrons has been thought of as infinite in capacity and the ionic conductor also has an abundance of ions in it; these tend to migrate up to the metal surface and appear there for the purpose of capturing some of the overflowing electrons.

What is being described here is an *electrochemical reaction*; i.e., it is *a chemical transformation involving a net electron transfer*, and it can be written in familiar style as

$$2H^+ + 2e \to H_2.$$

The hydrogen ions are discharged on the electrode and there is an evolution of hydrogen gas.

The simplicity of such a formulation should not obscure the fact that what has been described is a remarkable and distinctive part of chemistry. An *electric current*, a controllable electron stream, has been made in a controlled way to react with a *chemical substance* and produce another *new chemical substance*.[†] That is what a good deal of electrochemistry is about—*it is about the electrical path for producing chemical transformations*. Much of electrochemistry is also connected with the other side of this penny, namely, the production of electric currents *and therefore electric power* directly from changes in chemical substances, which is one of the newer methods of producing electrical energy (fuel cells, Section 11.2).

1.2.4. Electrochemical and Chemical Reactions

There is another aspect of the electrochemical reaction which has just been described. It concerns the happenings to the iodide ions of the hy-

---

[†] There is not much limitation on what kind of chemical substance; e.g., it does not have to be an ion. $C_2H_4 + 4H_2O \to 2CO_2 + 12H^+ + 12e$ is as much an electrochemical reaction as is $2H^+ + 2e \to H_2$.

drogen iodide, which must also have been present in the solution of HI in water. Where do they go while the hydrogen ions are being turned into hydrogen molecules?

The I⁻ ions have not yet appeared in the picture because only half of the picture has been shown. Electrical sources have two terminals. The consideration of a power source pumping electrons into a platinum plate in contact with an ionic solution is essentially a *thought experiment*. In the real situation, one immerses another electronic conductor in the same solution and connects this second electronic conductor to the other terminal of the power source (Fig. 1.5). Then, whereas electrons from the power source pour *into* the platinum plate, they would flow *away from* the second electronic conductor (made, e.g., of rhodium) and back to the power source. It is clear that, if we want a system which can carry on for some time with hydrogen ions receiving electrons from the platinum plate, then iodide ions have got to give up electrons *to* the rhodium plate at the same rate as electrons are given up *from* the platinum. Thus, the whole system can function smoothly without the loss of electroneutrality which would arise were the hydrogen ions to receive electrons from the platinum without a balancing event to occur at the other plate. Such a process would be required to remove the negatively charged ions which would become excess ones directly the positively charged hydrogen ions had been removed from the solution.

The assembly, or system, consisting of one electronic conductor which acts as an electron source for particles in the ionic conductor and another electronic conductor acting as an electron sink receiving electrons from the ionic conductor is known as an *electrochemical cell, or electrochemical system*.

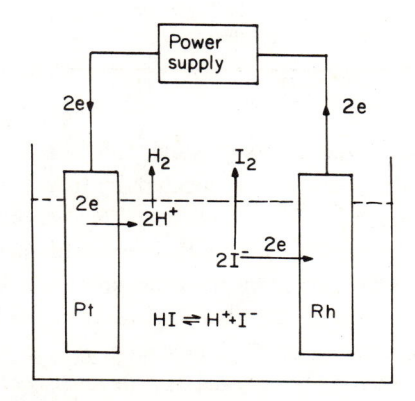

Fig. 1.5. The electrochemical reactor.

We have seen that, at one charged plate, electrochemical electron-transfer reactions can occur. What happens if one takes into account the second plate? There, the electron transfer is from the solution to the other electronic conductor. Thus, if we consider the two electronic conductor–ionic conductor interfaces, there is no *net* electron transfer. The electron outflow from one electronic conductor equals the inflow to the other; i.e., a *purely chemical reaction* (one not involving net electron transfer) *can be carried out in an electrochemical cell.* Such net reactions in an electrochemical cell are formally identical to the familiar thermally induced chemical reactions in which molecules collide with each other and form new species with new bonds. There are, however, very fundamental differences between the ordinary chemical way of effecting a reaction and the less familiar electrical or electrochemical way, in which the reactants collide not with each other but with separated "charge-transfer catalysts," as the two plates which serve as electron-exchange areas might be called. One of the differences, of course, pertains to the facility with which the rate of a reaction in an electrochemical cell can be controlled; all one has to do is to control electronically the power source. This facile control arises because the electrochemical reaction rate *is* the rate at which the power source pushes out and receives back electrons through the electrochemical cell.

Thus, one could write the electrochemical happenings as

$$2HI \xrightarrow[\text{solution}]{\text{in}} 2H^+ + 2I^-$$

$$2H^+ + 2e \xrightarrow[\text{plate}]{\text{at Pt}} H_2 \uparrow$$

$$2I^- \xrightarrow[\text{plate}]{\text{at Rh}} I_2 + 2e$$

$$\overline{2HI \longrightarrow H_2 + I_2}$$

Thus, from an *overall* point of view, this net cell reaction is *identical* to that which would occur if one heated hydrogen iodide and produced hydrogen and iodine by a *purely chemical, or thermal, reaction.*

There is another way in which the electrical method of carrying out chemical reactions is distinct from the other methods for achieving chemical changes (Fig. 1.6). "Ordinary" chemical reactions, like the homogeneous combination of H_2 and I_2 or the heterogeneous combinations of H_2 and O_2, occur because thermally energized molecules occasionally collide and, during the small time they stay together, change around some bonds to

```
H ——I                              H   I
  |                       H··I     |  +|
  | Thermal                :       H   I
  | collisional   ———→   Bond
  | activation           H··I breaking
  ↑
H —— I
```

<u>Chemical (or Thermal) Reaction</u>

```
HI      water
(gas)   ———→    H⁺ + I⁻

         Accepts electrons          Donates electrons
           from metal                   to metal
       ↙                                          ↘
     H₂                                            I₂
```

<u>Electrochemical (or Electric) Reaction</u>

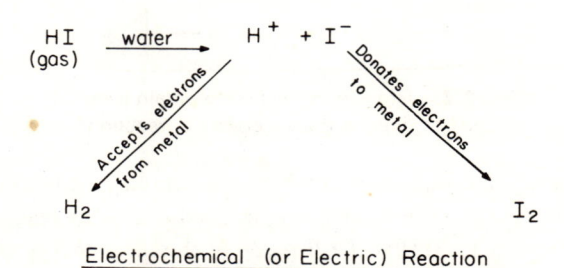

Fig. 1.6. The chemical and electrochemical ways of carrying out reactions. In the electrochemical way, the particles do not collide with each other but with separated sources and sinks of electrons.

form a new arrangement. Correspondingly, photochemical reactions occur by mechanisms in which photons strike molecules and give them extra energy so that they break up and form new compounds. In a similar way, the high-energy particles emanating from radioactive substances can energize molecules, which then react. The electrical method of causing chemical transformations is different from the other two general methods of provoking chemical reactivity in that the overall electrochemical cell reaction is composed of two separate electron-transfer reactions which occur at spatially separated electrode-electrolyte interfaces and are susceptible to electrical control.

If electrical energy provokes and controls chemical reactions, chemical reactions can presumably give rise to a flow of electricity. Thus, two reactant substances may be allowed to undergo spotaneous electron transfer at the separated (electrode) sites characteristic of the electrochemical way of bringing about chemical reactions, and then the electrons transferred in the two reactions at the two interfaces can be made to flow through an electrical load, for example, the circuit of an electric motor (Fig. 1.7). In this reverse

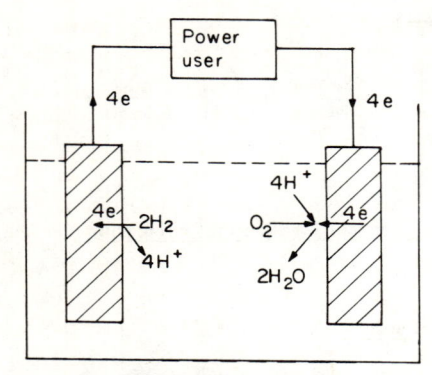

Fig. 1.7. Grove was the first to obtain electric power directly from a chemical reaction.

process, too, there is a unique aspect when one compares it with the production of available energy from thermally induced chemical reactions. It can be shown (*cf.* Section 11.3) that the *fraction* of the total energy of the chemical reaction which can be converted to mechanical energy turns out to be intrinsically much greater in the electrical than in the chemical way of producing electricity.

1.3. BASIC ELECTROCHEMISTRY

1.3.1. Electrochemistry before 1950

Certain fields develop logically, usually from an initial empirical discovery. Such a one is the production of energy from the heat evolved in nuclear reactions. The harnessing of nuclear energy through controlled nuclear reactions in a reactor (Fermi *et al.*, 1942) was the culmination of the work of Becquerel (1896) on radioactivity, of Einstein (1905) on mass–energy equivalence, of Rutherford (1911) on the nuclear atom model, of Cockroft and Walton (1930) on the artificial transmutation of elements, and of Hahn and Meitner (1939) on fission.

Electrochemistry made a much earlier and, indeed, even more promising, start than did nuclear chemistry. Luigi Galvani described the first observations in a field which, nearly 180 years later, has come again to focus upon its electrical aspects. Galvani wrote (1791):

> I had dissected a frog...and had placed it upon a table on which there was an electric machine, while I set about doing certain other things. The frog was entirely separated from the conductor of the machine and indeed was at no small distance away from it. While one of those who were as-

sisting me touched lightly and by chance the point of his scalpel to the internal crural nerves of the frog, suddenly all the muscles of its limbs were seen to be so contracted that they seemed to have fallen into tonic convulsions. Another of my assistants, who was making ready to take up certain experiments in electricity with me, seemed to notice that this happened only at the moment when a spark came from the conductor of the machine. He was struck with the novelty of the phenomenon, and immediately spoke to me about it, for I was at the moment occupied with other things and mentally preoccupied. I was at once tempted to repeat the experiment so as to make clear whatever might be obscure in it. For this purpose I took up the scalpel and moved its point close to one or the other of the crural nerves of the frog while at the same time one of my assistants elicited sparks from the electric machine. The phenomenon happened exactly as before. Strong contractions took place in every muscle of the limb, and at the very moment when the sparks appeared the animal was seized as it were with tetanus.

This observation of Galvani's was closely followed by an observation by Volta, who wrote thus in 1800:

I have the pleasure of communicating to you, and through you to the Royal Society, some striking results I have obtained in pursuing my experiments on electricity excited by the mere mutual contact of different kinds of metal, and even by that of other conductors, also different from each other, either liquid or containing some liquid, to which they are properly indebted for their conducting power. The principle of these results is the construction of an apparatus having a resemblance in its effects to the Leyden flask. The apparatus to which I allude, and which will, no doubt, astonish you, is only the assemblage of a number of good conductors of different kinds arranged in a certain manner. Thirty or more pieces of copper, or, better, silver, applied each to a piece of tin or zinc, which is much better, and as many strata of salt water or any other conducting liquid, or pasteboard, skin, etc., well soaked in these liquids; such strata interposed between every pair of two different metals and always in the same order are all that is necessary for constructing my new instrument.

To this apparatus, much more similar to the natural electric organ of the torpedo or the electric eel, etc., than to the Leyden flask, I would wish to give the name of the "Artificial Electric Organ."...

All the facts which I have related in this long paper in regard to the action which the electric fluid excited and, when moved by my apparatus, exercises on the different parts of our body which the current attacks and passes through; an action which is not instantaneous, but which lasts, and is maintained during the whole time that this current can follow the chain not interrupted in its communications; in a word, an action the effects of which vary according to the different degrees of excitability in the parts, as has been seen; all these facts, sufficiently numerous, and others which may still be discovered by multiplying and varying the experiments of this kind, will open a very wide field for reflection, and of views, not only curious, but particularly interesting to Medicine. There will be a great deal to occupy the anatomist, the physiologist, and the practitioner.

These discoveries by Galvani and Volta near the beginning of the nineteenth century not only emphasized the role of electrochemistry in the production of electricity but also demonstrated the importance of electrochemical phenomena in biology.

This brilliant and very early start to electrochemistry was carried on by the discovery by Faraday in 1834 of the laws which govern the quantity of a chemical substance formed by the passage of a certain amount of electricity across an interface such as that at the platinum–hydrogen ion-containing solution of Fig. 1.6. The laws of Faraday are among the most exactly verified laws of science and led to the realization by Stoney (1891) that electricity, like matter, is atomic in nature. Stoney proposed the name *electron* for the "atom," or unit, of electric charge.

A fourth major advance in electrochemistry at this very early time consisted of the first attainment by Grove of electrochemical energy conversion. In 1834, Grove produced electrical energy from the union of hydrogen and oxygen in an electrochemical cell (Fig. 1.7). The far-reaching significance of the achievement of Grove's was realized by Ostwald as early as 1894. At the Bunsen Gesellschaft meeting of that year, Ostwald pointed out that the *direct* conversion of chemical energy to electrical power could be achieved electrochemically with a possible conversion efficiency of nearly 90%. In contrast, the (indirect) thermal method (a chemical reaction produces heat which expands a gas which pushes a cylinder which drives a wheel which turns a generator), however, was subject to the Carnot restriction on the maximum possible efficiency with which the heat energy liberated in a chemical reaction can be converted into mechanical work. The restriction rests on intrinsic thermodynamic reasoning and argues for a practical upper limit of about 40% for the efficiency of conversion of chemical energy into electricity by heat (Fig. 1.8).

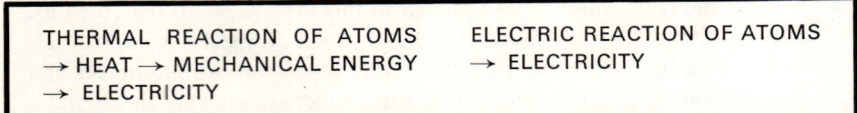

| THERMAL REACTION OF ATOMS → HEAT → MECHANICAL ENERGY → ELECTRICITY | ELECTRIC REACTION OF ATOMS → ELECTRICITY |

Fig. 1.8. Ostwald pointed out that electrically achieved chemical reactions could convert a higher fraction of the energy difference between their initial and final states to electricity than could thermally achieved reactions.

The progress of electrochemistry in these early times thus seemed excellent. For example, it appeared to have been relatively greater than the corresponding progress in the field of homogeneous reaction kinetics and had obtained a head start over, e.g., thermionics. Furthermore, fundamental advances were made in the early years of the present century by Le Blanc, who made an early application of ideas of nonstationary potentials across interfaces, and by Tafel, who, in 1905, established what could perhaps be called the most used law in electrochemistry, namely, that the potential difference $\Delta\varphi$ across the interface at which an electrochemical reaction is occurring changes linearly with the logarithm of the current density i, or

$$\Delta\varphi = a - b \ln i$$

where a and b are independent of $\Delta\varphi$, or

$$i = e^{a/b} e^{-\Delta\varphi/b}.$$

This empirical law of Tafel presents an early example of a series of experimental laws, involving the rate at which events occur, such as that found by Arrhenius for the relation of reaction velocity and flow velocity to temperature and that by Richardson for the dependence of electron emission on the thermionic work function of the electrode.

The discoveries by Galvani, Volta, Faraday, LeBlanc, and Tafel indicated a promising future for electrochemistry with a rapid development in the understanding of the *rates* of electrochemical cell reactions as a function of the potential differences across cells. But this promise was not quickly fulfilled. A somewhat stultifying influence settled upon electrochemistry. This influence arose from the great success achieved at about the turn of the century by the *thermodynamic* treatment of electrochemical cells *in equilibrium*, i.e., the treatment of the situation when no *net* current passes across the interfaces in a cell and, therefore, when no net cell reaction occurs. This thermodynamic view considered the electrical *energy* lost or gained when an electric charge was, in a thought experiment, taken around a circuit consisting of an electrochemical cell with its two electrode–solution interfaces. The electrical energy change is the algebraic sum of the potential differences in the cell, multiplied by the charge transferred in the reactions at each interface. Such an *electrical energy* sum could then be placed equal to the change in *free energy* which occurs in the net *chemical* reaction which takes place in an electrochemical cell.

This brilliant thermodynamic analysis made by Nernst (1891) opened up excellent ways of getting at the free energy changes in chemical reactions and solving the problem of distinguishing between the change in the *heat content* of reactants and products and the change in the *free energy* between them. Great as these advances were in the history of chemistry, they had also weighty, if indirect, negative consequences upon the development of electrochemistry. Two of these were:

1. They made the viewpoint in the treatment of electrochemical reactions across interfaces "*potential*-centric" and not "current (or rate[†])-centric." Thus, when the current was changed, it was thought that the corresponding (unrationalized) changes in electrode potential were due to some illicit[‡] malfunction somewhere in the cell. The change was sometimes, particularly by the polarographers, ascribed to a delay in transporting reactants through the ionically conducting solution up to the interfaces, so that the electrode surface sees a different concentration of ions from that which it would have seen at zero current. The thought that there was a net flow of electrons across the interfaces only *because* the potential difference across them had been made to differ from the equilibrium one of Nernst was suppressed in the zeal of the potential-centric and rate-indifferent thermodynamic approach.

2. Apart from making them regard the potential as the variable *dependent* on the current, the overwhelming success of the thermodynamic treatment—and its emphasis in all textbooks of physical chemistry for the next 50 years— seems to have awed electrochemists from about 1910 onward. They did not, with a very few exceptions, try to develop a rational *kinetic* treatment of the interfacial charge transfer. As late as 1947, most of them were still trying to do the impossible, i.e., to treat the highly (thermodynamically) irreversible electrode reactions by a series of misconceptions and approximations on the basis of reversible thermodynamics. This fundamental error and lack of conceptualization *held a dead hand on the electric mode of achieving chemical reactions* and on the direct conversion of chemical to electrical energy, for some 40 to 50 years. This period in electrochemistry could well be called "The Great Nernstian Hiatus" (Fig. 1.9).

[†] The current density across an interface is the *rate* of electron passage across it per unit area. Thus, it is usually written as amp cm^{-2}.

[‡] The attitude of post-Nernstian electrochemists toward changes in potential away from that described by Nernst's equations seems sometimes to have been tinged with a derogatory moral fervor. The word *polarized*, used to denote in a vague way the nonequilibrium condition of electrodes, had a downing and critical kind of flavor even in the late 1940's. It will be seen later, however, that polarization is an essential factor associated with electrochemical reactions out of equilibrium and, without polarization, the reactions cannot occur at a *net* rate.

Nernst's law:	$\Delta\phi = \Delta\phi^0 + \dfrac{RT}{F} \ln a_i$
Tafel's law in its potentiocentric form:	$\Delta\phi = P + Q \ln i$
Tafel's law in its current-centric form:	$i = A\, e^{B\eta}$

Fig. 1.9. The great Nernstian hiatus. Here $\Delta\phi$ is the potential across the interface; a_i is the activity of a species in the solution; and i is the current density ($\Delta\phi^0$, P, Q, A, and B are constants). Affected by the successful thermodynamic thinking of Nernst's theory of galvanic cells at equilibrium, electrochemists regarded the current passing in cells out of equilibrium as affecting the potential differences at the interfaces; they did not for a long time understand that the current flowed across them only because the potential changed away from that of equilibrium, and possessed an *over*potential, η.

One probable consequence of great socio-economic importance can be mentioned. It will be seen in Chapter 11 that the electric mode of bringing about reactions can be reversed so that electricity can be produced directly from chemical reactions in an electrochemical system or cell. The efficiency of the conversion of chemical energy into electricity and the power which an electrochemical converter produces depend on the nonthermodynamic phenomenon of gross departures of the potential difference across electrochemical cells from the equilibrium cell potential (the degree of this departure is called the *overpotential*). It was of course impossible to rationalize the overpotential in the early fuel cells, or electrochemical converters, so that a practical fuel cell was not produced until the 1960's, and not even then, one capable of producing more than a few kilowatts. It is of interest to speculate that, had the Nernstian hiatus not inhibited the development of the underlying subject of electrode kinetics, then, perhaps, practical electrochemical energy converters would have been produced several decades earlier. A consequence of such a development, had it become possible to build fuel cells with megawatt output, would then have been the absence of the atmospheric pollution which has arisen from the production of power by the chemical combustion reaction of hydrocarbons with oxygen to carbon dioxide, together with other more complex products, some of which are the base substances of smog.

1.3.2. The Treatment of Interfacial Electron Transfer as a Rate Process: The 1950's

By about 1950 onward, most electrochemists in Europe and America began generally turning toward the view that the immediate cause of the current flowing across the interface in an electrochemical cell was the fact that the outside power source caused the electric potential difference across the electric double layer at the interfaces to change from the values that would correspond, for given electrochemical reactions, to the zero-current situation of equilibrium, i.e., to assume values such that there was an overpotential.

Such a view had been implied in a rudimentary way earlier by Butler (1924) and stated explicitly by Volmer and Erdey–Gruz in 1930. Only in Russia,[†] under the leadership of Frumkin, had the new attitude been consistently adopted during the 1930's and 1940's. Most other electrochemists during the 1930's and 1940's thought of the potential at the interface as a quantity given by thermodynamics only. It took until the 1950's to get electrochemists generally moving along in the direction of relating current densities to deviations of the electrode potentials from the value which they possessed when the interface was at equilibrium, i.e.,

$$i = Ae^{B\eta}$$

where the *over*potential (the departure of the potential from that characteristic of equilibrium) η was the current-provoking quantity, and not as

$$\Delta \varphi = a - b \ln i$$

where the attitude was that the potential was basically that given by equilibrium thermodynamics but disturbed *by* the passage of electrons across the interface.

The change of attitude among electrochemists in the 1950's was more than a formal one because it compelled them to think about the interphasic region in a more realistic way. Electrochemistry, after a delay of 35 to 40 years, became molecular kinetic and structural in attitude. Its researchers began to talk in terms of the molecular structure at the interface and the effect of the interfacial field on the transfer of electrons between the electrode and the particles in the adjacent layer of solution. The understanding of the *kinetic* interpretation of equilibrium became widespread, i.e., equilibrium corresponds to electrons crossing the interface at the same rates in both directions; the methods were developed by which the various steps in electrochemical reactions could be elucidated. Electronics became an indispensable tool to the study and control of electrochemical reactions. There was sufficient work at the end of the decade for the first[‡] general

---

[†] Isolated examples of this attitude elsewhere certainly existed, e.g., in the work of Audubert in France and of Bowden in England.

[‡] A brief text in Russian, due to Frumkin's school, dates from 1952, but is almost unknown in the West.

ELECTROCHEMISTRY 19

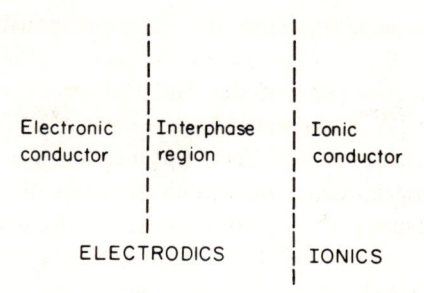

```
Electronic    | Interphase  | Ionic
conductor     | region      | conductor

        ELECTRODICS        | IONICS
```

Fig. 1.10. Ionics and electrodics.

textbook on the kinetics of electrochemical reactions to be written (Vetter, 1961) and to mark the end of the first stage of the rebirth and growth of electrochemistry (Figs. 1.10 and 1.11).

1.3.3. Quantum Electrochemistry: The 1960's

Electrochemistry by the mid-1950's showed much vigor arising from the new attitudes; workers in electrochemistry were returning from other occupations (see below) to the attractive task of increasing knowledge

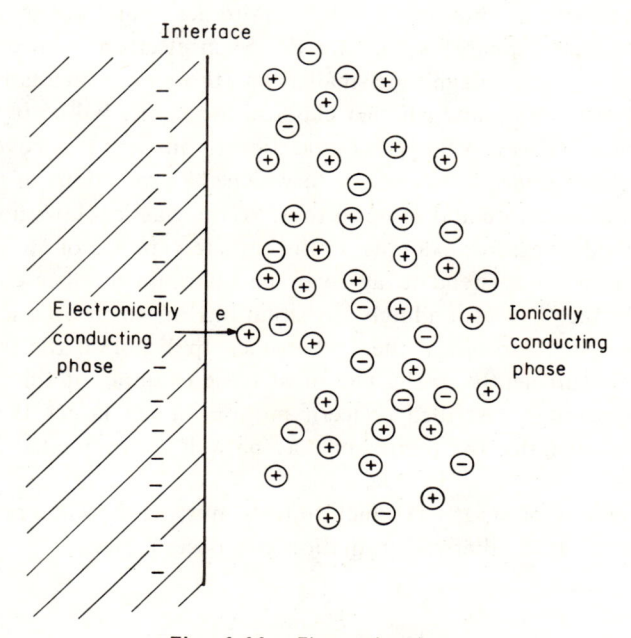

Fig. 1.11. Electrochemistry.

concerning interfacial charge transfer. Electrochemistry was no longer a frozen field.

But the electrochemists of the mid-1950's had not been sufficiently discerning in the *type* of electron transfer which they considered. In thermionic electron emission at high temperatures, the electrons have to climb up and over an energy cliff, and significant rates of transfer are obtained by raising the temperature, often to 1000 to 2000°K, so that, in Richardson's expression

$$i = AT^2 e^{-\Phi R/T}$$

(where Φ is the thermionic work function and A is a constant for a given substance), the exponential term is made sufficiently low.

In the situations treated by electrochemistry, the transfer of electrons across interfaces, the most common one is that of a metal in contact with an aqueous ion-containing solution. Here, the value of the work function Φ gets changed from that value which it has in a vacuum and which applies essentially to the thermionic emission case. But, when, in the 1960's, electrochemists tackled the more detailed task of considering what *kind* of electron transfer would correspond to the range of rates of charge transfer observed at the metal–solution interfaces, it turned out that *classical* electron emissions (electrons climbing *over* barriers to get in and out of the electronically conducting phase) would not do. Such electron transfers provided rates many orders of magnitude smaller than those experimentally observed.

How does one know whether classical mechanics will apply effectively to a problem? This is a fairly easy question to answer; it is covered by the *correspondence principle*, according to which the predictions of the quantal laws correspond more and more to those of the classical laws under certain limiting conditions. For example, in the emission spectra of atoms, classical and quantum theory tend to agree if the quantum number considered is sufficiently big. Correspondingly, in situations in which atomic vibrations are concerned, the classical theory becomes applicable if the frequency of vibration is sufficiently small. The most basic example of this rule is that if the happenings described concern motions much larger than their de Broglie wavelengths, the treatment may be well approximated by classical mechanics.

The energy of a particle which tunnels must be the same on both sides of the barrier (for otherwise, radiation will occur). Hence,

$$\frac{p_1^2}{2m} + U_1 = \frac{p_2^2}{2m} + U_2$$

where p_1 and p_2 are the momenta of the electrons inside the metal and the atom which receives the electron, respectively; U_1 is the potential energy of the electron inside the metal and U_2 that of the electron in an atom in the solution. The proper calculation of U_1 and U_2 is an exercise in the theory of electron transfer at interfaces (see Chapter VIII). Suppose $U_2 = 0$ and $U_1 = \Phi$, the thermionic work function. Then, with the de Broglie wavelength

$$\lambda = \frac{h}{p}$$

$$\lambda_{(e)} = \frac{h}{\sqrt{p_1^2 + 2m\Phi}}$$

But p_1 is the momentum of electrons in the Fermi level and $p_1^2/2m_e \approx$ kinetic energy $\approx$ the Fermi energy of an electron in the metal. But the Fermi energy of an electron differs from its work function only by the surface potential, which is much less than the work function. Hence,

$$\frac{p_1^2}{2m} \approx \Phi \quad \text{or} \quad p_1^2 \sim 2m\Phi$$

Hence,

$$\lambda_e = \frac{h}{(4m\Phi)^{1/2}}$$

Taking Φ as some 5×10^{-12} erg atom^{-1},

$$\lambda \approx 10^{-3} \text{ cm}$$

Thus, the relevant de Broglie wavelength is several thousand angstroms. However, at a given interface, the distance over which an electron jumps in an electrode reaction is always of the order of a few angstroms. Hence, the transition with which one is associated in an interfacial electron transfer is much less than the length of the relevant de Broglie wave. The transitions must be treated quantally.

It is of interest to note that the first paper to treat the electrode–solution interface quantally was published by Gurney in 1931, but the matter was then allowed to rest there for some 20 years instead of having been exploited as one of the first pieces of quantum chemistry to be published. Basic electrochemistry (and all that it implies in respect to the properties of materials, the direct conversion of chemical energy to mechanical work, and the functioning of biological systems) must be worked out in quantum, rather than classical, mechanics.

1.3.4. Ions in Solution, as well as Electron Transfer across Interfaces

From what has been said so far, it might be thought that the field encompassed by electrochemistry concerns the functions of interfacial charge transfer only. This is certainly not the impression one would get if one looked at other books on electrochemistry. In fact, in so far as most of these were published in the 1940's or earlier, one gets almost the reverse impression; electron transfer across interfaces was dealt with sparingly indeed in those days.

There is a good historical reason for this reversing disbalance. When the development of interfacial electrochemistry along modern lines became restricted by the overthermodynamic attitude of its adherents in the pre-1950 days, much attention was diverted to what had seemed previously to some extent the accompanying side issues, i.e., the physical chemistry of the bulk solution adjoining the double layer. This had concentrated upon an interest in deviations in the behavior of solutions from laws derived on the assumption that interactions between particles in them were negligible. Thus, the famous Debye-Hückel theory of such interactions—sophisticated and successful—attracted the attention of electrochemists away from the blocked interfacial studies. Thus, from about 1920 to 1950, the majority of research workers in the electrochemical field were occupied with determining, e.g., activity coefficients of salts in dilute aqueous solutions, the electrical conductance of molten salts, or electrostatic effects on the dissociation constants of acids or bases in aqueous solutions, etc. (Figs. 1.12 and 1.13).

The situation from the 1950's is that, with the opening up of the rate process and quantum mechanical approaches to interfacial reactions, most attention in electrochemistry concerns mechanisms at the interface. But we must not forget to research the fields concerned with ions in solution, too. It will be seen in Chapter 3, for example, that a great deal remains to be done before we can say that we have a sophisticated picture of ionic solutions at practical concentrations. The pure ionic systems, such as liquid sodium chloride or silicate, require deeper examination, particularly from a theoretical viewpoint, for at present it is only possible to provide a quantitative calculation of transport properties if knowledge of an experimental parameter, the melting point, is available. Besides these feelings of a desire to finish the job which was begun in 1930 and 1940, there is a reasoned case for including "ionics" in the field of electrochemistry. If electrochemistry is basically the study of the electrified interphasic region between an electronic and an ionic conductor and of the transfer of electrons across such

ELECTROCHEMISTRY 23

Period of Early Discovery	Nernstian Hiatus	Rise of Ionics	Electrochemistry Becomes Predominantly Concerned with Interfaces	Electrochemistry Becomes an Interdisciplinary Area
Muscles move when currents pass	Overthermodynamical approach prevents development of charge transfer kinetics at interfaces	Study of ions in solutions, adjunct to electrochemical studies, becomes identified with electrochemistry itself in England and America	The kinetics of electrical reactions at interfaces is formulated, and some basic progress in investigating its mechanisms is made	Electrochemistry plays a role in space power, stability of materials, functioning of biological cells, nylon synthesis, vehicular transportation
1791	1910–1940	1920–1940	1950's	1960's

Fig. 1.12. Electrochemistry, a perspective from afar.

Date	Period Concerned	Main Events
1791–1830	The initial period of great experimental discoveries	Muscular movements are associated with electric currents; there is a "profound connection" of biological and electrical events; a definite amount of current produces a definite weight of deposited material; electric energy can be produced directly from chemical reactions
1890–1905	Electrode kinetics makes a hesitant beginning, but is potentiocentric and soon fades	Tafel's equation connects current and interfacial potential
1891–1947	The great Nernstian hiatus	Nernst's first paper on the thermodynamics of galvanic cells (1891); thermodynamically oriented electrode "kinetics" still dominates Faraday meeting of 1947
1920–1940	The (substitutional) rise (and fall) of ionics dominates electrochemistry	The theory of Debye and Huckel is the first statistical-mechanical theory of solutions to give quantitative success. But it is shown to be intrinsically limited to dilute solutions. No one breaks through this barrier.
1940–1950	Ionics, as well as electrodics, becalmed	Ionics stuck in mathematical difficulties for concentrated solutions. Electrodics not yet fully unfrozen from Nernstian hiatus, except in Russia, where it is developed through 1940's.
1924–1941	Isolated papers give basis to awakening of 1950's	Butler interprets Nernstian potentials kinetically; Volmer gives a theoretical formulation of current–potential relation; Gurney introduces a quantum-mechanical approach to charge transfer, but it is attacked and forgotten; Frumkin connects current to double-layer structure; Horiuti formulates a statistical mechanics of interfacial relations; Eyring formulates a current–potential relation in terms of the theory of absolute reaction rates.

Fig. 1.13. Electrochemistry, a perspective from a medium distance.

Date	Period Concerned	Main Events
1949–1960	Electrochemistry awakes; fundamental research becomes widespread and intense. It is strongly oriented to kinetics of process at interfaces	International Committee of Electrochemical Thermodynamics and Kinetics (CITCE) formed, 1949; Understanding of concept of overpotential and exchange current density becomes general among fundamental electrochemists; methods of examining the mechanism of electrochemical reactions becomes established; two international journals of electrochemistry are founded; NASA chooses electrochemical path for space power; the silver–zinc cell provides high electrochemical power density; Vetter writes a general textbook on electrode kinetics.
1960	Electrochemistry becomes realized as an interdisciplinary area. Its applications spread widely.	Quantum mechanics receives increased attention: Gurney's view is revived. Space vehicles powered electrochemically; quantitative verification of electrochemical theory of failure under stress; Del Duca proposes an electrochemical theory of the biological cell; nylon synthesis by an electrochemical route commences; Ford announces intention to introduce electrochemically powered cars.

Fig. 1.13. *(continued).*

a region, then it is not possible to do much in the theoretical direction unless one knows the structure of both of the phases on either side of the interphase. The structure of the electronically conducting phase—the metal or semiconductor—is well taken care of in the physics of the solid state. Who is to study the solution side except the electrochemists? It is they, primarily, who need to know about it.

Shall such an attitude of keeping ionics as a part of electrochemistry be permanent?

There are some difficulties to it. The subjects of electron transfer across interfaces and of the interactions of ions in solutions differ considerably. They have different parentage, the latter in the theory of solutions and bulk properties of liquids, the former in interfacial chemistry and electrostatics. Although ionics is a field of great breadth of application, its fundamentals are not interdisciplinary. It is a conceptually straightforward part of the applications of statistical mechanics to the interactions of particles in liquids, and its difficulties are largely mathematical. It might better be treated under the "physical chemistry of solutions."

Finally, what turns out to be regarded as a field in a given period of time cannot be decreed. It works itself out. Let us wait another decade or two to determine the wisdom of keeping ionics as well as electrodics within electrochemistry. Here, we take the attitude that we need a knowledge of ionically conducting fluid phases if we are to understand the structure of electrified interfaces and deal with interfacial charge-transfer reactions. Nearly half of what we shall describe here as electrochemistry will be about ions in solutions and in pure ionic liquids. But, it should be understood, we regard ionics as the supportive aspect in electrochemistry, and we wish to solicit the reader's support in thinking of our subject primarily as the study of the structural and other properties of the interphasic region and of charge transfer across interfaces.

1.4. THE RELATIONS OF ELECTROCHEMISTRY TO OTHER SCIENCES

1.4.1. Some Diagrammatic Presentations

Let us look at Figure 1.14 to see something of the parentage of *conventional* electrochemistry, in both the ionic and electrodic aspects.

We could also look at these relations in a different way and derive electrochemistry with the dominant emphasis on charge transfer at inter-

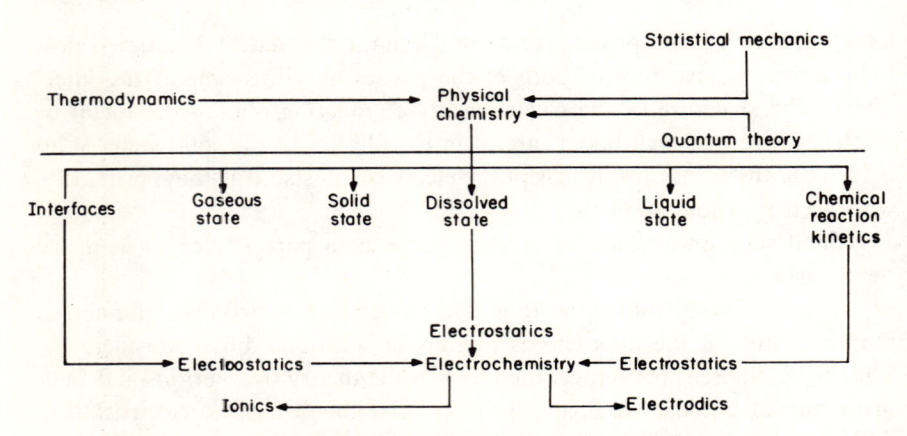

Fig. 1.14. Physical chemistry and electrochemistry

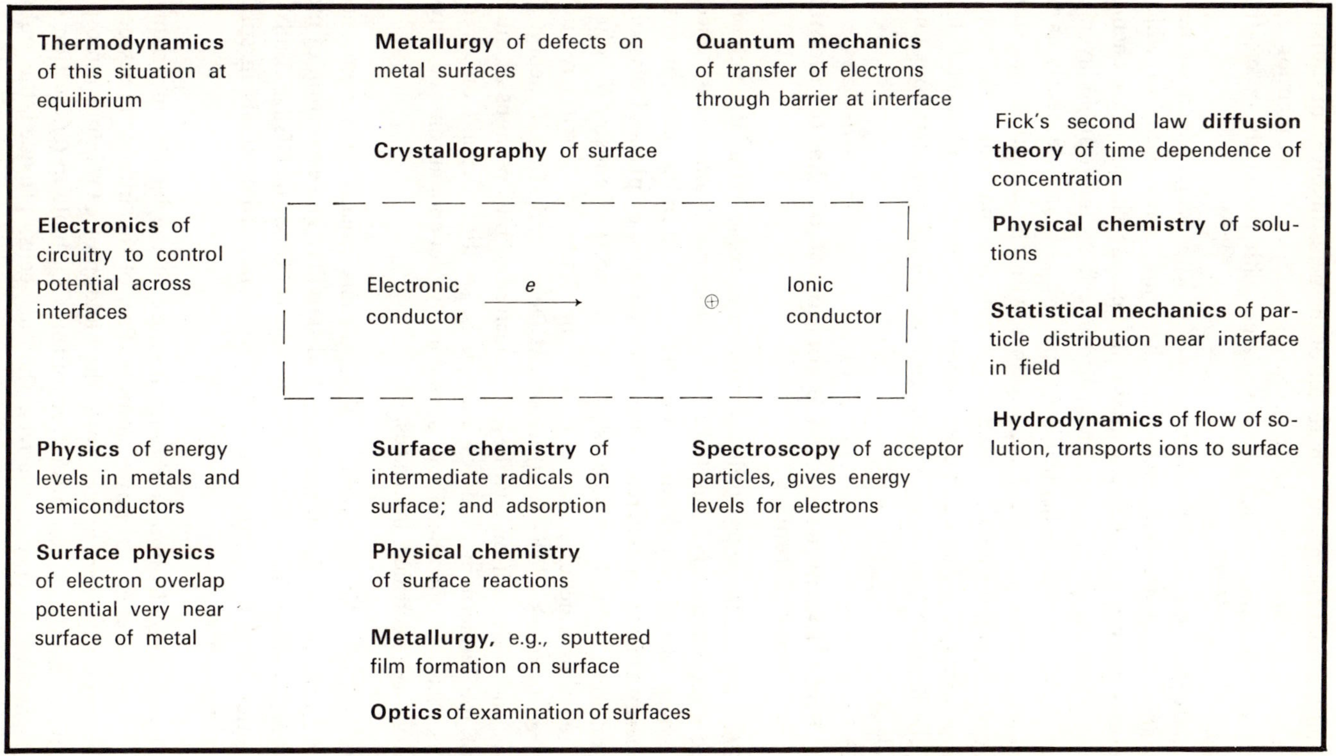

Thermodynamics of this situation at equilibrium

Electronics of circuitry to control potential across interfaces

Physics of energy levels in metals and semiconductors

Surface physics of electron overlap potential very near surface of metal

Metallurgy of defects on metal surfaces

Crystallography of surface

Electronic conductor — $e \rightarrow$ — $\oplus$ — Ionic conductor

Surface chemistry of intermediate radicals on surface; and adsorption

Physical chemistry of surface reactions

Metallurgy, e.g., sputtered film formation on surface

Optics of examination of surfaces

Quantum mechanics of transfer of electrons through barrier at interface

Fick's second law **diffusion theory** of time dependence of concentration

Physical chemistry of solutions

Statistical mechanics of particle distribution near interface in field

Hydrodynamics of flow of solution, transports ions to surface

Spectroscopy of acceptor particles, gives energy levels for electrons

Fig. 1.15. Some disciplines invoved in the study of charge transfer at interfaces.

faces and the interdisciplinary character of the fields involved in studying it. Such a view is implied in Figure 1.15, where space limits mention of the number of disciplines which are associated with the study of electrified interfaces.

Apart from the large number of areas of knowledge which are needed to advance modern electrochemistry, there are many areas to which it contributes or even plays the essential role. Because much surface chemistry under real conditions concerns electrified interfaces and double layers, the systems to which electrochemical concepts are relevant are nearly as wide as surface chemistry itself. This, together with the fact that the subject embraces interactions between electric currents and materials—i.e., between two large areas of physics and chemistry—, signifies the widespread character of the fields of knowledge which electrochemical considerations underly (Fig. 1.15).

1.4.2. Some Examples of the Involvement of Electrochemistry in Other Sciences

Chemistry. Chemistry itself contains much basic material which originated in electrochemistry. The third law of thermodynamics was conceived on the basis of the temperature variations observed in the heat-content and free-energy change in cell reactions. The concepts of pH and dissociation constant were formerly studied as part of the electrochemistry of solutions. Ionic reaction kinetics in solution and the Bronsted–Bjerrum relation is expressed in terms of the electrochemical Debye and Hückel theory. Electrolysis, metal deposition, syntheses at electrodes, plus perhaps as much as half of the modern methods of analysis in solution depend on electrode kinetics. All colloid chemistry is essentially dependent on the physical chemistry of electrified interfaces.

Metallurgy. An electrochemical aspect has already been mentioned in respect to the properties of materials and stability. The extraction of metals from ores dissolved in molten salts, the separation of metals from mixtures in solution, and the formation of single crystals from metals in solution are among others.

Engineering. Examples abound because of the interest in driving vehicles by electrochemical power sources. Part of this interest arises because electrochemical power sources avoid the buildup of CO_2 in our atmosphere and thus avoid the deleterious effects of this CO_2 buildup (*cf.* Section 11.2). Electrochemical energy converters, the fuel cells, provided the onboard power for the first space vehicles, and there are general prospects of evolution away from the thermal to the electrochemical method of realizing the energy of chemical reactions. One of the more certain aspects of the future is the cheapening of electric power, with a sharp fall in the cost when controlled hydrogen fusion becomes practical. Such changes would clearly favor economically the increasing adoption of electrochemical processes. Electrochemical engineering systems include those of a large proportion of the nonferrous metals industries, in particular, the production of aluminum by deposition from a molten salt containing the oxide.

Biology. Food is converted to mechanical work by biochemical mechanisms which have an efficiency much greater than that from some corresponding form of energy conversion involving the heat-engine principle. The mechanism involves electrochemical reactions. Correspondingly, the mechanism of the transmission of impulses through nerves, as well as that of the stability of blood, and probably many of the macromolecules involved in biological processes, depend on aspects of electrochemistry which concern electrochemical charge transport and double-layer interaction.

Geology. An example of electrochemistry in geology concerns certain types of soil movements. The movement of earth under stress depends on the viscosity of the earth regarded as a slurry, i.e., a viscous mixture of suspended solid in water with a consistency of very thick cream. Such mixtures of material exhibit thixotropy which depends on the interactions of the double layers between colloidal particles. These depend on the concentration of ions which affect the field across the double layer and cause the colloidal structures upon which the soil's consistency depends to repel each other and remain stable. Thus, addition of ionic solutions to soils in certain conditions may cause a radical increase in their ability to flow.

1.4.3. Electrochemistry as an Interdisciplinary Field, Apart from Chemistry?

All fields in chemistry, e.g., that of the liquid state or of reaction kinetics, have connections to other parts of the general subject; and, indeed, all subjects treated under chemistry tend to be subject, as time goes on, to treatments at the more advanced degree of sophistication attained under the aspect of the science we call physics. Chemists dare to make approximate treatments of relatively complicated subjects not yet simple enough for physicists to tackle in a more exact way. But the involvement, e.g., of the study of liquids and gaseous reaction kinetics is largely with the parent

Electronics:
Concerns interfacial charge transfer between two electronic conductors.

Electrodics:
Concerns interfacial charge transfer between an electronic and an ionic conductor.

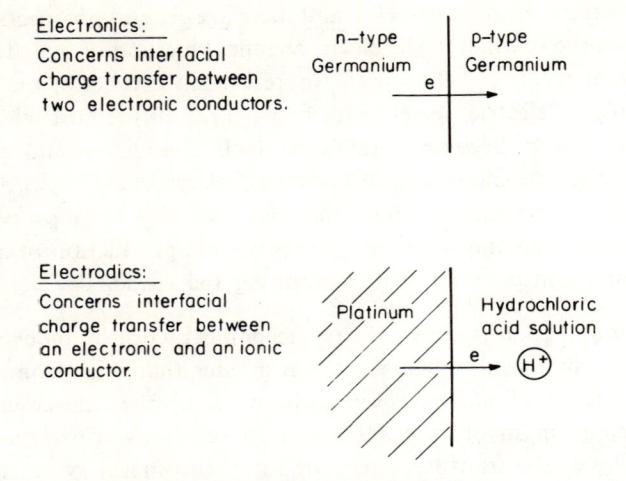

Fig. 1.16. Electronics and electrodics.

areas of chemistry itself (see Fig. 1.14)—e.g., statistical and quantum mechanics. A direct connection to areas right outside chemistry does not exist for the two examples of other fields of physical chemistry given above.

In electrochemistry, however, there is an immediate definitional connection to the physics of current flow and electric fields. Further, it is difficult to pursue interfacial electrochemistry without an acquaintance with some principles of theoretical structural metallurgy and electronics as well as of hydrodynamic theory. Conversely (see Section 1.4.2), the range of fields in science in which the important steps are controlled by the electrical properties of interfaces and the flow of charge across them is great and exceeds the range to which other areas of what is specifically physical chemistry[†] maintain relevance. In fact, so great is it that a worker who is concerned, e.g., with the creation of passive films on metals and their resistance to environmental attack is scarcely in intellectual contact, e.g., with the man who is interested in attempting to find the model for why blood clots or with him who is seeking to extract a metal from sea water.

These widespread involvements with other areas of science, in respect to both the family of disciplines which compose the considerations involved in electrochemical thoughts, and the breadth of applications of electrical aspects of surfaces and surface reactions throughout the sciences connected

---

[†] As apart from areas of basic science, e.g., quantum mechanics, which primarily originate in physics and underlie all chemistry, including, of course, electrochemistry.

with materials, may suggest that electrochemistry be handled increasingly in the future as an interdisciplinary area rather than a branch of physical chemistry.

Correspondingly, there exists a general tendency at the present time to break down the older formal disciplines of physical, inorganic, and organic chemistry and to make new groupings. That of materials science —the solid-state aspects of metallurgy, physics, and chemistry—is one. Energy conversion—the energy-producing aspects of nuclear fission, electrochemistry, photovoltaics, thermionic emission, magnetohydrodynamics, etc.—is another. Electrochemistry would be concerned with the part played by electrically charged interfaces and interfacial charge transfer in chemistry, metallurgy, biology, engineering, etc. *Perhaps* it would also involve areas which are directly related to such charge transfer, as is the physical chemistry of ionic solutions (see Section 1.3.4).

The precise subject matter considered covered by the term electrochemistry will evolve. But the divisions of science into physics, chemistry, metallurgy, biology, etc., are purely arbitrary and were made with a nineteenth century perspective. It may be that a time for regrouping and for the recognition of new areas, homogenized by a common aspect of sufficient importance and breadth of relevance in other sciences and to technology, is here—or may be overdue.

1.5. ELECTRODICS AND ELECTRONICS

Aristotle wrote in his *Poetics*: "What is important to master is metaphor. It is the mark of genius." The direct metaphor seeks things which may be likened to each other. Much may be learned from such likenesses. Electrochemical systems are like electronic systems. What are *electrodic* [*sic*] systems? They are two-phase systems in which one phase, an electronic conductor, is connected by an electric field (across which electrons sometimes pass) to the other phase, an *ionic* conductor. What are *electronic* systems? They are two-phase systems, *both* phases electronically conducting, between which an electron passes, as shown in Figure 1.16.

In both subjects, interfacial electron transfer is the essential event. In one, electrodics, it takes place from a phase in which electrons are the principal charge carriers to a phase in which ions are the principal charge carriers. In the other, electronics, the transfer of charge is to a phase in which the electron does not remain localized in one particle but is free to continue movement in the new phase. Superficially, there is not a great deal of difference between the initial theory of electronics and electrodics. This can be seen by comparing the basic equation in both fields. In electrodics, it is

$$i = Ac_{ion}e^{-\beta\eta F/RT}$$

where i is the rate of passage of electrons across the interface, A and β are constants for the system, c_{ion} is the concentration of ions in contact with the surface of the electronic conductor, and η is the disturbance of the potential difference across the interface from that corresponding to electron-transfer equilibrium. In electronics, the initial equations is that for thermionics, i.e.,

$$i = A'e^{-\Delta\Phi/RT}$$

where the terms have analogous meanings, $\Delta\Phi$ being the difference in energy of the electron inside and outside the electronic conductor from which it is ejected.

A similar analogy exists between the behavior of electrons and holes in semiconductors and the behavior of H^+ and OH^- ions in electrolytes. Insights of this type are important to make and should be sought because it is clear that "all Science is one," and, in a remote sense, all phenomena take place in consistency with the same general laws of physics. It is helpful, then, to try to break down the barriers between fields distinguished largely by having different names and, unfortunately, different terminologies. Often, one finds parts of one's problem have been solved in another field or the other field has a problem to which one's own field has a solution, or nearly so. Hence, electronics should be regarded as a neighboring subject to electrodics not only because electronic circuitry external to the electrodic system is essential in helping electrodics but because there is a fundamental closeness in electrodic, and some ionic, systems to those studied under the terms electronics and thermionics.

1.6. TRANSIENTS

Hitherto, we have referred to the interfacial transfer of electrons as though this were always a steady-state process, i.e., one occurring at a rate invariant with time. The corollary is, of course, that, because of the law (cf. Fig. 1.9)

$$i = Ae^{B\eta}$$

the interfacial potential difference across the double layer must be constant.

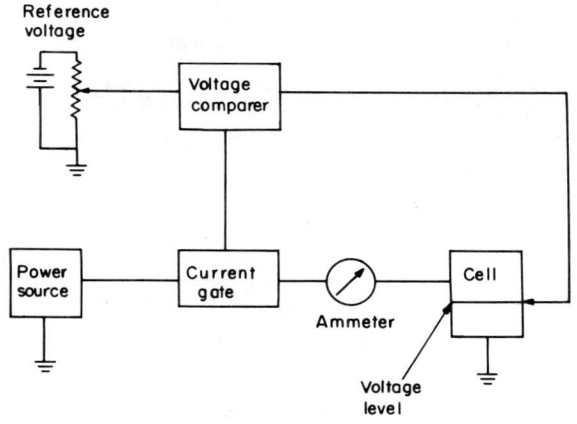

Fig. 1.17. An electronic circuit which produces transient behavior in an electrodic circuit.

It follows therefore that, if, between the power source and the cell in the circuit in Figure 1.17, we add a device which enables us to vary the η (the variable part of the potential difference across the interface) with time, the rate of charge transfer will vary. Further, the delay time between a change of potential and the resulting current is (1) small and (2) characteristic of the details of the type of process occurring across the interface. (For example, does the rate at which the electrons are being transferred control itself in a relatively simple way, e.g., just vary at the behest of the potential as the above equation indicates, or is the charge-transfer reaction rate indirectly controlled by something else, such as the adsorption of some entity from solution which might conceivably partly block the transfer site, with the degree of blockage also a function of potential but a different function from that of charge transfer?)

In the last section, an analogy was made between the fundamental processes involved in charge transfer at an electronic-conductor–ionic-conductor interface and those at an electronic-conductor–electronic-conductor interface. It has not yet been shown, however, that electronics is connected with electrochemistry in another way, i.e., as a tool, the value of which can hardly be overestimated. Thus, consider the matter of transients. By electronic circuits of appropriate design, one may impose upon the metal electrode electric-energy pulses of virtually any shape or size and with widely varying time durations. One can make the interfacial current dance to the tune being played by the electronic circuit. The time ranges in which one can operate are as low as tens of nanoseconds.

Such ability to program circuitry to make metal–solution interfaces do things at command and very quickly has great significance for research work on the electrode–solution interface. It allows us to change the electrical interfacial conditions from a given condition so quickly that, before the particles on the surface (intermediate radicals in some reaction, say) have had time to adjust to the new order, they have been hit by a change of potential which promotes a reaction which, e.g., electrically burns them up and wipes them off the surface. Then, by finding out the current passing during this process and integrating it with time, one has a measurement of the total *amount* of electricity equivalent to the radicals on the surface and hence a measure of the degree of coverage of the surface with radicals. Such a quantity is often a vital piece of knowledge in research on reaction mechanisms and eventually, e.g., on energy converters.

Such a use of transient techniques makes a good example of why the mechanism of electrical interfacial reactions has better prospects of being understood than have classical heterogeneous "chemical" reactions. Electrochemists have known of transient techniques to obtain the concentration of radicals on noble metal surfaces for about thirty years, but the corresponding technique of the heat pulse in thermal kinetics has had a much shorter history. Flash photolysis, a transient technique in gas kinetics, was introduced twenty years after the effective and more powerful transient technique available for radical investigation and control in electrochemical processes. The advantage of the electrochemical measurements is that one has really just to count electrons, and, for that, there are available a large number of rather sophisticated techniques.

1.7. ELECTRODES ARE CATALYSTS

Here again, as with electrodics and electronics, there is a comparison which is informative. When chemical reactions occur on surfaces—*heterogeneous reactions*, as they are called—, there is generally a great dependence of the rate upon the nature of the solid surface concerned. On some surfaces, the rate may be so small that the reaction is said not to take place at all. On others—they may be metals but are often also oxides—the rate is large; such surfaces are said to *catalyze* the reaction concerned, i.e., they make it go faster without themselves undergoing any change in the reaction.

The electronic conductors which are the source and sink of electrons in electrochemical interfacial electron-transfer reactions are called electrodes, and a few experimental facts quickly show that electrodes are catalysts. For example, see Table 1.1 of rates (expressed as the current passing equally in both directions across the surface when this is held at

TABLE 1.1

The Rate of the Hydrogen Evolution Reaction on a Series of Metals at the Equilibrium Potential

Metal	log rate, amp cm^{-2}	Metal	log rate, amp cm^{-2}	Metal	log rate, amp cm^{-2}
Pd	−3.0	Ag	−6.1	Sn	− 8.2
Pt	−3.1	Nb	−6.4	Tl	−10.2
Rh	−3.2	Mo	−6.5	Cd	−10.7
Ir	−3.7	Cu	−6.7	Mn	−10.7
Ni	−5.2	Ta	−7.0	Ga	−11.0
Fe	−5.2	Bi	−8.1	Pb	−11.3
Au	−5.7	Al	−8.0	Hg	−12.0
W	−5.9	Ti	−8.2		

the potential which corresponds to equilibrium) for the electrochemical reaction of hydrogen evolution

$$2H^+ + 2e \rightarrow H_2.$$

Thus, for this reaction, there is a very great dependence of the reaction rate upon the surface chosen.[†] Electrodes, then, are catalysts; but the science of electrocatalysis is in its infant stages because, like so much else in electrochemistry, the concept of electrodes as catalysts is only just becoming a concept—it was first written of in 1963 by Grubb. The rate at which a reaction occurs for a given departure from equilibrium in the electric potential applied across the interface depends on the constant A of the equation $i = Ae^{B\eta}$, and A must therefore be a kind of *chemical* rate constant, within the electrochemical reaction, upon which is superimposed the influence of the interfacial electric field.

1.8. THE ELECTROMAGNETIC THEORY OF LIGHT AND THE EXAMINATION OF ELECTRODE SURFACES

It will be recalled that electrochemistry is the study of electrified interfaces with special reference to electron transfer across them. A well-established piece of physics deals with the band structure of the energy levels

---

[†] In fact, the variation of some 10^{10} times is far greater than the range of variation of rates observed for a given reaction on various catalysts in thermal catalytic reactions.

in the electronic conductor. A good picture of the solution side is also available for dilute solutions (though some of the theory of the acceptor states in the molecules in solution had to be developed by electrochemists). What is little understood is the *state of the electrode surface* in contact with a solution. Of course, the surface of the liquid metal mercury is fairly well understood, but work on mercury surfaces in electrochemistry has become increasingly relegated to chemists interested less in fundamental advances in, e.g., electrocatalysis and more in using electrochemistry as a tool, perhaps for the analysis of compounds dissolved in solution. The modern trend is toward the study of solid surfaces. To work with them without knowing their structure is a little like attempting to determine the mechanism of decomposition of a solid without first knowing the crystal structure of the substance concerned.

It has been pointed out that there is an advantage in the electrochemical way of studying the mechanism of reactions compared with the chemical way because, in the former, one can make a determination of the degree of coverage of the surface by radicals and sometimes distinguish one radical from another. However, this advantage would be counteracted if one had no way of *looking* at surfaces on electrodes *in situ* (one can look at catalysts used *outside* solutions, e.g., those used in classical heterogeneous reactions, by means of electron microscopy). The necessity for *in situ* examination of electrode surfaces gains added significance because the presence of adsorbed layers and oxide films on surfaces is the rule rather than the exception for reactions on electrode catalysts.

It is of interest to remind ourselves why high-energy radiations, e.g., electron beams or X rays, cannot be used for looking at electrode surfaces *in situ*. It is because they interact strongly with the solution and thus get absorbed. The energy necessary to get electrons to pass through even a few millimeters of aqueous solution and still have energy to send back a message after reflection from the electrode is too much to make such methods practical. Thus, there is the significant difficulty that the electron-optical and X-ray methods used for surface studies in solid-state chemistry cannot be used in the case of electrode catalysts, the seat of the charge-transfer reactions which are the foundation of electrochemistry.

This difficulty can be compensated by using visible light, which can pass with facility through the solution, be reflected by the electrode surface, and emerge out from the solution (Fig. 1.18). If the incident light is polarized, then the process of reflection produces changes in the state of polarization. These changes are determined essentially by the state of the surface—the existence and nature of the film on the electrode—and can be

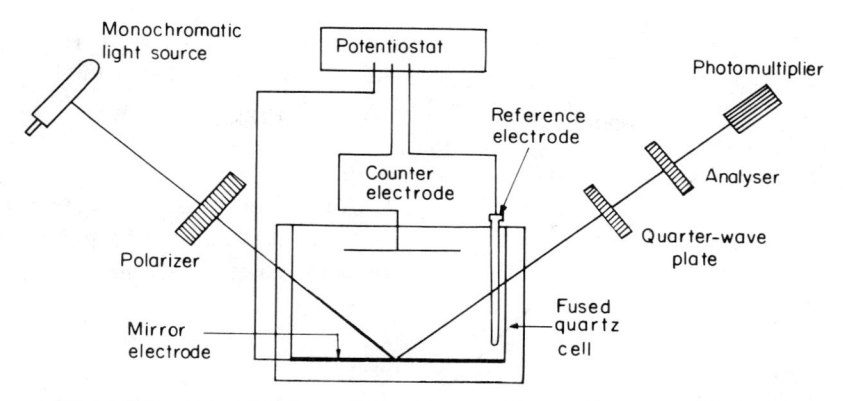

Fig. 1.18. Determination of certain properties of film-covered surfaces in solution by ellipsometry.

understood by using the electromagnetic theory of light and, in particular, Fresnel's equations for the reflection of polarized light.

This technique of ellipsometry was originated by Tronstad as early as 1937, but it has only been in the past few years that new and powerful ways of using ellipsometry to study electrode surfaces have developed. It seems likely that its use will spread. Applied to the detection and measurement of adsorbed films, it should give an optical means of following the concentration of radicals on the surface while the electronic instrumentation follows the rate of the charge transfer. Further, by using different frequencies of light, a kind of ellipsometric spectroscopy may be devised.

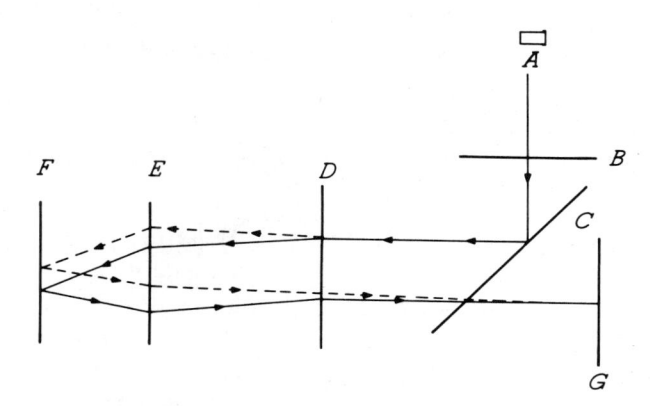

Fig. 1.19. Schematic of polarized-light microscopy on a surface in solution; A, light source; B, polarizer; C, semi-silvered mirror; D, Wolloston prism; E, microscope ens; F, object; G, analyzer.

Other methods of using polarized light involve interferometry with polarized light by which irregularities of a few hundred angstroms in the surface structure can be observed. Interference contrast microscopy, also using polarized light, lets us observe entities on the surface which are about half the wavelength of light in extent (Fig. 1.19). Both these kinds of microscopy use visible light and not electrons; hence, they can be used in the presence of solutions (but, conversely, do not resolve the picture as well as electrons because the wavelength, and hence resolving power, of visible light is in the range of thousands of angstroms, and that of electrons may be made to be a few angstroms.)

Electron microscopy can be and is used for obtaining the state of the surface of an electrode if the latter is removed suddenly from the solution while the current is still passing. Replicas of the surface structure which correspond to that at the moment of removal can then be made. If the current were just switched off and the electrode left in the solution, it would have a surface which would be difficult to associate with any clear condition of potential across the interface because of the growth which may continue *after* cessation of current. As this further growth involves strictly charge-transfer mechanisms and these demand the presence of a solution, the removal of the electrode from the solution leads to a cessation of current and prevention of further growth.

The use of electron microscopy in this way (with the electrode surface frozen at a certain moment and then examined) is of course hardly a convenient way of following changes on the electrodes surface, e.g., with potential.

1.9. SCIENCE, TECHNOLOGY, ELECTROCHEMISTRY, AND TIME

1.9.1. Do Interfacial Charge-Transfer Reactions Have a Wider Significance than Has Hitherto Been Realized?

It is informative, in this chapter, to make some attempt to fit electrochemistry among the sciences to obtain the relative measure of its significance. It is difficult to do this for a field, the modern phase of which dates back only 10 to 15 years. Let us try, but let us realize that what we are doing here is making a *speculation*, although we shall give some of the reasoning which makes it seem, at this time anyway, a probable one.

Firstly, we can ask a *relative* question: Is interfacial electrochemistry simply a special aspect of reaction kinetics, somewhat analogous to photochemistry? There, one might say, one studies the effect of energy packets striking molecules by means of radiations; and, in electrochemistry, one studies the effect of striking molecules dissolved in solution with electrons emitted from electrically charged conductors.

Or is the significance of interfacial electrochemistry something more? The evidence for a point of view tending to this latter direction is as follows:

1. Interfacial electrochemistry has a relatively high degree of ubiquity compared with those of other branches of knowledge outside physics. A way to appreciate this is to realize how often one is concerned with electrochemical phenomena outside the laboratory. For example, one shaves, starts the car, and listens to its radio on battery power; television pictures of the moon are transmitted from space vehicles to the earth by fuel-cell power; the office building may be partly made of electrochemically extracted aluminum; the water which is in the coffee may be obtained by electrochemical deionization from impure or brackish water; one may ride a car with a magnesium-alloy engine block produced from seawater by an electrochemical process; one uses clothes of nylon produced from adiponitrile, which is electrochemically synthesized; the gas in one's car contains lead tetraethyl, probably produced electrochemically; one may take a tranquilizer which exerts an effect by an electrochemical mechanism; one adds an inhibitor to the radiator fluid of a car to reduce electrochemical corrosion. Finally, one thinks by electrochemical mechanisms, and one's blood remains functional insofar as the electrochemical potential at the interface between its corpuscles and their solution remains sufficiently high.

2. This ubiquitous role of electrified interfaces throughout many aspects of science suggests that electrochemistry should not be regarded as a *branch* of chemistry. Rather, while most chemists have concentrated upon *thermally activated* reactions and their mechanisms, with electrochemical reactions as some special academic subcase, there is a parallel type of chemistry not based on the collisions of molecules and the energy transfers which underlie these collisions but on interfacial electron transfer. It is this chemistry of interfacial electron-transfer processes that seems to underlie much of what goes on around us.

Some examples of alternative thermal and electrical approaches to common chemical happenings may illustrate this (Table 1.2).

1.9.2. The Relation between Three Major Advances in Science, and the Place of Electrochemistry in the Developing World

If one stands sufficiently away from one's specialization in the sciences and looks as far back as the nineteenth century, then three great scientific

TABLE 1.2

Examples of the Alternative Thermal and Electrochemical Paths in Chemical Happenings

Phenomenon or process	Thermal	Electrochemical
The determination of free-energy changes and equilibrium constants in chemical reactions	Determine equilibrium constant and use $\Delta G^0 = -RT \ln K$	Determine thermodynamic cell potential and use $\Delta G_0 = -nFE$
Synthesis, e.g., water from hydrogen and oxygen	Occurs heterogeneously presumably by noncharge-transfer collisional processes $H_2 + \frac{1}{2}O_2 \rightarrow H_2O$	Occurs in electrochemical cell by reactions $H_2 \rightarrow 2H^+ + 2e$ $\frac{1}{2}O_2 + 2H^+ + 2e \rightarrow H_2O$ $\overline{H_2 + \frac{1}{2}O_2 \rightarrow H_2O}$
Biochemical digestion	Series enzyme-catalyzed chemical reactions	Some enzymatic reactions may act through electrochemical mechanisms analogous to local cell theory of corrosion
Many so-called chemical reactions, e.g., chemical synthesis of Ti	$TiCl_4 + 2Mg \rightarrow Ti + 2MgCl_2$ (apparently a thermal collisional reaction)	$2Mg \rightarrow 2Mg^{++} + 4e$ $Ti^{4+} + 4e \rightarrow Ti$ $4Cl^- \rightarrow 4Cl^-$
Production of electrical energy	$H_2 + \frac{1}{2}O_2$ explodes, produces heat, expands gas, causes piston to move, and drives generator	H_2 and O_2 ionize on electrodes, as above in this column, and produce current
Storage of electrical energy	Electricity pumps water up to height and allows it to fall on demand to drive generator	Allow to cause some electrochemical change, e.g., $Cd^{++} + 2e \rightarrow Cd$, $Ni^{++} \rightarrow Ni^{4+} + 2e$, which will be reversed on demand
Deionize water	Distill it and leave behind nonvolatile ions	Electrodialyze through membrane
Syntheses of inorganic and organic material, e.g., Al, adiponitrile	$2Al_2O_3 + 3C \rightarrow 3CO_2 + 6Al$; Tetrahydrofuran $\rightarrow$ 1,4-dichlorobutane $\rightarrow$ adiponitrile	$Al^{3+} + 3e \rightarrow Al$ $H_2C = HC - CN \xrightarrow[\text{coupling}]{\text{cathodic}}$ $NC - (CH_2)_4 - CN$

TABLE 1.2 *(continued)*

Phenomenon or process	Thermal	Electrochemical
Spreading of cracks through metal	Amount of stress at the apex of crack per unit area is so high that crack is propagated into metal bulk	Bottom of crack dissolves anodically, obtaining current from local cell formed with surface (which is an electron donor, probably to O_2 from air)
Practical enhancement of stability of metal	Cover surface with coating; prevent access of air	Appyl electron-donating field, reduce electron acceptance, and hence reduce $M \rightarrow M^{2+} + 2e$

contributions stand out in the matter of their impact on science and technology. They are:

1. The electromagnetic theory of light due to Maxwell (nineteenth century)
2. The theory of relativistic mechanics due to Einstein (nineteenth to twentieth centuries)[†]
3. The theory of quantum mechanics originating with the work of Planck, Einstein, and Bohr and developed by Schrödinger, Heisenberg, Born, and Dirac (twentieth century)

It is important to recall why these contributions to physics and chemistry are regarded as so outstandingly important. Maxwell's nineteenth century theory is the basis for a large fraction of twentieth century technology. It provides the basis for the transfer of energy and communication across distances, and the delivery of mechanical power at command as a consequence of the controlled application of electric currents. The significance of relativistic mechanics is that its theory of mass–energy equivalence constitutes the basis of the generation of nuclear power which, in association with electrical and electronic devices, makes it possible in decades rather than centuries to end man's necessity of having to work to be able to live. The quantum theory is the basis of the solid-state device and transistor technology and has given us such a great revolution in thinking, e.g., the

---

[†] Einstein's first paper was of 1905, but the thought mode and many consequences were based on thoughts typical of the nineteenth century, in particular, the time necessary for the propagation of radiation.

knowledge that macroscopic and microscopic systems behave in fundamentally different ways and insight into many phenomena at a molecular level (particularly the understanding of the permanence of genes), that it takes its place with the greatest of the contributions to knowledge.

For how is weighed the eventual magnitude of a contribution in science? Is it not the degree to which the applications which arise from it eventually change everyday life? Is not the essence of our present civilization the controlled development of the surroundings at man's command?

It is in this light that one may estimate the relative significance of the theory of electrified interfaces and thus of electrochemistry. It is of interest to note how interfacial charge-transfer theories are based on a combination of the electric currents of Maxwell's theory and the quantum-mechanical tunneling of electrons through energy barriers.

A number of illustrations have been given to support the statement that electrochemical mechanisms are relevant in many fields of science. The nineteenth century contributed to physics the theory of electromagnetism. The twentieth century contributed to physics the relativistic theory, an implication of which is the provision of limitless cheap energy, and the quantum theory, which underlies solid-state electronics and has promoted a fundamental revision of thinking in chemistry and biology. In the twenty-first century, it seems reasonable to assume that the major preoccupations will be in the direction of taking up and working out the consequences of these developments toward a predominantly electrical postindustrial technology from which noncreative work will have been eliminated.

Two very general types of probable advances can be made out. One is in the direction noted, for example, the development of practical photogalvanic energy-conversion devices for the direct conversion of light to electricity, or the development of replacement of parts in the body by artificial devices connected with body circuits.

The other type will be those developments necessary as corollaries of the interference with nature of the last 50 years, for example, electrically powered vehicular transportation to avoid increasing the CO_2 content of the atmosphere and processes for the reduction of the presently polluted state of much useful water, and of the present atmosphere.

Electrochemistry is the basic science upon which many of these electrically oriented advances of the next few decades will be founded. It underlies electrically powered synthesis, extractions (including fresh from brackish water), machining, stabilization of materials, storage of energy in the form of electricity, efficient conversion to electricity of the remaining fossil fuels, and the basis for practical developments in molecular biology.

It is worthwhile thinking, too, that the mode of organization of urban living is likely to develop as a function of available electricity from atomic sources. Corresponding to such a basic development will tend to come an increasing need to invest resources in recovering from atmospheric and aqueous pollution (even of the sea) and an increasing exhaustion of supplies.

In such a situation, the town will tend to be built around the reactor. The only form of energy used will be electrical. The town will be very largely self-contained. Little material mass will leave or enter it. The processes on which it will be run will be all electrical, and those involving matter, therefore, electrochemical. Transportation will use electrochemically stored electricity; manufacturing and machining processes and recovery of materials used or degraded will all be electrochemical. Polluted liquids will be electrodialytically regenerated. Sewage will be processed electrochemically. Medical electronics—the electronic–electrodic combination in medical research—will be highly developed.

Thus, by the end of the present century, it seems reasonable to expect the achievement of several electrochemically based innovations: a great extension in the use of cordless devices; the provision of cheap heat electrically from storers charged during off-peak times; electrochemically powered vehicles and perhaps ships; a viable photogalvanic solar conversion system; extensive use of electrochemical machining and electrochemically based tools; the electrochemical production of metals of great strength; an internal fuel-cell-powered heart; and electrometallurgical extractions of materials on a large scale from the moon. Many more areas of probable advance could be mentioned. A large, immediately developable area lies in the electrochemical aspects of molecular biology and in the development of circuitry which will join the brain and its probably electrochemical mechanisms to artificial limbs with their electrochemical functions and, perhaps, even to circuits not connected to the body.

Thus, a press of advances in the electrical aspects of matter, and therefore in electrochemistry, seems likely to develop, along with the transition of the principal attention of material scientists from the bulk to the surface. During the next few decades an increasing fraction of our surroundings will have an electrochemical aspect.

Let us, therefore, read this book with the realization that it is written early in an era in which the electrical aspects of existence will bear great increase, that, often in that time, the detailed interpretive positions described below as probable will be modified and improved, and that all those prospects which lie before us in electrochemical applications depend greatly for their realization upon the fundamental work by which the relevant basic mechanisms become understood.

To realize this vista, contributions from scientists of varied fields of original training will be necessary. They have to learn about the field of electrochemistry. It will be necessary to train electrochemists and electrochemical engineers in far greater numbers than is now done. The stimulus for the writing of this book is the need to provide these workers with material which assumes a background knowledge only of freshman physics and chemistry and yet takes the student in most of the areas considered to a position in which he can comprehend the specialist monograph in electrochemistry.

CHAPTER 2
ION–SOLVENT INTERACTIONS

2.1. INTRODUCTION

An electrochemical system (Fig. 2.1) includes two interfaces, at each one of which an electronic conductor is in contact with an ionic conductor (or electrolyte[†]). The electronic conductor is generally a metal but may well be a semiconductor. The ionic conductor, as the term suggests, is a material which consists of mobile ions.

How does one produce a medium of mobile ions? One method is based on the fact that certain substances, which, in the pure form, do not contain any significant concentration of ions, are able to interact to produce ions. This is how neutral, i.e., nonionic, molecules of water and of acetic acid interact to give an electrolytic solution of hydrogen ions and acetate ions (Fig. 2.2 and Table 2.1). This *chemical* method of producing an ionic conductor will be studied in Chapter 5.

Another approach is based on starting off with a *solid* ionic crystal and reducing the forces which hold the ions together. A stage is reached when the cohesive forces are so weakened that the ions, which could only

---

[†] The term *electrolyte* is used in electrochemistry to refer not only to the *ionically conducting medium* through which electricity is passed but also to the substances which, when dissolved (or melted), give rise to a conducting medium.

46 CHAPTER 2

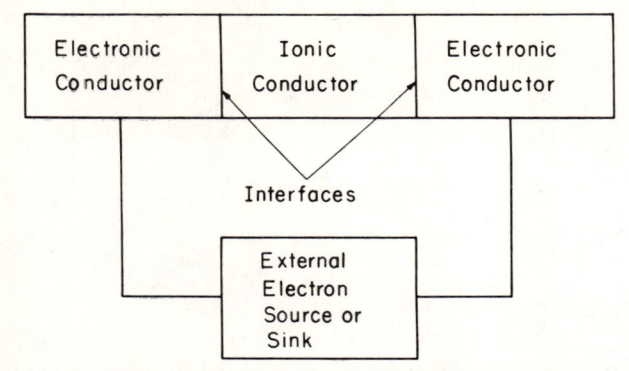

```
┌─────────────┐  ┌─────────────┐  ┌─────────────┐
│  Electronic │  │    Ionic    │  │  Electronic │
│  Conductor  │  │  Conductor  │  │  Conductor  │
└─────────────┘  └─────────────┘  └─────────────┘
                        ↓ ↓
                    Interfaces
              ┌──────────────────┐
              │     External     │
              │     Electron     │
              │    Source or     │
              │       Sink       │
              └──────────────────┘
```

Fig. 2.1. The essential parts of an electrochemical system.

vibrate in the solid, acquire a new degree of freedom—the freedom of translational motion.

There are two distinct ways in which the interionic forces in a crystal can be overcome. One method is based simply on an agitational effect. Heat energy is used to increase the tempo of the ionic vibrations in the solid until thermal forces prevail and the long-range order of the ionic arrangement in the crystal lattice is wiped out—the ionic crystal "melts" (see Chapter 6).

One is left with a *pure liquid electrolyte*, a molten material teeming with positive and negative ions and with free space which is far more plentiful than in the solid. These ions are in ceaseless random motion and ready to respond to applied electric fields by conducting electricity. What has been described is a *thermal* method of obtaining a pure liquid electrolyte.

There is, however, another way of overcoming the interionic forces in an ionic crystal and producing mobile ions. This is with the aid of a *solvent*. A crystal of potassium chloride, e.g., is placed in water. Soon it becomes apparent that the crystal *as an entity* has disappeared. The solvent has

$$H_3C-C\genfrac{}{}{0pt}{}{=O}{O-H} + O\genfrac{}{}{0pt}{}{H}{H} \xrightarrow{\text{Proton-transfer reaction}} \left[H_3C-C\genfrac{}{}{0pt}{}{=O}{O}\right]^- + \left[H-O\genfrac{}{}{0pt}{}{H}{H}\right]^+$$

Neutral acetic   Neutral   Acetate ion   Hydrogen ion
acid molecule   water
                                    molecule

Fig. 2.2. The chemical method of producing ionic solutions.

TABLE 2.1

Ionic Concentrations in Pure Water, Pure Acetic Acid, and Acetic Acid Solution

	Ionic concentration g-ions liter^{-1} at 25 °C
Pure water	10^{-7}
Pure acetic acid	$10^{-6.5}$
0.1N acetic acid solution	10^{-3}

enticed the ions out of the solid so that they can wander off into the solvent (Fig. 2.3). (The Greek word for *wanderer* is *ion*.) One has witnessed the process of *dissolution of an ionic crystal*.

What are the influences which the solvent brings to bear upon the ions of the crystal? What are the ion–solvent forces which overcome the ion–ion forces holding together the crystal?

It is obvious that questions such as these are of central significance to the understanding of ionic solutions and, hence, the electrochemical processes which occur in them. For the questions imply that ions in solution are constantly affected by ion–solvent forces, and that, to understand the behavior of ions inside an electrolytic solution, one has to reckon with the forces arising from the presence of the solvent. One must understand *ion–solvent interactions*.

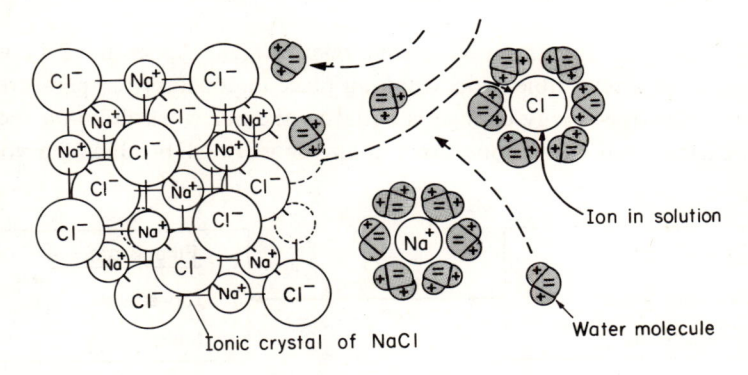

Fig. 2.3. Dissolution of an ionic crystal by the action of a solvent.

2.2. THE NONSTRUCTURAL TREATMENT OF ION–SOLVENT INTERACTIONS

2.2.1. A Quantitative Measure of Ion–Solvent Interactions

A field of study often undergoes a qualitative change when the concepts used can be associated with numbers and made quantitative. The problem, therefore, is to develop a quantitative measure of ion–solvent interactions. This type of problem is a common one in chemistry. It is often solved by considering two situations or states, one where the interactions operate (are "switched on") and the other where they do not exist (are "switched off"), and then computing the free-energy difference ΔG_{I-S} between the two states (Fig. 2.4).

In the case of ion–solvent interactions, the state in which the interactions exist is an obvious one; it is the situation in which ions are inside the solvent. Ions are charged particles, and charges interact with other charges. So there will also be ion–ion, as well as ion–solvent, interactions in the solution. But the former are excluded in the quantitative analysis of ion–solvent interactions; they will be given separate consideration later on (Chapter 3).

Now, what is a situation in which there are no ion–solvent interactions? Obviously, one in which there is no solvent. Hence, one must consider an initial state in which there are large spaces between individual ions, and nothing else present. The initial state, therefore, is that of *ions in vacuum* at an infinitely low pressure.

The problem, therefore, is to consider the free-energy change for the transfer of ions from vacuum to solution (Fig. 2.5).

$$\text{Ions in vacuum} \rightarrow \text{Ions in solution.}$$

Recall, however, the thermodynamic relation (*cf.* Appendix 2.1) which states that, in a reversible process taking place at constant temperature and pressure, the free-energy change is equal to the *net* work done *on* the system, i.e., the total work done other than the work of producing a volume change.

Initial state	Free energy change, ΔG_{I-S}	Final state
No ion-solvent Interactions	→	Ion-solvent Interactions

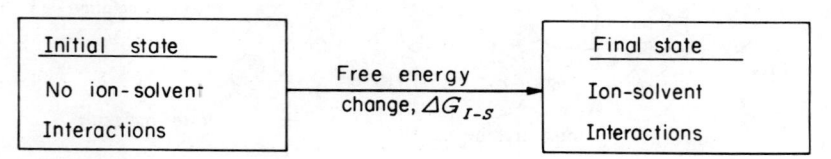

Fig. 2.4. The free-energy change arising from ion–solvent interactions.

ION–SOLVENT INTERACTIONS 49

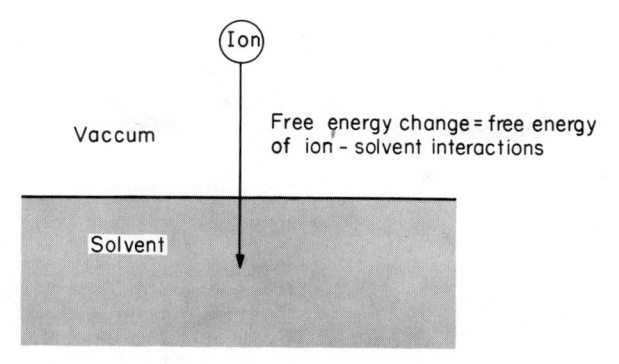

Fig. 2.5. The free energy of ion–solvent interactions is the free-energy change resulting from the transfer of ions from vacuum into solution.

Hence, the basic problem of deriving an expression for the free energy of ion–solvent interaction can be defined as follows. What is the work done when one transfers an ion from vacuum into a position deep inside the solvent? This work will include the energy of all the interactions between the ion and the surrounding solvent, for example, water.

2.2.2. The Born Model: A Charged Sphere in a Continuum

A moment's thought will reveal that, to work out exactly all the ion–solvent interactions, one must know the structure of the solvent, i.e., the dispositions of all the particles constituting the solvent and the forces between the ion and these particles. But the solvent, e.g., water, may have a fairly complex structure. To understand this structure, one must be able to answer a vast number of questions. For example, are there discrete solvent molecules, or are they associated to such an extent that one should not speak of separate molecules? What do the ions do to the solvent structure? Do they disrupt it, or are there spaces inside the structure so that ions can be smuggled in but cause little damage to it?

The problem seems insuperable, but one can resort to modelistic thinking. Models are simplified representations of the real microstructure of nature, often as mental pictures derived from the macroscopic world. They are intended to reproduce approximately the essential features of the real situation. The better they are able to predict experimental quantities, the better do they serve as aids to thinking about how nature really works.

An example of a *very* crude and approximate model for ion–solvent interactions is that suggested by Born in 1920. In the Born model, an ion

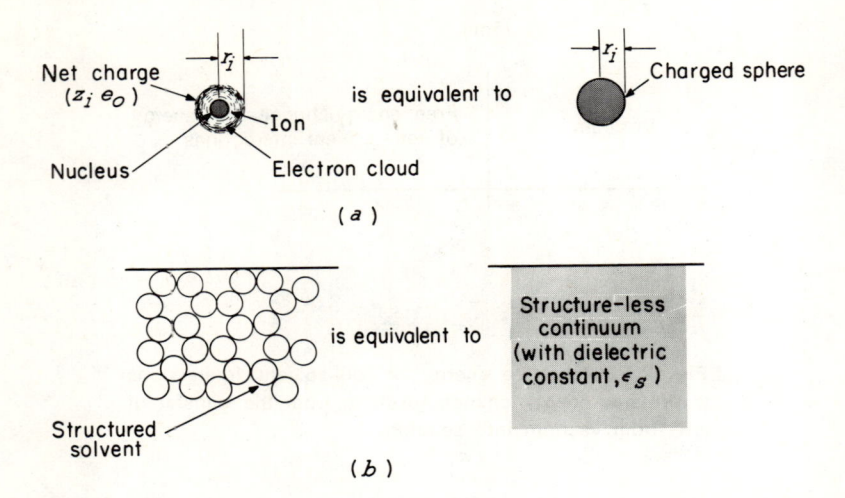

Fig. 2.6. The Born model for ion–solvent interactions considers (a) an ion equivalent to a charged sphere and (b) the structured solvent equivalent to a structureless continuum.

is viewed as a rigid sphere (of radius r_i) bearing a charge $z_i e_0$ (e_0 is the electronic charge), and the solvent is taken to be a structureless continuum (Fig. 2.6). Thus, the problem of ion–solvent interaction assumes the following form: What is the work done in transferring a charged sphere from vacuum into a continuum (Fig. 2.7)?

By considering a charged sphere equivalent to an ion, the Born model is assuming that it is only the *charge* on the ion (or charged sphere) that is responsible for ion–solvent interactions. The interactions between the solvent and the ion are considered to be solely *electrostatic* in origin.

The Born model suggests a simple thought process for calculating the free energy ΔG_{I-S} of ion–solvent interactions, i.e., the work of transferring an ion from vacuum into the solvent (Fig. 2.8). One uses a thermodynamic cycle. The basic idea behind a thermodynamic cycle is the law of the conservation of energy. If one starts with a certain system (say, an ion in vacuum) and then goes through a hypothetical cycle of changes, ending up with the starting condition (i.e., the ion in vacuum), then the algebraic sum of all the energies involved in the various steps must be zero. The particular cycle that will be used is the following: (1) The ion (or charged sphere) is first considered in a vacuum, and the work W_1 of stripping it of its charge $z_i e_0$ is computed. (2) This uncharged sphere is slipped into the solvent; this process will involve no work, i.e., $W_2 = 0$ because the only

ION–SOLVENT INTERACTIONS 51

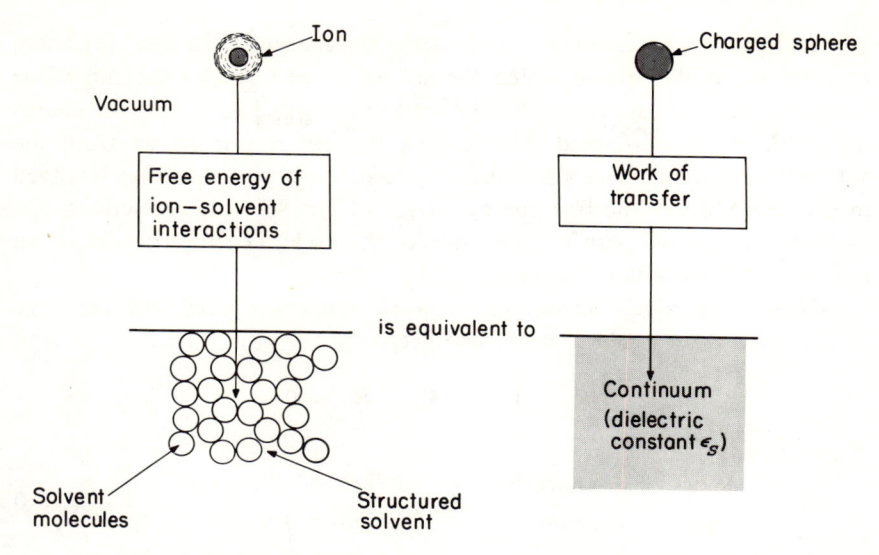

Fig. 2.7. The Born model views the free energy of ion–solvent interactions as equal to the work of transferring a charged sphere (of radius r_i and charge $z_i e_0$) from vacuum into a continuum (of dielectric constant ε_S).

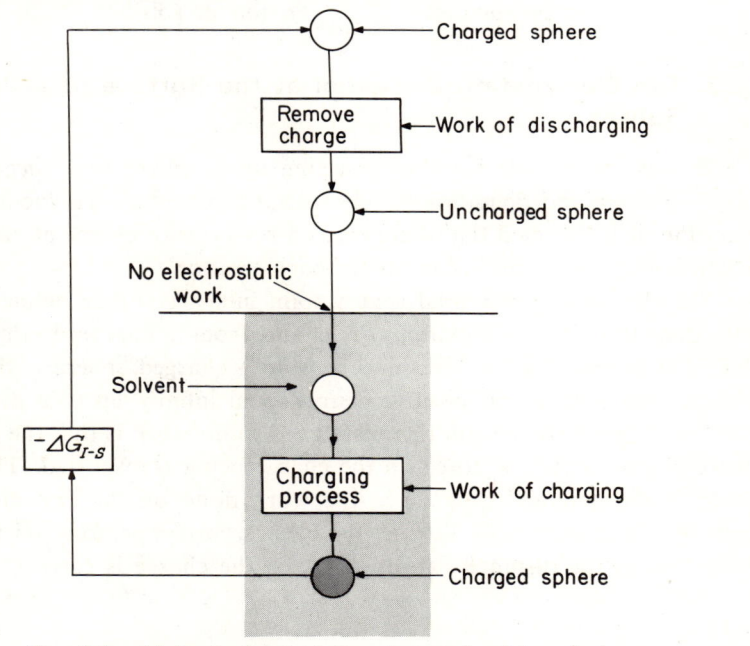

Fig. 2.8. Method of calculating the work of transferring a charged sphere from vacuum into the solvent by a thermodynamic cycle.

interactional work is assumed to arise from the charge on the ion.[†] (3) Then, the charge on the sphere inside the solvent is restored to the full value $z_i e_0$—one says, the sphere is charged up to the value $z_i e_0$—, and the charging work W_3 is computed. (4) Finally, the ion is transferred from the solvent to vacuum. Since this transfer process is opposite to that involved in the definition of the free energy ΔG_{I-S} of ion–solvent interactions, the work W_4 associated with this last step of the cycle, i.e., the transfer of an ion from the solvent to vacuum, yields $-\Delta G_{I-S}$.

Now, if the algebraic sum of the work terms associated with the steps of the cycle is set equal to zero, one gets

$$W_1 + W_2 + W_3 + W_4 = 0$$

or

$$\begin{matrix}\text{Work of discharging}\\\text{ion in vacuum}\end{matrix} + 0 + \begin{matrix}\text{Work of charging}\\\text{ion in solvent}\end{matrix} - \Delta G_{I-S} = 0$$

i.e.,

$$\Delta G_{I-S} = \begin{matrix}\text{Work of discharging}\\\text{ion in vacuum}\end{matrix} + \begin{matrix}\text{Work of charging}\\\text{ion in solvent}\end{matrix} \quad (2.1)$$

2.2.3. The Electrostatic Potential at the Surface of a Charged Sphere

In considering the work of charging up a sphere *in a vacuum*, one starts off from the definition of electrostatic potential. To facilitate the definition, it is assumed that there exists a reservoir of charge at an infinite distance away from the sphere under consideration.

The electrostatic potential ψ at a point in space is then defined as the work done to transport a unit *positive* charge from infinity up to that point. Thus, the potential ψ_r at a distance r from a charged sphere is the work done to transport a unit positive charge from infinity up to a distance r from the sphere. The reason there is a need to do work is that the charged sphere exerts an electric force on the charge being transported. The *magnitude* of the potential, $|\psi_r|$, i.e., the work done on the unit charge, is given by the electric field X_r (i.e., the electric *force* operating on the unit charge) times the distance r through which the charge is carried

$$|\psi_r| = X_r \cdot r \quad (2.2)$$

[†] What happens when charges cross interfaces is discussed in Section 7.2.

ION–SOLVENT INTERACTIONS 53

←──────────── Direction of increasing potential r

Positively charged sphere ←── Direction of transport of charge ── (+) ── Direction of field ──→
                                                              ↑
                                                       Unit positive charge

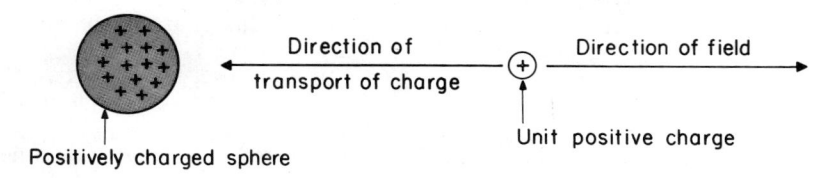

Fig. 2.9. The relative directions of the field due to a charged sphere, of the movement of the test charge, and of increasing electrostatic potential.

The *sign* of the potential ψ_r is thought out as follows. Suppose the sphere is charged positively. Then it exerts a repulsive force on the unit *positive* charge, and the potential ψ_r, the work which has to be done *by* an external agency in transporting the unit positive charge, i.e., overcoming the repulsive interaction, will be taken to be positive. But the electric force of the charged sphere on the unit positive charge, i.e., the field, acts in a direction *opposite* to that in which the charge is being moved (Fig. 2.9). Since both the field and the direction of transport are vectors (quantities with direction and magnitude) and since the vectors point in opposite ways, their product is negative.[†] Hence, to relate the *positive* potential ψ_r to the product of the field and distance, it is necessary to state that

$$\psi_r = -X_r \cdot r \tag{2.3}$$

Equation (2.2) for the electrostatic potential is valid only if the field X_r acting on the unit positive charge remains the same, independent of the distance of the unit charge from the source of the field. Suppose, however, as will be seen to be the case with the field due to a charged sphere, the field varies with distance from the source of the field. Then one must allow for the inconstancy of the field in the definition of the potential at a point. What one does is to take the field X_r as a constant over an infinitesimally short distance dr. In this case, the electrostatic potential ψ_r at a point r is obtained by summing up all the little bits of work, $X_r\,dr$, as the unit charge is carried from infinity up to the point r in steps of length dr, i.e.,

$$\psi_r = -\int_{\infty}^{r} X_r\,dr \tag{2.4}$$

[†] The product of two vectors A and B is $AB \cos\theta$, where θ is the angle between the two vectors. If the vectors are in opposite directions, $\theta = \pi$ and $\cos\theta = -1$ and the product is $-AB$.

By inserting an upper limit of r_i in the integration, one can indicate that the unit charge has been brought up to the surface of the sphere. Thus, the electrostatic potential at the surface of the charged sphere is

$$\psi_{r_i} = -\int_{\infty}^{r_i} X_r \, dr \tag{2.5}$$

The electric force X_r operating on a unit charge *in vacuum* is obtained from Coulomb's law for the electric force F between two charges q_1 and q_2, i.e.,

$$F = \frac{q_1 q_2}{r^2} \tag{2.6}$$

where r is the distance between the charges. Thus, by setting $q_1 = q$ and $q_2 = 1$, the electric force per unit charge (i.e., the electric field X_r) due to a charge q becomes

$$X_r = \frac{q}{r^2} \tag{2.7}$$

Substituting for X_r in equation (2.5), one gets for the potential at the surface of the sphere

$$\psi_{r_i} = -\int_{\infty}^{r_i} \frac{q}{r^2} \, dr$$

$$= +\frac{q}{r_i} \tag{2.8}$$

2.2.4. On the Electrostatics of Charging (or Discharging) Spheres

The electrostatic potential at the surface of the sphere pertains to the work of transporting a *unit* charge to the sphere; hence, the work done in transporting a charge of any other magnitude is simply given by the product of the potential and the magnitude of that charge. It will be noticed, however, that the electrostatic potential at the surface of the sphere varies with the charge q on the sphere. So the work of adding on any charge to the sphere depends upon *how much charge q the sphere already has*. This is awkward. So the best thing to do is start with an uncharged sphere ($q = 0$) and add charge onto it in little driblets or infinitesimal amounts, dq, each of which requires an infinitesimal amount of work, dw, given by the product of the potential and the infinitesimal charge dq, i.e.,

$$dw = \psi_{r_i} \, dq \tag{2.9}$$

This procedure is known as a *charging process*.

ION–SOLVENT INTERACTIONS 55

If, therefore, one starts with an *uncharged* sphere of radius r_i in a *vacuum* and slowly builds up the charge from zero to a final value which can be taken as $z_i e_0$, corresponding to a charge on an ion containing z_i electronic charges, then the total work consists of all the little elements of work, dw, i.e.,

$$W = \int dw = \int_0^{z_i e_0} \psi_{r_i} dq$$

$$= \int_0^{z_i e_0} \frac{q}{r_i} dq = \left[\frac{q^2}{2r_i}\right]_0^{z_i e_0}$$

$$= \frac{(z_i e_0)^2}{2r_i} \tag{2.10}$$

Obviously, the work of discharging a charged sphere in a vacuum is the negative of the charging work because, in the discharging process, one is taking away charge from a charged sphere, i.e.,

$$W_{\text{disch}} = -\frac{(z_i e_0)^2}{2r_i} \tag{2.11}$$

Now that the process of discharging a sphere in a *vacuum* has been analyzed, one can consider the charging process when the sphere is placed inside the *solvent*. The question is: can one use the vacuum formula for the electrostatic potential at the surface of the sphere, i.e., Eq. (2.8)?

$$\psi_{r_i} = \frac{q}{r_i}$$

The answer is no, because this formula was obtained from the expression for the electric force between two charges in a vacuum and it is known that the electric force between two charges depends on the medium between them. The electric force in the presence of a material medium is less than that which operates when only a vacuum is present. A simple explanation of this phenomenon is given later on (*cf.* Section 2.5). The ratio of the force in vacuum to the force in the medium is a characteristic of the medium and is known as its *dielectric constant* ε (Fig. 2.10)

$$\varepsilon = \frac{\text{Electric force in vacuum}}{\text{Electric force in medium}} \tag{2.12}$$

Hence, the coulombic force between two charges in a medium of dielectric constant ε is

$$F = \frac{q_1 q_2}{\varepsilon r^2} \tag{2.13}$$

56 CHAPTER 2

| Electric force in medium | $= \dfrac{q_+ q_-}{\varepsilon r^2}$ | → | Is ϵ times less than | → | Electric force in vacuum | $= \dfrac{q_+ q_-}{r^2}$ |

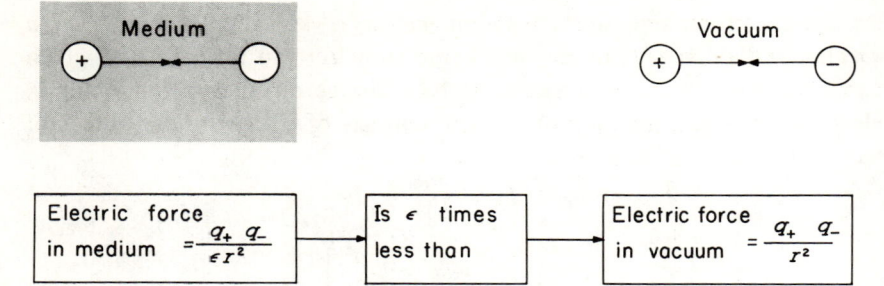

Fig. 2.10. The electric force between two charges q_+ and q_- in vacuum and in a medium.

and the electric field becomes

$$X = \frac{q}{\varepsilon r^2} \qquad (2.14)$$

Hence, the potential at the surface of the sphere of radius r_i placed in a medium of dielectric constant ε is

$$\psi = \frac{q}{\varepsilon r_i} \qquad (2.15)$$

In terms of this expression for the electrostatic potential at the surface of a charged sphere situated in a medium of dielectric constant ε, the electrostatic work of charging a sphere becomes

$$\begin{aligned} W_{\text{ch.}} &= \int_0^{z_i e_0} \psi_{r_i} \, dq \\ &= \int_0^{z_i e_0} \frac{q}{\varepsilon r_i} \, dq \\ &= \frac{(z_i e_0)^2}{2\varepsilon r_i} \end{aligned} \qquad (2.16)$$

2.2.5. The Born Expression for the Free Energy of Ion–Solvent Interactions

Now that the basic electrostatics of charging and discharging spheres has been presented, it can be immediately applied to the model suggested by Born for the calculation of the free energy of ion–solvent interactions.

It has been argued in Section 2.2.2 (see also Fig. 2.8) that the free energy of ion–solvent interactions, ΔG_{I-S}, is given by the sum of the work done to discharge an ion in vacuum and the work done to charge it up in the solvent of dielectric constant ε_s. Since, according to the Born model, a sphere (of radius r_i and charge $z_i e_0$) is considered to be equivalent to an ion of radius r_i and charge $z_i e_0$, it follows that the work of discharging an ion in vacuum is equal to the work of discharging the equivalent sphere in vacuum and the work of charging an ion in the solvent is equal to the work of charging the equivalent sphere in the solvent. Hence [cf. Eq. (2.1)],

$$\Delta G_{I-S} = \text{Work of discharging equivalent sphere in vacuum}$$
$$+ \text{Work of charging equivalent sphere in solvent}$$

$$= -\frac{(z_i e_0)^2}{2r_i} + \frac{(z_i e_0)^2}{2\varepsilon_s r_i} \quad \text{per ion}$$

$$= -\frac{(z_i e_0)^2}{2r_i}\left(1 - \frac{1}{\varepsilon_s}\right) \quad \text{per ion}$$

$$= -N_A \frac{(z_i e_0)^2}{2r_i}\left(1 - \frac{1}{\varepsilon_s}\right) \quad \text{per mole of ions} \quad (2.17)$$

where N_A is the Avogadro number (Fig. 2.11).

Thus, by considering that the free energy of ion–solvent interactions is given by the net electrostatic work of discharging a sphere (of the same size and charge as the ion) in a vacuum, of transferring the discharged

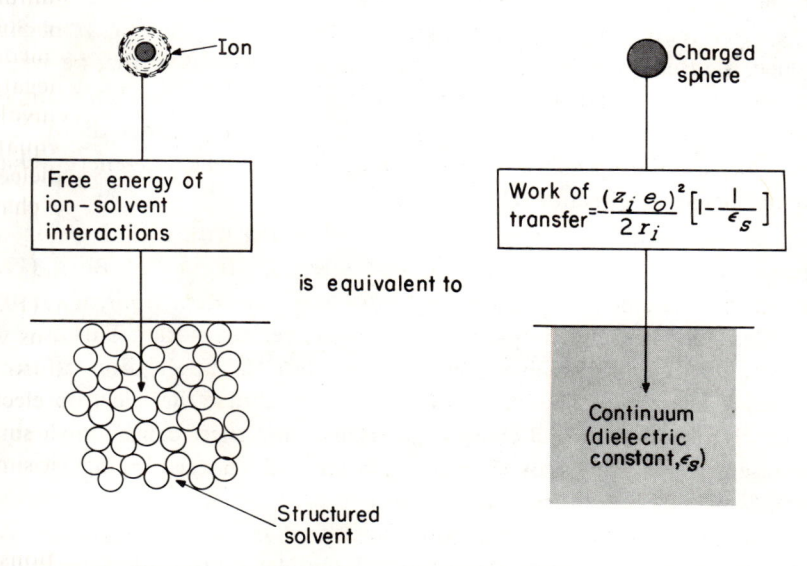

Fig. 2.11. The free-energy change resulting from the transfer of an ion from a vacuum into the solvent.

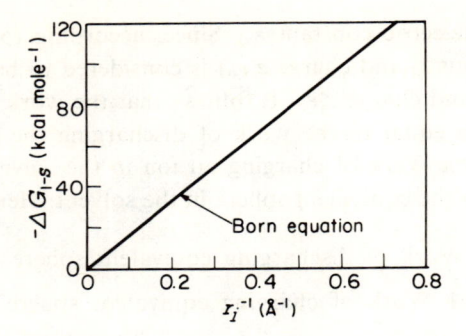

Fig. 2.12. The free energy of ion–solvent interactions as a function of the reciprocal of the ionic radius.

sphere into a medium with the same dielectric constant as the solvent, and of then charging the sphere till it has the same charge as the ion, the Born model has yielded the free-energy change resulting from the transfer of ions from a vacuum to solvent.

What is the importance of this free-energy change? The importance derives from the fact that systems in nature try to attain a state of minimum free energy. Thus, if the ΔG_{I-S} is negative, then ions exist more stably in the solvent than in vacuum. Since the dielectric constant of any medium is greater than unity, $1 > 1/\varepsilon_s$, and, therefore, ΔG_{I-S} is always negative; hence, the Born equation (2.17) shows that all ions would rather be involved in ion–solvent interactions than be left in vacuum. The Born equation predicts that the smaller the ion (smaller r_i) and the larger the dielectric constant ε_s, the greater will be the magnitude of the free-energy change in the negative direction (Fig. 2.12).

If one stands back and looks at the situation with regard to ions and their existence in solvents before and after the theory of Born (1920), several points emerge. One set out to discover the interactions of ions with a solvent, and one ended up doing a problem in electrostatics. This illustrates a feature of electrochemistry—it often involves the application of electrostatics to chemistry. The basis of this link is of course that electrochemistry is involved with ions and charged interfaces, and these can be most simply represented in models by charged spheres and charged plates, the stuff with which electrostatics deals.

One has also seen in the Born theory of ion–solvent interactions an example of very simple thinking based on models. A complicated situation has been reduced to a simple one by the choice of a simple model. In the case of ion–solvent interactions, once the analogies between an ion and a charged sphere and between a structured solvent and a dielectric continuum are stressed, the rest is easy.

It will be shown later that, not unexpectedly, the Born model oversimplifies the problem, but one must see the model in its historical perspective. It was proposed at a time when the very existence of charged particles in solution was questioned. Indeed, the Born approach to ion–solvent interactions and the fact that it gave answers of the same order of magnitude as experiment (*cf.* Section 2.2.9) helped to confirm the hypothesis that ions exist in solution. Seen in historical perspective, the simple Born model may be recognized as an important step forward.

2.2.6. The Enthalpy and Entropy of Ion–Solvent Interactions

Before finding out about the experimental testing of the Born theory, it is preferable to recover from the theoretical expression for the free energy ΔG_{I-S}, the enthalpy (heat) and entropy changes associated with ion–solvent interactions. This is because it is the *heat* of ion–solvent interactions, rather than the free energy, which is obtained directly from the experimentally measured heat changes observed to occur when solids containing ions are dissolved in a solvent, i.e., when ion–solvent interactions are provoked.

By making use of the combined first and second laws ($dE = T\,dS - p\,dV$) in $G = H - TS = E + PV - TS$, one gets

$$dG = V\,dP - S\,dT \qquad (2.18)$$

and, at constant pressure,

$$\left(\frac{\partial G}{\partial T}\right)_P = -S \qquad (2.19)$$

Thus, applying (2.19) to a transformation from state 1 to state 2 results in

$$\left(\frac{\partial G_2}{\partial T}\right)_P - \left(\frac{\partial G_1}{\partial T}\right)_P = -(S_2 - S_1) \qquad (2.19a)$$

Hence

$$\frac{\partial \Delta G}{\partial T} = -\Delta S \qquad (2.19b)$$

since $\Delta G = G_2 - G_1$ and $\Delta S = S_2 - S_1$.

Hence, all one has to do to get the entropy changes associated with ion–solvent interactions is to differentiate ΔG_{I-S} [given by Eq. (2.17)] with respect to temperature. During this differentiation, the question arises whether the dielectric constant should be treated as a constant or as a variable with temperature. At this stage of the presentation, one does not have a feel for dielectric constants to be able to answer the question (see, however, Section 2.5); so one has to appeal to experiment. It turns out that the dielectric constant does vary with temperature (Table 2.2) and must therefore be treated as a variable in differentiating Eq. (2.17) with respect to temperature.

Thus, the entropy change due to ion–solvent interactions is

$$\Delta S_{I-S} = -\left(\frac{\partial \Delta G_{I-S}}{\partial T}\right)_P = \frac{N_A(z_i e_0)^2}{2r_i} \frac{1}{\varepsilon_s^2} \frac{\partial \varepsilon_s}{\partial T} \qquad (2.20)$$

and from

$$\Delta H_{I-S} = \Delta G_{I-S} + T \Delta S_{I-S} \qquad (2.21)$$

one has for the heat change:

$$\Delta H_{I-S} = -\frac{N_A z_i^2 e_0^2}{2r_i}\left[1 - \frac{1}{\varepsilon_s} - \frac{T}{\varepsilon_s^2}\frac{\partial \varepsilon_s}{\partial T}\right] \qquad (2.22)$$

Now that one has a theoretical expression for a *heat* change, it is time to think of comparing the predictions of the Born theory with experiment. There are, however, a few conceptual questions first to be considered.

TABLE 2.2
Variation of Dielectric Constant of Water with Temperature

Temperature, °C	Dielectric constant ε_W	Temperature, °C	Dielectric constant ε_W
0	87.74	50	69.91
10	83.83	60	66.81
20	80.10	70	63.85
25	78.30	80	61.02
30	76.54	90	58.31
40	73.15	100	55.72

2.2.7. Can One Experimentally Study the Interactions of a Single Ionic Species with the Solvent?

The Born theory has indicated an elementary, modelistic, and oversimplified way of calculating the free energy, entropy, and enthalpy changes associated with the interaction of a *single* ionic species with the solvent. But how can these quantities be experimentally measured so that one can check on the degree of correctness of the Born model?

It is in connection with the actual process of measuring the heat of the ion–solvent interactions that there arises a difficulty which is best brought out by the following thought experiment. Suppose one imagines that a beam of any *one* species of ions (Fig. 2.13), say K^+ ions, is introduced into a solvent. What will be the result? The solvent will become positively charged. So the work of introducing an ion into a charged solution will consist not only of the work arising from ion–solvent interactions but also of the coulombic work of bringing an ion into a *charged* medium. One has to conclude that, even if one could experimentally measure the heat change in the imaginary process of shooting a beam of one ionic species into a solvent, it would not correspond to the calculated heat of ion–solvent interactions [*cf*. Eq. (2.22)] because the experimental value would include the extra energy arising from the interactions of the charged ions with the charged solution, a serious complicating factor.

In contrast to the thought experiment referred to in the preceding paragraph the conventional method of introducing ions into a solvent is to dissolve a *salt* which must contain both positive and negative ions. Thus, one can dissolve KCl to introduce K^+ ions into the solvent. Since one introduces a negatively charged Cl^- ion along with every positively

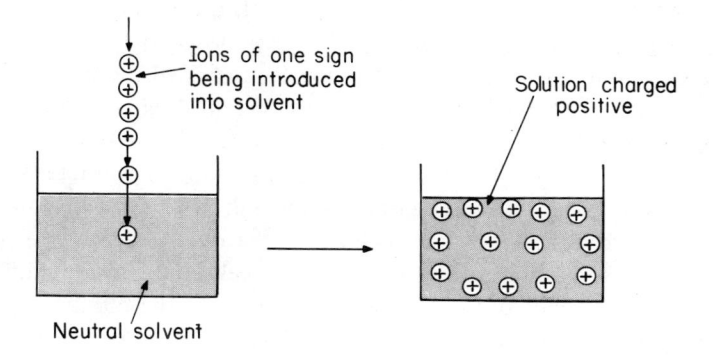

Fig. 2.13. The result of introducing ions of one sign into a neutral solvent is a **charged** solution.

charged K$^+$ ion, the solution never gets charged; it maintains its electroneutrality.

At the same time, however, one has generated another problem. *Two ionic species have been introduced into the solvent, and each one is going to have its own work of interactions so what is experimentally measured is actually the *sum* of the contributions of the two ionic species to ion–solvent interactions. How, then, can the sum of the heats of ion–solvent interactions of the positive and negative ions, i.e., the heat of *salt*–solvent interactions, be separated into the individual contributions of each ionic species?

This matter of attempting to extract from an experimental measurement on a salt (which consists of both positive and negative ions) the value of the property concerned for *one* of the ions constituting the salt is a frequent one in electrochemistry. Thus, for the kind of reason shown up by the thought experiment described above and pictured in Fig. 2.13, only the combined effects of the contributions of positive *and* negative ions can be experimentally measured, the separation of the effects into the contributions of the individual ionic species being an experimentally insoluble problem (*cf.*, for example, the question of individual ionic activity coefficients, Section 3.4).

A safe but rather timid way out of the situation is to remain content with determining only the effects of electrolytes (salts). One could even seek to justify this attitude by saying that one should only think about things which can be measured without ambiguity. A corollary of this viewpoint is that experimental heats of ion–solvent interactions of *individual* ionic species are neither meaningful nor significant.

There is, however, entirely objective evidence, free from any taint of the inaccuracies of a given model, for individual heats of ion–solvent interactions. One can tabulate heats of salt–solvent interactions (their experimental determination is described in Section 2.2.8) for *pairs* of salts with a common ion, e.g., KCl and NaCl, KBr and NaBr, and KI and NaI. It is seen (Table 2.3) that there are almost constant *differences* in the heats of interactions of these different salts with water. This is just what one would expect if each of the ionic species was making a characteristic contribution to the heats of interaction of the salt with the solvent. Then, because, in each pair of salts in Table 2.3, the differing ions are lithium and sodium or sodium and potassium, respectively, the constant difference would be the difference of the heats of interaction of these individual ionic species with the solvent. Thus, it *is* meaningful to try to obtain experimental values of the heats of solvation of the individual ionic species.

There are several approaches which could be made to the problem

TABLE 2.3

Constant Differences in the Heats of Solvation of Pairs of Salts with a Common Ion

Salt	ΔH_{S-H_2O}, kcal mole^{-1}	Difference
LiF	−245.2	−27.4
NaF	−217.8	
LiCl	−211.2	−27.4
NaCl	−183.8	
LiBr	−204.7	−27.4
NaBr	−177.3	
LiI	−194.9	−27.4
NaI	−167.5	
. . . .
NaCl	−183.8	−20.0
KCl	−163.8	
NaBr	−177.3	−20.0
KBr	−157.3	
NaI	−167.5	−20.0
KI	−147.5	

of obtaining heats of interaction of *individual* ions with the solvent from the experimental measurements (*cf.* Section 2.2.8) of heats of *salt*–solvent interactions. A crude one is based on the realization that, according to the Born model, the heat of ion–solvent interactions is inversely proportional to the ionic radius [*cf.* Eq. (2.22)]. Hence, if one considers a salt in which the positive and negative ions are of equal radii, then the heat of the *salt*–solvent interactions can be split equally between the two species. The usual pair chosen for this purpose is KF because the radii of the K$^+$ and F$^-$ ions are almost equal (Table 2.4). Of course, once the heat of ion–solvent interactions is obtained for any single ionic species, then, one can quickly get the heats for all other ionic species. For example, if the heat of ion–solvent interactions is known for K$^+$ (because of the above procedure applied to KF), then the heat for Cl$^-$ can be obtained from the measured heat for KCl because the heat of interaction of KCl with the solvent is the sum of that due to the K$^+$ and Cl$^-$ ions.

TABLE 2.4
Some Ionic Radii

Ion	Ionic radius, Å	Ion	Ionic radius, Å
Li^+	0.60	Fe^{++}	0.76
Na^+	0.95	Co^{++}	0.74
K^+	1.33	Ni^{++}	0.72
Rb^+	1.48	Cu^{++}	0.72
Cs^+	1.69	Al^{3+}	0.50
Ag^+	1.26	Sc^{3+}	0.81
Be^{++}	0.31	La^{3+}	1.15
Mg^{++}	0.61	Fe^{3+}	0.64
Ca^{++}	0.99	F^-	1.36
Sr^{++}	1.13	Cl^-	1.81
Ba^{++}	1.35	Br^-	1.95
Zn^{++}	0.74	I^-	2.16
Mn^{++}	0.80		

The approach just described for the separation of the heat of salt–solvent interactions into its component heats of ion–solvent interactions is a very simple one. More complicated methods have been suggested. But almost all of them have one important defect in common: they start with ideas based on a specific model (e.g., the idea following from Born's simple view that ions of the same radius will have the same heat of interaction with the solvent) and then split the experimental heat of salt–solvent interactions into heats of interaction of individual ions with the solvent. Hence, one generally has only so-called "experimental" values of the heat of ion–solvent interactions. The heats of interaction with the solvent for *single ionic species* can never be as "clean" and unambiguous as those for *electrolytes* (salts); the latter are given by pure experiment, untainted by any (always approximate) model-oriented speculations.

2.2.8. The Experimental Evaluation of the Heat of Interaction of a Salt and Solvent

It remains therefore to describe how the heat of salt–solvent interaction is experimentally measured. If one takes a solid salt and dissolves it in a solvent, it is only in exceptional cases that marked temperature changes, i.e., thermal effects, are noticed (Table 2.5). In general, the dissolution process is accompanied by only a slight heating or cooling (usually < 10 kcal mole^{-1}).

TABLE 2.5

Heat of Solution ΔH_{soln}, the Heat Absorbed During Dissolution of Some Ionic Crystals at 25°C

kcal mole^{-1}

Cations	Anions			
	F$^-$	Cl$^-$	Br$^-$	I$^-$
Li$^+$	+1.1			
Na$^+$	+0.1	+0.9	−0.2	
K$^+$	−4.2	+4.1	+4.8	+4.9
Rb$^+$	−6.3	—	+5.2	+6.2
Cs$^+$	−9.0	—	—	+7.9

These heat changes are very small compared with the heat of ion–solvent interactions (of an order of magnitude of 100 kcal mole^{-1} for singly-charged ions) calculated by the Born model. What is the reason for the big difference? The answer is not difficult to see. The Born model calculates the heat change when isolated ions are transferred *from vacuum* into the solvent, in contrast to the heat of solution which is the measure of the heat change when ions *from an ionic crystal* are transferred into the solvent. But, in the crystal lattice, the ions are engaged in interactions with each other. This means that two processes are occurring simultaneously when a crystal is dissolved. Firstly, the lattice is being dismantled and the ions separated from each other, and, secondly, the ions are entering into interactions with the solvent. It is only this second part of the dissolution process in which ion–solvent interactions play a part.

To get at this ion–solvent part of the process, one must know the ion–ion interactions in the crystal lattice. In other words, one must know how much energy is required to dismantle the lattice and take the ions so far from each other that they do not interact any more. Then, the difference between the heat change associated with breaking up the lattice and producing a gas of ions extremely far from each other and that associated with the dissolution process is equal to the heat of interaction of the salt with the solvent.

To be specific, consider a potassium fluoride crystal. First, this ionic lattice must be conceptually disassembled and a very dilute (to cut down ion–ion interaction) *gas* of K$^+$ and F$^-$ ions produced. The heat-content

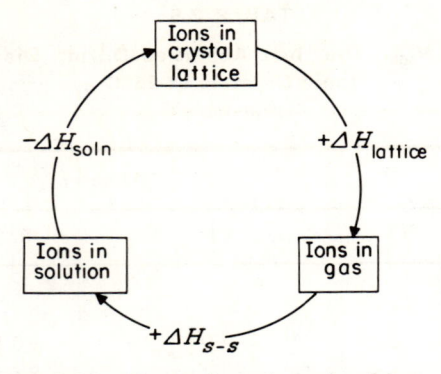

Fig. 2.14. Thermodynamic cycle of hypothetical changes to obtain the heat of salt–solvent interactions.

change associated with this process is called the *lattice energy* $\Delta H_{\text{lattice}}$. Then, the ions, which are infinitely far apart, are introduced into the solvent, in which process the heat-content change is the heat of salt–solvent interactions, ΔH_{S-S}. Finally, the ions in the solvent are assembled back into the crystal; this is the opposite of the process of dissolution and is therefore associated with *minus* the heat of solution, $-\Delta H_{\text{soln}}$. The crystal is back where it started (Fig. 2.14), a thermodynamic cycle has been completed, and the various heat changes must all algebraically add up to zero, i.e.,

$$\Delta H_{\text{lattice}} + \Delta H_{S-S} - \Delta H_{\text{soln}} = 0 \tag{2.23}$$

or

$$\Delta H_{S-S} = \Delta H_{\text{soln}} - \Delta H_{\text{lattice}} \tag{2.24}$$

Both the quantities on the right-hand side are experimentally accessible (*cf.* Table 2.6), and, hence, one can obtain experimental values of the heat of salt–solvent interactions (Table 2.7).

If, now, the ΔH_{S-S} for KF is taken and divided equally (*cf.* Section 2.2.7) between K^+ and F^- ions because the radii of these ions are almost equal, one gets the heat of interaction of these two individual ions with the solvent. Once these values are obtained, one can construct a table of so-called[†] "experimental" heats of ion–solvent interactions (Table 2.8).

[†] So-called because their evaluation has involved the postulate that ions which have equal crystallographic radii also have equal interaction with the solvent.

ION–SOLVENT INTERACTIONS 67

TABLE 2.6
Lattice Energy $\Delta H_{\text{lattice}}$ of Ionic Crystals and Heat of Solution,[†] ΔH_{soln} kcal mole^{-1} at 25°C

Salt	$\Delta H_{\text{lattice}}$	ΔH_{soln}
LiF	+246.3	+1.1
NaF	+217.9	+0.1
KF	+193.6	−4.2
RbF	+186.4	−6.3
CsF	+177.9	−0.9
NaCl	+184.7	+0.9
KCl	+167.9	+4.1
NaBr	+177.1	−0.2
KBr	+162.1	+4.8
RbBr	+157.4	+5.2
KI	+152.4	+4.9
RbI	+148.6	+6.2
CsI	+144.5	+7.9

[†] These are the measured molar heats of solution extrapolated to infinite dilution to avoid ion–ion interactions (*cf.* Chapter 3).

TABLE 2.7
Experimental Values of the Heats of Interaction between a Salt and Water ΔH_{S-S} kcal mole^{-1} at 25°C

	F$^-$	Cl$^-$	Br$^-$	I$^-$
Li$^+$	−245.2	−211.2	−204.7	−194.9
Na$^+$	−217.8	−183.8	−177.3	−167.5
K$^+$	−197.8	−163.8	−157.3	−147.5
Rb$^+$	−192.7	−158.7	−152.2	−142.4
Cs$^+$	−186.9	−152.9	−146.4	−136.6

TABLE 2.8

The Experimental Heats ΔH_{I-H_2O} of Interaction between Individual Ions and Water, Assuming $\Delta H_{K^+-H_2O} = \Delta H_{F^--H_2O} = \frac{1}{2}\Delta H_{KF-H_2O}$

Ion	Heat of ion–water interactions, kcal mole^{-1} at 25 °C
Li$^+$	−146.3
Na$^+$	−118.9
K$^+$	− 98.9
Rb$^+$	− 93.8
Cs$^+$	− 88.0
F$^-$	− 98.9
Cl$^-$	− 64.9
Br$^-$	− 58.4
I$^-$	− 48.6

2.2.9. How Good Is the Born Theory?

In making numerical calculations based on Eq. (2.22) so that a comparison between the predictions of the Born theory and the results of experiment can be made, the first question to decide is: What value of the radius of the ions shall be substituted into the Born equation (2.22)? Since the experimental values of the heat of ion–solvent interactions are based on dismantling an ionic crystal and then plunging the ions so produced into the solvent, the obvious radius to choose is the radius of ions obtained from X-ray measurements on the corresponding ionic crystal, i.e., the crystallographic radius.

When the crystal radii are inserted into the Born expressions for the free energy and enthalpy of ion–solvent interactions, the resulting values turn out to be of the correct order (Table 2.9). An important conclusion can be drawn from this order of magnitude comparison: Ion–solvent interactions arise largely (at least for the ions concerned) from coulombic forces for those are the only forces reckoned with in the Born model.

However, a detailed examination of Table 2.9 reveals that the Born values for the heats of ion–solvent interactions are numerically too high, in some cases nearly 50% too high.

It also turns out that the experimental heats of interaction between ions and solvent do not vary inversely as the radius, as predicted by the

TABLE 2.9

Ionic Radii and the Born Free Energy ΔG_{I-H_2O} and Enthalpy ΔH_{I-H_2O} of Ion–Water Interactions at 25°C

Ion	r_i, Å	ΔG_{I-H_2O}, kcal mole^{-1}	ΔH_{I-H_2O}, kcal mole^{-1} Calc. from Eq. (2.22)	ΔH_{I-H_2O}, kcal mole^{-1} Exp. from Table 2.8
Li$^+$	0.60	−273.2	−277.7	−146.3
Na$^+$	0.95	−172.6	−175.5	−118.9
K$^+$	1.33	−123.2	−125.3	− 98.9
Rb$^+$	1.48	−110.8	−113.1	− 93.8
Cs$^+$	1.69	− 97.0	− 98.6	− 88.0
F$^-$	1.36	−120.5	−122.6	− 98.9
Cl$^-$	1.81	− 90.6	− 92.1	− 64.9
Br$^-$	1.95	− 84.1	− 85.5	− 58.4
I$^-$	2.16	− 75.9	− 77.2	− 48.6

Born equation (2.22) (Fig. 2.15). Of course, one can arbitrarily adjust the values of the radii to differ from the crystallographic radii and then obtain better fit between theory and experiment. Indeed, it was found some years ago by Latimer, Pitzer, and Slansky that, by adding 0.85 Å to the radii of the positive ions and 0.1 Å to those of the negative ions, one can "remove" the discrepancy between the calculated and observed values (Fig. 2.16 and

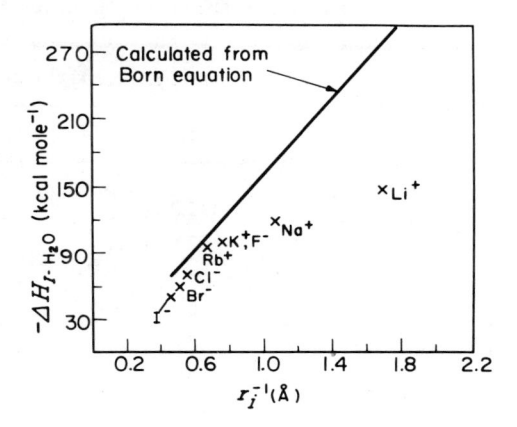

Fig. 2.15. Experimental heats of ion–water interactions do not vary inversely as the ionic radius.

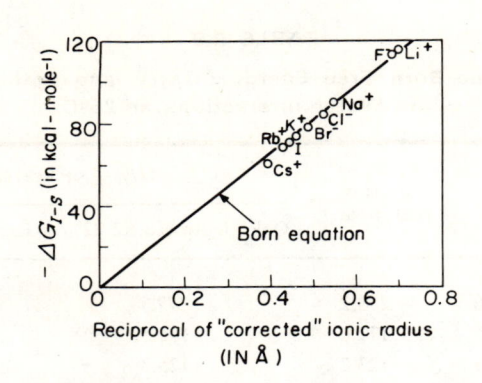

Fig. 2.16. By adding 0.85 and 0.10 Å to the crystallographic radii of positive and negative ions, respectively, the calculated Born free energies of ion–water interactions vary inversely with the corrected ionic radii.

Table 2.10). But then one has to give a theory of where the magic numbers 0.85 Å and 0.1 Å come from. Pending the proposal of some such theory, one is back at the starting point, i.e., the Born theory suggests that ion–solvent interactions are much stronger than experiment shows them to be (Table 2.9).

TABLE 2.10

Calculated Born Values of ΔH_{I-H_2O} after Adding 0.85 Å to the Radii of the Positive Ions and 0.10 Å to Those of Negative Ions

Ion	Corrected radius, Å	ΔH_{I-H_2O}, kcal mole^{-1} Calc. with corrected radii	Exp. from Table 2.8
Li$^+$	1.45	−115.8	−146.3
Na$^+$	1.80	− 92.6	−118.9
K$^+$	2.18	− 98.5	− 98.9
Rb$^+$	2.33	− 90.1	− 93.8
Cs$^+$	2.54	− 65.6	− 88.0
F$^-$	1.46	−114.1	− 98.9
Cl$^-$	1.91	− 87.3	− 64.9
Br$^-$	2.05	− 81.3	− 58.4
I$^-$	2.26	− 73.8	− 48.6

Another approach at explaining the discrepancy between the Born theory and experiment revealed in Fig. 2.15 centers around the value chosen for the dielectric constant ε_s to be used in numerical calculations with Eq. (2.22). The Born theory uses the experimentally measured value for the bulk solvent, e.g., 80 for water. But this bulk value is what one would obtain if one put the solvent (e.g., water) between the plates of a capacitor and measured the reduction factor for the electric force compared with the value in vacuum (Fig. 2.17). What one has found out in this measurement is the *average* dielectric constant of the solvent, i.e., of the solvent taken in bulk quantities. It is this *bulk* value ε_s which has been used in the Born equations, but it is the *effective* value ε_{eff} *near the ion* which can be considered of greater significance to the charging process (*cf.* Section 2.2.4). Now, if ε_{eff} near the ion is not equal to the bulk value but is much less than it, they, by using ε_{eff} instead of ε_s in the Born equation (2.22), the ΔH_{I-S} will be pulled down in the right direction, i.e., toward the experimental value.

But at this stage of the present discussion of ion–solvent interactions, there is no rationale for considering to what extent the values of dielectric constants near an ion may be different from the bulk values. In fact, the atomistic origins of dielectric constants have yet to be explored. The ε_s which has appeared in the Born theory has been taken from experiment, and, in this aspect, the theory has a somewhat nonstructural, thermodynamic flavor.

To go further, one has to think of the structure of the solvent near an ion. After all, how can one discuss ion–solvent interactions if, as prescribed by the Born model, one turns a blind eye to the structure of the solvent?

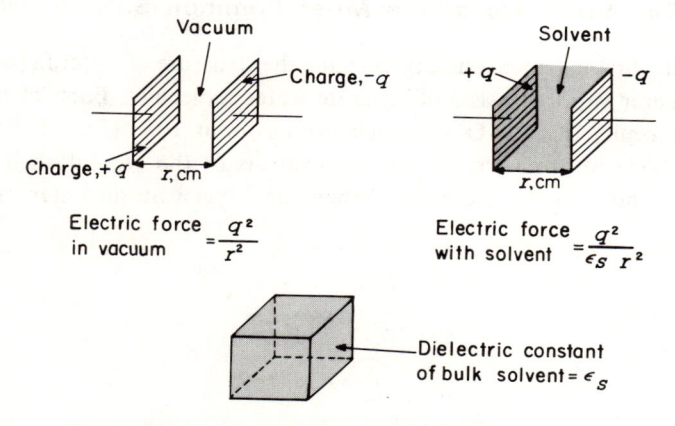

Fig. 2.17. The bulk dielectric constant of a solvent.

If one knew the arrangement of the particles in the solvent and the forces operating between them and the ion, then one could make at least approximate calculations of the ion–solvent interaction on a particulate basis rather than on the continuum basis used by Born. Thus, a new strategy for understanding ion–solvent interactions must be mapped. At first, one must understand the structure of the solvent in the bulk far away from the ion; then, one must understand the structure near the ion. It is a mix of the ion–solvent forces from both regions which determines the energy of the ion–solvent interactions.

Further Reading

1. M. Born, *Z. Physik*, **1**: 45 (1920).
2. W. M. Latimer, K. S. Pitzer, and C. M. Slansky, *J. Chem. Phys.*, **7**: 108 (1939).
3. E. J. W. Verwey, *Rec. Trav. Chim.*, **60**: 887 (1961); **61**: 127 (1942).
4. B. E. Conway and J. O'M. Bockris, "Ionic Solvation" in: J. O'M. Bockris, ed., *Modern Aspects of Electrochemistry*, No. 1, Butterworth's Publications, Inc., London, 1954.
5. K. J. Laidler and C. Pegis, *Proc. Roy. Soc.* (*London*), **A241**: 80 (1957).
6. R. H. Stokes, *J. Am. Chem. Soc.*, **86**: 979, 982, and 2337 (1964).
7. E. Glueckauf, *Trans. Faraday Soc.*, **60**: 572 (1964).
8. W. A. Millen and D. W. Watts, *J. Am. Chem. Soc.*, **89**: 6051 (1967).
9. K. Ross, *Aust. J. Phys.*, **21**: 597 (1968).

2.3. STRUCTURAL TREATMENT OF THE ION–SOLVENT INTERACTIONS

2.3.1. The Structure of the Most Common Solvent, Water

In the first instance, one can examine the structure of water in its gaseous form. Water vapor consists of separate water molecules. Each of these is a bent molecule, the H—O—H angle being about 105° (Fig. 2.18). In the gaseous oxygen atom, there are six electrons in the second shell (two $2s$ electrons and four $2p$ electrons). When the oxygen atoms enter into bond

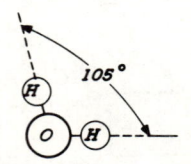

Fig. 2.18. A water molecule is nonlinear.

formation with the hydrogen atoms, there is a blurring of the distinction between the *s* and *p* electrons. The six electrons from oxygen and the two from hydrogen interact, and it is found that four pairs of electrons tend to distribute themselves so that they are most likely to be found in four approximately equivalent directions in space. Since the motion of electrons is described by quantum mechanics, according to which one cannot specify precise *orbits* for the electrons, one talks of the regions where the electrons are likely to be found as *orbitals*, or blurred orbits. The electron orbitals in which the electron pairs are likely to be found are arranged approximately along the directions joining the oxygen atom to the corners of a tetrahedron (Fig. 2.19). The eight electrons around the oxygen are neither *s* nor *p* electrons; they are sp^3 hybrids. Of the four electron orbitals, two are used for the O—H bond, and the remaining two remain as free orbitals for the lone pair of electrons. Because of the repulsion of the electron pairs, the H—O—H angle is not exactly equal to the tetrahedral angle (109° 28′) but is less than that.

The free orbitals in which are found the electron lone pairs confer an interesting property (Fig. 2.20) on the water molecule. The center of "gravity" of the negative charge in the water molecule does not coincide with the center of gravity of the positive charge. In other words, there is a charge separation within the electrically neutral water molecule; it can be considered an electric dipole. The moment of a dipole is defined by the product of the full electronic charges at either end times the distance between the centers of electrical charge. The dipole moment of water is 1.87×10^{-18} esu in the gas phase (but becomes larger when the water molecule is associated with other water molecules).

The availability of the free orbitals (with lone electron pairs) on the oxygen atom contributes not only to the dipolar character of the water molecule but also to another interesting consequence. The two lone pairs can be used for electrostatic bonding onto two other hydrogen atoms. This is what happens in a crystal of ice. The oxygen atoms lie in layers with each layer consisting of a network of open, puckered hexagonal rings (Fig. 2.21). Each oxygen atom is tetrahedrally surrounded by four other oxygen atoms (Fig. 2.22). In between any two oxygen atoms is a hydrogen atom which provides a hydrogen bonding (Fig. 2.23). At any instant, the hydrogen atoms are not situated exactly halfway between two oxygens. Each oxygen has two hydrogen atoms near it (the two hydrogen atoms of the water molecule) at an estimated distance of about 1.75 Å. Such a network structure of associated water molecules contains interstitial regions (between the tetrahedra) which are larger than the dimensions of a water molecule (Fig.

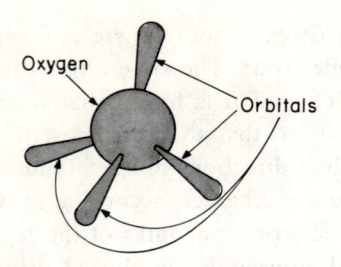

Fig. 2.19. The hybrid orbitals of an oxygen atom.

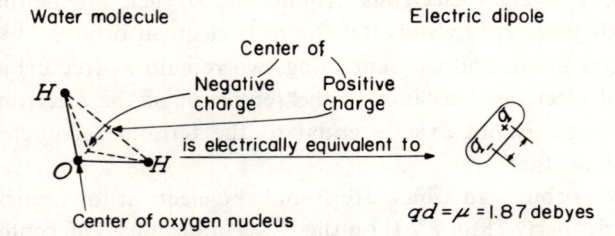

Fig. 2.20. A water molecule can be considered electrically equivalent to a dipole.

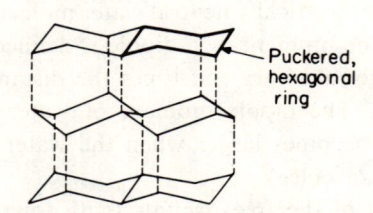

$qd = \mu = 1.87$ debyes

Fig. 2.21. The oxygen atoms in ice, which are located at the intersections of the lines in the diagram, lie in a network of open, puckered hexagonal rings.

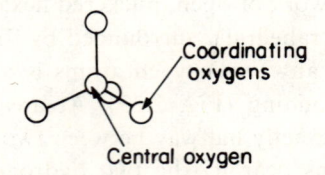

Fig. 2.22. Each oxygen atom in ice is tetrahedrally coordinated by four other oxygen atoms. The hydrogen atoms are not shown in the diagram.

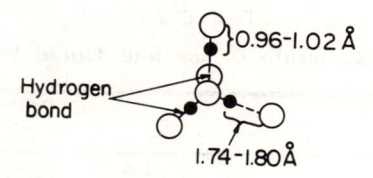

Fig. 2.23. The hydrogen bond between two oxygen atoms (the oxygen and hydrogen atoms are indicated by ○ and ●, respectively).

2.24). Hence, a free nonassociated water molecule can enter the interstitial regions with little disruption of the network structure.

Structural research, originating from a classic paper by Bernal and Fowler, has shown that *liquid water*, under most conditions, is best described as *a somewhat broken-down, slightly expanded* (Table 2.11) *form of the ice lattice*, but this statement must not be taken to mean that there is no association of water molecules in water. X-ray and other techniques indicate that, in water, there is a considerable degree of short-range order characteristic of the tetrahedral bonding in ice. Thus, liquid water *partly* retains

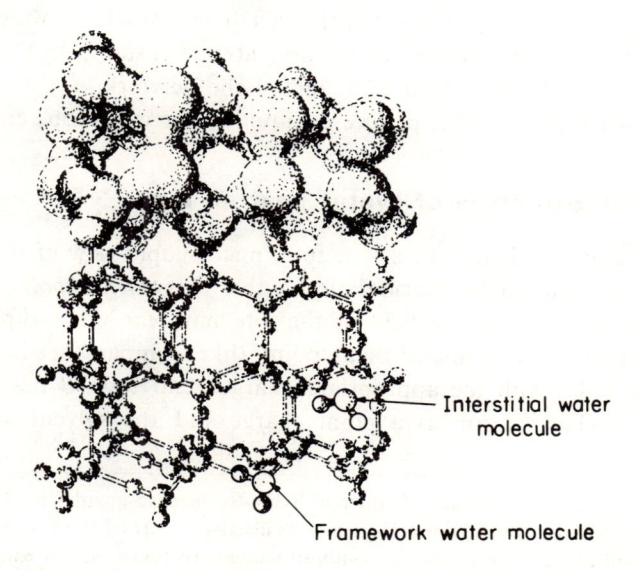

Fig. 2.24. The structure of ice with interstitial spaces large enough to accommodate a free, unassociated water molecule (based on a diagram from *General Chemistry* by Linus Pauling).

76　CHAPTER 2

TABLE 2.11

The Structure of Ice and Liquid Water

	Ice	Liquid water
Mean O–O distance	2.76 Å	2.92 Å
Number of oxygen nearest neighbors	4	4.4–4.6

the tetrahedral bonding and resulting network structure characteristic of the crystalline structure of ice.

In addition to the water molecules which are part of the network, there can be a certain fraction of structurally free, unassociated water molecules in interstitial regions of the network (Fig. 2.25). When a network water molecule breaks its hydrogen bonds with the network, it can move into interstitial regions as an interstitial water molecule which can rotate freely. Thus, the classification of the water molecules into network water and free (or interstitial) water is not a static one. It is dynamic. As argued by Frank, clusters of water molecules are cooperating to form networks, and, at the same time, the networks can break down. A water molecule may be free in an interstitital position at one instant, and, in the next instant, it may become held as a unit of the network.[†]

What happens to this picture of liquid water when ions enter it?

2.3.2. The Structure of Water near an Ion

The aim here is to take a structural microscopic view of the ion inside the solvent. The central consideration is that *ions orient dipoles*. The spherically symmetrical electric field of the ion may tear water dipoles out of the water lattice and make them point (like compass needles toward a magnetic pole) with the appropriate charged end toward the central ion. Hence, viewing the ion as a point charge and the solvent molecules as

---

[†] Knowledge of the structure of water is likely to increase greatly in the near future. There is need to regard the present view—as usual—as part of an evolving picture, one which will become more complete with an increase of research. For example, there is evidence which suggests that normal water, described here, contains a small quantity of a second form of water, which has a much higher boiling point and lower freezing point than normal water. The new form of water may be separated from normal water by passage through some types of capillary tubes.

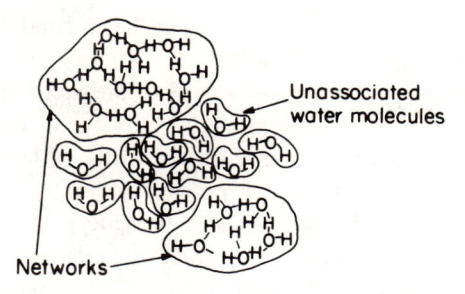

Fig. 2.25. Schematic diagram to show that, in liquid water, there are networks of associated water molecules and also a certain fraction of free, unassociated water molecules.

electric dipoles, one comes out with a picture of ion–dipole forces as the principal basis of ion–solvent interactions.

Due to the operation of these ion–dipole forces, a certain number of water molecules in the immediate vicinity of the ion (more about just how many, later) may be trapped and oriented in the ionic field. Such water molecules cease to associate with other water molecules to form the networks characteristic of water (*cf.* Section 2.3.1). They are *immobilized* except in so far as the ion moves, in which case the sheath of immobilized water molecules moves with the ion. In other words, the ion and its water sheath are a single kinetic entity. (More discussion of this is in Section 2.4.3). Thus, the picture (Fig. 2.26) is of *ions enveloped by a solvent sheath* of oriented, immobilized water molecules.

What about the situation far away from the ion? At a sufficient distance away from the ion, its influence is negligible because the ionic fields

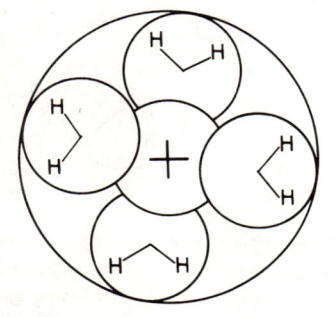

Fig. 2.26. An ion enveloped by a sheath of oriented solvent molecules.

have become attenuated virtually to zero. The normal structure of water is undisturbed; it is that of *bulk water*.

In the region between the *solvent sheath* (where the ionic influence determines the water orientation) and the bulk water (where the ionic influence has ceased to affect the orientation of water molecules), the orienting influences of the ion *and* the water network operate; the former tries to align the water dipoles parallel to the spherically symmetrical ionic field, and the water network tries to make the water in the in-between region continue the tetrahedral arrangement (Fig. 2.27). Caught between the two types of influences, the in-between water adopts a compromise structure that is neither completely oriented nor disoriented. The compromising water molecules are not close enough to the ion to become oriented perfect-

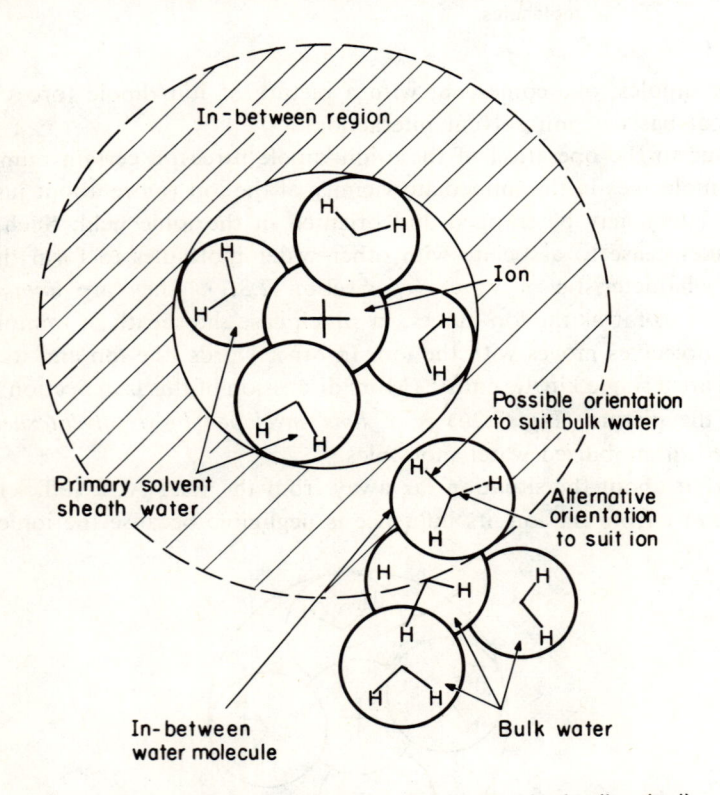

Fig. 2.27. Schematic diagram to indicate that, in the (hatched) region between the primary solvated ion and bulk water, the in-between water molecules must compromise between an orientation which suits the ion (oxygen-facing ion) and an orientation which suits the bulk water (hydrogen-facing ion).

ly around it, and neither are they sufficiently far away from the ion to be part of the structure of bulk water; hence, depending on their distance from the ion, they orient out of the water network to varying degrees. In this intermediate region, the water structure is said to be *partly broken down*.

One can summarize this description of the structure of water near an ion by referring to three regions (Fig. 2.28). In the *primary*, or structure-enhanced, region next to the ion, the water molecules are immobilized and oriented by the ionic field; they move as and where the ion moves. Then, there is a *secondary*, or structure-broken, region, in which the normal bulk structure of water is broken down to varying degrees. The in-between water molecules, however, do not partake of the translational motion of the ion. Finally, at sufficient distance from the ion, the water structure is unaffected by the ion and displays the tetrahedrally bonded networks characteristic of bulk water.

The three regions just described differ in their degree of sharpness. The primary region—to be discussed in greater detail below—in which there are (at least for some ions) water molecules which share the translational motion of the ion, is a fairly sharply defined region. In contrast, the secondary region, stretching from the termination of the primary region to

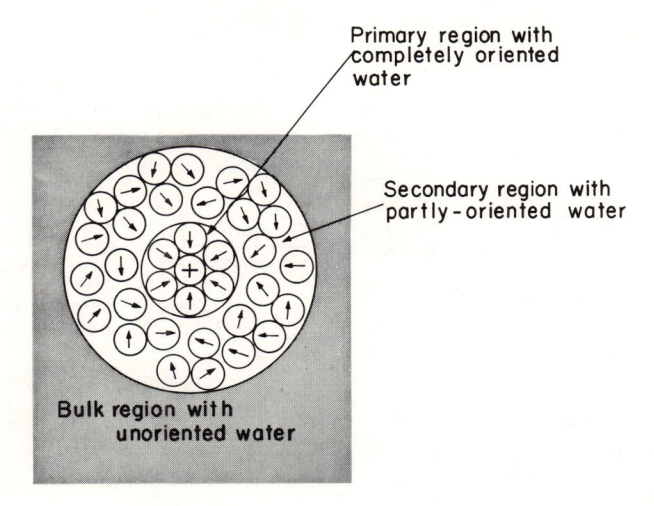

Fig. 2.28. The neighborhood of an ion may be considered to consist of three regions with differing solvent structures: (1) the primary or structure-forming region, (2) the secondary or structure-breaking region, and (3) the bulk region.

80 CHAPTER 2

the resumption of the normal bulk structure, cannot be sharply defined; the bulk properties and structure are asymptotically approached.

These structural changes in the primary and secondary regions are generally referred to as *solvation* (or as *hydration* when water is the solvent). Since they result from interactions between the ion and the surrounding solvent, one often uses the terms solvation and ion–solvent interactions synonymously; the former is the structural result of the latter.

2.3.3. The Ion–Dipole Model of Ion–Solvent Interactions

The above description of the solvent surrounding an ion can now be used as the basis of a *structural* treatment of ion–solvent interactions, initiated by Bernal and Fowler (1933).

Consider an isolated ion in the gas phase above the solvent. The total work done to transfer this ion from a very dilute gas of ions to the inside of the solvent defines the free energy of solvation (*cf.* Section 2.2.1), i.e., the free-energy change arising from ion–solvent interactions, ΔG_{I-S}. This free-energy change is composed of both enthalpy changes and entropy changes. The latter arise from changes in the degrees of freedom (translational, rotational, and vibrational) experienced by the water molecules as they come out of the water structure and associate with an ion. In this simplified treatment, only the enthalpy changes will be treated.

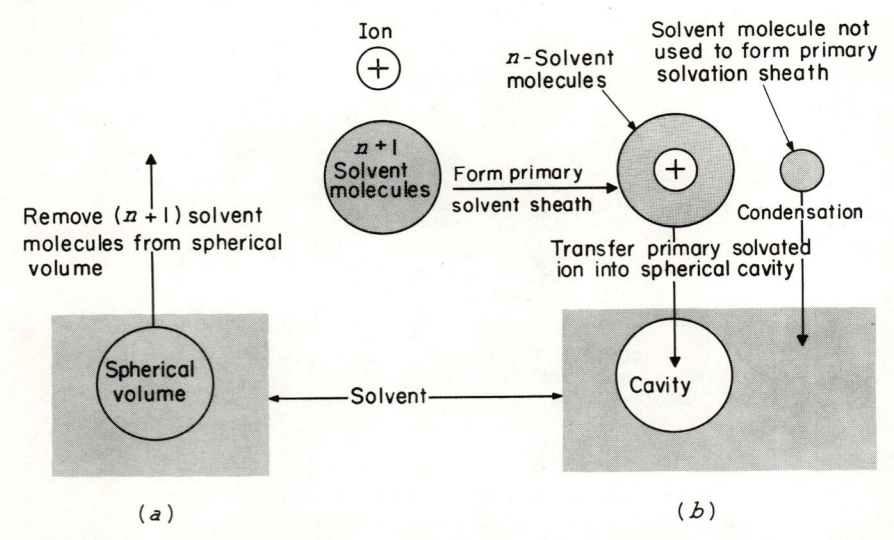

Fig. 2.29. A thought experiment to separate out various aspects of ion–solvent interactions.

ION–SOLVENT INTERACTIONS 81

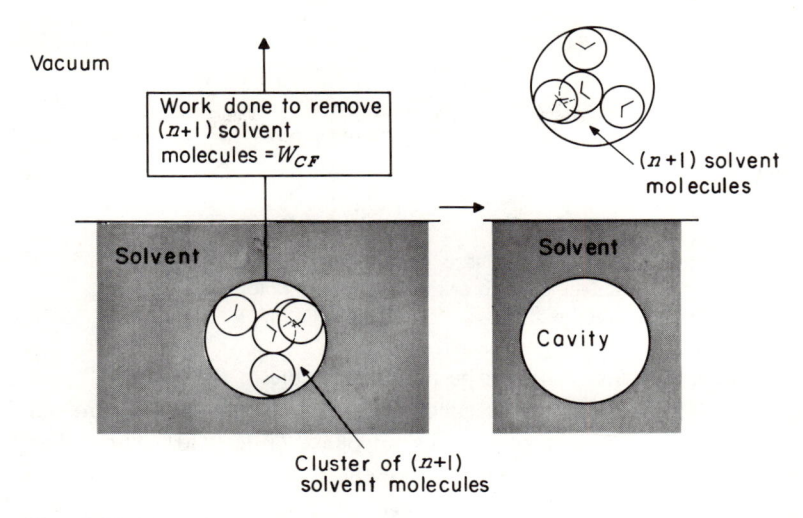

Fig. 2.30. The formation of a cavity in the solvent by the removal of $n + 1$ solvent molecules.

The ion–solvent interactions consist of several contributions. There is, for example, the interaction between the ion and the n nearest neighbors which surround the ion and make up the primary solvent sheath (see Fig. 2.28). Then, there is the energy used up for the structure breaking in the secondary region (Fig. 2.28), etc.

To separate out the various aspects of the total interaction, one can consider a thought experiment (Fig. 2.29) proposed by Eley and Evans in which the proper number of solvent dipoles are taken from the solvent to the gas phase and there oriented around the ion (by ion–dipole forces). Finally, the primary solvated ion is transferred into the solvent, upon which structure breaking, etc., occurs.

The steps of this thought experiment will now be described more elaborately.

1. One starts the thought experiment with the knowledge gained from various types of experiment that the primary solvated ion will occupy a volume corresponding to the volume of n primary solvent molecules *plus* one more to make room for the bare ion.[†] This volume corresponding to

---

[†] Note that, as a first approximation, it is assumed that the volume of a water molecule is the same as that of a bare ion. For some ions, this is a reasonable approximation. Thus, the radius of a water molecule is 1.38 Å and that of K^+ is 1.33 Å.

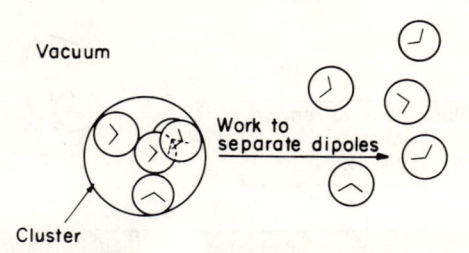

Fig. 2.31. Dissociation of a cluster of $n + 1$ molecules by breaking the bonds holding them together.

$n + 1$ solvent molecules must be made available in the solvent for immersing a primary solvated ion. Hence, $n + 1$ molecules will be removed from the solvent and taken into the vacuum phase (Fig. 2.30). Thus, the cavity which is left in the solvent will be large enough to accommodate an ion plus n molecules in its *primary solvation sheath*. Let this work of cavity formation be represented by W_{CF}.

2. Before the $n + 1$ solvent molecules just removed from the solvent can orient around the ion in the gas phase, they must be detached from the cluster of $n + 1$ molecules and made free to orient around the ion. To make this feasible, the bonds holding together these $n + 1$ solvent dipoles in the cluster are broken asunder (Fig. 2.31), i.e., the group is dissociated in the gas phase into $n + 1$ separate molecules. This dissociation will involve an amount of work represented by W_D.

3. Next, ion–dipole bonds are forged between the ion and n out of the $n + 1$ solvent dipoles, and, thus, the primary solvent sheath is formed. The work of interaction between an ion and a dipole (of moment μ_s and radius r_s) for the configuration shown in Fig. 2.32 is approximately

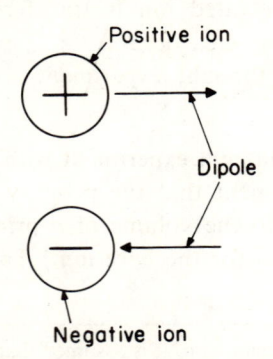

Fig. 2.32. The minimum interaction-energy orientation of a dipole to an ion.

ION–SOLVENT INTERACTIONS 83

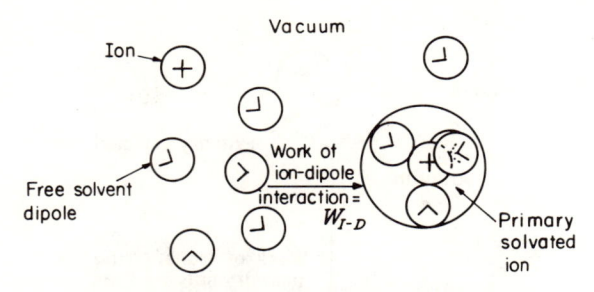

Fig. 2.33. Formation of a primary solvated ion.

given by (*cf.* Appendix 2.2)†

$$-\frac{z_i e_0 \mu_s}{(r_i + r_s)^2}$$

But, it is n solvent molecules that are involved in the primary solvent sheath. Hence, per mole of ions, the ion–dipole interaction work (Fig. 2.33) is‡

$$W_{I-D} = -\frac{N_A n z_i e_0 \mu_s}{(r_i + r_s)^2} \qquad (2.25)$$

4. Now, the ion together with its primary solvent sheath is transferred from vacuum into the cavity in the solvent (Fig. 2.34). What work is involved in this transfer? A simple way to look at it is to imagine that the solvated ion in the gas phase is discharged and then, still preserving its solvent sheath, is sneaked into the cavity formed in step 1 of the thought experiment (*cf.* Fig. 2.30), whereafter the discharged but still solvated ion is charged up to its normal value $z_i e_0$. What has been described is simply a Born charging process (*cf.* Section 2.2.5). There is, however, an important difference between the Born charging done here and that previously described (Fig. 2.35). It is not a bare ion but a primary solvated ion which undergoes the charging process. Hence, the radius to be used in the Born expression (2.22) is no longer the crystallographic radius r_i but the radius of a solvated ion, i.e., $r_i + 2r_s$.

Since it has been decided to deal only with enthalpies (or heat-content

---

† Note that the dielectric constant does not appear in this expression because there is only vacuum between the dipole (i.e., the water molecule) and the (adjacent) ion.
‡ Note that the ion–dipole work always contributes a negative quantity to the heat of solvation, independently of the sign of z_i, because the dipole always orients so that that pole is in contact with the ion which makes the interaction attractive.

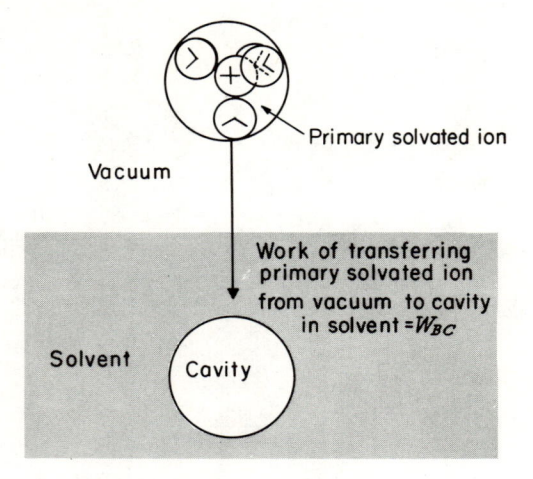

Fig. 2.34. Transfer of a primary solvated ion from vacuum into a cavity in the solvent.

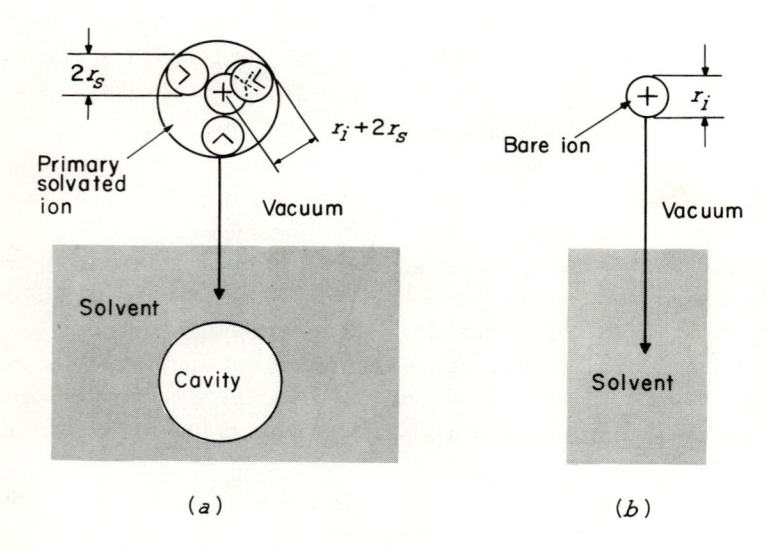

Fig. 2.35. The difference between the Born charging process in (a) the ion–dipole model of solvation in which a primary solvated ion of radius $r_i + 2r_s$ is transferred into the solvent (Section 2.3.3) and in (b) the nonstructural model of Born (*cf.* Section 2.2.5) in which a bare ion of radius r_i is involved.

changes), one can set the work of transferring a solvated ion from vacuum into a cavity in the solvent equal to the Born *heat* of solvation. This contribution to the total heat of ion–solvent interactions shall be called the *Born charging contribution*, W_{BC}. Thus, per mole of ions,

$$W_{BC} = -\frac{N_A(z_i e_0)^2}{2(r_i + 2r_s)}\left(1 - \frac{1}{\varepsilon_s} - \frac{T}{\varepsilon_s^2}\frac{\partial \varepsilon_s}{\partial T}\right) \quad (2.26)$$

Is it reasonable to use here an equation based on the Born model even though one motivation for this structural treatment of solvation is to get away from the Born nonstructural approach? The justification is as follows. The radius $r_i + 2r_s$ has been precisely defined (its ambiguity was a problem in the Born model), the water outside the cavity is, at this stage of the thought experiment, normal and undisturbed, and, therefore, its dielectric constant (another ambiguity of the Born model) should be that of *bulk* water. Thus, by considering the process of ion–solvent interactions as occurring in steps with corresponding heat-content changes, one of the steps, namely, the introduction of a primary solvated ion into an *undisturbed* solvent, has been made to resemble a Born charging process.

5. Once the cavity is filled up with the solvated ion and the Born charging is carried out, one must ask whether the solvated ion leaves the surrounding water undisturbed. It does not (Figs. 2.28 and 2.36). The introduction of the primary solvated ion into the cavity does lead to some

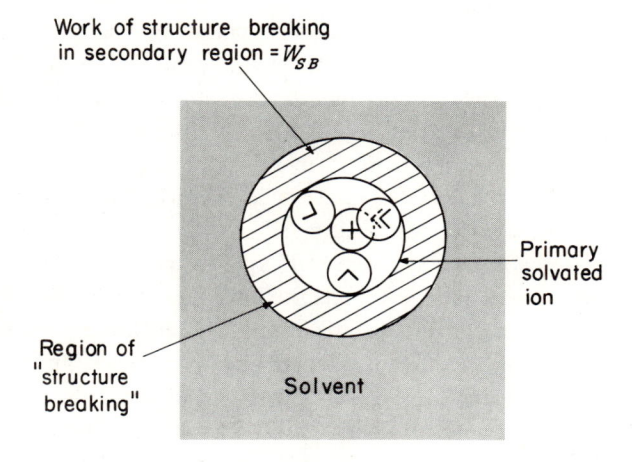

Fig. 2.36. The introduction of a primary solvated ion into the cavity causes disturbance to the structure of the solvent in the immediate vicinity of the solvated ion.

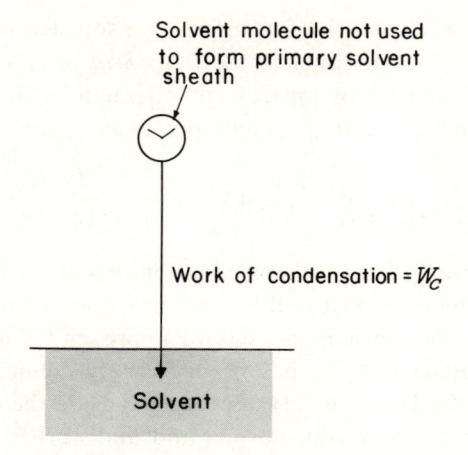

Fig. 2.37. The condensation of a water molecule left behind in vacuum because it was not used to form the primary solvent sheath.

disturbance of the structure of the surrounding solvent. In fact, this is the structure breaking that has been referred to in dealing with the secondary region, between the primary solvent sheath and the bulk water far away from the ion (Fig. 2.28). Let this work of structure breaking be represented by W_{SB}.

6. One must check up on the cycle now. Have all the solvent molecules (taken out of the solvent into vacuum to create the cavity) been returned to the solvent? Of the $n + 1$ solvent molecules removed, only n have returned in the company of the ion as members of its solvation sheath. The one water molecule which did not become part of the solvation sheath of the ion and which has been left behind in vacuum, has to be returned to the solvent to complete the cycle (Fig. 2.37). The work involved in this process is equal to the work of condensation, W_C.

Now, all the solvent molecules which were removed from the solvent in the thought experiment have returned to the solvent. In addition, the ion which was in vacuum at the beginning of the thought experiment has been transferred into the solvent. Hence, any work resulting from plunging the ion into the solvent must result purely from ion–solvent interactions. This work (or heat)[†] of solvation or ion–solvent interactions is therefore

---

[†] Note that, as already stated, there is an approximation being made here: It is the free-energy change which is exactly equal to the work done (Appendix 2.1). One has

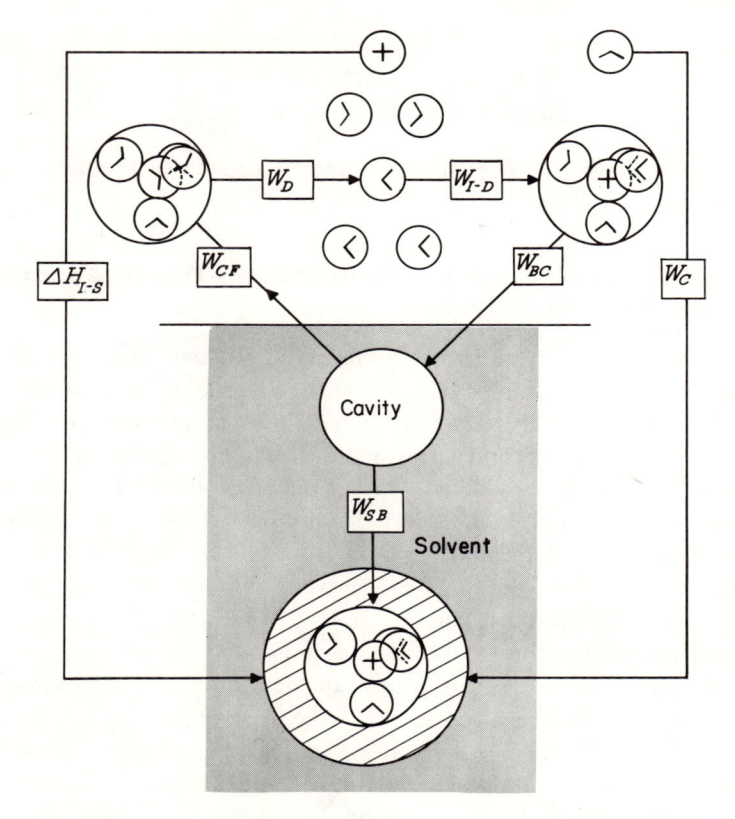

Fig. 2.38. How the total heat ΔH_{I-S} of ion–solvent interactions has been separated in a thought experiment into the various steps of cavity formation W_{CF}, cluster dissociation W_D, formation of primary solvated ion W_{I-D}, Born charging W_{BC}, structure breaking W_{SB}, and condensation, W_C.

given by (Fig. 2.38) the sum of all the pieces of work performed in each step, i.e.,

$$\Delta H_{I-S} = W_{CF} + W_D + W_{I-D} + W_{BC} + W_{SB} + W_C \quad (2.27)$$

$$= W + W_{I-D} + W_{BC} \quad (2.28)$$

where

$$W = W_{CF} + W_D + W_{SB} + W_C \quad (2.29)$$

---

neglected $T\Delta S$, where ΔS is the change of entropy during the solvation process. Structural theories of the entropy of hydration are known but will not be discussed here. The error introduced by the approximation is about 10%.

Hence,

$$\Delta H_{I-S} = W - \frac{N_A n z_i e_0 \mu_s}{(r_i + r_s)^2} - \frac{N_A (z_i e_0)^2}{2(r_i + 2r_s)} \left(1 - \frac{1}{\varepsilon_s} - \frac{T}{\varepsilon_s^2} \frac{\partial \varepsilon_s}{\partial T}\right) \quad (2.30)$$

where the Avogadro number has been introduced to get the heat of solvation per mole of ions.

2.3.4. Evaluation of the Terms in the Ion–Dipole Approach to the Heat of Solvation

The Born term, i.e., the last term in Eq. (2.30), can be easily calculated. One uses the crystallographic radius r_i of the ion, the radius r_s of the solvent molecules, and the bulk dielectric constant ε_s of the solvent. The ion–dipole term, i.e., the second term in the expression (2.30) for the heat of solvation, can also be calculated without difficulty *provided* one knows—or estimates—the number n of solvent molecules which coordinate (or are nearest neighbors to) the ion.

The first term in Eq. (2.30), however, is more awkward. It will be recalled [Eq. (2.29)] that it consists of W_{CF}, the work of forming a cavity in the solvent by the removal into the gas phase of a cluster of $n + 1$ solvent molecules; W_D, the work of splitting up the cluster and separating to infinity the $n + 1$ solvent molecules; W_{SB}, the work of altering the orientation of the solvent molecules in the solvent around the primary solvated ions; and W_C, the work of condensing the one solvent molecule (from the cluster) which is not used in the solvation of the ion.

The work W_D of breaking the cluster and separating the $n + 1$ solvent molecules can be considered either as the work of separating dipoles, i.e., the work arising from dipole–dipole forces or, in the case of hydrogen-bonded liquids such as water, the work of breaking hydrogen bonds (Fig. 2.39). Since about 5 kcal mole^{-1} is required to break hydrogen bonds, the value of W_D depends on the value of n in the cluster of $n + 1$ solvent mole-

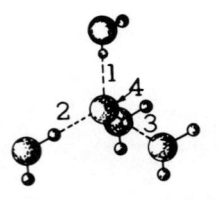

Fig. 2.39. Four hydrogen bonds (which are numbered) must be broken to separate the cluster of $4 + 1 = 5$ water molecules.

cules which are removed from the solvent to make room for an ion and its n nearest neighbors. If, for example, an ion surrounds itself with four water molecules in a tetrahedral configuration, then the cluster consists of $4 + 1 = 5$ water molecules, and four hydrogen bonds must be broken per cluster to separate the water molecules. Since 1 mole of cluster must be removed from the solvent for the solvation of 1 mole of ions, it is necessary to break 4 moles of hydrogen bonds per mole of ions. This requires $4 \times 5 = 20$ kcal mole^{-1}.

The work W_C of condensing one solvent molecule per ion, or 1 mole of solvent molecules per mole of ions, can be taken from the experimental latent heat of condensation (Fig. 2.40); it is about -10 kcal mole^{-1}.

The cavity formation work W_{CF} and the structure-breaking work W_{SB} can be only roughly calculated. When the $n + 1$ water molecules are removed to form the cavity, a certain number of hydrogen bonds linking these molecules to those outside the cavity are broken (Fig. 2.41). When the primary solvated ion is introduced into the cavity, some of the solvent molecules surrounding the solvated ion have to reorient. This reorientation leads to the breakage of some hydrogen bonds and the formation of others. Thus, if one considers the combined steps of cavity formation *and* structure breaking, a certain *net* number of hydrogen bonds will be broken. Once this number is known, one can easily get $W_{\text{CF}} + W_{\text{SB}}$ by multiplying the net number of hydrogen bonds broken by 5 kcal mole^{-1}.

A simple way of getting this number is to look at the water structure before and after the solvated ion is introduced into the cavity. A careful study of Fig. 2.42 shows that, whereas 12 hydrogen bonds are broken in the cavity formation step involving the removal of $4 + 1 = 5$ water mole-

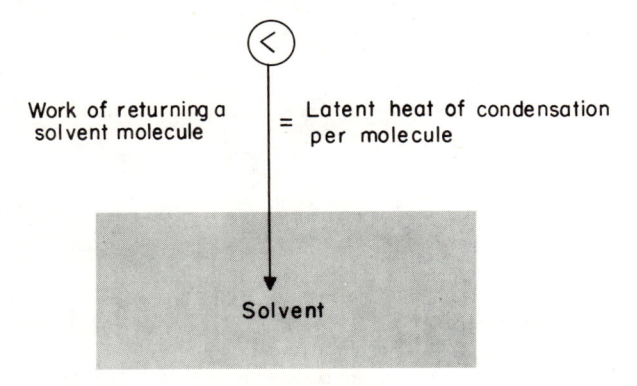

Fig. 2.40. The work of condensing a water molecule is equal to the latent heat of condensation per molecule.

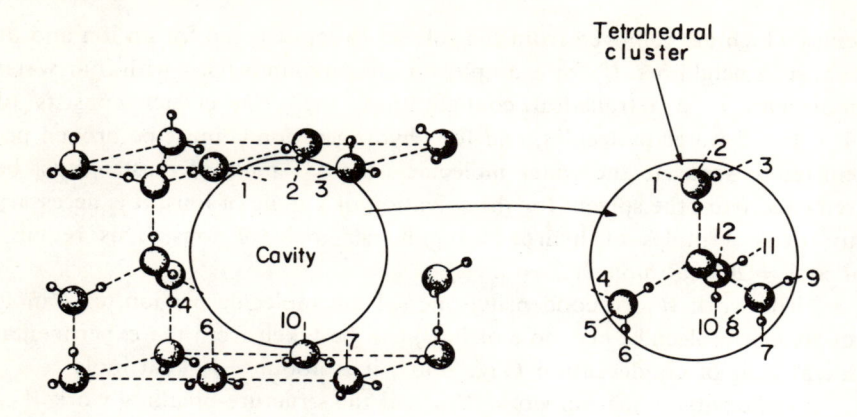

Fig. 2.41. A total of 12 hydrogen bonds are broken when a tetrahedral cluster of water molecules is removed from the solvent to form the cavity (numbers represent broken hydrogen bonds).

Cluster with its broken hydrogen bonds
(a)

Solvated positive ion
(b)

Two hydrogen bonds not remade

Solvated positive ion

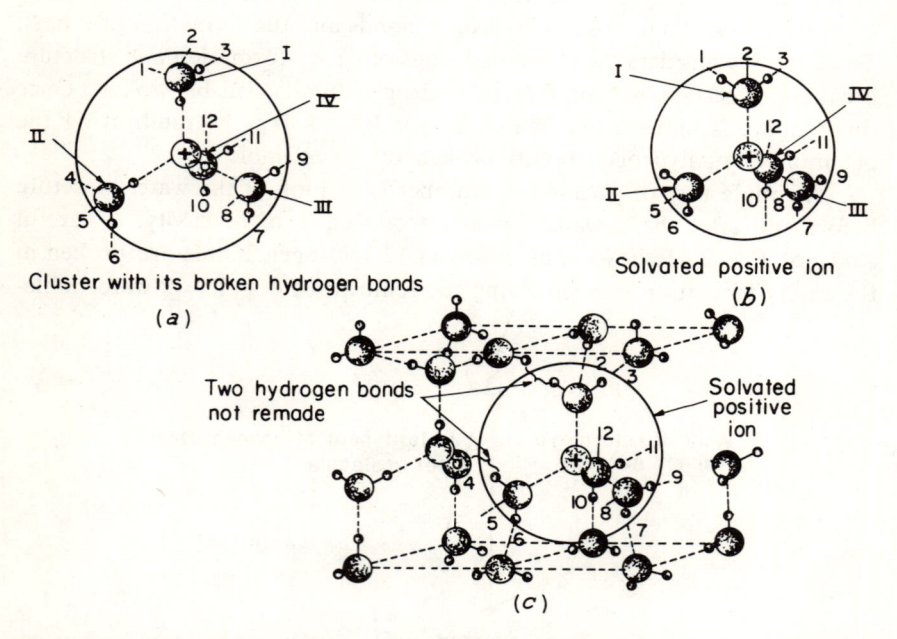

(c)

Fig. 2.42. Schematic diagram to show that, out of four coordinating water molecules [I, II, III, and IV in (a)] in a tetrahedral cluster removed from the cavity, two water molecules [I and II in (b)] reorient in the formation of a primary solvated positive ion, and, therefore, only 10 H bonds [see (a)] are remade when the solvated *positive* ion is introduced into the cavity.

cules, only 10 hydrogen bonds are remade when the primary solvated positive ion is introduced into the cavity. That is, a net number of 2 hydrogen bonds are broken per ion in the combined process of cavity formation and structure breaking. The corresponding heat change $W_{CF} + W_{SB}$ is $2 \times 5 = 10$ kcal mole^{-1} of ions.

It is now possible to write down for a tetrahedrally coordinated positive ion an approximate value for the work term [cf. Eq. (2.29)] $W = W_D + W_C + W_{CF} + W_{SB}$. Using the arguments just presented, i.e., $W_D = 20$ kcal mole^{-1}, $W_C = -10$ kcal mole^{-1}, and $W_{CF} + W_{SB} = 10$ kcal mole^{-1}, one has for four-coordinated positive ions

$$W = W_D + W_C + W_{CF} + W_{SB}$$
$$= 20 - 10 + 10$$
$$= 20 \text{ kcal mole}^{-1} \qquad (2.31)$$

Now consider negative ions. If, once again, tetrahedral coordination is considered, then W_D continues to be the work required to break up a cluster of five water molecules, i.e., it is 20 kcal mole^{-1}. The latent heat of condensation of a water molecule obviously remains the same (-10 kcal mole^{-1}) for positive and negative ions. But a perusal of Fig. 2.43 shows that the negative ion differs from the positive ion in that more water molecules have to reorient when the primary solvated ion is introduced into the cavity. In other words, the orientation of water molecules *around* a primary solvated ion is less compatible with the water molecules *in* the primary solvation shell of negative ions than with those of positive ions. Thus, of the 12 hydrogen bonds broken in forming the cavity, only 8 are remade when the cavity is filled up with a solvated ion (see Fig. 2.43). That is, the net number of hydrogen bonds broken in the combined process of cavity formation and structure breaking is four in the case of tetrahedrally coordinated negative ions; and the corresponding work $W_{CF} + W_{SB}$ is $4 \times 5 = 20$ kcal mole^{-1}. Consequently, the work W [cf. Eq. (2.29)] for four coordinated negative ions is given by

$$W = W_D + W_C + W_{CF} + W_{SB}$$
$$= 20 - 10 + 20$$
$$= 30 \text{ kcal mole}^{-1} \qquad (2.32)$$

Now that the work W has been evaluated, it can be introduced into Eq. (2.30) for the heat of hydration. Thus, for four coordination, one has

Cluster with its broken H-bonds
(a)

Solvated negative ion
(b)

Four hydrogen bonds not remade

(c) Solvated negative ion

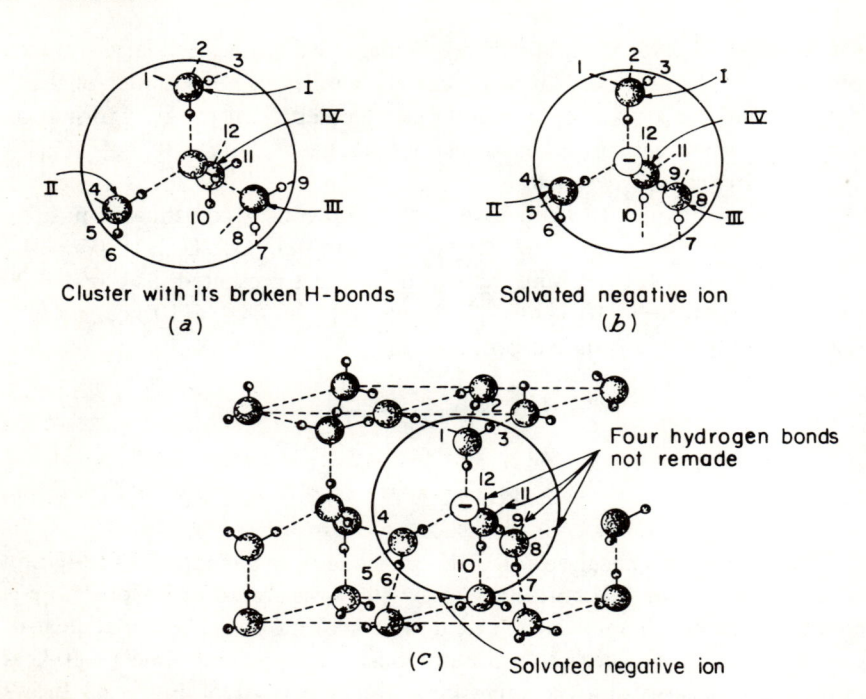

Fig. 2.43. Schematic diagram similar to Fig. 2.42 except that a *negative* ion is being considered here. Thus, two water molecules [III and IV in Fig. 2.42 (b)] reorient in the formation of a primary solvated negative ion; and, therefore, only 8 H bonds [1 to 7 and 10 in Fig. 2.42 (c)] out of 12 H bonds [see Fig. 2.42 (a)] are remade when the solvated ion is introduced into the cavity.

$$\Delta H_{I-S} = 20 - \frac{4N_A z_i e_0 \mu_W}{(r_i + r_W)^2} - \frac{N_A(z_i e_0)^2}{2(r_i + 2r_W)} \left(1 - \frac{1}{\varepsilon_W} - \frac{T}{\varepsilon_W^2} \frac{\partial \varepsilon_W}{\partial T}\right) \quad (2.33)$$

for positive ions, and

$$\Delta H_{I-S} = 30 - \frac{4N_A z_i e_0 \mu_W}{(r_i + r_W)^2} - \frac{N_A(z_i e_0)^2}{2(r_i + 2r_W)} \left(1 - \frac{1}{\varepsilon_W} - \frac{T}{\varepsilon_W^2} \frac{\partial \varepsilon_W}{\partial T}\right) \quad (2.34)$$

for negative ions.

By analyzing the structure breaking for octahedral ($n = 6$) coordination, one can develop expressions for the heat of solvation of ions with six solvent molecules in their primary solvent shells. In this case, the expression is

$$\Delta H_{I-S} = 15 - \frac{6N_A z_i e_0 \mu_W}{(r_i + r_W)^2} - \frac{N_A(z_i e_0)^2}{2(r_i + 2r_W)} \left(1 - \frac{1}{\varepsilon_W} - \frac{T}{\varepsilon_W^2} \frac{\partial \varepsilon_W}{\partial T}\right) \quad (2.35)$$

2.3.5. How Good Is the Ion–Dipole Theory of Solvation?

The heats of ion–solvent interactions calculated on the basis of the ion–dipole approach [Eq. (2.33) and (2.34)] can now be compared with the experimental values used to test the Born theory. The comparison is therefore made with values obtained by equally dividing the experimental heat of solvation of the salt KF between K^+ and F^- ions and then using the individual values thus gained in data for the experimental hydration heats of other salts (*cf.* Section 2.2.7).

The comparison (Table 2.12) shows that the ion–dipole model is a considerable improvement over the rudimentary Born continuum model. The improvement indicates that the ion–dipole model is on the right track in considering that an ion sees the solvent in contact with it as consisting of discrete water dipoles which orient around it. It is only the solvent lying farther out which the ion views Born-wise as a dielectric continuum. Thus, by assuming that the solvent has the bulk dielectric constant right up to the surface of the ion, the Born model missed the work of orientation of water dipoles around the ion and the related change of dielectric constant of water near the ion.[†]

When one considers numerically the various contributions to the heats of ion–solvent interactions calculated from Eqs. (2.33) and (2.34), it can be seen (Table 2.13) that the main contributions come from the ion–dipole and Born charging terms, i.e.,

$$-\frac{N_A n z_i e_0 \mu_W}{(r_i + r_W)^2}$$

and

$$-\frac{N_A (z_i e_0)^2}{2(r_i + 2r_W)} \left(1 - \frac{1}{\varepsilon_W} - \frac{T}{\varepsilon_W^2} \frac{\partial \varepsilon_W}{\partial T}\right)$$

respectively. This fact must be taken to mean that ion–solvent interactions are *essentially electrostatic in origin*. The ion behaves like a charged sphere to the water outside the primary solvent sheath and like an orienting attracting charge to the water molecules inside the primary solvent shell.

The approximately ± 10% agreement between the calculated and experimental values for the heats of solvation of ions should normally be cause for jubilation, but the situation here is abnormal. The so-called experimental values have been obtained by splitting the unambiguous experi-

---

[†] The connection between the orientation of water dipoles around ions and the dielectric constant of the medium will be looked into much further in Section 2.5.

TABLE 2.12

Comparison between the Experimental Heats of Ion–Solvent Interactions and Those Calculated on the Basis of the Ion–Dipole Approach

Ion	ΔH_{I-S} calc. with Eqs. (2.33) and (2.34), kcal mole^{-1}	Exp. ΔH_{I-S} with $\Delta H_{K^+-H_2O} = \Delta H_{F^--H_2O} = \tfrac{1}{2}\Delta H_{KF-H_2O}$
Li$^+$	−160.1	−146.3
Na$^+$	−119.2	−118.9
K$^+$	− 90.5	− 98.9
Rb$^+$	− 81.9	− 93.8
Cs$^+$	− 71.8	− 88.0
F$^-$	− 78.6	− 98.9
Cl$^-$	− 56.8	− 64.9
Br$^-$	− 51.6	− 58.4
I$^-$	− 44.7	− 48.6

mental value of the heat of solvation of KF equally between K$^+$ and F$^-$ ions, i.e., by assuming that ions of equal radius but opposite signs have equal heats of solvation. This assumption, based on the Born nonstructural model, *taints* the so-called experimental values of the individual heats of ionic solvation, which are being used to test the numerical values calculated

TABLE 2.13

Various Contributions to the Ion–Solvent Interactions

Ion	Heat of Born charging, kcal mole^{-1}	Heat of ion–dipole interactions, kcal mole^{-1}
Li$^+$	−49.6	−130.5
Na$^+$	−45.0	− 94.2
K$^+$	−40.8	− 69.9
Rb$^+$	−39.3	− 62.6
Cs$^+$	−37.5	− 54.3
F$^-$	−40.5	− 68.1
Cl$^-$	−36.5	− 50.3
Br$^-$	−35.4	− 46.2
I$^-$	−33.9	− 40.8

ION–SOLVENT INTERACTIONS

by the structural ion–dipole theory. One has therefore to scrutinize the assumption more carefully in order to assess the progress made in the understanding of the heats of solvation of individual ions.

2.3.6. The Relative Heats of Solvation of Ions on the Hydrogen Scale

Consider the unambiguous experimental value ΔH_{HX-H_2O}, the heat of interaction between HX and water. It is made up of the heats of solvation of H$^+$ ions and X$^-$ ions[†]

$$\Delta H_{HX} = \Delta H_{H^+} + \Delta H_{X^-} \quad (2.36)$$

The heat of solvation of X$^-$ ions *relative* to that of H$^+$ ions, i.e., ΔH_{X^-} (rel) can be defined by considering ΔH_{H^+} an arbitrary zero in Eq. (2.36)

$$\Delta H_{X^-}(\text{rel}) = \Delta H_{X^-}(\text{abs}) + \Delta H_{H^+}(\text{abs}) = \Delta H_{HX} \quad (2.37)$$

where the notation (*rel*) and (*abs*) has been inserted to distinguish between the relative ΔH_{X^-} value of X$^-$ ions on a arbitrary scale of ΔH_{H^+} (abs) = 0 and the absolute or true ΔH_{X^-} values. From equation (2.37),

$$\Delta H_{X^-}(\text{rel}) = \Delta H_{HX} \quad (2.38)$$

and, since ΔH_{HX} can be experimentally obtained to a precision determined by measuring techniques, one can see that the relative heats of solvation, ΔH_{X^-} (rel), are clear-cut experimental quantities.

Relative heats of solvation can also be defined for positive ions. One writes

$$\Delta H_{MX} = \Delta H_{M^+}(\text{abs}) + \Delta H_{X^-}(\text{abs}) \quad (2.39)$$

and substitutes for ΔH_{X^-} (abs) from Eq. (2.37). Thus, one has

$$\Delta H_{X^-}(\text{abs}) = \Delta H_{X^-}(\text{rel}) - \Delta H_{H^+}(\text{abs}) \quad (2.40)$$

$$= \Delta H_{HX} - \Delta H_{H^+}(\text{abs}) \quad (2.41)$$

[†] One should, strictly speaking, write

$$\Delta H_{HX-H_2O} = \Delta H_{H^+-H_2O} + \Delta H_{X^--H_2O}$$

but the –H$_2$O will be dropped out in the subsequent text to make the notation less cumbersome.

TABLE 2.14

Relative Heats of Hydration of Individual Ions, ΔH_{H^+}(abs) = 0

Ion	Relative heats of hydration
Li$^+$	+136.34
Na$^+$	+163.68
K$^+$	+183.74
Rb$^+$	+188.80
Cs$^+$	+194.60
F$^-$	−381.50
Cl$^-$	−347.50
Br$^-$	−341.00
I$^-$	−331.20

which, when inserted into Eq. (2.39), gives

$$\Delta H_{MX} = \Delta H_{M^+}(\text{abs}) + \Delta H_{HX} - \Delta H_{H^+}(\text{abs}) \qquad (2.42)$$

or

$$\Delta H_{M^+}(\text{abs}) - \Delta H_{H^+}(\text{abs}) = \Delta H_{MX} - \Delta H_{HX} \qquad (2.43)$$

Taking ΔH_{H^+}(abs) as an arbitrary zero in this equation permits the definition of the relative heat ΔH_{M^+}(rel) of solvation of positive ions

$$\Delta H_{M^+}(\text{rel}) = \Delta H_{M^+}(\text{abs}) - \Delta H_{H^+}(\text{abs}) \qquad (2.44)$$

$$= \Delta H_{MX} - \Delta H_{HX} \qquad (2.45)$$

Since ΔH_{MX} and ΔH_{HX} are unambiguous experimental quantities, so are the relative heats of solvation, ΔH_{M^+}(rel), of positive ions.

On this basis, a table of *relative* heats of solvation of individual ions can be drawn up (*cf.* Table 2.14). These relative heats will now be used to gauge the assumption that ions of equal radii and opposite charge have equal heats of solvation.

2.3.7. Do Oppositely Charged Ions of Equal Radii Have Equal Heats of Solvation?

Consider two ions M_i^+ and X_i^- of *equal radius r_i but opposite charge*. If their *absolute* heats of solvation are equal, then, one expects that

$$\Delta H_{M_i^+}(\text{abs}) - \Delta H_{X_i^-}(\text{abs}) = 0 \qquad (2.46)$$

But, from the definition of the *relative* heats of solvation of positive ions [Eq. (2.45)] and of negative ions [Eq. (2.40)], one has by subtraction

$$\Delta H_{M_i^+}(\text{abs}) - \Delta H_{X_i^-}(\text{abs}) = [\Delta H_{M_i^+}(\text{rel}) - \Delta H_{X_i^-}(\text{rel})] + 2\Delta H_{H^+}(\text{abs}) \tag{2.47}$$

If, therefore, the left-hand side is zero, then one should find, since $\Delta H_{H^+}(\text{abs})$ is a constant, that

$$\Delta H_{M_i^+}(\text{rel}) - \Delta H_{X_i^-}(\text{rel}) = \text{a constant} \tag{2.48}$$

This prediction can easily be checked. One makes a plot of the experimentally known *relative* heats of solvation of positive and negative ions as a function of ionic radius. By erecting a perpendicular at a radius r_i, one can get the difference $[\Delta H_{M_i^+}(\text{rel}) - \Delta H_{X_i^-}(\text{rel})]$ between the relative heats of solvation of positive and negative ions *of radius r_i*. By repeating this procedure at various radii, one can make a plot of the differences $[\Delta H_{M^+}(\text{rel}) - \Delta H_{X^-}(\text{rel})]$ as a function of radius. If oppositely charged ions of the same radius have the same absolute heats of hydration, then, $[\Delta H_{M^+}(\text{rel}) - \Delta H_{X^-}(\text{rel})]$ should have a constant value independent of radius. It does not (Fig. 2.44).

When, however, one examines the terms in Eqs. (2.33) and (2.34) for the heat of ion–solvent interaction, it is clear that neither the Born charging term nor the ion–dipole term depends on the *sign* of the charge on the ion. The only term which does depend on the sign of the ionic charge is the first term W, which is 20 kcal mole^{-1} for positive ions and 30 kcal

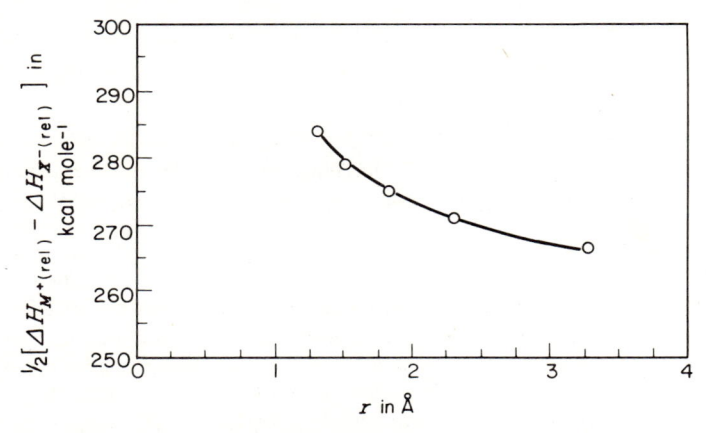

Fig. 2.44. Plot of the difference between the relative heats of hydration of oppositely charged, equiradii ions *versus* ionic radius.

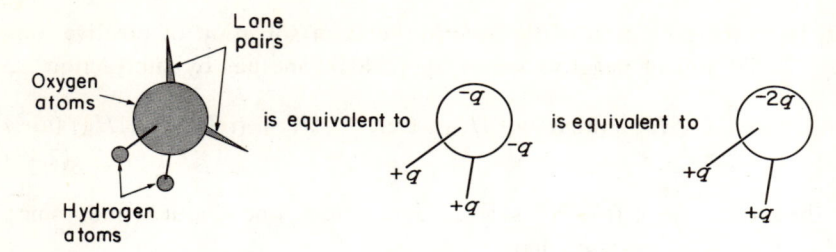

Fig. 2.45. The electrical equivalence between a water molecule and a quadrupole.

mole^{-1} for negative ions having $n = 4$. Since this argues for a constant difference between the heats of solvation of positive ions and negative ions, W cannot explain why $[\Delta H_{M^+} \text{(rel)} - \Delta H_{X^-} \text{(rel)}]$ is not a constant but varies with radius.† One has to seek an alternative explanation.

2.3.8. The Water Molecule Can Be Viewed as an Electrical Quadrupole

The structural approach to ion–solvent interactions has been developed so far by considering that the electrical equivalent of a water molecule is an idealized dipole, i.e., two charges of equal magnitude but opposite sign separated by a certain distance. Is this an adequate representation of the charge distribution in a water molecule?

Consider an ion in contact with the water molecule; this is the situation obtained in the primary hydration sheath. The ion is close enough to see one positively charged region near each hydrogen nucleus and two negatively charged regions corresponding to the lone pairs near the oxygen atom. In fact, from this intimate viewpoint, the charge distribution in the water molecule can be represented (Fig. 2.45) by a model with four charges of

---

† The assumption has been made that W depends only on the sign of the charge on the ion and not on the radius of the ion. While this assumption is reasonable in an approximate evaluation of W, a closer examination of the expression for W, i.e., $W = W_D + W_C + (W_{CF} + W_{SB})$, reveals that the volume of the cavity that has to be formed and the extent of structure breaking do depend on ionic radius. However, the work of cavity formation has a sign opposite to that of structure breaking, and, therefore, there should be some cancellation of these radius-dependent effects. Further, if W_+ for positive ions has a radius dependence, so has W_- for negative ions, and it is likely that, in taking the difference $W_+ - W_-$ for two ions of opposite charge but equal radius, there is no significant radius dependence of $W_+ - W_-$. This, however, is an assumption which has to be substantiated by exact analysis. A similar statement applies to the assumption that n is independent of radius, and this must clearly break down for sufficiently large ions.

equal magnitude q—a charge of $+q$ near each hydrogen atom, and two charges each of value $-q$ near the oxygen atom. Thus, rather than consider that the water molecule can be represented by a dipole (an assembly of two charges), a better approximation, suggested by Buckingham (1957), is to view it as a *quadrupole*, i.e., an assembly of four charges. What may this increase in realism of model do to the remaining discrepancies in the theory of ion–solvent interactions?

2.3.9. The Ion–Quadrupole Model of Ion–Solvent Interactions

It will be recalled that the structural calculation of the heat of ion–solvent interactions (*cf.* Sections 2.3.3 and 2.3.4) involved the following cycle of hypothetical steps: (1) A cluster of $n+1$ water molecules is removed from the solvent to form a cavity; (2) the cluster is dissociated into $n+1$ independent water molecules; (3) n out of $n+1$ water molecules are associated with an ion in the gas phase through the agency of ion–*dipole* forces; (4) the primary solvated ion thus formed in the gas phase is plunged into the cavity; (5) the introduction of the primary solvated ion into the cavity leads to some structure breaking in the solvent outside the cavity; and (6), finally, the water molecule left behind in the gas phase is condensed into the solvent. The heat changes involved in these six steps are W_{CF}, W_D, W_{I-D}, W_{BC}, W_{SB}, and W_C, respectively, where, for $n=4$.

$$W = W_{CF} + W_D + W_{SB} + W_C = +20 \text{ for positive ions}$$
$$= +30 \text{ for negative ions} \quad (2.29)$$

$$W_{I-D} = -\frac{4N_A z_i e_0 \mu_W}{(r_i + r_W)^2} \quad (2.25)$$

$$W_{BC} = -\frac{N_A (z_i e_0)^2}{2(r_i + 2r_W)} \left(1 - \frac{1}{\varepsilon_W} - \frac{T}{\varepsilon_W^2} \frac{\partial \varepsilon_W}{\partial T}\right) \quad (2.26)$$

and the total heat of ion–water interactions is

$$\Delta H_{I-H_2O} = W + W_{I-D} + W_{BC} \quad (2.28)$$

If one scrutinizes the various steps of the cycle, it will be realized that only for one step, namely, step 3, does the heat content change [Eq. (2.25)] depend upon whether one views the water molecule as an electrical dipole or quadrupole. Hence, the expressions for the heat changes for all steps *except* step 3 can be carried over as such into the theoretical heat of ion–

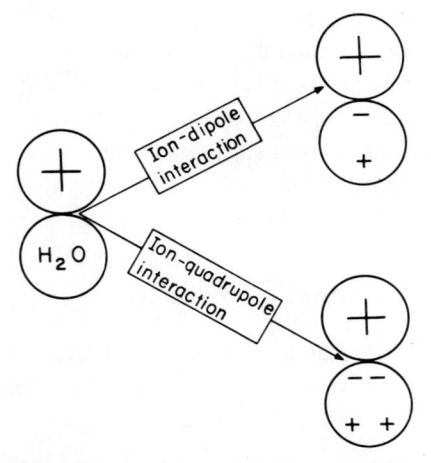

Fig. 2.46. Improvement in the calculation of the ion–water molecule interactions by altering the model of the water molecule from a dipole to a quadrupole.

water interactions, ΔH_{I-H_2O}, derived earlier. In step 3, one has to replace the heat of ion–dipole interactions, W_{I-D} [Eq. (2.25)] with the heat of ion–quadrupole interactions (Fig. 2.46).

But what is the expression for the energy of interaction between an ion of charge $z_i e_0$ and a quadrupole? The derivation of a general expression requires sophisticated mathematical techniques, but, when the water molecule assumes a symmetrical orientation (Fig. 2.47) to the ion, the ion–quadrupole interaction energy can easily be shown to be (Appendix 2.3)

$$E_{I-Q} = -\frac{z_i e_0 \mu_W}{r^2} \pm \frac{z_i e_0 p_W}{2r^3} \tag{2.49}$$

$$E_{I-Q} = -\frac{z_i e_0 \mu_W}{r^2} + \frac{z_i e_0 p_W}{2r^3}$$

$$E_{I-Q} = -\frac{z_i e_0 \mu_W}{r^2} - \frac{z_i e_0 p_W}{2r^3}$$

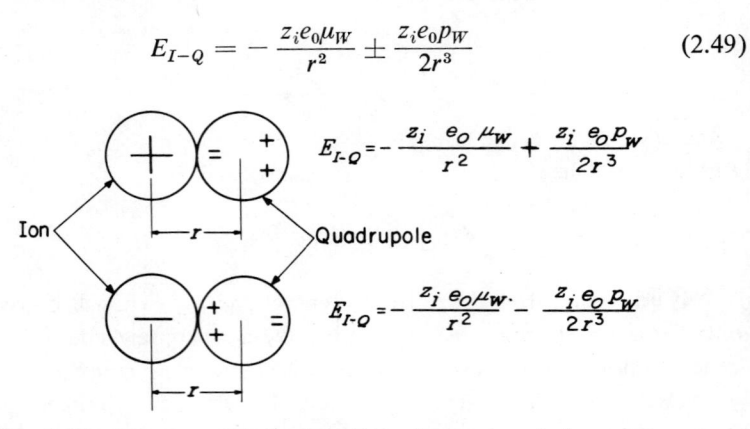

Fig. 2.47. The symmetrical orientation of a quadrupole to an ion.

where the $+$ in the $\pm$ is for positive ions, and the $-$ is for negative ions, and the p_W is the quadrupole moment (3.9×10^{-26} esu) of the water molecule. It is at once clear that a difference will arise for the energy of interaction of positive and negative ions with a water molecule, a result hardly forseeable from the rudimentary Born viewpoint and hence probably accountable for the result of Fig. 2.44.

The first term in this expression [Eq. (2.49)] is the dipole term, and the second term is the quadrupole term. It is obvious that, with increasing distance r between ion and water molecule, the quadrupole term becomes less significant. Or, in other words, the greater the value of r, the more reasonable it is to represent the water molecule as a dipole. But, as the ion comes closer to the water molecule, the quadrupole term becomes significant, i.e., the error involved in retaining the approximate dipole model becomes more significant.

When the ion is in contact with the water molecule, as is the case in the primary solvation sheath, the expression (2.49) for the ion–quadrupole interaction energy becomes

$$E_{I-Q} = -\frac{z_i e_0 \mu_W}{(r_i + r_W)^2} \pm \frac{z_i e_0 p_W}{(r_i + r_W)^3} \qquad (2.50)$$

The quantity E_{I-Q} represents the energy of interaction between one water molecule and one ion. If, however, four water molecules surround one ion and one considers a mole of ions, the heat change W_{I-Q} involved in the formation of a primary solvated ion through the agency of ion–quadrupole forces is given by

$$W_{I-Q} = 4N_A E_{I-Q} = -\frac{4N_A z_i e_0 \mu_W}{(r_i + r_W)^2} \pm \frac{4N_A z_i e_0 p_W}{2(r_i + r_W)^3} \qquad (2.51)$$

where, as before, the $+$ in the $\pm$ refers to positive ions and the $-$ to negative ions.

Substituting this expression for W_{I-Q} in place of W_{I-D} in expression (2.28) for the heat of ion–water interactions, one has

$$\Delta H_{I-H_2O} = 20 - \frac{4N_A z_i e_0 \mu_W}{(r_i + r_W)^2} + \frac{4N_A z_i e_0 p_W}{2(r_i + r_W)^3} - \frac{N_A (z_i e_0)^2}{2(r_i + 2r_W)}$$

$$\times \left(1 - \frac{1}{\varepsilon_W} - \frac{T}{\varepsilon_W^2} \frac{\partial \varepsilon_W}{\partial T}\right) \qquad (2.52)$$

for positive ions and

$$\Delta H_{I-H_2O} = 30 - \frac{N_A z_i e_0 \mu_W}{(r_i + r_W)^2} - \frac{4N_A z_i e_0 p_W}{2(r_i + r_W)^3} - \frac{N_A (z_i e_0)^2}{2(r_i + 2r_W)}$$
$$\times \left(1 - \frac{1}{\varepsilon_W} - \frac{T}{\varepsilon_W^2} \frac{\partial \varepsilon_W}{\partial T}\right) \quad (2.53)$$

for negative ions.

2.3.10. Ion–Induced-Dipole Interactions in the Primary Solvation Sheath

If one compares Eqs. (2.52) and (2.53) with Eqs. (2.33) and (2.34), it is clear that the ion–quadrupole calculation of ΔH_{I-H_2O} differs from the ion–dipole calculation of the same quantity in only one respect: In representing the water molecule by a quadrupole, one is making a more refined assessment of the interactions between the ion and the water molecules of the primary solvation sheath. At this level of sophistication, one wonders whether there are other subtle interactions which one ought to consider.

For instance, when the water molecule is in contact with the ion, the field of the latter tends to distort the charge distribution in the water molecule. Thus, if the ion is positive, the negative charge in the water molecule tends to come closer to the ion and the positive charge to move away. This implies that the ion tends to induce an extra dipole moment in the water molecule over and above its permanent dipole moment. For small fields, one can assume that the induced dipole moment μ_{ind} is proportional to the inducing field X

$$\mu_{\text{ind}} = \alpha X \quad (2.54)$$

where α, the proportionality constant, is known as the *deformation polarizability* and is a measure of the "distortability" of the water molecule along its permanent dipole axis.

Thus, one must consider the contribution to the heat of formation of the primary solvated ion, i.e., step 3 of the cycle used in the theoretical calculation presented above, arising from interactions between the ion and the dipoles induced in the water molecules of the primary solvent sheath. The interaction energy between a dipole and an infinitesimal charge dq is $-\mu \, dq/r^2$, or, since dq/r^2 is the field dX due to this charge, the interaction energy can be expressed as $-\mu \, dX$. Thus, the interaction energy between the dipole and an ion of charge $z_i e_0$, exerting a field $z_i e_0/r^2$ can be found by performing the integration $-\int_0^{z_i e_0/r^2} \mu \, dX$. In the case of permanent dipoles, μ does not depend on the field X and one gets the result (*cf.* Appendix 2.2)

$$-\int_0^{z_i e_0/r^2} \mu \, dX = -\frac{z_i e_0}{r^2} \mu \qquad (2.55)$$

For induced dipoles, however, $\mu_{\text{ind}} = \alpha X$, and, hence,

$$-\int_0^{z_i e_0/r^2} \mu \, dX = -\int_0^{z_i e_0/r^2} \alpha X \, dX = -\left[\frac{\alpha X^2}{2}\right]_0^{z_i e_0/r^2} = -\alpha \frac{(z_i e_0)^2}{2r^4} \qquad (2.56)$$

Considering a mole of ions and four water molecules in contact with an ion, the heat of ion–induced-dipole interactions is

$$-\frac{4N_A \alpha (z_i e_0)^2}{2(r_i + r_W)^4}$$

Introducing this induced dipole effect into the expression for the heat of ion–solvent interactions [Eqs. (2.52) and (2.53)], one has

$$\Delta H_{I-H_2O} = 20 - \frac{4N_A z_i e_0 \mu_W}{(r_i + r_W)^2} + \frac{4N_A z_i e_0 p_W}{2(r_i + r_W)^3}$$
$$- \frac{N_A (z_i e_0)^2}{2(r_i + 2r_W)} \left(1 - \frac{1}{\varepsilon_W} - \frac{T}{\varepsilon_W^2} \frac{\partial \varepsilon_W}{\partial T}\right) - \frac{4N_A \alpha (z_i e_0)^2}{2(r_i + r_W)^4} \qquad (2.57) \leftarrow$$

for **positive** ions, and

$$\Delta H_{I-H_2O} = 30 - \frac{4N_A z_i e_0 \mu_W}{(r_i + r_W)^2} - \frac{4N_A z_i e_0 p_W}{2(r_i + r_W)^3}$$
$$- \frac{N_A (z_i e_0)^2}{2(r_i + 2r_W)} \left(1 - \frac{1}{\varepsilon_W} - \frac{T}{\varepsilon_W^2} \frac{\partial \varepsilon_W}{\partial T}\right) - \frac{4N_A \alpha (z_i e_0)^2}{2(r_i + r_W)^4} \qquad (2.58) \leftarrow$$

for **negative** ions.

2.3.11. How Good Is the Ion–Quadrupole Theory of Solvation?

A simple test for the validity of these theoretical expressions (2.57) and (2.58) can be constructed. Consider two ions M_i^+ and X_i^- of equal radius but opposite charge. The difference $\Delta H_{M_i^+}$ (abs) $- \Delta H_{X_i^-}$ (abs) in their absolute heats of hydration is obtained by subtracting Eq. (2.58) from Eq. (2.57). Since the signs of the dipole term, namely,

$$\frac{4N_A z_i e_0 \mu_W}{(r_i + r_W)^2}$$

the Born charging term, namely,

$$\frac{N_A(z_ie_0)^2}{2(r_i + 2r_W)} \left(1 - \frac{1}{\varepsilon_W} - \frac{T}{\varepsilon_W^2} \frac{\partial \varepsilon_W}{\partial T}\right)$$

and the induced dipole term, namely,

$$\alpha \frac{4N_A(z_ie_0)^2}{2(r_i + r_W)^4}$$

are invariant with the sign of the charge of the ion, they cancel out in the subtraction (so long as the orientation of a dipole near a cation is simply the mirror image of that near an anion). The quadrupole term, however, does not cancel out because it is positive for positive ions, and negative for negative ions. Hence, one obtains[†]

$$\Delta H_{M_i^+}(\text{abs}) - \Delta H_{X_i^-}(\text{abs}) = -10 + \frac{4N_A z_i e_0 p_W}{(r_i + r_W)^3} \qquad (2.59)$$

It is seen from this equation that the quadrupolar character of the water molecule would make oppositely charged ions of equal radii have radius-dependent differences in their heats of hydration (cf. Fig. 2.44). Further, Eq. (2.47) has given

$$\Delta H_{M_i^+}(\text{abs}) - \Delta H_{X_i^-}(\text{abs}) = \Delta H_{M_i^+}(\text{rel}) - \Delta H_{X_i^-}(\text{rel}) + 2\Delta H_{H^+}(\text{abs}) \qquad (2.47)$$

By combining Eqs. (2.47) and (2.59), the result is

$$\Delta H_{M_i^+}(\text{rel}) - \Delta H_{X_i^-}(\text{rel}) = -2\Delta H_{H^+}(\text{abs}) - 10 + \frac{4N_A z_i e_0 p_W}{(r_i + r_W)^3} \qquad (2.60)$$

Thus, the ion–quadrupole model of ion–solvent interactions predicts that, if the experimentally available differences $\Delta H_{M_i^+}(\text{rel}) - \Delta H_{X_i^-}(\text{rel})$ in the relative heats of solvation of oppositely charged ions of equal radii r_i are plotted against $(r_i + r_W)^{-3}$, one should get a straight line with a slope $+4N_A z_i e_0 p_W$. From Fig. 2.48, it can be seen that the experimental points do give a straight line except as the ionic radius falls below about 1.3 Å. Further, the theoretical slope (1078 kcal mole^{-1} Å^3) is in fair agreement with the experimental slope (909 kcal mole^{-1} Å^3).

[†] The expression (2.59) is based on the assumption of the radius independence of $W_+ - W_- = 20 - 30 = -10$ and the constancy of n with radius over the interval concerned (cf. footnote on p. 98).

It can therefore be concluded that, by considering a quadrupole model for the water molecule, one can not only explain why oppositely charged ions of equal radius have differing heats of hydration (*cf.* Fig. 2.44) but also predict quantitatively the way these differences in the heats of hydration vary with radius.

It remains to test the values of the absolute heat of hydration calculated from Eqs. (2.57) and (2.58). The question is: With what experimental values must the calculated values be compared? Obviously, the theoretical values cannot be checked with so-called experimental values obtained—following the rudimentary Born view of Eq. (2.22)—by dividing ΔH_{KF-H_2O} by two to get $\Delta H_{K^+-H_2O}$ and $\Delta H_{F^--H_2O}$ because this procedure assumes incorrectly that oppositely charged, equiradii ions should have equal heats of interaction with water molecules. Hence, one has to develop a set of absolute heats of hydration which is based on the fact that oppositely charged ions of equal radii have differences in their heats of hydration and that these differences are proportional to $(r_i + r_W)^{-3}$.

An elegant method of obtaining such experimental values is available. Starting from the experimentally proved linearity of $\Delta H_{M_i^+}(\text{rel}) - \Delta H_{X_i^-}(\text{rel})$ versus $(r_i + r_W)^{-3}$ (*cf.* Fig. 2.48), one can take Eq. (2.60)

$$\Delta H_{M_i^+}(\text{rel}) - \Delta H_{X_i^-}(\text{rel}) = -2\Delta H_{H^+}(\text{abs}) - 10 + \frac{4N_A z_i e_0 p_W}{(r_i + r_W)^3}$$

and, following Halliwell and Nyburg (1963), extrapolate the $\Delta H_{M_i^+}(\text{rel}) - \Delta H_{X_i^-}(\text{rel})$ versus $(r_i + r_W)^{-3}$ plot to infinite radius, i.e., to $(r_i + r_W)^{-3} \to 0$. The intercept which is 522 kcal mole^{-1} is then equal to $-2\Delta H_{H^+}(\text{abs}) - 10$, or

$$\Delta H_{H^+}(\text{abs}) = 266 \text{ kcal mole}^{-1}$$

Once one has obtained thus the experimental absolute heat of hydration of the proton, one has the heat of hydration of one individual species but on a much better basis than that gained by splitting the heats of hydration of KF in half. This semiabsolute value can now be introduced into the heats of hydration of HX compounds to yield individual heats of hydration of X$^-$, which can then be used with the experimental heats of hydration of MX to give the individual heats of hydration of other ions (Table 2.15).

When these experimental values are used to check the values of the absolute heat of hydration calculated by theory, i.e., by Eqs. (2.57) and (2.58), it is seen (Table 2.16) that there is agreement between theory and experiment, with an average disagreement of about 5%.

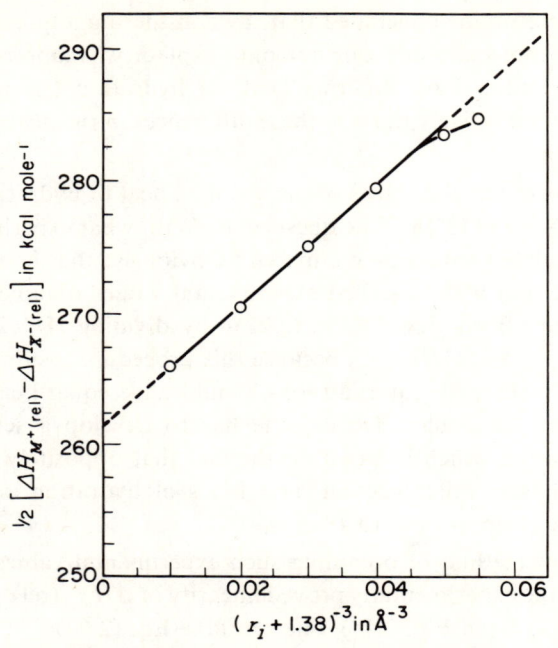

Fig. 2.48. The plot of half the difference in the relative heats of hydration of positive and negative ions of the same radii *versus* $(r_i + 1.38)^{-3}$. The solid line is through experimental points and the dotted line is the extrapolation of the straight line going through the experimental points.

TABLE 2.15

Quasi-Experimental Absolute Heats of Hydration of Various Individual Ions

Ion	Absolute heat of hydration
Li$^+$	−129.7
Na$^+$	−102.3
K$^+$	− 82.3
Rb$^+$	− 77.2
Cs$^+$	− 71.4
F$^-$	−115.5
Cl$^-$	− 81.5
Br$^-$	− 75.0
I$^-$	− 65.2

TABLE 2.16
Comparison of ΔH_{I-H_2O} Calculated from Equations (2.57) and (2.58) with Experimental Values[†]

Ion	Born term	Ion-dipole term	Ion-quadrupole term	Ion-induced-dipole term	Total[‡]	Experimental	Deviation, %
Li+	−49.6	−130.5	+69.5	−62.4	−153.0[§]	−129.7	−18
Na+	−45.0	− 94.2	+42.6	−32.7	−109.3	−102.3	− 6.8
K+	−40.8	− 69.7	+27.1	−19.2	− 82.6	− 83.3	− 0.4
Rb+	−39.3	− 62.6	+23.1	−14.6	− 73.4	− 77.2	+ 5
Cs+	−37.5	− 54.3	+18.7	−10.5	− 63.6	− 71.4	+11
F−	−40.5	− 68.1	−26.2	−16.5	−121.3	−115.5	− 5
Cl−	−36.5	− 50.3	−16.6	− 9.4	− 82.8	− 81.5	− 2
Br−	−35.4	− 46.2	−14.6	− 7.9	− 74.1	− 75.0	+ 1.2
I−	−33.9	− 40.8	−12.2	− 6.4	− 63.3	− 65.2	+ 3

[†] All values in kilocalories per mole.
[‡] These totals include the +20 kcal mole^{-1} for positive ions and +30 kcal mole^{-1} for negative ions [cf. Eqs. (2.57) and (2.58)].
[§] Some authors [cf. A. D. Buckingham, *Discussions Faraday Soc.*, **24**: 151 (1957)] have used a Li+ ionic radius of 0.78 Å, in which case the calculated $\Delta H_{I-H_2O} = 144.3$ kcal mole^{-1}, corresponding to a deviation of 11%. The figure of −153.0 corresponds to an ionic radius of 0.60 Å.

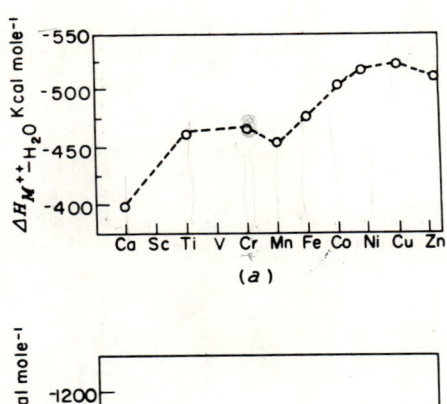

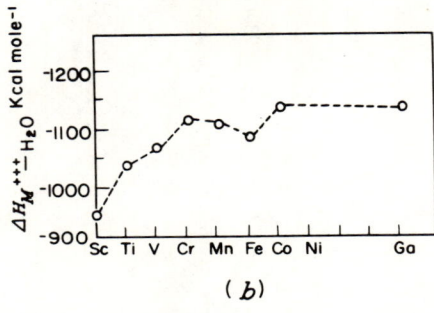

Fig. 2.49. Plot of the heats of hydration of transition-metal ions (and their immediate neighbors) *versus* atomic number; (a) divalent ions and (b) trivalent ions.

2.3.12. The Special Case of Interactions of the Transition-Metal Ions with Water

Theoretical considerations on the heats of ion–solvent interactions have been restricted thus far to alkali-metal and halide ions. For these ions, it has been shown that the ion–quadrupole theory is in good agreement with experiment.[†] In the case of the transition-metal ions, some complications arise.

A simple way of presenting these complications is to plot the experimental heats of hydration of transition-metal ions *versus* their atomic number. It is seen that, in the case of both divalent and trivalent ions, the heats of hydration lie on double-humped curves [Fig. 2.49(a) and (b)].

Now, if the transition-metal ions had spherical charge distributions, then one would expect that, with increasing atomic number, there would

---

† The theory is almost, but not quite, as good for doubly charged alkaline-earth ions.

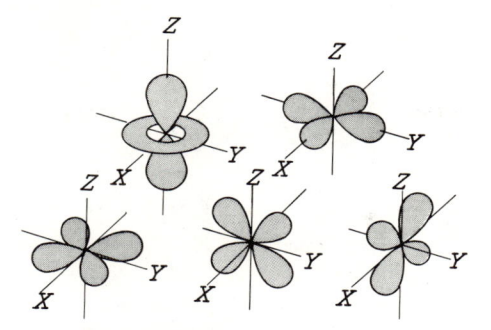

Fig. 2.50 The five 3d orbitals.

be a decreasing ionic radius† and thus a smooth and monotonic increase of the heat of hydration in the negative direction. The double-humped curve implies therefore the operation of factors which make transition-metal ions deviate from the behavior of charged spheres. What are these factors?

In the case of transition-metal ions, it is the shapes of the 3d orbitals which contribute the special properties. The 3d orbitals are not spherically symmetrical; in fact, they are as shown in Fig. 2.50.

In a gaseous ion (i.e., a free unhydrated ion), all the 3d orbitals are equally likely to be occupied because they all correspond to the same energy. Now, consider what happens when the ion becomes hydrated by six‡ water molecules situating themselves at the corners of an octahedron enveloping the ion. The lone electron pairs of the oxygen atoms (of the water molecules) exert a repulsive force on the valence electrons of the ion (Fig. 2.51).

This repulsive force acts *to the same extent* on all the *p* orbitals, as may be seen from Fig. 2.52. The *d* orbitals, however, can be classified in two types: (1) those that are directed *along* the X, Y, Z axes—these are known as the d_ε orbitals—, and (2) those that are directed *between* the axes—these are known as the d_y orbitals. It is clear (*cf.* Fig. 2.53) that the repulsive field of the lone electron pairs of the oxygen atoms acts more strongly on the d_ε orbitals than on the d_y orbitals. Thus, under the electrical

---

† The radius of an ion is determined mainly by the principal quantum number and the effective nuclear charge. As the atomic number increases in the transition-metal series, the principal quantum number remains the same, but the effective charge seen by the valence electrons increases; hence, the ionic radius should decrease with atomic number.

‡ The figure of six, rather than four, is used because of the experimental evidence that transition-metal ions undergo six coordination.

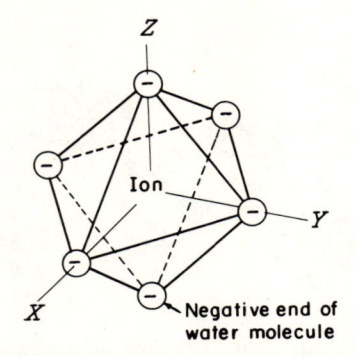

Fig. 2.51. Schematic diagram to show that the valence electrons of the positive ion are subject to the repulsion of the negative ends (⊖) of the octahedrally coordinating water molecules. The negative charge arises from the presence of lone electron pairs on the oxygen atoms of the water molecules.

influence of the water molecules of the primary solvation sheath, all the $3d$ orbitals do not correspond to the same energy. They are differentiated into two groups: The d_ε orbitals correspond to a higher energy and the d_γ orbitals to a lower energy. It will now be shown that this splitting of the $3d$ orbitals into two groups (with differing energy levels) affects the heat of hydration and hence makes it deviate from the values expected on the basis of theory developed earlier in this chapter, which neglected interactions of the water molecules with the electron orbitals in the ion.

Consider a free vanadium ion V^{++} and a hydrated vanadium ion.

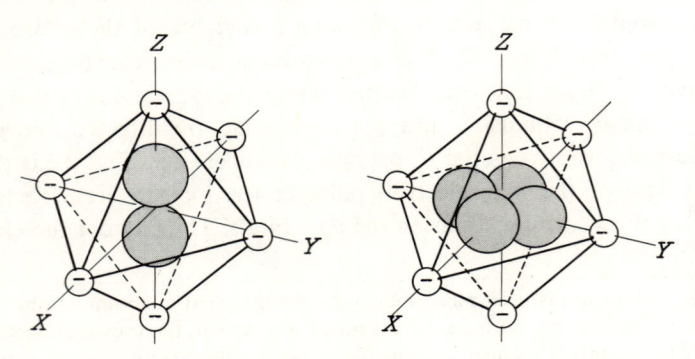

Fig. 2.52. Schematic diagram that shows that the three p orbitals (directed along the axes) are equally affected by the repulsive field of octahedrally coordinating water molecules.

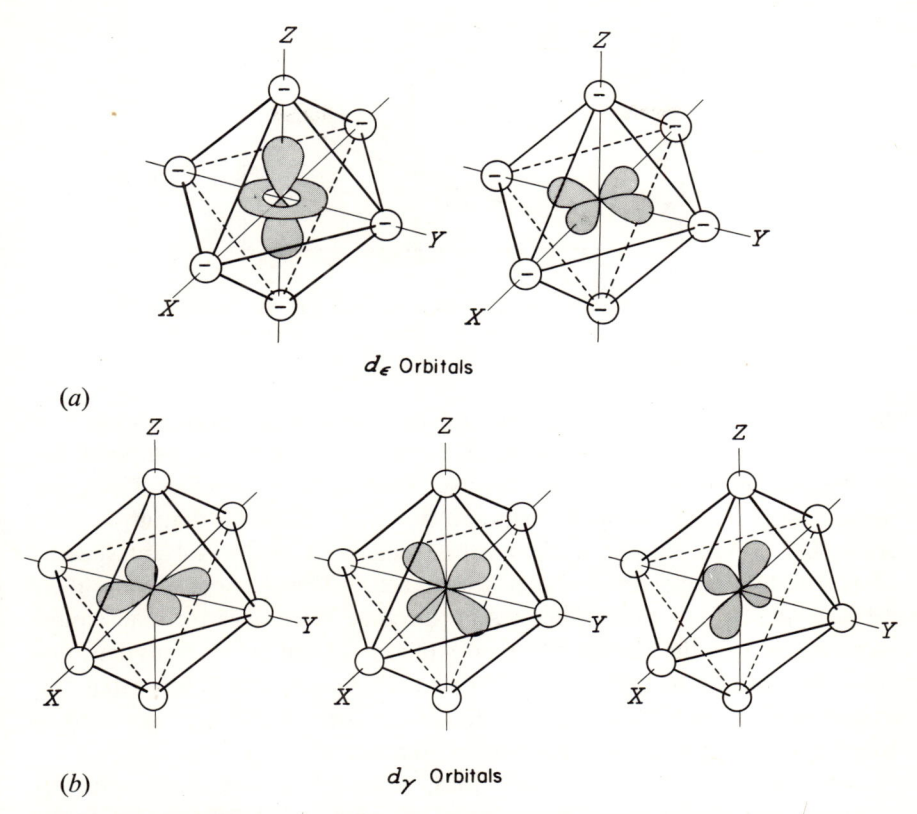

Fig. 2.53. Because the d_ε orbitals [see (a)] are directed along the axes and toward the negative ends of the water molecules, they correspond to a higher energy than the d_γ orbitals [see (b)] which are directed between the axes.

In the case of the free ion, all the five $3d$ orbitals (the two d_ε and the three d_γ orbitals) are equally likely to be occupied by the three $3d$ electrons of vanadium; the reason is that, in the free ion, all the five $3d$ orbitals correspond to the same energy. In the hydrated V^{++} ion, however, the d_γ orbitals, corresponding to a lower energy, are more likely to be occupied than the d_ε orbitals. This implies that the mean energy of the ion is less when the ion is subject to the electrical field of the solvent sheath than when it is free. Thus, the change in the mean occupancy of the various $3d$ orbitals, arising from the electrical field of the water molecules coordinating the ion, has conferred an extra stabilization (lowering of energy) on the ion–water system, and, to that extent, the heat of hydration is made more negative.

In the case of the hydrated divalent manganese ion, however, its five

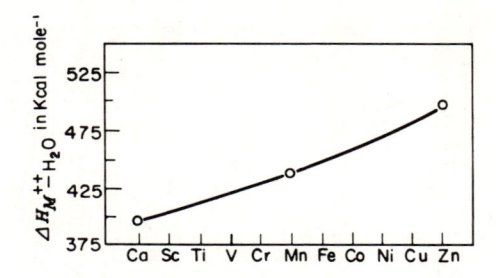

Fig. 2.54. The plot of the heat of hydration of Ca^{++}, Mn^{++}, and Zn^{++} *versus* atomic number.

$3d$ electrons are distributed[†] among the five $3d$ orbitals, and the *decrease* in energy of three electrons in the d_γ orbitals is exactly compensated for by the *increase* in energy of the two electrons in the d_ε orbitals. Thus, the mean energy of the ion in the hydrated state is the same as that in the free state, and there is no extra stabilization produced by the solvent sheath. Similarly, for Ca^{++} with no $3d$ electrons and Zn^{++} with a completely filled $3d$ shell, the heat of hydration does not become more negative than would be expected from the electrostatic theory of ion–solvent interactions developed in Section 2.3.10 and earlier. It can be concluded, therefore, that the experimental heats of hydration of these three ions should vary in a monotonic manner with atomic number, as, indeed, they do (Fig. 2.54).

All the other transition-metal ions, however, should have contributions to their heats of hydration from the energy stabilization produced by the field of the water molecules. It is these contributions which produce the double-humped curve of Fig. 2.55. If, however, for each ion, the energy[‡] corresponding to the water-field stabilization is subtracted from the experimental heat of hydration, then the resulting values should lie on the same smooth curve yielded by plotting the heats of hydration of Ca^{++}, Mn^{++}, and Zn^{++} *versus* atomic number. This reasoning is found to be true (*cf.* Fig. 2.55).

The argument has been presented here for divalent ions, but it is equally valid (Fig. 2.56) for trivalent ions. Here, it is Sc^{+++}, Fe^{+++}, and Ga^{+++} which are similar to manganese in that they do not acquire any

---

[†] The five electrons tend to occupy five different orbitals for the following reason: In the absence of the energy required for electrons with opposite spins to pair up, electrons with parallel spins tend to occupy different orbitals because, according to the Pauli principle, two electrons with parallel spins cannot occupy the same orbital.

[‡] This energy can be obtained spectroscopically.

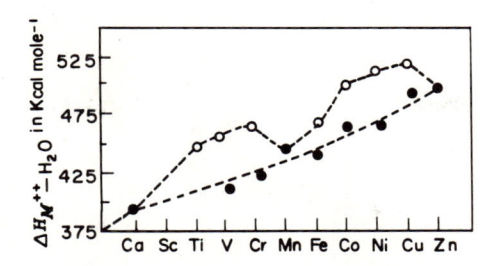

Fig. 2.55. The plot of the heat of hydration of the divalent transition-metal ions *versus* atomic number (○, experimental values; ●, values after subtracting water-field stabilization energy).

stabilization energy from the field of the water molecules, i.e., water-field stabilization energy.

In conclusion, therefore, it is the contribution of the water-field stabilization energy to the heat of hydration which is the special feature distinguishing transition-metal ions from the alkali-metal, alkaline-earth-metal, and halide ions in their interactions with the solvent.

2.3.13. Some Summarizing Remarks on the Energetics of Ion–Solvent Interactions

The first aspect of the ion–solvent interactions considered in this chapter has been the *energetics* of these interactions, i.e., the theory of the free energy and heat of solvation.

In an initial attempt to calculate these energies, a continuum approach was adopted with the ion likened to a charged sphere and the structured

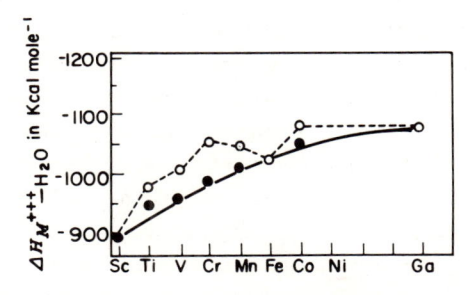

Fig. 2.56. The same as Fig. 2.55 but for trivalent ions.

solvent to a continuum having the bulk dielectric constant of the solvent. According to this first model of Born, the free energy of solvation is equal to the electrostatic work of discharging the sphere (which represents the ion) in vacuum, transferring the uncharged sphere into the solvent, and then charging it up again.

To compare with experiment the free energies and heats of solvation thus calculated, it was necessary to find a way of knowing the experimental heats of solvation of *individual* ions. This provided a fair problem because experimental measurements of the lattice energies and heats of dissolution only gave the *sum* of the heats of solvation of both ions of a *salt* and not those of individual ions. As a first attempt, therefore, it was decided to split the heat of solvation of KF equally between K^+ and F^- ions, which have almost equal radii. The rationale behind this step was the Born model, which argued for ions of equal radii having equal heats of solvation.

When these KF-derived experimental heats were compared with the Born heats of solvation, it turned out that the latter were of the right order of magnitude. This demonstrated the essentially electrostatic character of ion–solvent interactions. The calculated values were however too high which necessitates further theoretical considerations. Rather than be tempted into tampering with the ionic radii and dielectric constant and making them into adjustable parameters, it was decided to take detailed, structural view of the process of ion–solvent interactions.

Confining oneself to water as the solvent, a sketch was presented of the structure of water. Emphasis was laid on the fact that, while water is an uncharged molecule, its centers of the negative and positive charges do not coincide and thus confer a polar character on the molecule. The charge distribution in the water molecule was, as a first approximation, taken as equivalent to an electric dipole. Another feature of the structure of water is the linking up of individual water molecules into tetrahedral networks consisting of many water molecules in each unit.

When an ion enters water, the structure of the latter is disturbed to an extent which depends on the distance from the ion. The ion is able to wrench water molecules out of their continuous network and make them orient around the ion by coulombic forces which arise from the polar nature of the water molecule. Thus is formed the primary solvated ion. The water structure just outside the primary hydration sheath is broken to some extent, but, further away, the normal network structure obtains.

This picture of the structure of the solvent around ions was used as the basis of the ion–dipole theory of solvation. The procedure was to view the total heat of solvation as consisting of three contributions. The first contribution was considered to arise from the interactions between the ion and the water molecules which are members of the primary solvent sheath; and the calculation involved the heat of ion–dipole interactions. The second contribution to the heat of solvation pertained to the interaction between a primary solvated ion and the surrounding solvent, which, as a first step, was reckoned to have the bulk structure of water. The calculation of this second part of the total heat of solvation followed the procedure of the Born model except that one was dealing with a primarily solvated ion rather than a bare ion. Since the solvent in the immediate neighborhood of the primary solvent ion does not have the bulk structure, a correction to the second contribution was introduced to reckon with the structure breaking. The change in heat content due to the structure breaking around a primary solvated ion constituted the third contribution to the total heat of solvation of ions.

When the values of the heat of solvation calculated by the ion–dipole theory were compared with the so-called experimental KF-derived values, it became clear that the structural picture of solvation was in far better accord with the experimental than the Born model was.

The presentation would have been terminated at that point but for an interesting check on the validity of the "experimental" values of ionic solvation heats which were being used as the test of theory. Clear-cut experimental evidence was produced to demonstrate that the difference in the absolute heats of solvation of ions of opposite charge but equal radius was neither zero nor independent of ionic radius. This difference should be zero according to the so-called experimental KF-derived heats, and it should be radius independent according to the ion–dipole theory. Thus arose the need for a deeper probing into the structural picture of ion–solvent interactions and the development of the ion–quadrupole theory of solvation.

The one important advance of the ion–quadrupole theory over the ion–dipole theory lay in the more careful assessment of the interactions between the ion and the water molecules of the primary solvation sheath. It was realized that, while the charge distribution in the water molecule may appear equivalent to an electrical dipole, to an ion situated far away, the charge distribution is better represented as an electrical quadrupole, from the point of view of an ion in contact with the water molecule. Thus, what needed calculation was the heat of ion–quadrupole interactions in the primary solvated ion.

A confidence in the ion–quadrupole approach was gained immediately when the theory showed that ions of opposite charge but equal radii should have differing heats of solvation—as found experimentally. So, after also taking into account interactions between the ions and dipoles induced in the water molecules, the final expression for the heat of solvation according to the ion–quadrupole theory was used to calculate values for the various ions.

These calculated values obviously could not be compared with KF-derived experimental values derived on the assumption that oppositely charged ions of equal radii have equal heats of solvation. Fortunately, it was possible to isolate experimentally unambiguous heats of hydration of ions relative to the hydrogen ion and use these relative heats to yield an absolute heat of hydration of the hydrogen ion and, thus, the absolute heats of other ions. Comparison of these reliable experimental absolute heats with those calculated by the ion–quadrapole theory showed close correspondence between theory and experiment.

In the case of the transition-metal ions, a special factor had to be considered. The electric field exerted by the solvent sheath alters the distribution of the $3d$ electrons of the transition-metal ions among the various $3d$ orbitals and thus confers an extra stabilization energy upon the ion–solvent system. The heats of hydration of these ions is made more negative than it would be if no interactions with directed orbitals were considered.

This, then, is the basic theory of the energetics of ion–solvent interactions. It is clear that the polar nature of the solvent molecule is of fundamental importance to the heat of solvation. Viewing the water molecule as a dipole gives reasonable heats of solvation; viewing it as a quadrupole gives good results. One can, of course, go further. For instance, one can consider lateral interactions between the water molecules of the primary solvent sheath, or one can decide whether there are chemical as opposed to primarily electrostatic interactions between the ion and the contiguous water molecules. But these are matters of greater detail, which will be left to the tomes.

Further Reading

1. J. D. Bernal and R. H. Fowler, *J. Chem. Phys.*, **1**: 515 (1933).
2. D. D. Eley and M. G. Evans, *Trans. Faraday Soc.*, **34**: 1093 (1938).
3. H. S. Frank and M. Evans, *J. Chem. Phys.*, **13**: 507 (1945).
4. L. E. Orgel, *J. Chem. Soc.*, **1952**: 4756 (1952).
5. H. S. Frank and W. Y. Wen, *Discussions Faraday Soc.*, **24**: 133 (1957).
6. A. D. Buckingham, *Discussions Faraday Soc.*, **34**: 151 (1957).
7. O. G. Holmes and D. S. McClure, *J. Chem. Phys.*, **26**: 1686 (1957).
8. H. S. Frank, *Proc. Roy. Soc.* (*London*), **A247**: 481 (1958).

9. L. Pauling, *The Nature of the Chemical Bond*, 3rd ed., Cornell University Press, Ithaca, N.Y., 1960.
10. H. S. Frank and A. S. Quist, *J. Chem. Phys.*, **34**: 604 (1961).
11. G. Némethy and H. A. Scheraga, *J. Chem. Phys.*, **36**: 3882 and 3401 (1962).
12. G. R. Choppin and K. Buijs, *J. Chem. Phys.*, **39**: 2035 and 2042 (1963).
13. D. J. G. Ives, *Some Reflections on Water*, J. W. Ruddock, London, 1963.
14. H. F. Halliwell and S. C. Nyburg, *Trans. Faraday Soc.*, **58**: 1126 (1963).
15. J. P. Hunt, *Metal Ions in Aqueous Solution*, W. A. Benjamin, Inc., New York, 1963.
16. J. Lee Kavanau, *Water and Solute–Water Interactions*, Holden–Day Inc., San Francisco, 1964.
17. R. P. Marchi and H. Eyring, *J. Phys. Chem.*, **68**: 221 (1964).
18. R. P. Feynman, R. B. Leighton, and M. Sands, *The Feynman Lectures on Physics*, Addison-Wesley Publishing Company, Inc., Reading, Mass., 1964.
19. B. E. Conway, "Proton Solvation and Proton Transfer Processes in Solution," in: J. O'M. Bockris, ed., *Modern Aspects of Electrochemistry*, No. 3, Butterworth's Publications, Inc. London, 1964.
20. O. Ya. Samoilov, *Structure of Aqueous Electrolyte Solutions and the Hydration of Ions*, Consultants Bureau, New York, 1965.
21. S. Golden and C. Guttmann, *J. Chem. Phys.*, **43**: 1894 (1965).
22. D. R. Rosensteig, *Chem. Rev.*, **65**: 467 (1965).
23. B. E. Conway and M. Salomon, in: B. E. Conway and R. G. Barradas, eds., *Chemical Physics of Ionic Solutions*, John Wiley & Sons, Inc., New York, 1966.
24. V. I. Klassin and Yu. Zinovev, *Kolloid. Zh.* (English translation), **29** (5): 561 (1967).
25. B. V. Deryagin, Z. M. Zorin, and N. V. Churaev, *Kolloid. Zh.* (English translation), **30** (2): 232 (1968).
26. A. K. Covington and P. Jones, *Hydrogen-Bonded Solvent Systems*, Taylor and Francis Ltd., London, 1968.

2.4. THE SOLVATION NUMBER

2.4.1. How Many Water Molecules Are Involved in the Solvation of an Ion?

Mathematically speaking, the electric force originating from an ion becomes zero only at infinity. In effect, however, the force fades out to a negligible value after quite a short distance (of the order of tens of angstroms). Beyond this cutoff distance, solvent molecules may be regarded as unaware of an ion's presence. There is therefore a certain effective volume around the ion within which its influence operates. How many solvent molecules are inside this volume and could therefore be said to be partici-

TABLE 2.17
Hydration Number Ascribed to the Sodium Ion in According to Different Experimental Methods

Ion	Hydration numbers reported
Na$^+$	1, 2, 2.5, 4.5, 6–7, 16.9, 44.5, 71

pants in the solvation of the ion? This number may be termed the *solvation number*[†] (or *hydration number* when water is the solvent).

The question of the value of the solvation number is an interesting one. It is no surprise, therefore, that a large number of different methods have in the past been used to determine the solvation number (more about these methods later). But, the alarming thing is that exceedingly discrepant results are obtained by the various methods. For instance, widely varying hydration numbers ranging from 1 to 71 (Table 2.17) have been ascribed to the sodium ion. Are some of the methods wholly incorrect, or is there a confusion as to what constitutes a hydration number?

The answer can be approached, if not attained precisely, by the following considerations. What value of hydration number a particular method gives depends on what types of ion–solvent interactions the method senses. If it can pick up the interactions of an ion with water molecules several molecular diameters away in the secondary region, it will report that a large number of water molecules are involved in solvation, i.e., a high hydration number. If, however, the method only detects how many water molecules an ion takes along in the course of its thermal motions through the solution (i.e., those tightly bound to it), then it will report a small hydration number.

To avoid ambiguity, it is best to define a *primary solvation number* as the number of solvent molecules which surrender their own translational freedom and remain with the ion when it moves relative to the surrounding solvent. Of course, a solvent molecule loses its independent translational motions only when it is overwhelmed by the ionic force field into adopting

---

[†] This total effective number of solvent molecules involved in interactions should not be confused with the number *n* used in the structural treatment of the energetics of solvation. The latter number was meant to represent the number of solvent molecules in contact with the ion and assumed to be aligned in its field.

TABLE 2.18

Hydration Numbers

Ion	Hydration number	Number of independent methods on which result is based
Li$^+$	5 ± 1	5
Na$^+$	5 ± 1	5
K$^+$	4 ± 2	4
Rb$^+$	3 ± 1	4
F$^-$	4 ± 1	3
Cl$^-$	1 ± 1	3
Br$^-$	1 ± 1	3
I$^-$	1 ± 1	2

a minimum-energy orientation to the ion. Thus, the primary solvation number can also be defined as the number of solvent molecules which are aligned in the force field of the ion.

This definition provides a criterion for discussing the different methods of determining solvation numbers. The primary solvation number should be determined by only those methods which register the number of water molecules which are associated with the ion in its travels through the solution.

When, however, these methods are used (and they will be presented in Section 2.4.5), it turns out (Table 2.18) that the number of water molecules determined by some of them are *less* than what geometry says the number of water molecules in contact with the ion should be. This latter number is a coordination number,[†] i.e., the number of nearest-neighbor water molecules which are in contact with or coordinate or surround an ion. The question, therefore, arises: Why does not all the coordinated water join the ion in its zig-zag motions through the solution? In fact, why is the solvation number not always equal to the coordination number? Further,

---

[†] In the structural treatment of the heats of solvation, it was tacitly assumed that the number n of primary solvent molecules aligned in the ionic field is equal to the coordination number. In other words, the structural treatment slurs over the distinction between the number that are oriented in the ionic field (i.e., move with the ion) and the number in contact with the ion.

what happens when coordinating solvent molecules desert their positions in the coordination shell of an ion as soon as it begins its voyage through the solvent, say, in response to an electric field? Do the missing solvent molecules leave voids in the coordination shell of a moving ion? The concept of solvation number will become clear only when such questions are answered.

2.4.2. Static and Dynamic Pictures of the Ion–Solvent Molecule Interaction

Suppose that, in a thought experiment, a bare ion is made to stop during its movements through the solution. At that instant, the hypothetical *stationary* ion will be surrounded or coordinated by water molecules still associated in a network structure (Fig. 2.57). What will happen? The ionic force field will operate on the neighboring water dipoles. The forces, which are essentially ion–dipole in nature, will cause some of the water molecules to break away from the water network and attach themselves to the ion.

What is the consideration on the basis of which a particular water molecule decides to embrace the ion by aligning into its field or to shun it and remain in the water network? The consideration is simple: Is the ion–dipole interaction energy greater in magnitude than the hydrogen-bond energy keeping the particular water molecule in the network? If the ion-dipole energy is greater in magnitude, the water molecule should link itself with the ion and form part of the primary solvation sheath. If not, the water molecule should remain in the water network.

The whole thought experiment described above is a *static* one. All that has been done is to consider the energies in the initial state (a water

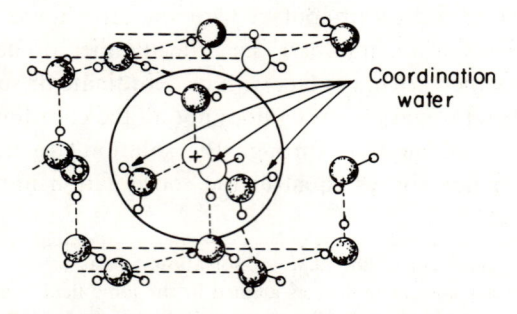

Fig. 2.57. A hypothetical stationary ion coordinated by water molecules still associated into a network structure.

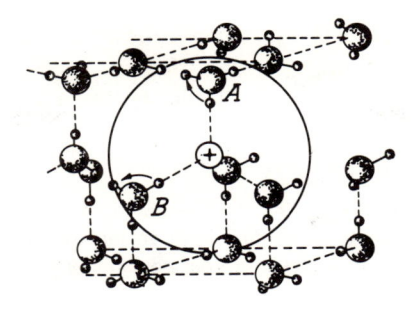

Fig. 2.58. Schematic diagram to show that, of four water molecules which coordinate an ion, two water molecules, *A* and *B* must reorient from positions in which one of their H atoms faces the ion to positions in which the same H atoms are away from the ion. (The required reorientation is shown by an arrow.)

molecule in the water network and an ion nearby) and in the final state (water bound to the ion by ion–dipole forces).

But ions can be kept stationary only in thought experiments. In reality, they exist in a state of ceaseless motion (see Chapter 4). So time and movement must come into the picture of ions interacting with water molecules. One must abandon a static view for a dynamic view.

One can develop a dynamic view along the following lines (Samoilov). Consider a water molecule bound by hydrogen bonds to the water network. Suppose that, at a time taken as zero ($t = 0$), an ion suddenly appears next to the water molecule. If the net force on the water molecule is in favor of its association with the ion rather than with the water network, it will try to get into an equilibrium position around the ion, i.e., the water molecule will try to align into a minimum-energy orientation. This usually means that the water molecule has to reorient (or jump through a small distance or both) from the position it had in the water structure to the new position of alignment in the ionic field (Fig. 2.58).

But these reorienting or *jumping movements* to be made by the water molecule will *require a finite time*, the value of which depends on the critical activation energy required for the reorientation or jumping process. Let this time required for the orientation of a water molecule into the coordination sheath around an ion be $\tau_{\text{water orient}}$ (Fig. 2.59). This orientation time will not have a unique value because it will depend on how far the ion is situated from and on how the ion is located with respect to the water

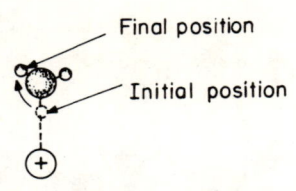

Fig. 2.59. The time required for a water molecule in contact with an ion to reorient from an initial to a final position (shown in figure) is τ_{orient}.

network holding the water molecule. So one is talking about an average water-orientation time.

Now, instead of considering the ion suddenly placed next to the water molecule at $t = 0$, one can visualize the ion resting or waiting near the water molecule in between its hops from location to location in the solvent (*cf.* Section 4.2). Of course, if a water molecule belonging to the water network is to orient toward the ion, it must do so when the ion is within a certain small distance of the water molecule. But how long does the ion stay within this jumping range? That depends on how long the ion pauses next to the water molecule in the course of its jumps through the solvent. The longer the ion waits near the water molecule, the longer is the time available for the water molecule to break out of the water lattice and swing into that intimate ion–dipole relationship with the ion which characterizes

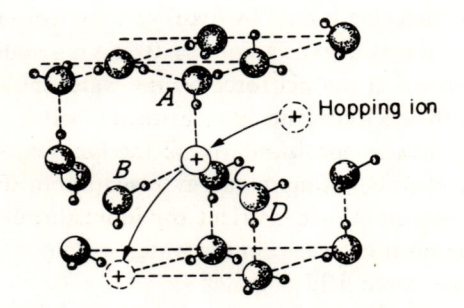

Fig. 2.60. In the course of its hops through the solution, the hopping ion can be considered to spend a time $\tau_{\text{ion wait}}$ in contact with the neighboring water molecules. Will one of these orient itself into a position of minimum interaction energy with the ion before the latter has jumped to a new position?

a seat in the primary hydration sheath. The hopping ion spends a certain time in "contact" with the particular water molecule under discussion. Call this contact time $\tau_{\text{ion wait}}$ (Fig. 2.60).

2.4.3. The Meaning of Hydration Numbers

Now, an interesting qualitative conclusion becomes clear. If the time an ion waits near a water molecule is *long* compared with the average time a water molecule takes to orient into association with an ion, then the probability of the water molecule's being captured by the ion is high. That is, the probability of an ion's capturing a water molecule depends on the ratio $\tau_{\text{ion wait}}/\tau_{\text{water orient}}$.

If $\tau_{\text{ion wait}}/\tau_{\text{water orient}}$ is large, then the ion will be surrounded by the full geometrically permitted complement of bonded water molecules *during all its zig-zag motions* through the solution. Under these circumstances, the hydration number (i.e., the number of water molecules which participate in the translational motions of the ion) will be equal to the coordination number.

If, however, $\tau_{\text{ion wait}}/\tau_{\text{water orient}}$ is of the order of unity, then the situation is interesting. The time an ion spends in the neighborhood of a water molecule is of the order of the water reorientation time, and, hence, though the ion is not sure to capture a water molecule, there is a certain probability, less than unity. At the same time, one must consider the opposite process: An ion with a bound water molecule collides with a water molecule belonging to the water network. There will be a certain probability that the ion will lose its water to the water network. But there are plenty of water molecules all around and the ion has a chance of making up its loss. Thus, over a period of time which is long compared with the period of contact between a moving ion and a specific water molecule, the ion has aligned and trapped in its field a certain number of water molecules which is less than the number of water molecules which geometrical close packing makes possible, i.e., the coordination number.

The collisions between ions and water molecules linked to the water network are analogous to any other collision process. Consider, for example, the collisions between neutrons and U^{238} nuclei, in which slow neutrons stand a better chance of being captured than fast neutrons. One says that there is a large capture cross section for slow neutrons. It is as if a slow moving neutron sees a bigger target than a fast moving one.

What happens if the ions wait for so short a time that, even before a water molecule has had time to break out of the water structure and turn around, the ion has hopped away? Then, the probability of a water molecule's being captured by the ion is zero, and, on a time average, the ion will not have any *aligned* water molecules in contact with it, i.e., its primary hydration sheath is empty. This does not mean that such ions are not surrounded by interacting solvent molecules or that they would have no coordination water. It only means that, because $\tau_{\text{ion wait}}/\tau_{\text{water orient}} \ll 1$, the ion does not wait long enough at any particular site for the contiguous water molecules to swing out of the water network into minimum-energy orientation with the ion. Even if the ion does capture a water molecule, it is bound to lose it soon. It also means that the moving ion exchanges water molecules so easily with the surrounding solvent that, in effect, the moving ion does not carry its sheath along with it. Its solvation number is zero, though its coordination number is that dictated by geometry.

The picture of solvation numbers presented here is a dynamic one. The solvation number refers to the number of water molecules which remain aligned with the ion *during* its jumps through the medium. But it is not necessary that the same individual water molecules serve in the solvation sheath for an indefinitely long time. A given water molecule may serve the ion for some time, but it is not imprisoned for life in its hydration shell. A chance collision, and the particular water molecule may link up again with the water network, get left behind by the hopping ion, and watch another water molecule yield to the attraction of the ionic field and be incorporated in the primary solvation sheath.

2.4.4. Why Is the Concept of Solvation Numbers Useful?

In all this dynamic exchange of solvent molecules between the coordination region and the main bulk of solvent, has the concept of solvation number any utility? Yes, the solvation number can be considered the *effective* number of solvent molecules to be "permanently" bound to the ion and to follow its motion from site to site. The kinetic entity is not the bare ion but the ion plus the solvation number of water molecules.

The concept of solvation number permits one to suppress the dynamic nature of the primary solvation sheath from many modelistic considerations of ions in solution. This is important particularly in situations where one would overcomplicate an analysis by considering the details of the constant exchange of water molecules between the ionic primary hydration shell and the solvent. The overall total action of the ion on the water may be replaced conceptually by a strong binding between the ion and some effective number (the solvation number) of solvent molecules; this effective number may well be almost zero in the case of large ions, e.g., iodide, cesium, and tetraalkylammonium. The solvation number clearly diminishes with increase of ionic radius because, with increasing ionic radius, the distance to the coordinating water molecules increases and, thus, the ionic force field which aligns the ion diminishes so that the water molecules have less inclination to reorient away from their solvent-structure positions.

Of course, there may be situations where, quite independent of the ratio $\tau_{\text{ion wait}}/\tau_{\text{water orient}}$, there are thermodynamic restrictions against the association of solvent molecules with the ion, e.g., the ion–solvent molecule interaction energy may be less in magnitude than the solvent molecule–solvent molecule energy. In such cases, the solvation number will be zero on static considerations alone.

2.4.5. On the Determination of Solvation Numbers

All this discussion would be pointless if there were no agreement between the different methods of measuring primary solvation numbers. Fortunately, it turns out that there *is* some degree of agreement between the values reported by different methods so long as they are methods which determine the *primary* solvation number (Section 2.4.1), as opposed to the vague and asymptotic concept of total solvation number (see Table 2.19).

TABLE 2.19

Comparison between the Hydration Numbers Determined by Different Methods

Ion	Compressibility	Mobility	Entropy	Theoretical calc.
Li$^+$	5–6	6	5	6
Na$^+$	6–7	2–4	4	5
Mg$^+$	16	14	13	
Ca^{++}	—	7.5–10.5	10	
Zn^{++}	—	10 –12.5	12	
Cd^{++}	—	10 –12.5	11	
Fe^{++}	—	10 –12.5	12	
Cu^{++}	—	10.5–12.5	12	
Pb^{++}	—	4 – 7.5	8	
K$^+$	6–7	—	3	3
F$^-$	2	—	5	5
Cl$^-$	0.1	0.9	3	3
Br$^-$	0	0.6	2	2
I$^-$	0	0.2	1	0

A detailed discussion of the various methods of determining solvation numbers is not intended in this treatment. Nevertheless, it is illustrative to present two examples.

Consider, for example, the compressibility method. The compressibility β is defined by the expression

$$\beta = -\frac{1}{V}\left(\frac{\partial V}{\partial p}\right)_T \qquad (2.62)$$

If a pure solvent is considered, then its compressibility may be written thus

$$\beta_{\text{solv}} = -\frac{1}{V}\left(\frac{\partial V}{\partial p}\right)_T \qquad (2.63)$$

Now suppose that an ionic solution is considered. Will its compressibility be the same as that of the pure solvent, i.e., β_{solv}? A physical picture of why a solvent is compressible will provide a qualitative answer.

Let water be the solvent. It has been described (see Section 2.3) as having quite an open framework structure with many holes in it. When a pressure is applied, the water molecules can break out of the tetrahedral framework and enter the interstitial spaces; the water molecules become packed more closely (Fig. 2.61). Thus, the volume decreases.

This is not the only way of compressing water. When ions are introduced into the water, they are capable of wrenching water molecules out of the water framework so as to envelop themselves with solvent sheaths. Because the molecules are oriented in the ionic field, the water is more compactly packed in the primary solvation shell as compared to the packing if the

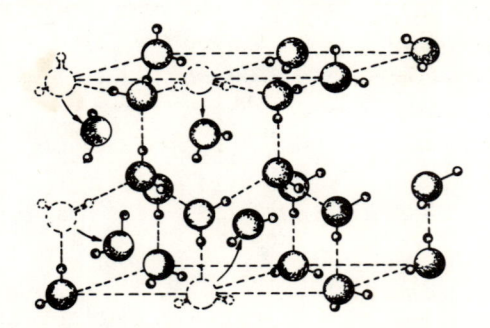

Fig. 2.61. Schematic diagram to show that, when an external pressure is applied to water, water molecules break out of the networks and occupy interstitial spaces.

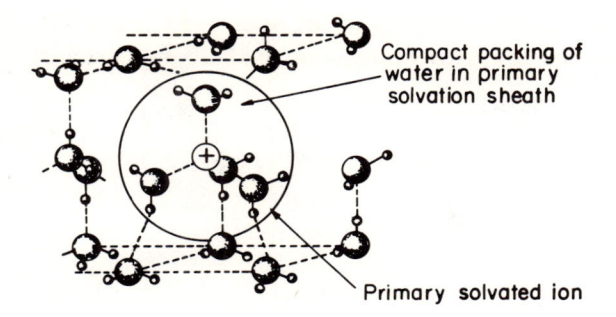

Fig. 2.62. Schematic diagram to illustrate the principle of electrostriction; owing to the ionic field, water molecules are more compactly packed in the primary solvation sheath than in the field.

ion were not there (Fig. 2.62). The water has become compressed by the introduction of the ion. But what is the origin of the influence of the ion? The origin is the electric field of the ion. Thus, electric fields cause compression of the material medium upon which they exert their influence; this phenomenon is known as *electrostriction*.

Since the introduction of ions into a solvent causes the solvent molecules *in the primary solvent shell* to be highly compressed, these water molecules may be supposed not to respond to any further pressure which may be applied. Thus, the compressibility of an ionic solution is less than that of the pure solvent because of the incompressibility of the primary solvation sheath.[†]

It is easy to calculate the ratio of the compressibility of a solvent β_{solv} to that of the solution β_{soln}. Suppose the primary hydration number is n_h. Then, n_i moles of ions are solvated with $n_i n_h$ moles of incompressible water. Now, if n_w moles of water correspond to a total volume V of solution, $n_i n_h$ moles of incompressible water would correspond to a volume $V n_h n_i / n_w$ of incompressible solution. Defining the symbol y thus

$$y = \frac{n_h n_i}{n_w} \qquad (2.64)$$

[†] Outside the primary solvent sheath, the water molecules are not oriented to the same degree as those inside the primary solvation sheath because the orienting ionic field is less. This means that the nonprimary water molecules are less electrostricted and free to respond to pressure. One can, to good approximation, say that the water outside the primary solvation shell has the same compressibility as the pure solvent.

the volume of the incompressible part of the solution is yV. This volume must be excluded from the expression for the compressibility of the ionic solution. Thus,

$$\beta_{\text{soln}} = -\frac{1}{V}\left[\frac{\partial}{\partial p}(V - yV)\right]_T \tag{2.65}$$

$$\beta_{\text{solv}} = -\frac{1}{V}\left[\frac{\partial}{\partial p}V\right]_T \tag{2.63}$$

$$\frac{\beta_{\text{soln}}}{\beta_{\text{solv}}} = 1 - y \tag{2.66}$$

Hence, from (2.64),

$$\frac{\beta_{\text{soln}}}{\beta_{\text{solv}}} = 1 - \frac{n_h n_i}{n_w} \tag{2.67}$$

$$n_h = \frac{n_w}{n_i}\left(1 - \frac{\beta_{\text{soln}}}{\beta_{\text{solv}}}\right) \tag{2.68}$$

This equation can be used to obtain the hydration number by determining the compressibility of the pure solvent and the ionic solution. There are several methods available for studying compressibilities. The ultrasonic method, for example, depends on the fact that sound travels by a compression–rarefaction process, and, thus, the velocity of an ultrasonic wave can be used to determine the compressibilities of solvent and solution, needed for Eq. (2.68) (Fig. 2.63).

The *mobility* method of measuring hydration numbers is based on the following argument (*cf.* also Section 4.4.8). Suppose an ion is made to

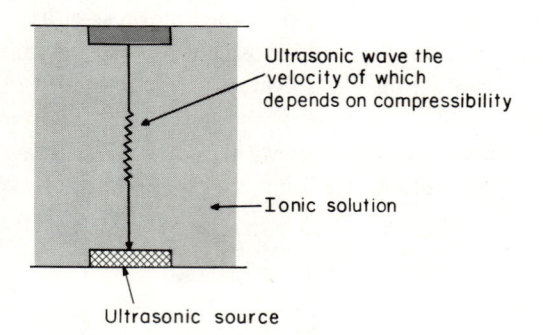

Fig. 2.63. The ultrasonic method of determining hydration numbers is based on the fact that the velocity of the ultrasonic wave depends, by the compressibility of the solution, on the extent of primary hydration in the ionic solution.

ION–SOLVENT INTERACTIONS   129

drift by the application of an external electric field. The motion of the ion is opposed by the viscous resistance of the solution. When a steady-state velocity is reached, the electric force is equal to the hydrodynamic viscous force (Fig. 2.64). The former is simply $z_i e_0 X$, where X is the electric field (or potential gradient) in the solution applied by two electrodes placed in solution (X is often measured in volts per centimeter). The latter is expressed by a famous classical formula of hydrodynamics called *Stokes' law*. This law, which describes the force experienced by a sphere moving in a viscous medium, states that

$$\text{Viscous force} = 6\pi r \eta v \tag{2.69}$$

where r is the radius of the moving ion and η is the viscosity of the medium. Thus,

$$z_i e_0 X = 6\pi r \eta v \tag{2.70}$$

or

$$r = \frac{z_i e_0 X}{6\pi \eta v}$$

$$= \frac{z_i e_0}{6\pi \eta u} \tag{2.71}$$

where $u(=v/X)$, i.e., the velocity under unit electric field, is a measurable quantity and is often called the *electrical mobility* of the ion (*cf.* Section 4.4.3).

Once the radius r of the solvated ion is obtained from Eq. (2.71),

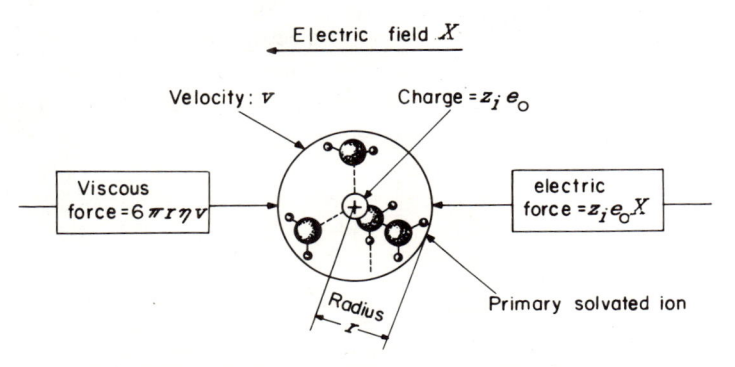

Fig. 2.64. The mobility method of determining hydration numbers is based on finding out the radius of a primary solvated ion from the fact that, when an ion in solution attains a steady-state velocity, the electric force $z_i e X_0$ is exactly balanced by opposing viscous force $6\pi \eta r v$.

one can calculate the hydration number n_h by a simple geometric argument (see Fig. 2.65)

$$n_h = \frac{\frac{4}{3}\pi r^3 - \frac{4}{3}\pi r_{cryst}^3}{\frac{4}{3}\pi r_{H_2O}^3}$$

$$= \frac{(r^3 - r_{cryst}^3)}{r_{H_2O}^3} \qquad (2.72)$$

where r_{cryst} is the crystallographic radius of the ion and r_{H_2O} is the radius of the water molecule, both of which are known from independent data.

Both the compressibility and the mobility methods of determining primary hydration numbers are based on quite loose approximations. The compressibility method assumes that the solvent *inside* the primary solvation sheath is completely incompressible and the water *outside* has the same compressibility as the pure solvent.

It will be recalled, however, that, in the secondary region (see Fig. 2.28 and Section 2.3.2) between the primary solvation sheath and the bulk water, there is structure breaking and partial alignment of the water molecules. Hence, instead of a sharp change of compressibilities at the boundary of the primary solvation shell, it is likely that there will be a smooth variation in compressibility from the ion out into the bulk solvent.

The mobility method, on the other hand, ignores the fact that, because the secondary region does not have the structure of the bulk solvent, the viscosity of the medium constituting the immediate neighborhood of the moving primary solvated ion is not the viscosity of the bulk solvent. It should be the *local* viscosity of the region surrounding the primary solvated ion. Such local viscosities are uncertain in value. Another approximation in the mobility method is that it neglects electrostrictional compression in computing the volume occupied by the water molecules in the primary hydration sheath (and, in the rudimentary version of the theory given here, also free space between water molecules).

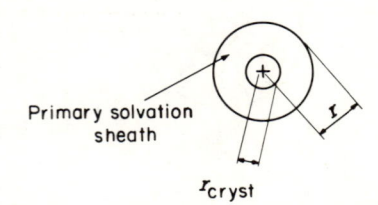

Fig. 2.65. The calculation of the hydration number from the radius of the primary solvated ion.

These approximations in the mobility method are offset by one big advantage. The method gives directly the hydration number of one ionic species, e.g., sodium ions, and not the *sum* of the hydration numbers of positive *and* negative ions. In the compressibility method, however, one only obtains the hydration number of the *salt*, and thus one has all the problems of resolving the value for the salt into the individual ionic values that were encountered in getting individual ionic heats of hydration from heats of hydration of salts (*cf.* Section 2.2.7). One has to depend on some independent (and sometimes somewhat circular) argument, for instance, that the relatively large iodide ion should have a hydration number of zero wholly to the positive ion. Of course, once one is certain of the hydration number of one ion, one can then get out those of other ions by taking the appropriate salts.

There are in all about five experimental methods which yield primary hydration numbers. The results show approximate agreement (± 1). Each method involves some doubts and approximations, and, in some cases, it is difficult to estimate with even a tolerance of $\pm 25\%$ what effect the approximations would have on the hydration numbers. Nevertheless, when one recalls the wild spread (*cf.* Table 2.17) of the values of hydration numbers obtained by not distinguishing between methods which determine primary and total hydration numbers, it must be accepted that the results of Table 2.20 hang together at least very much better than those in which a distinction between primary and other types of solvation is neglected. The results permit one to conclude the basic correctness of the picture of an ion influencing quite a bit of the surrounding solvent but actually succeeding

TABLE 2.20

Primary Hydration Numbers

Ion	From compressibility	From entropies	From apparent molal vol	From mobility	Most probable integral value
Li$^+$	5–6	5	2.5	3.5–7	5 ± 1
Na$^+$	6–7	4	4.8	2.4	4 ± 1
K$^+$	6–7	3	1.0	—	3 ± 2
F$^-$	2	5	4.3	—	4 ± 1
Cl$^-$	0–1	3	0	—	2 ± 1
Br$^-$	0	2	—	—	2 ± 1
I$^-$	0	1	—	—	1 ± 1

in trapping in its field only a certain number of water molecules which become the baggage of the ion in its travels through the solution.

There is, however, one surprising thing in the extent of agreement between the various methods. The mobility method is based on the nonequilibrium process of conduction, and the compressibility method, for example, is based on the system's being in equilibrium. Yet, the two methods yield fairly concordant results. The point, however, is that, in considering hydration numbers, one is not concerned with whether the whole system (the assembly of ions and solvent particles) is in static equilibrium or dynamic change. One is concerned with the state of the individual ions. But these are in ceaseless motion irrespective of whether the whole assembly is in equilibrium or not. Thus, even methods, such as the compressibility method, which involve measurements of the solution at equilibrium, concern in fact *ions* in a very dynamic state and should therefore give nearly the same hydration and solvation numbers as one would expect when the ions are drifting under nonequilibrium conditions, e.g., under an electric field.

Further Reading

1. H. Ulich, *Z. Elektrochem.*, **36**: 497 (1930).
2. H. Ulich, *Z. Physik. Chem. (Leipzig)*, **168**: 141 (1934).
3. A. Passynsky, *Acta Physicochim. URSS*, **8**: 385 (1938).
4. J. O'M. Bockris, *Quart. Rev. (London)*, 3: 173 (1949).
5. B. E. Conway and J. O'M. Bockris. "Ionic Solvation," in: J. O'M. Bockris, ed., *Modern Aspects of Electrochemistry*, Vol. I, Butterworth's Publications, Inc., London. 1954.
6. J. Padova, *J. Chem. Phys.*, **40**: 391 (1964).
7. R. Zana and E. Yeager, *J. Phys. Chem.*, **71**: 521 (1967); **71**: 4241 (1967).
8. J. F. Hinton and E. S. Amis, *Chem. Rev.*, **67**: 367 (1967).

2.5. THE DIELECTRIC CONSTANT OF WATER AND IONIC SOLUTIONS

2.5.1. An Externally Applied Electric Field Is Opposed by Counterfields Developed within the Medium

The solvation of ions arises from the interactions between solvent molecules and ions. These interactions result in the *orientation* of the, e.g., water molecules toward the *ions*. It follows that, as the ionic concentration increases, the fraction of the water in a solution which is trapped by ionic fields in the solvation sheaths also increases. Is this concentration-dependent extent of ion–water interactions revealed in any macroscopic property of the electrolyte? The dielectric constant of an electrolyte is such a property, but, to understand the relationship between dielectric constant and hydration and, thus, the dependence of dielectric constants on ionic concentration, one has to have a picture of the atomistic basis of the dielectric constant of a medium.

The dielectric constant always appears in the expression for the capacity of an electrical condenser, so why not start to try to understand dielectric constants in terms of the behavior of capacitors?

Consider a capacitor with plates which are A cm² in area and d cm apart (Fig. 2.66). Let there be a vacuum between the plates. To set up a potential difference V between the plates, a charge of q_{vac} has to be supplied to the plates from an external source. The capacity C_{vac} is defined as the charge that the capacitor can store per unit of potential difference between the plates, i.e.,

$$C_{\text{vac}} = \frac{q_{\text{vac}}}{V} \tag{2.73}$$

Now suppose that one placed between the plates some material which does not conduct electronically, i.e., a *dielectric* material such as paraffin. (If the material is an electronic conductor, it will allow a current to flow between the plates. This current would hamper the storage of charge and, hence, would frustrate simple reasoning.) In the presence of a dielectric material between the plates, how much charge must be supplied to the plates to set up the same potential difference V across the capacitor? It is found experimentally that a greater amount of charge is required in the presence of the dielectric than without it. Let this charge be q_{diel} (Fig. 2.67). Then,

$$C_{\text{diel}} = \frac{q_{\text{diel}}}{V} \tag{2.74}$$

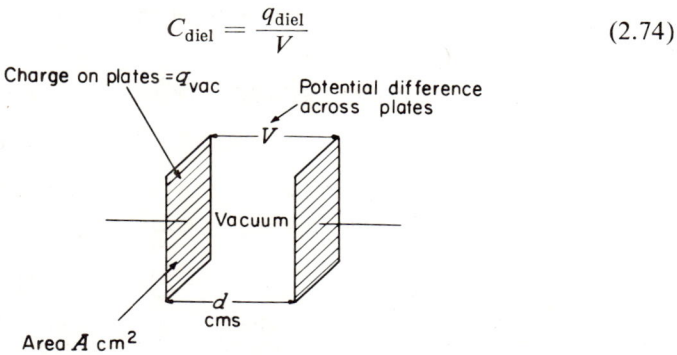

Fig. 2.66. A capacitor with vacuum between the plates.

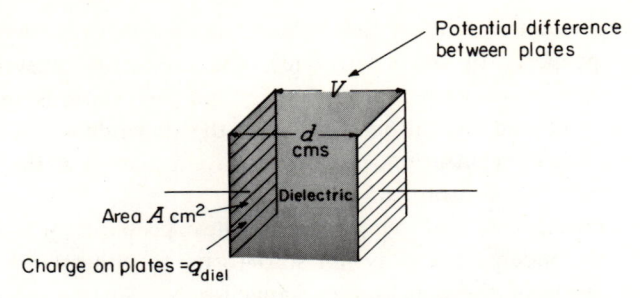

Fig. 2.67. A capacitor with a dielectric material between the plates.

The ratio of the capacity C_{diel} in the presence of the dielectric to the vacuum value C_{vac} defines the dielectric constant ε

$$\frac{C_{\text{diel}}}{C_{\text{vac}}} = \varepsilon \tag{2.75}$$

But why does the presence of the dielectric material affect the charge necessary to sustain a potential difference V between the capacitor plates? It is here that dipoles enter the picture. The point is that the electric field (due to the charge at the plates) *orients* the dipoles in the material (Fig. 2.68). Even if there are no permanent dipoles in the material (e.g., in paraffin) the electric field has the ability to displace the centers of negative charge (the electron clouds) from the centers of positive charge (the atomic nuclei) in the molecules of the material (Fig. 2.69). This deformation of the atoms and molecules then makes them behave as dipoles as long as the field is *on*. These induced dipoles do not have to be oriented; they are born aligned parallel to the field.

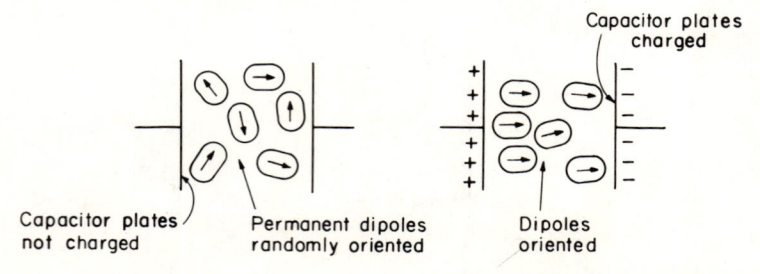

Fig. 2.68. The charge on the capacitor plates orients the permanent dipoles of a dielectric which consists of polar molecules.

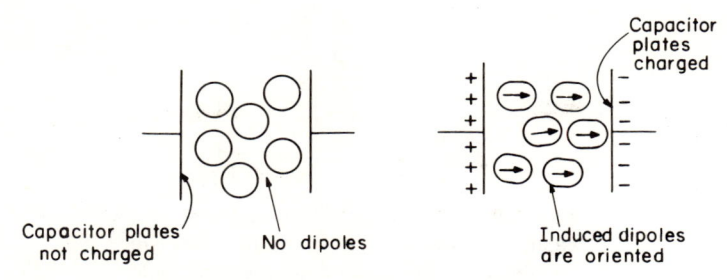

Fig. 2.69. The charge on the capacitor plates induces dipoles in the molecules of a dielectric.

The effect of the externally applied field on the dielectric material can now be qualitatively understood. In the absence of the field, the permanent dipoles are all arranged higgledy-piggledy; the switching-on of the field produces a net orientation of permanent dipoles and induces temporary dipoles. These oriented dipoles (and the induced dipoles) generate an internal field which is directed *counter* to the external field, i.e., to the field which is applied from an external source via the capacitor plates. Hence, the *net* field between the capacitor plates is less than the external field to the extent that an internal opposing field is produced in the dielectric material (Fig. 2.70). It follows that a given charge on the plates produces a smaller potential difference (field times distance between plates) compared with the value in the absence of the dielectric, or, conversely, a larger charge is required to produce unit potential difference. From Eq. (2.74), it is clear that the capacity of the condenser has been increased by interposing a dielectric material between the plates.

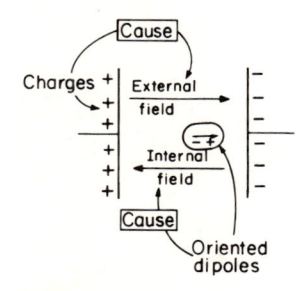

Fig. 2.70. The orientation of dipoles in the dielectric sets up an internal field which is directed counter to the external field produced by the charges on the plates.

Charge density, q

Oriented dipole layer

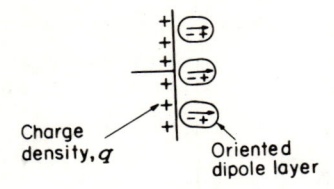

Fig. 2.71. At the interface between the capacitor plates and the dielectric, there are charges on the plates and an oriented dipole layer.

2.5.2. The Relation between the Dielectric Constant and Internal Counterfields

Now a relation will be derived between dielectric constant (i.e., the increase in the capacity over the vacuum value) and the internal counterfield which depends on the susceptibility of the dielectric material to having its dipoles oriented and its molecules distorted into dipoles.

Consider the interface between the metallic plates of the capacitor and the dielectric material (Fig. 2.71). Let there be a charge density q on the plates. The externally applied field has produced a dipole layer adjacent to the capacitor plates. This dipole layer is equivalent to two sheets of charge (Fig. 2.72), i.e., there will be a charge density q_{dipole} on the plane going through one end of the oriented dipoles and an equal but opposite charge density on the plane through the other end of the dipoles.

The next step in the argument is to compute the net field which is set up in the dielectric material as a result of the external field and the internal counterfield. For this purpose, use will be made of Gauss's law which states that the electric field normal to the surface (called the *Gaussian surface*) of *any* volume is equal to 4π times the charge enclosed in that volume.

Dipole layer is electrically equivalent to two sheets of charge

q

$+q_{\text{dipole}}$

$-q_{\text{dipole}}$

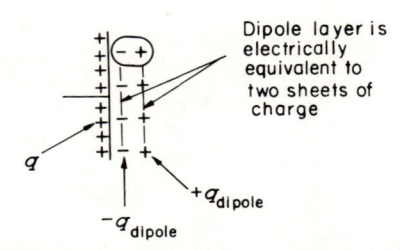

Fig. 2.72. The oriented dipole layer at the plate–dielectric interface is equivalent to two sheets of charge.

ION–SOLVENT INTERACTIONS 137

So, all that needs to be done is to choose a Gaussian surface which will yield the *net* electric field between the capacitor plates and then use Gauss's law.

For this purpose, a brick-shaped volume is considered (Fig. 2.73). Its location and size are of importance. Two of its faces are considered of *unit area* and parallel to the capacitor plates. The dimensions of the volume and its location are such that it encloses the charge q on the plate and the counter charge q_{dipole} on *one* end of the oriented dipoles—the end nearest the plates. Thus, the net charge inside the brick surface is $q - q_{\text{dipole}}$.

Hence, by Gauss's law, the field X_{ext} directed from the capacitor plate toward the dielectric is

$$X_{\text{ext}} = 4\pi(q - q_{\text{dipole}}) \tag{2.76}$$

The question is: What is q_{dipole}? In other words, how does the extent to which dipoles are oriented and induced in the dielectric depend on the magnitude of the applied electric field? In general, q_{dipole} can be expressed by a power series

$$q_{\text{dipole}} = A_1 X_{\text{ext}} + A_2 X^2_{\text{ext}} + \cdots \tag{2.77}$$

where $A_1, A_2, \ldots$ are constants independent of X_{ext}, but, for small fields, only the first term may be taken as significant. Thus,

$$q_{\text{dipole}} = A_1 X_{\text{ext}} \tag{2.78}$$

The constant A_1 must depend on the number n of dipoles per cubic centimeter and a constant which expresses how susceptible the material is to

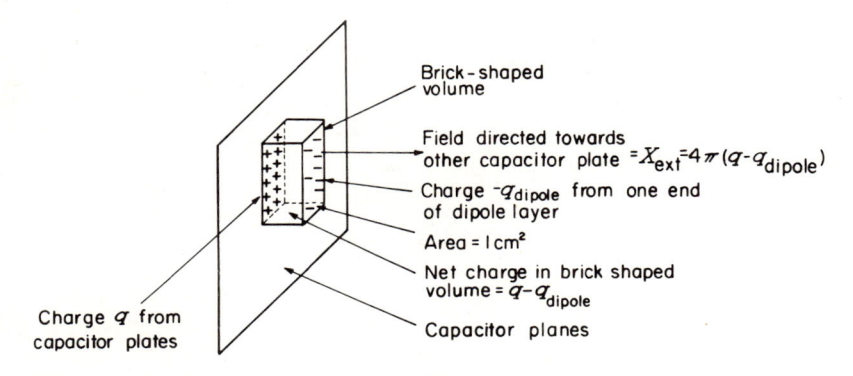

Fig. 2.73. The brick-shaped Gaussian surface used to compute the field that is directed from the capacitor plates into the dielectric.

having its molecules deformed into dipoles and its polar molecules oriented by the electric field. This electric susceptibility of the material is known as the *polarizability* of a molecule and designated by the symbol α. Thus,

$$q_{\text{dipole}} = n\alpha X_{\text{ext}} \tag{2.79}$$

where n is the number of dipoles per square centimeter.

Introducing this expression for q_{dipole} into the Eq. (2.76) for the field, one has

$$X_{\text{ext}} = 4\pi q - 4\pi n\alpha X_{\text{ext}}$$

or

$$X_{\text{ext}} = \frac{4\pi q}{1 + 4\pi n\alpha} \tag{2.80}$$

But the electric field between the plates is the gradient of the potential, so, since the potential varies linearly between the plates of a parallel plate condenser,

$$X_{\text{ext}} = \frac{V}{d} \tag{2.81}$$

Hence, Eq. (2.80) becomes

$$\frac{q}{V} = \frac{1 + 4\pi n\alpha}{4\pi d} \tag{2.82}$$

or, by multiplying both sides by A, the area of the plates

$$\frac{Aq}{V} = \frac{(1 + 4\pi n\alpha)A}{4\pi d} \tag{2.83}$$

But Aq (area of the plates times charge *density* on the plates) is the total charge on the plates. Hence Aq/V is the capacity C of the parallel-plate condenser, which is shown in elementary books on physics to be equal to $\varepsilon A/4\pi d$. Hence, from Eq. (2.83), one has

$$\frac{Aq}{V} = C = \frac{\varepsilon A}{4\pi d} = \frac{(1 + 4\pi n\alpha)A}{4\pi d}$$

or

$$\varepsilon - 1 = 4\pi n\alpha \tag{2.84}$$

Thus, one has obtained the fundamental relation between the dielectric constant and the polarizability of the stuff between the capacitor plates.

Charge on capacitor plates tends to orient dipole and thermal collisions tend to disorient dipole

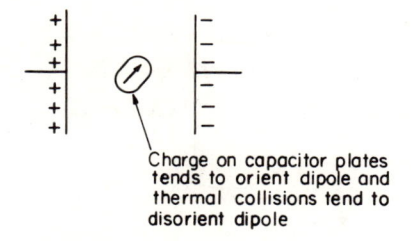

Fig. 2.74. The permanent dipoles of the dielectric are subject to an electrical orienting force and a thermal disorienting force.

2.5.3. The Average Dipole Moment of a Gas-Phase Dipole Subject to Electrical and Thermal Forces

The picture so far has been macroscopic to the extent that the atomistic basis of the dielectric constant is concealed in the polarizability α. But how is α related to atomistic quantities?

As a prelude to tackling a condensed medium such as water, a simpler system will first be analyzed. Consider polar molecules (permanent dipoles) which are so far from each other that their mutual interactions can be considered negligible. For example, consider a dilute gas of dipoles. Now, the electric field arising from the charge on the plates tends to line up the dipoles with their positive heads oriented toward the negative plate of the condenser, but, at the same time, thermal collisions between the dipoles are trying to knock them out of alignment (Fig. 2.74). Hence, the dipoles strike a compromise between the electrical orienting force and the thermal disorienting force. The compromise is described, following Debye, in the Boltzmann distribution law

$$n_\theta = R e^{-W/kT} \tag{2.85}$$

where n_θ is the number of dipoles per unit solid angle at an angle θ to the applied field (Fig. 2.75), R is a proportionality constant, and W is the work

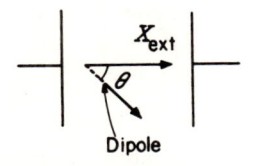

Fig. 2.75. An example of a dipole at an angle θ to the external field.

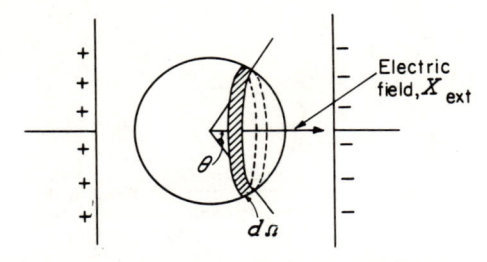

Fig. 2.76. The infinitesimal solid angle $d\Omega$.

done by the molecule in aligning with the field, i.e.,

$$W = -\mu X_{\text{ext}} \cos\theta \tag{2.86}$$

Hence,

$$n_\theta = Re^{\mu X_{\text{ext}} \cos\theta / kT} \tag{2.87}$$

or the number dn_θ of dipoles in an infinitesimal solid angle $d\Omega$ to the external field (Fig. 2.76) is

$$dn_\theta = Re^{\beta \cos\theta} \, d\Omega \tag{2.88}$$

where β has been written instead of $\mu X_{\text{ext}}/kT$.

Consider the dipoles oriented at an angle θ_1 (Fig. 2.77). The component of their dipole moment in the direction of the field is $\mu \cos\theta_1$. Hence, these dipoles are behaving *as if* they only have the moment $\mu \cos\theta_1$ in the *same* direction as the field. Similarly, other dipoles oriented at an angle θ_2 appear as if they have a moment $\mu \cos\theta_2$ in the direction of the field. Thus, the average moment $\langle\mu\rangle$ of a molecule under the combined influence of electrical and thermal forces is (by the standard formula for finding an average quantity)

$$\langle\mu\rangle = \frac{\int_0^\pi \mu \cos\theta \, Re^{\beta \cos\theta} \, d\Omega}{\int_0^\pi Re^{\beta \cos\theta} \, d\Omega} \tag{2.89}$$

Fig. 2.77. If a dipole is oriented at an angle θ to the external field, it has a moment $\mu \cos\theta$ in the direction of the field.

ION–SOLVENT INTERACTIONS 141

Now, it can be seen from Fig. 2.78 that $d\Omega = 2\pi \sin\theta\, d\theta = -2\pi\, d(\cos\theta)$. Hence,

$$\langle\mu\rangle = \frac{\int_\pi^0 \mu\cos\theta\, e^{\beta\cos\theta}\, d(\cos\theta)}{\int_\pi^0 e^{\beta\cos\theta}\, d(\cos\theta)} \quad (2.90)$$

or, by writing $x = \cos\theta$ (so that $x = 1$ when $\theta = 0°$ and $x = -1$ when $\theta = \pi$),

$$\langle\mu\rangle = \frac{\mu\int_{-1}^{1} e^{\beta x} x\, dx}{\int_{-1}^{1} e^{\beta x}\, dx} \quad (2.91)$$

For small fields, $\beta x = \mu X_{ext} x / kT \ll 1$. Hence, one can expand $e^{\beta x}$ by a Taylor's series

$$e^{\beta x} = 1 + \beta x + \frac{(\beta x)^2}{2!} + \cdots \quad (2.92)$$

and omit terms higher than the second

$$e^{\beta x} = 1 + \beta x \quad (2.93)$$

Thus,

$$\frac{\langle\mu\rangle}{\mu} = \frac{\int_{-1}^{1} x\, dx + \int_{-1}^{1} \beta x^2\, dx}{\int_{-1}^{1} dx + \int_{-1}^{1} \beta x\, dx}$$

$$= \frac{[x^2/2]_{-1}^{1} + [\beta x^3/3]_{-1}^{1}}{[x]_{-1}^{1} + [\beta x^2/2]_{-1}^{1}}$$

$$= \frac{\beta}{3} \quad (2.94)$$

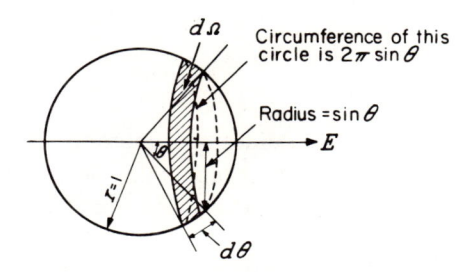

Fig. 2.78. Arriving at an expression for the infinitesimal solid angle $d\Omega$.

142   CHAPTER 2

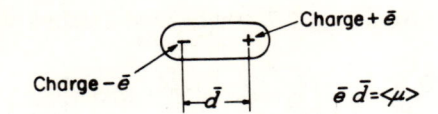

Fig. 2.79. The average moment $\langle\mu\rangle$ of a gas dipole.

or

$$\langle\mu\rangle = \frac{\mu^2 X_{ext}}{3kT} \tag{2.95}$$

This is the average or *effective* moment which a gas dipole exhibits in the direction of the weak[†] external field when it is subject to electrical orienting and thermal randomizing forces.

2.5.4. The Debye Equation for the Dielectric Constant of a Gas of Dipoles

The average moment $\langle\mu\rangle$ of a gas dipole in the direction of the field may be considered to arise from charges $+\bar{e}$ and $-\bar{e}$ separated by a distance $\bar{d}$ (Fig. 2.79)

$$\langle\mu\rangle = \bar{e}\bar{d} \tag{2.96}$$

Suppose that one considers (Fig. 2.80) a brick-shaped volume in the bulk of the dielectric. Let the volume be 1 cm² in area and d cm in thickness. If the gas contains n dipoles per cubic centimeter, then the brick-shaped volume ($\bar{d}$ cm³) can enclose $n\bar{d}$ dipoles. Further, the charge density q_1 on a *plane* going through the dipole ends will be equal to the number of dipoles per square centimeter ($= n\bar{d}$) times the charge per dipole ($= \bar{e}$); i.e.,

$$q_1 = n\bar{e}\bar{d} = n\langle\mu\rangle \tag{2.97}$$

and, inserting the expression (2.95) for $\langle\mu\rangle$, one has

$$q_1 = \frac{n\mu^2}{3kT} X_{ext} \tag{2.98}$$

An identical argument is valid at the interface between the capacitor plate and the dielectric, which means that the orientation of dipoles under electrical forces and thermal motions sets up a charge density q_1

[†] The qualifying word *weak* is introduced here in the light of the approximation contained in Eq. (2.93).

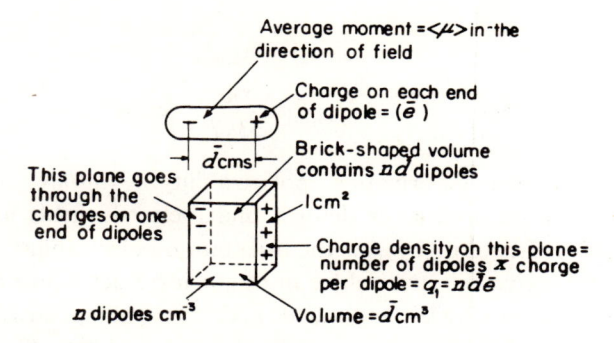

Fig. 2.80. The net orientation of dipoles produces a charge density q on a plane in the dielectric.

on a plane parallel to the capacitor plate. This charge q_1 is *opposite* in sign to that externally set up on the plates, i.e., to the charge which causes q_1.

It will be recalled however that the basic equation for the dielectric constant [Eq. (2.84)] was derived on the basis that a charge density of q_{dipole} was produced on a plane *adjacent* to the capacitor plates. Is $q_1 = q_{\text{dipole}}$? No, because one must also consider the contribution from the *induced* dipoles produced by the distortion of molecules subject to an electrical field; the charge q_1 only takes into account the effect of the *permanent* dipole moment of the molecules of the dielectric. The average moment *induced* in a nonpolar molecule will be considered to contribute a charge density q_2 to the plane adjacent to the capacitor plates, so that

$$q_{\text{dipole}} = q_2 + q_1 \tag{2.99}$$

or, from Eq. (2.79),

$$\alpha = \frac{q_{\text{dipole}}}{nX_{\text{ext}}} = \frac{q_2}{nX_{\text{ext}}} + \frac{q_1}{nX_{\text{ext}}} \tag{2.100}$$

One can substitute for q_1 in this equation Eq. (2.98), obtaining

$$\alpha = \frac{q_2}{nX_{\text{ext}}} + \frac{\mu^2}{3kT} \tag{2.101}$$

In this expression, the total polarizability α and the number of molecules per cubic centimeter are constant for a given field; so is $\mu^2/(3kT)$ at a fixed temperature. Hence, $q_2/(nX_{\text{ext}})$ must also be a constant, and one can write

$$\alpha = \alpha_{\text{deform}} + \alpha_{\text{orient}} \tag{2.102}$$

$$= \alpha_{\text{deform}} + \frac{\mu^2}{3kT} \tag{2.103}$$

where α_{deform} is the deformation polarizability ($= q_2/nX_{\text{ext}}$) and is a measure of how the molecules deform under electric fields and become induced dipoles. The term α_{orient} is the orientation polarizability and reveals how successful a given electric field is in producing a net dipole orientation.

The expressions (2.102) and (2.103) for the total polarizability α of the dipole gas can be inserted into the fundamental relation (2.84) between the polarizability and the dielectric constant. Thus,

$$\varepsilon - 1 = 4\pi n \alpha_{\text{deform}} + 4\pi n \alpha_{\text{orient}} \tag{2.104}$$

$$= 4\pi n \alpha_{\text{deform}} + \frac{4\pi n \mu^2}{3kT} \tag{2.105}$$

Thus did Debye link the macroscopic property of dielectric constant to a molecular property, namely, the dipole moment of the gas molecule (*cf.* Fig. 2.81).

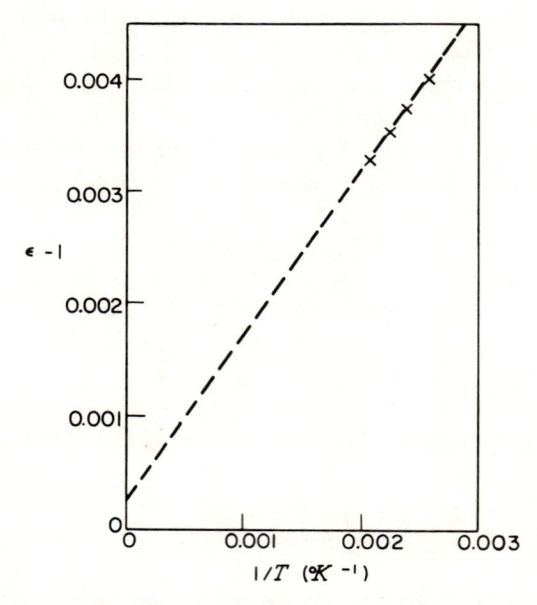

Fig. 2.81. The variation of the dielectric constant of water vapor with temperature is as predicted by the Debye equation.

There are two important features of this expression (2.105). Firstly, temperature affects the dielectric constant by affecting the orientation of dipoles; the higher the temperature, the less successful is the externally applied field in opposing the thermal disorienting motions and producing a net orientation of the dipoles. So it was a wise decision to treat the dielectric constant as a function of temperature in differentiating the Born free energy of solvation to obtain the entropy of solvation (see Section 2.2.5). Secondly, as the temperature increases, the orientation polarization $4\pi n\mu^2/(3kT)$ and, thus, the dielectric constant decreases. Eventually, when the temperature is large enough,

$$4\pi n\alpha_{\text{deform}} \gg \frac{4\pi n\mu^2}{3kT}$$

and

$$\varepsilon - 1 \approx 4\pi n\alpha_{\text{deform}} \qquad (2.106)$$

Things have become too hot for the dipoles to have any net orientation at all; they become randomly oriented. But there still remains deformation polarization due to induced dipoles, and this is what then decides the dielectric constant.

2.5.5. How the Short-Range Interactions between Dipoles Affect the Average Effective Moment of the Polar Entity Which Responds to an External Field

The Debye theory for the dielectric constant of a gas of permanent dipoles invites an obvious comment. It may be satisfactory for a gas of water molecules[†] which are not involved in mutual interactions, but what has it to do with the dielectric constant of *liquid* water with its network structure and of ionic *solutions* containing solvated ions? One must review some of the basic assumptions in the treatment of a gas of dipoles and see to what extent they would have to be changed in discussion of the dielectric constant of an associated liquid.

Firstly, it was assumed that there were no interactions between the dipoles in the vapor. This assumption is obviously untenable in liquid water. Water is a broken-down form of ice. It is quasi crystalline in the sense that there are in liquid water large groups ("icebergs") of water molecules associated by hydrogen bonding. In each of these large groups,

---

[†] For the consideration of the dielectric constant of water, it is not necessary to take into account the quadrupole character of a water molecule; it is quite adequate to consider a water molecule as it were an electrical dipole.

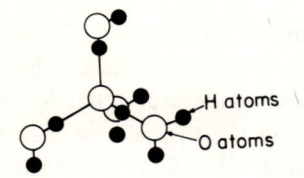

Fig. 2.82. The tetrahedral unit of a central water molecule linked to four neighboring water molecules by hydrogen bonds.

an important structural unit may be distinguished. This structural unit or subgroup consists of a central water molecule tetrahedrally linked to four[†] other molecules by hydrogen bonds (Fig. 2.82).

How does such a subgroup respond to an electric field? When the electric field tries to align the central molecule of a tetrahedral unit, it has to reckon not only with randomizing thermal motions but also with the coordinating water molecules which seek to maintain definite orientation relationships with the central molecule, i.e., the orientations characteristic of the tetrahedral subgroup with hydrogens facing oxygen lone pairs, etc. So, if any aligning occurs, the whole subgroup has to align[‡] (Fig. 2.83). What matters, therefore, is the moment μ_{group} of a subgroup of water molecules, not the dipole moment μ of an *isolated* water molecule. The effective moment of the group as a whole is equal to the dipole moment of the central molecule plus the components of the dipole moments of the four neighboring water molecules of the tetrahedral unit (these components being taken along the moment of the central molecule). In other words, the effective moment μ_{group} is the vector sum of the dipoles in the group, i.e.,

$$\mu_{\text{group}} = \mu + g(\mu \overline{\cos} \gamma)$$
$$= \mu(1 + g \overline{\cos} \gamma) \qquad (2.107)$$

where g is the number of nearest-neighbor water molecules linked with the central molecule and $\overline{\cos} \gamma$ is the average of the cosines of the angles

---

[†] In ice, the number of coordinating water molecules is exactly four; in water, it is 4.4 to 4.6 (*cf.* Table 2.11). For simplicity, this distinction will not be stressed.

[‡] It has been suggested that the icebergs present in water contain (at 20 °C) about 50 to 60 water molecules with hydrogen bonding effective throughout the group. Nevertheless, the *sub*group of four neighboring water molecules plus a central water molecule is considered in this discussion the orienting entity because the central molecule is regarded as exerting short-range forces, apart from hydrogen bonding, only on its nearest neighbors, i.e., on four water molecules.

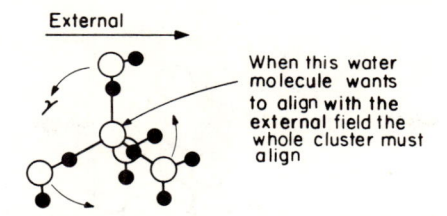

Fig. 2.83. When the central molecule of the tetrahedral group tries to align with the external field, the whole group aligns.

between the dipole moment of the central water molecule and those of its bonded neighbors.

For example, if the cluster consists of the tetrahedral group, then, $g = 4$ and $\overline{\cos \gamma} = 1/3$, in which case $1 + g \overline{\cos \gamma} = 7/3$. That is, owing to the interlinking of the water dipoles, the effective dipole moment of the aligning group is 7/3 times the dipole moment of a free water molecule.

Once the effective moment of a subgroup is computed, one can go through the Debye argument saying that thermal motions oppose the alignment of subgroups. After writing $\beta = \mu_{\text{group}} X/(kT)$, the average moment of a dipole cluster is given by [cf. Eq. (2.95)]

$$\langle \mu_{\text{group}} \rangle = \frac{\mu^2 (1 + g \overline{\cos \gamma})^2}{3kT} X \qquad (2.108)$$

A simple test of this equation is to examine what happens when the subgroups break up, say, by a temperature increase. The correlated alignments of the water molecules are diminished, and the dipoles become independent. The quantity $\overline{\cos \gamma}$ tends to zero,[†] i.e., the various neighboring dipoles contribute a zero average component along the particular central dipole. One gets back the simple Debye equation (2.95) for the average dipole moment.

2.5.6. The Local Electric Field in a Condensed Polar Dielectric

From the effective moment of a group of dipoles which orient as one unit (Eq. 2.108), one can obtain the orientation polarizability thus

$$\alpha_{\text{orient}} = \frac{q_1}{nX_{\text{ext}}} = \frac{\langle \mu_{\text{group}} \rangle}{X_{\text{ext}}} = \frac{\mu^2 (1 + g \overline{\cos \gamma})^2}{3kT} \frac{X}{X_{\text{ext}}} \qquad (2.109)$$

[†] The quantity $\overline{\cos \gamma}$ is averaged over those neighboring molecules which hold an orientation relationship with the central molecule. When thermal forces destroy the orientation relationship, the average of $\cos \gamma$ is zero, i.e., the neighboring water molecules are orienting independent of the central molecule.

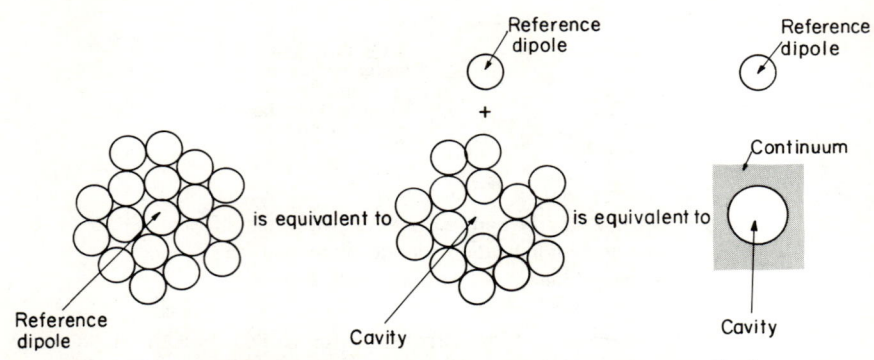

Fig. 2.84. The approximately spherical cavity into which a reference dipole or group of dipoles is caged.

where X_{ext} is the externally applied field. But what is X, the electric field operating on the central molecule and the group? Is it simply equal to the external field X_{ext}?

These questions bring up the second main difference between a gas of free, noninteracting dipoles and a liquid with dipoles associated into subgroups which are part of networks. In the gas phase, all that a water molecule feels is the external field emanating from the capacitor plates—the other water molecules are too far away to exercise any influence. In the liquid phase, however, any water molecule of group of molecules is in "contact" with other water molecules. So, what matters to a given water molecule (or group of water molecules) is the true *local* field X_{loc}, which is the sum of the external field and the field of the surrounding particles on the given molecule (or group). Thus, the electric field X in the expression for the orientation polarizability is the local field and not the external field and should therefore be replaced by X_{loc}.

Onsager showed an interesting way of calculating the local field X_{loc}. The thinking proceeds thus: To a close approximation, a reference dipole or a group of interlinked dipoles is caged by the surrounding particles in a spherical hole (Fig. 2.84). So the local field operating on the reference dipole (or group) is the field in the empty spherical cavity owing to the externally applied field and the surrounding molecules which are partly oriented by the field.

Now, to compute the field in the cavity, consider two regions (Fig. 2.85): one outside the cavity (region I) and one inside it (region II). Let the center of the cavity be arbitrarily taken as the origin of a polar coordinate system and as the point to which all potentials are referred, i.e., as the zero of potentials.

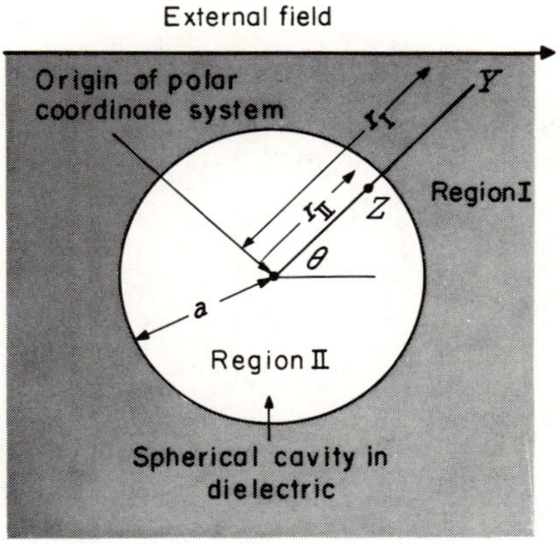

Fig. 2.85. The computation of the field in the spherical cavity.

The procedure will consist in obtaining expressions for the potential in region I and region II. At the surface of the sphere, the spatial coordinates are the same, and the potentials, as expressed for region I and region II, must be the same and can hence be equated.

The potential φ_{II} at a point Z (with coordinates r_{II} and θ) *inside* the cavity, i.e., in region II, is simply equal to $-X_{loc}r_{II}\cos\theta$, where X_{loc} is the local field[†]

$$\varphi_{II} = -X_{loc}r_{II}\cos\theta \qquad (2.110)$$

The potential φ_I at a point $Y(r_I, \theta)$ is a more interesting matter. It is equal to the sum of the potentials due to the external field (this potential is $-X_{ext}r_I\cos\theta$) and, what may appear surprising, that field due to the *empty* spherical cavity.

Why should an empty cavity exert any influence on the surroundings? Because, whenever a dielectric material is placed in an electric field, charges appear on the boundary (or surface) of the material. The cavity is an internal surface, so there are charges on its surface. It is these charges which exert an electrical influence on the surroundings.

Another point about these charges is that they are always set up so as to oppose the externally applied field. The distribution of these charges

---

[†] The minus sign comes in because the potential decreases in the direction of an electric field, i.e., "down" the field. Thus, any point has a potential which is negative relative to that at a point farther "up" the field.

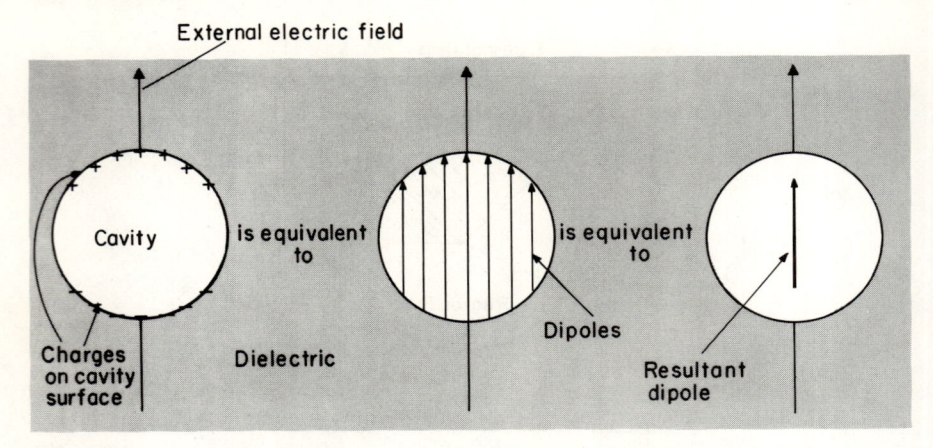

Fig. 2.86. The electrical effect of a cavity is equivalent to the effect of a single dipole.

looks like a series of dipoles all pointing in the direction of the field (Fig. 2.86). In fact, one can conceptually replace all these dipoles by one single dipole of moment m located at the center of the cavity. In conclusion, therefore, the electrical effect of an empty spherical cavity (in a dielectric which is placed in an externally applied field) is equivalent to the effect of a hypothetical dipole of moment m located at the center of the cavity. Thus, the spherical cavity sets up a potential $m(\cos\theta)/r_\mathrm{I}^2$ (the usual expression for the dipole potential) at the point $Y(r_\mathrm{I}, \theta)$ in region I.

The total potential φ_I at Y (in Fig. 2.85) is given by the sum of the potential due to the external field, which is $-X_\mathrm{ext} r_\mathrm{I} \cos\theta$, plus that due to the cavity, which at Y is $m(\cos\theta)/r_\mathrm{I}^2$,

$$\varphi_\mathrm{I} = -X_\mathrm{ext} r_\mathrm{I} \cos\theta + \frac{m \cos\theta}{r_\mathrm{I}^2} \qquad (2.111)$$

Now that expressions for φ_I and φ_II have been obtained, the situation at the surface of the cavity can be considered. Here, $r_\mathrm{I} = r_\mathrm{II} = a$, the radius of the sphere, and, thus, $\varphi_\mathrm{I} = \varphi_\mathrm{II}$, i.e.,

$$-X_\mathrm{ext} a \cos\theta + \frac{m \cos\theta}{a^2} = -X_\mathrm{loc} a \cos\theta$$

or

$$X_\mathrm{loc} = X_\mathrm{ext} - \frac{m}{a^3} \qquad (2.112)$$

There is a further condition that must be satisfied at the cavity surface. The electric field normal to the hole surface in the two opposite directions

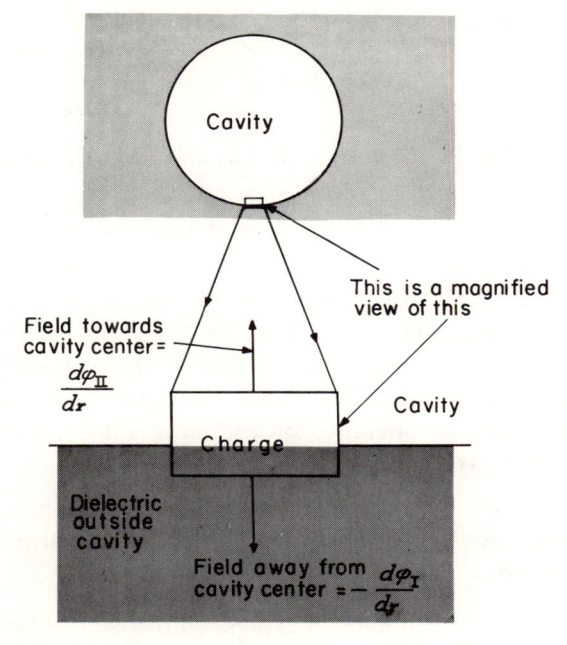

Fig. 2.87. A Gaussian surface for obtaining the electric fields normal to the cavity surface.

(toward and away from the cavity center) must be equal. To get at these fields, one invents an imaginary Gaussian surface shaped like a pillbox and containing some charge q (Fig. 2.87). The field away from the cavity is $-d\varphi_I/dr$, and, from Gauss's law, one has

$$\frac{-d\varphi_I}{dr} = \frac{4\pi q}{\varepsilon}$$

or

$$-\varepsilon \frac{d\varphi_I}{dr} = 4\pi q \qquad (2.113)$$

The field toward the cavity center (Fig. 2.87) is

$$-\frac{d\varphi_{II}}{dr} = 4\pi q \qquad (2.114)$$

(ε being unity because, inside the cavity, there is vacuum). Hence, at the surface of the cavity,

$$\left(\frac{d\varphi_{II}}{dr}\right)_{r=a} = \varepsilon \left(\frac{d\varphi_I}{dr}\right)_{r=a} \qquad (2.115)$$

which, from Eqs. (2.110) and (2.111), is

$$X_{\text{loc}} = \varepsilon X_{\text{ext}} + \frac{2\varepsilon m}{a^3} \tag{2.116}$$

From the simultaneous Eqs. (2.112) and (2.116), one has

$$X_{\text{loc}} = \frac{3\varepsilon}{2\varepsilon + 1} X_{\text{ext}} \tag{2.117}$$

Hence, the local field in the cavity, which operates on the dipole cluster, has been evaluated. Since $\varepsilon > 1$ (i.e., $2\varepsilon + 1 < 2\varepsilon + \varepsilon$ or $2\varepsilon + 1 < 3\varepsilon$), it is obvious that the local field in condensed media is always greater than the externally applied field. This means, then, that a given dipole is subject to a stronger electrical orienting force when it is inside a liquid than when it is in a dilute gas.

2.5.7. The Dielectric Constant of Liquids Containing Associated Dipoles

It is now easier to obtain an expression for the dielectric constant of a liquid, such as water, in which the polar molecules are associated into subgroups and networks.

One starts from the fundamental expression

$$\varepsilon - 1 = 4\pi n \alpha \tag{2.84}$$

which, after substituting for α with

$$\alpha = \frac{q_{\text{dipole}}}{nX_{\text{ext}}} = \frac{q_1}{nX_{\text{ext}}} + \frac{q_2}{nX_{\text{ext}}} \tag{2.100}$$

becomes

$$\varepsilon - 1 = 4\pi n \left(\frac{q_1}{nX_{\text{ext}}} + \frac{q_2}{nX_{\text{ext}}} \right) \tag{2.118}$$

The q_1 term which arises from orientation polarization is elaborated as follows. Since [cf. (2.97)]

$$q_1 = n \langle \mu_{\text{group}} \rangle \tag{2.119}$$

one can use Eq. (2.108) to write

$$q_1 = \frac{n\mu^2 (1 + g \overline{\cos \gamma})^2}{3kT} X_{\text{loc}} \tag{2.120}$$

or, introducing the expression (2.117) for the local field,

$$q_1 = \frac{n\mu^2(1 + g\overline{\cos\gamma})^2}{3kT} \frac{3\varepsilon}{2\varepsilon + 1} X_{\text{ext}} \qquad (2.121)$$

Hence,

$$\frac{q_1}{nX_{\text{ext}}} = \frac{\mu^2(1 + g\overline{\cos\gamma})^2}{3kT} \frac{3\varepsilon}{2\varepsilon + 1} \qquad (2.122)$$

The second term in Eq. (2.118) is connected with the deformation polarization. Following the procedure of Section 2.5.4, one argues that q_2 is proportional to the number n per unit volume of deformable molecules and to the electric field operating on them (the proportionality constant being α_{deform}). In the gas phase, the field operating on the deformable molecules is the external field X_{ext}; in condensed media, it will be the local field X_{loc}. Thus,

$$q_2 = n\alpha_{\text{deform}} X_{\text{loc}} \qquad (2.123)$$

Once again, the expression (2.117) for X_{loc} can be introduced to give

$$q_2 = n\alpha_{\text{deform}} \frac{3\varepsilon}{2\varepsilon + 1} X_{\text{ext}} \qquad (2.124)$$

or, by rearrangement,

$$\frac{q_2}{nX_{\text{ext}}} = \alpha_{\text{deform}} \frac{3\varepsilon}{2\varepsilon + 1} \qquad (2.125)$$

Now, the expressions for q_1/nX_{ext} and q_2/nX_{ext} can be inserted into Eq. (2.118) to give

$$\varepsilon - 1 = 4\pi n \frac{3\varepsilon}{2\varepsilon + 1} \left[\alpha_{\text{deform}} + \frac{\mu^2(1 + g\overline{\cos\gamma})^2}{3kT} \right] \qquad (2.126)$$

It is conventional, however, to write this in a slightly altered form as

$$\frac{(\varepsilon - 1)(2\varepsilon + 1)}{9\varepsilon} = \frac{4\pi n}{3} \left[\alpha_{\text{deform}} + \frac{\mu^2(1 + g\overline{\cos\gamma})^2}{3kT} \right] \qquad (2.127)$$

This is the Kirkwood equation for the dielectric constant of a condensed medium. It takes into account the short-range interactions between polar molecules which lead to the formation of molecular groups orienting as a unit under the influence of electric fields. It also considers the actual local field, as distinct from the externally applied electric field, operating on the orienting entities.

Table 2.21
Dielectric Constant of Liquid Water at Various Temperatures

Temperature,	Calc	Obs
0	84.2	88.0
25	78.2	78.5
62	72.5	66.1
83	67.5	59.9

From the Kirkwood equation, it is clear that the *association of dipoles arising from short-range forces* is a very important factor in determining the dielectric constant of a liquid. The linking together of dipoles increases g, the number of dipoles which are nearest neighbors to a reference dipole, and thus increases the dielectric constant.

In using the Kirkwood equation (2.127), it is necessary to know g, the number of dipoles which are nearest neighbors or coordinate in particular to a reference dipole. This quantity g can be obtained from X-ray data, and, for liquid water, it is 4.4 to 4.6. Based on this value, there is good agreement (Table 2.21) between the observed values of dielectric constant and those calculated by the Kirkwood equation.

To illustrate the influence of dipole association on the dielectric constant, consider liquid water and SO_2. The dipole moments of the SO_2 and H_2O molecules are almost the same, but their dielectric constants are quite different (Table 2.22). This difference arises because, in water, there are groups of strongly interacting, hydrogen-bonded water molecules, whereas such association of SO_2 molecules does not occur.

In the light of the above discussion, hydrogen-bonded liquids should in general be expected to have high dielectric constants compared with those of nonassociated liquids, and, indeed, this is the case (Fig. 2.88).

TABLE 2.22
The Dipole Moments and Dielectric Constants of Liquid Water and Sulfur Dioxide

	Dipole moment, debyes	Dielectric constant, at 25 °C
H_2O	1.85	78.5
SO_2	1.67	12.35

ION–SOLVENT INTERACTIONS   155

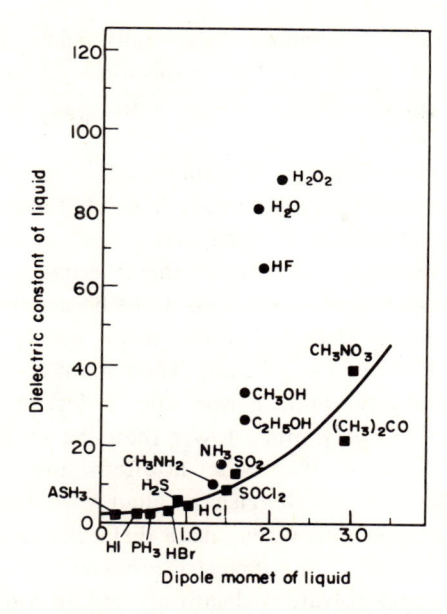

Fig. 2.88. The dielectric constants of liquids as a function of their dipole moments (■'s represent unassociated liquids, and ●'s represent H-bonded liquids).

2.5.8. The Influence of Ionic Solvation on the Dielectric Constant of Solutions

Some indication has now been given of how the association of dipoles and the local field can increase the dielectric constant into the range of several tens compared with values of 5 to 10 for unassociated liquids. An elementary, quantitative picture of the dielectric constant of pure water has also been presented. The next question is: What happens to the dielectric constant when ions enter an associated liquid such as water? In other words, how does the dielectric constant of an *ionic solution* differ from that of *pure water*?

Recall the picture of the structure of water in the presence of ions (Section 2.3.2). Far from an ion, the water structure is normal and the usual bulk dielectric constant would obtain. In the secondary region, the existence of structure breaking implies that there is a decrease of the parameter g of the Kirkwood equation (2.127), i.e., there is a decrease in the number of water molecules linked to any particular water molecule. But, as g decreases, the dielectric constant also decreases. This implies that the dielectric constant in the secondary region falls below the bulk value. It will be, depending on the distance from the ion, somewhere between the bulk value ($\sim$80) and the value in the solvation sheath, which turns out (see below) to be about 6.

The situation in the primary solvation sheath is interesting. The solvent dipoles here are oriented to the ion and are so firmly fixed in this orientation that they are almost insensitive to the external field and to the field of the surrounding water molecules. As far as the external and local fields are concerned, they are so bound to the ions as to be unorientable. Thus, the water molecules in the primary solvation sheath hardly contribute to the orientation polarizability of the solution. Their main contribution is to the deformation polarizability, and, hence, the dielectric constant of this primary solvent water is very much lower than the bulk value. Measurements of the dielectric constant of water at alternating field frequencies so high that the dipoles are too sluggish to align with the alternating field suggest that completely bound water has a dielectric constant of about 6.

So one has a picture of the dielectric constant of water being very low ($\sim$6) in the primary solvation sheath around an ion and then rapidly increasing till the bulk value ($\sim$80) is attained outside the structure-breaking region (Fig. 2.89). It will be seen later (Section 7.4.24b) that one should reckon with a similar variation of dielectric constant near a charged electrode adjacent to which there is a layer of oriented water.

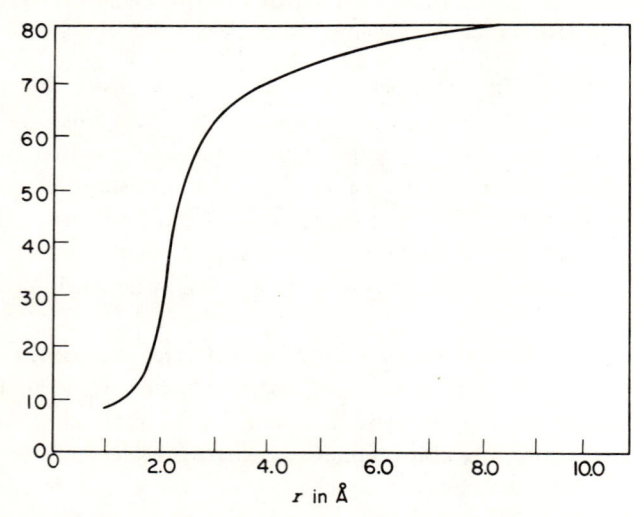

Fig. 2.89. The variation of the dielectric constant around an ion.

TABLE 2.23
Dielectric Constants of Aqueous Solutions of Electrolytes[†]

Electrolyte	Dielectric constant
LiCl	66 ± 2
NaCl	69 ± 2
KCl	70 ± 2
RbCl	70 ± 2
NaF	68 ± 2
KF	67 ± 2
NaI	65 ± 2
KI	64 ± 2
$MgCl_2$	50 ± 2
$BaCl_2$	52 ± 2
$LaCl_3$	36 ± 2
Na_2SO_4	58 ± 2

[†] In all cases, the values pertain to solutions containing a mole of electrolyte.

A simple conclusion follows from the variation of dielectric constant in the neighborhood of an ion. The dielectric constant of an ionic solution must be determined partly by the dielectric constant of the water which is unaffected in its structure by the presence of ions and partly by the dielectric constant of the water in the primary and secondary solvation sheaths around ions. But, the dielectric constant of the water both in the primary solvation sheaths and in the structure-breaking secondary region is lower than the value for the pure solvent. Hence, the dielectric constant of a solution must be lower than that of the pure solvent (Table 2.23). Further, the depression of the dielectric constant depends on the concentration of the ions—the higher the ionic concentration, the lower the dielectric constant is (Fig. 2.90).

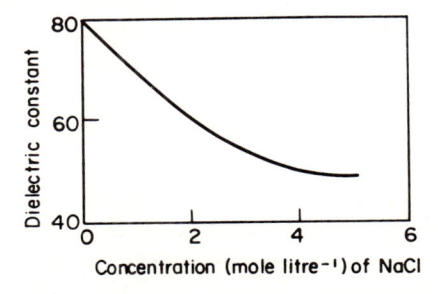

Fig. 2.90. The variation of the dielectric constant of NaCl solutions with electrolyte concentration.

Further Reading

1. L. Onsager, *J. Am. Chem. Soc.*, **58**: 1486 (1936).
2. J. G. Kirkwood, *J. Chem. Phys.*, **7**: 911 (1939).
3. G. Oster, and J. G. Kirkwood, *J. Chem. Phys.*, **11**: 175 (1954).
4. L. Pauling, *The Nature of the Chemical Bond*, 3rd ed., Chap. 12, Cornell University Press, Ithaca, N.Y., 1960.
5. A. Prock and G. McConkey, *Topics in Chemical Physics*, Chap. 1, Elsevier Publishing Co., New York, 1962.
6. R. H. Cole, *J. Chem. Phys.*, **39**: 2602 (1963).
7. N. E. Hill, *Trans. Faraday Soc.*, **69**: 344 (1963).
8. R. P. Feynmann, R. B. Leighton, and M. Sands, *The Feynmann Lectures on Physics*, Vol. II, Chap. 10, Addison-Wesley Publishing Co., Inc., Reading, Mass., 1964.
9. W. Dannhauser, and L. W. Bahe, *J. Chem. Phys.*, **41**: 2666 (1964).
10. B. E. Conway, and R. G. Barradas, *Chemical Physics of Ionic Solutions*, John Wiley & Sons. Inc., New York, 1966.
11. C. P. Smyth, *Ann. Rev. Phys. Chem.*, **17**: 433 (1966).

2.6. ION–SOLVENT–NONELECTROLYTE INTERACTIONS

2.6.1. The Problem

The picture that has emerged in this chapter is of ions interacting with the solvent and producing the interesting effects that go under the name *solvation*. The quasi-lattice structure of the solvent is not the same after the ions have entered it. Some of the water molecules are wrenched out of the quasi lattice and appropriated by the ions as part of their primary solvation sheaths. Even farther away in the secondary solvation sheaths, the ions produce telltale effects of structure breaking.

What happens if, in addition to ions and water molecules, molecules of nonelectrolyte are also present in the system? Or, what will occur if, to a solution already containing nonelectrolyte molecules at saturation concentration, ions are added?

One thing is certain: there will be less free water to dissolve the nonelectrolyte, because some of it will be removed from the solvent into the solvation sheath of the ion. This means that the nonelectrolyte molecules find themselves suddenly having much less water to associate with, and some of them will hence shun the loneliness imposed by the water's preference for ions, and reassociate themselves with their parent lattice, i.e., precipitate out. This is the origin of a term from technological organic chemistry—"salting out." Occasionally, however, the ions are deviants, and associate preferably with the nonelectrolyte solute, shunning the water. In the rarer instances where these deviants appear, there is a rapid departure of the nonelectrolyte from the parent lattice, and thus the solubility of the former is enhanced. One speaks of *salting in*.

Two aspects of the theory of salting out—the normal case—are considered below. Firstly, the effects of the *primary* solvation sheath have to be taken into account—how the requisition of water by the ions causes the nonelectrolyte's solubility to decrease. Secondly, the effects of secondary solvation (interactions outside the solvation sheath) are calculated.

2.6.2. The Change in Solubility of a Nonelectrolyte Due to Primary Solvation

It is very easy to calculate this change of solubility, so long as there are data available for the solvation numbers of the electrolyte concerned. Let it be assumed that the normal case holds true, and the ions are solvated entirely by water molecules. Then (*cf.* Section 3.6.2) recalling that one liter of water contains 55.55 moles, the number of water molecules left free to dissolve nonelectrolyte after the addition of ions is

$$55.55 - c_i n_s, \tag{2.128}$$

where c_i is the number of gram moles of the electrolyte l^{-1}, and n_s is the solvation number of the electrolyte concerned.

Assuming at present that the solubility of the nonelectrolyte is simply proportional to the number of water molecules outside the hydration sheath, then

$$\frac{S}{S_0} = \frac{55.55 - c_i n_s}{55.55} \tag{2.129}$$

or

$$S = S_0 - \frac{S_0 c_i n_s}{55.55} \tag{2.130}$$

where S_0 is the solubility of the nonelectrolyte before addition of electrolyte and S that after it. Then, suppose $n_s = 6$, $c_i = 1M$; then

$$\frac{S - S_0}{S_0} = -0.11 \tag{2.131}$$

Correspondingly, S/S_0 is 0.89. For comparison, with $KClO_4$ as electrolyte, and 2,4-dinitrophenol as nonelectrolyte, the corresponding experimental value (extrapolated) is about 0.7. Some further effect must be taken into account.

2.6.3. The Change in Solubility Due to Secondary Solvation

It has been stressed in Section 2.4 that solvation is a far reaching phenomenon, although only the primary solvation number can be determined significantly (if by no means accurately, cf. Table 2.19). Correspondingly, there certainly are effects of ions on the properties of their solvent which lie outside the radius of the primary hydration sheath. These effects must now be accounted for, in so far as they relate to solubility of a nonelectrolyte. Let the problem be tackled as though no primary solvation had withdrawn water from the solution. One can write

$$n_{NE,r} = n_{NE,b} e^{-\Delta G^0/RT} \tag{2.132}$$

where $n_{NE,r}$ is the *number* per unit volume of nonelectrolyte molecules at a distance r from the ion, $n_{NE,b}$ is the same number in the bulk, and the free-energy change ΔG^0 is the work W_r done to remove a mole of water molecules and insert a mole of nonelectrolyte molecules at a distance r from the ion outside the primary solvation sheath.

A naïve calculation would run thus: The field X_r of the ion at a distance r falls off with distance according to Coulomb's law[†]

$$X_r = \frac{z_i e_0}{\varepsilon r^2} \tag{2.133}$$

If, now, a dipole (aligned parallel to the ionic field) is moved from infinity where the field $X_r = 0$ through a distance dr to a point where the corresponding field is dX, the *elementary* work done is $-\mu\, dX$.

Thus, the work to bring a mole of nonelectrolyte molecules from infinity to a distance r is $-N_A \int_0^{X_r} \mu_{NE}\, dX$, and the work done to remove a water molecule to infinity is $N_A \int_0^{X_r} \mu_W\, dX$. The net work of replacing a water molecule by a nonelectrolyte molecule would therefore be given by

$$W_r = N_A \left(\int_0^{X_r} \mu_W\, dX - \int_0^{X_r} \mu_{NE}\, dX \right) \tag{2.134}$$

Now, the reader will probably be able to see there is a flaw here. Where? The error is easily recognized if one recalls the Debye argument for the average moment of a gas dipole. For what is the guarantee that a

---

[†] Notwithstanding the considerations of Section 2.5.8, the use of the *bulk* dielectric constant of water for dilute solutions of nonelectrolyte is not very inaccurate in the region *outside* the primary solvation sheath. The point is that, in this region (i.e., at distances > 5 to 10 Å from the ion's center), there is negligible structure breaking and, therefore, negligible decrease of dielectric constant from the bulk value.

ION–SOLVENT INTERACTIONS   161

water dipole far from the ion is aligned parallel to the ionic field? What about the thermal motions which tend to knock dipoles out of alignment? So what matters is the *average* dipole moment of the molecules in the direction of the ionic field. Thus, one has to follow the same line of reasoning as in the treatment of the dielectric constant of a polar liquid and think in terms of the average moment $\langle \mu \rangle$ of the individual molecules, which will depend in Debye treatment on the interplay of electrical and thermal forces and in Kirkwood treatment also on possible short-range interactions and association of dipoles. One has therefore

$$W_r = N_A \left(\int_0^{X_r} \langle \mu_W \rangle \, dX - \int_0^{X_r} \langle \mu_{NE} \rangle \, dX \right) \quad (2.135)$$

and, by making use of relations (2.79) and (2.97), which results in[†]

$$\langle \mu \rangle = \alpha X \quad (2.136)$$

one has

$$W_r = N_A \left(\int_0^{X_r} \alpha_W X \, dX - \int_0^{X_r} \alpha_{NE} X \, dX \right)$$

$$= \frac{N_A (\alpha_W - \alpha_{NE}) X_r^2}{2} \quad (2.137)$$

This expression for the work of replacing a water molecule by a non-electrolyte molecule at a distance r from an ion can now be introduced into Eq. (2.132) to give (in number of nonelectrolyte molecules per unit volume)

$$n_{NE,r} = n_{NE,b} \exp \left[\frac{(\alpha_{NE} - \alpha_W) X_r^2}{2kT} \right] \quad (2.138)$$

The exponent of Eq. (2.138) is easily shown to be less than unity at 25°C for most ions. Thus, for distances outside the primary hydration shell of nearly all ions, one can expand the exponential and retain only the first two terms, i.e.,

$$n_{NE,r} = n_{NE,b} \left[1 + \frac{(\alpha_{NE} - \alpha_W) X_r^2}{2kT} \right] \quad (2.139)$$

Thus, the *excess number* (not moles) per unit volume of nonelectrolyte molecules at a distance r from the ion is (again, in number of molecules

---

[†] The polarizability α used here refers to the orientation polarizability (see Section 2.5.4). Far away from the ion, the factor of deformation polarizability can be ignored.

per unit volume)

$$n_{NE,r} - n_{NE,b} = \frac{n_{NE,b}(\alpha_{NE} - \alpha_W)X_r^2}{2kT} \tag{2.140}$$

But this is only the excess number of nonelectrolyte molecules per unit volume *at a distance r* from the ion. What is required is the *total* excess number per unit volume throughout the region outside the primary solvation sheath, i.e., in region 2. One proceeds as follows: The excess *number*, *not* per unit volume, but in a spherical shell of volume $4\pi r^2 \, dr$ around the ion is

$$(n_{NE,r} - n_{NE,b})4\pi r^2 \, dr$$

and, therefore, the total excess number of nonelectrolyte molecule per ion in region 2 is (with r_h equal to the radius of the primary hydration sheath)

$$n_{NE,2} - n_{NE,b} = \int_{r_h}^{\infty} (n_{NE,r} - n_{NE,b})4\pi r^2 \, dr$$

$$= \int_{r_h}^{\infty} \frac{n_{NE,b}(\alpha_{NE} - \alpha_W)X_r^2}{2kT} 4\pi r^2 \, dr \tag{2.141}$$

If one sets $X_r = z_i e_0/(\varepsilon r^2)$, then, the excess number of nonelectrolyte molecules caused to be in solution per ion is

$$\frac{4\pi(z_i e_0)^2 n_{NE,b}(\alpha_{NE} - \alpha_W)}{2\varepsilon^2 kT} \int_{r_h}^{\infty} \frac{1}{r^2} \, dr = \frac{4\pi(z_i e_0)^2 n_{NE,b}}{2\varepsilon^2 kT r_h} (\alpha_{NE} - \alpha_W) \tag{2.142}$$

Hence, the excess number of nonelectrolyte molecules in a real solution containing c_i moles liter^{-1} of binary electrolyte is given by Eq. (2.142), multiplied by the number of moles per cubic centimeter in the solution, namely $N_A c_i/1000$. The expression is[†]

$$n_{NE} - n_{NE,b} = \frac{N_A c_i}{1000} \left[\frac{4\pi(z_i e_0)^2 n_{NE,b}(\alpha_{NE} - \alpha_W)}{\varepsilon^2 kT r_h} \right] \tag{2.143}$$

where n_{NE} is the number of nonelectrolyte molecules per cubic centimeter in the solution after addition of the electrolyte.

[†] A factor of 2 has been removed from the denominator of this equation, compared with Eq. (2.142), because there are *two* ions in the binary electrolyte, each of which is assumed to give the same effect on the solubility.

ION–SOLVENT INTERACTIONS   163

Hence

$$\frac{n_{NE} - n_{NE,b}}{n_{NE,b}} = \frac{N_A c_i}{1000}\left[\frac{4\pi(z_i e_0)^2(\alpha_{NE} - \alpha_W)}{\varepsilon^2 k T r_h}\right] \quad (2.144)$$

The terms n_{NE} and $n_{NE,b}$ are *numbers* of nonelectrolyte molecules per cubic centimeter. Hence, $N_A n_{NE} = S$ and $N_A n_{NE,b} = S_0$, where these terms have been defined in Section 2.6.2. It follows that

$$S = S_0 - \frac{S_0 N_A c_i}{1000}\left[\frac{4\pi(z_i e_0)^2(\alpha_W - \alpha_{NE})}{\varepsilon^2 k T r_h}\right] \quad (2.145)$$

This equation shows up well the sign of the effect of the secondary solvation. From (Section 2.5.4), and Eqs. (2.79) and (2.103) it is seen that the orientation polarizabilities, α_{NE} and α_W, are largely dependent on the square of the permanent dipole moments of the molecules. Water has, in comparison with many nonelectrolytes, the highest dipole moment. When, thus, $\alpha_W > \alpha_{NE}$, S is *less* than S_0 and there is salting out. HCN is an example of a substance the dipole moment of which is greater than that of water. (It masquerades as a nonelectrolyte because it is little dissociated in aqueous solution.) Appropriately, HCN is often salted in.

2.6.4. The Net Effect on Solubility of Influences from Primary and Secondary Solvation

Equation (2.130) can be written as

$$S_{\text{prim}} = S_0 - K_1 c_i \quad (2.146)$$

Equation (2.145) can be written as

$$S_{\text{sec}} = S_0 - K_2 c_i \quad (2.147)$$

The treatment of the effect of the secondary solvation has assumed that the primary solvational effects do not exist. In fact, the secondary solvational effects work on the diminished concentration of nonelectrolyte which arose because of the primary solvation. Hence,

$$S = S_{\text{prim}} - K_2 c_i \quad (2.148)$$

$$= S_0 - K_1 c_i - K_2 c_i \quad (2.149)$$

$$= S_0 - \frac{S_0 c_i n_s}{55.55} - \frac{N_A c_i S_0}{1000}\left[\frac{4\pi(z_i e_0)^2(\alpha_W - \alpha_{NE})}{\varepsilon^2 k T r_h}\right] \quad (2.150)$$

164   CHAPTER 2

S as written in (2.148) has taken into account the primary and secondary solvation and can be identified with the solubility of the nonelectrolyte after addition of ions to the solution. Hence,

$$\frac{S_0 - S}{S_0} = \frac{n_s c_i}{55.55} + \frac{N_A c_i}{1000}\left[\frac{4\pi(z_i e_0)^2(\alpha_W - \alpha_{NE})}{\varepsilon^2 kTr_h}\right] \quad (2.151)$$

If one writes

$$\frac{\Delta S}{S_0} = k_1 c_i \quad (2.152)$$

the k_1 is called Setchenow's constant.

There is fair agreement between Eq. (2.152) and experiment. If the nonelectrolyte has a dipole moment less than that of water, it salts out. In the rare cases in which there is a dipole moment in the nonelectrolyte greater than that of water, the nonelectrolyte salts in.

Salting out has practical implications. It is part of the electrochemistry of everyday industrial life. One reclaims solvents such as ether from aqueous solutions by salting them out with NaCl. Salting *out* enters into the production of soaps and the manufacture of dyes. Detergents, emulsion polymerization (rubber), and the concentration of antibiotics and vitamins from aqueous solutions, all depend in some part of their manufacture upon salting *in*. But the salting in with which they are associated is not the rare deviant phenomenon arising when the dipole moment of the nonelectrolyte is greater than that of water. It possesses the characteristic of always being associated with organic electrolytes in which the ions are *big*. Why do such ions (the dipole moment of which is *less* than that of water) cause "anomalous salting in"?

2.6.5. The Case of Anomalous Salting in

The picture given above seems satisfactory as a first approximation and for dilute solutions, but it has this one disturbing feature, namely, there are situations where theory predicts salting out, but experiment shows salting in. In such cases, the theory seems to favor ions' being surrounded by water, whereas, in fact, nonelectrolyte seems to accumulate near ions. Now, it will be recalled that only ion–dipole forces have been reckoned with in treating the interactions among the particles populating the primary solvation shell. Thus, the ion–water and ion–nonelectrolyte forces have been considered to be of the ion-dipole type of directional forces which orient polar particles along the ionic field. Perhaps this restriction is too severe, for there are also nondirectional forces, namely, dispersion forces.

These dispersion forces can be seen classically as follows: The *time-average* picture of an atom may show spherical symmetry because the charge due to the electrons orbiting around the nucleus is smoothed out in time. But an *instantaneous* picture of, say, a hydrogen atom would show a proton "here" and an electron "there"—two charges separated by a distance. Thus, every atom has an *instantaneous* dipole moment; of course, the time average of all these oscillating dipole moments is zero.

Then, an instantaneous dipole in one atom will induce an instantaneous dipole in a contiguous atom, and an instantaneous dipole–dipole force arises. When these forces are averaged over all instantaneous electron configurations of the atoms and thus over time, it is found that the time-averaged result of the interaction is finite, *attractive*, and nondirectional. Forces between particles, which arise in this way, are called *dispersion forces*.

The dispersion forces give rise to an interaction energy which has the same distance dependence as that due to dipole–dipole interactions. That is, the potential energy varies as r^{-6} and may be written as λ/r^6, where λ is a constant independent of r. The rapid decrease of such forces with increase of distance from the origin makes it unnecessary to consider dispersion interactions outside the primary solvation shell; they are by then already too feeble to warrant consideration. Inside the primary hydration sheath, the dispersion interaction can be treated in the same way as the ion–dipole interaction. That is, in the replacement of a water molecule by a nonelectrolyte molecule, one must take into account not only the difference in ion–dipole interactions, but also the difference between the dispersion interactions. Thus, one must now reconsider the situation in which the ion interacts with water molecules. The picture of the last three sections was that the water would usually be the entity to be predominantly attracted to the ion, so that the amount of water available for the dissolution of the nonelectrolyte would be *reduced*, and its solubility consequently would fall. The only exceptions which were recognized for this were those unusual cases in which the nonelectrolyte had a dipole moment greater than that of water. As to the ion, only its radius and charge played a part in the matter: it did not influence the situation in any more structural way, e.g., in terms of its polarizability, etc.

Now, however, with the introduction of dispersive force interactions in the competition of the water and the nonelectrolyte for ions, these considerations may have to be modified. On what molecular features do dispersion forces depend? What relative attractiveness has a given ion, on the one hand for a water molecule, and on the other for a nonelectrolyte?

Equations for the dispersive force interactions have been worked out for interactions in the gas phase. A simple equation would be

$$U = \frac{h\nu \alpha_{\text{d-ion}} \alpha_{\text{d-mol}}}{2R^6}$$

where the ν is the frequency of vibration of the electron in its lowest energy state, the α's are the *distortion polarizabilities* of the entities indicated, and R is their distance apart. It must be noted at once that the polarizability indicated here, the distortion polarizability, differs from that which enters into the equations for the dipole effects, which has simply been termed α. This latter α, which influences the theory of the effects of the secondary solvation upon salting out and salting in (Sections 2.6.3 and 2.6.4), is that due to the orientation of dipoles against the applied field; distortion polarizability, α_d, is that due to the stretching of the molecule under the influence of a field. The former polarizability, α, is simply connected to the dipole moment of the molecule according to a formula such as Eq.(2.103). But the latter, α_d, is more complexly connected to the *size* of the molecule. There is a parallelism with the radius, and in spherical symmetry cases it is found that α_d approximately follows r^3, where r is the radius of the molecule.

Now, if the size of the nonelectrolyte is greater than that of the water molecule (and for organic nonelectrolytes this is often so), it is clear from the above equation that the dispersive interaction of a given ion is going to be greater with the nonelectrolyte than with the water. This is a reversal of the behavior regarded as usual when only the permanent dipoles of the water and the nonelectrolyte are taken into account (for the dipole moment of water is higher than that of most nonelectrolytes). But it may be asked, in view of the above situation, why is not salting in (that which happens when the nonelectrolyte out-competes the water molecule in its attraction to the ion) the normal case? In this section, it is the dispersive interactions which have been the center of attention: one has suddenly, isolatedly considered them. But, one has to ask whether the dispersive ion–water (and dispersive ion–nonelectrolyte) interactions will dominate over the ion–dipole interactions which have been at the center of the stage in Sections 2.6.2–2.6.4. If the ion–dipole interactions dominate the ion-dispersive interactions, the considerations of those earlier sections are applicable, and salting out is the norm, with salting in a rare exception. When the dispersive interactions predominate, it is the other way around—salting in becomes the norm.

What factors of structure tend to make the dispersive forces dominate the situation? Clearly, they will be more likely to have the main influence upon the situation if the nonelectrolyte is big (because then the distortion polarizability is big); but there are many situations where quite big nonelectrolytes are still salted *out*. The dispersive interaction contains the *product* of the polarizability of both the ion and the nonelectrolyte (or the ion and water, depending upon which interaction one is considering), so that it is when *both* the ion *and* the nonelectrolyte are big (hence, both the α's large) that the dispersive situation has the likelihood of being that which dominates the issue, rather than the ion–dipole interaction.

In accordance with this it is found that, for example, if one maintains the nonelectrolyte constant[†] and varies the ion in size, though keeping it of the same type, salting in begins to predominate when the ion size exceeds a certain value. A good example is in the case of the ammonium ion, and a series of the tetra-alkyl ammonium ions with increasing size, i.e., NR_4^+, where R is CH_3, C_2H_5, etc. Here, the salting in already begins with the methyl ammonium ion (its α_d is evidently big enough), the ammonium ion alone giving salting out. The degree of salting in increases with increase of the size of the tetra-alkyl ammonium cation. Thus, the observations made at the end of Section 2.6.4 concerning the salting in of detergents, emulsions, and antibiotics by organic ions is, in principle, rationalized. An attempt has been made to make these considerations quantitative. Within the rough demands which the approximate nature of the equation for the dispersive energy allows, there is agreement between theory and experiment.

These considerations of the effect of ions upon the solubility of nonelectrolytes are sufficiently complicated to merit a little summary. The field is divided into two parts. The first part concerns systems in which the dispersive interactions are negligible compared with the dipole interactions. Such systems tend to contain relatively small ions and molecules. Here salting out is the expected phenomenon, and salting in occurs only in the rare case in which the nonelectrolyte dipole moment is greater than the dipole moment of the solvent. In the other division of the phenomena of solubility effects caused by ions, the substances concerned tend to be large, and, because distortion polarizability increases with size, this makes the dispersive energy interactions dominate. Then, salting in becomes the more expected situation.

[†] For example, it may be benzoic acid, considered a nonelectrolyte because it dissociates to a very small degree.

Further Reading

1. P. Debye and J. McAulay, *Z. Physik*, **26**: 22 (1927).
2. J. O'M. Bockris, J. Bowler-Reed, and J. A. Kitchener, *Trans. Faraday Soc.*, **47**: 184 (1951).
3. R. McDevit and F. Long, *Chem. Rev.*, **51**: 119 (1952).
4. B. E. Conway and J. O'M. Bockris, "Solvation," in: J. O'M. Bockris, ed., *Modern Aspects of Electrochemistry*, No. 1, Butterworth's Publications, Inc., London, 1954.
5. E. L. McBain and E. C. Hutchison, *Solubilization and Related Phenomena*, Academic Press, New York, 1955.
6. R. M. Diamond, *J. Phys. Chem.*, **67**: 2513 (1963).
7. B. E. Conway, J. E. Desnoyers, and A. C. Smith, *Phil. Trans. Roy. Soc. London*, **A256**: 389 (1964).
8. J. E. Desnoyers, C. Jolicoeur, and G. E. Pelletier, *Can. J. Chem.*, **42**: 3232 (1965).
9. W. W. Drost Hansen, *Advanced Chem.*, Ser. No. 67, 70–120 (1967).
10. W. Drost-Hansen, *Chem. Phys. Letters*, **2** (8): 647 (1968).

Appendix 2.1. Free Energy Change and Work

The free-energy change ΔG which a system undergoes in a process can be written quite generally as

$$\Delta G = \Delta E + p\,\Delta V + V\,\Delta p - T\,\Delta S - S\,\Delta T \qquad (A2.1.1)$$

If the process occurs at constant pressure and temperature,

$$\Delta p = 0 = \Delta T \qquad (A2.1.2)$$

and, therefore,

$$\Delta G = \Delta E + p\,\Delta V - T\,\Delta S \qquad (A2.1.3)$$

If, further, the process is reversible, the heat Q put *into* the system is related to the entropy change through

$$Q = T\,\Delta S \qquad (A2.1.4)$$

and, from the first law of thermodynamics,

$$\begin{aligned}\Delta E &= Q - W \\ &= T\,\Delta S - W\end{aligned} \qquad (A2.1.5)$$

where W is the total work done *by* the system.

Substituting for $T\,\Delta S$ from Eq. (A2.1.5) in Eq. (A2.1.3), one has

$$\Delta G = -(W - p\,\Delta V) \qquad (A2.1.6)$$

Since W is the total work (including mechanical work) and $p\,\Delta V$ is the mechanical work of volume expansion,

$$\Delta G = -(\text{work other than mechanical work done } by \text{ the system}) \qquad (A2.1.7)$$

or

$$\Delta G = \text{work other than mechanical work done } on \text{ the system} \qquad (A2.1.8)$$

Appendix 2.2. The Interaction between an Ion and a Dipole

The problem is to calculate the interaction energy between a dipole and an ion placed at a distance r from the dipole center, the dipole being oriented at an angle θ to the line joining the centers of the ion and dipole (Fig. A2.2.1). (By convention, the direction of the dipole is taken to be the direction from the negative end to the positive end of the dipole.)

The ion–dipole interaction energy U_{I-D} is equal to the charge $z_i e_0$ of the ion times the potential ψ_r due to the dipole at the site P of the ion

$$U_{I-D} = z_i e_0 \psi_r \qquad (A2.2.1)$$

Thus, the problem reduces to the calculation of the potential ψ_r due to the dipole. According to the law of superposition of potentials, the potential due to an assembly of charges is the sum of the potentials due to each charge. Thus, the potential due to a dipole is the sum of the potentials $+q/r_1$ and $-q/r_2$ due to the charges $+q$ and $-q$ which constitute the dipole and are located at distances r_1 and r_2 from the point P. Thus,

$$\psi_r = \frac{q}{r_1} - \frac{q}{r_2}$$

$$= q\left(\frac{1}{r_1} - \frac{1}{r_2}\right) \qquad (A2.2.2)$$

From Fig. A2.2.2, it is obvious that

$$r_1^2 = Y^2 + (z+d)^2 \qquad (A2.2.3)$$

and, therefore,

$$\begin{aligned}\frac{1}{r_1} &= [Y^2 + (z+d)^2]^{-\frac{1}{2}} \\ &= [(Y^2 + z^2) + d^2 + 2zd]^{-\frac{1}{2}} \\ &= (r^2 + d^2 + 2zd)^{-\frac{1}{2}} \\ &= \frac{1}{r}\left[1 + \left(\frac{d}{r}\right)^2 + \frac{2dz}{r^2}\right]^{-\frac{1}{2}}\end{aligned} \qquad (A2.2.4)$$

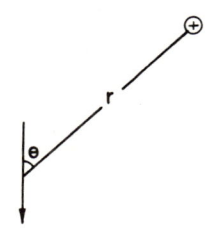

Fig. A2.2.1.

170 CHAPTER 2

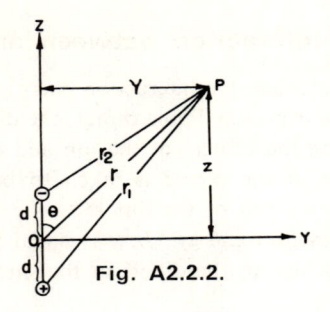

Fig. A2.2.2.

At this stage, an important approximation is made, namely, that the distance $2d$ between the charges in the dipole is negligible compared with r. In other words, the approximation is made that

$$1 + \left(\frac{d}{r}\right)^2 + \frac{2dz}{r^2} \sim 1 + \frac{2dz}{r^2} \tag{A2.2.5}$$

It is clear that the validity of the approximation decreases the closer the ion comes toward the dipole, i.e., as r decreases.

Making the above approximation, one has [see Eq. (A.2.2.4)]

$$\frac{1}{r_1} \sim \frac{1}{r}\left(1 + \frac{2dz}{r^2}\right)^{-\frac{1}{2}} \tag{A2.2.6}$$

which, by the binomial expansion taken to two terms, gives

$$\frac{1}{r_1} = \frac{1}{r}\left(1 - \frac{dz}{r^2}\right) \tag{A2.2.7}$$

By similar reasoning,

$$\frac{1}{r_2} = \frac{1}{r}\left(1 + \frac{dz}{r^2}\right) \tag{A2.2.8}$$

By using Eqs. (A2.2.7) and (A2.2.8), Eq. (A2.2.2) becomes

$$\psi_r = -\frac{2dq}{r^2}\frac{z}{r} \tag{A2.1.9}$$

Since $z/r = \cos\theta$ and $2dq$ is the dipole moment μ.

$$\psi_r = -\frac{\mu\cos\theta}{r^2} \tag{A2.2.10}$$

or the ion–dipole interaction energy is given by

$$U_{I-D} = \frac{-z_i e_0 \mu \cos\theta}{r^2} \tag{A2.2.11}$$

ION–SOLVENT INTERACTIONS   171

Appendix 2.3. The Interaction between an Ion and a Water Quadrupole

Instead of presenting a sophisticated general treatment for ion–quadrupole interactions, a particular case of these interactions will be worked out. The special case to be worked out is that corresponding to the water molecule being oriented with respect to a positive ion so that the interaction energy is a minimum.

In this orientation (see Fig. A2.3.1), the oxygen atom and a positive ion are on the Y axis which bisects the H—O—H angle. Further, the positive ion, the oxygen atom, and the two hydrogen atoms are all considered in the XY plane. The origin of the XY coordinate system is located at the point Q, which is the center of the water molecule. The ion is at a distance r from the origin.

The ion–quadrupole interaction energy U_{I-Q} is simply given by the charge on the ion times the potential ψ_r at the site of the ion due to the charges of the quadrupole,

$$U_{I-Q} = z_i e_0 \psi_r \qquad (A2.3.1)$$

But the potential ψ_r is the sum of the potentials due to the four charges $q_1, q_2, q_3,$ and q_4 in the quadrupole (1 and 2 are the positive charges at the hydrogen, and 3 and 4 are the negative charges at the oxygen). That is,

$$\psi_r = \psi_1 + \psi_2 + \psi_3 + \psi_4 \qquad (A2.3.2)$$

Each one of these potentials is given by the usual coulombic expression for the potential

$$\psi_r = \frac{q_1}{r_1} + \frac{q_2}{r_2} - \frac{q_3}{r_3} - \frac{q_4}{r_4} \qquad (A2.3.3)$$

where the minus sign appears before the third and fourth terms because q_3 and q_4

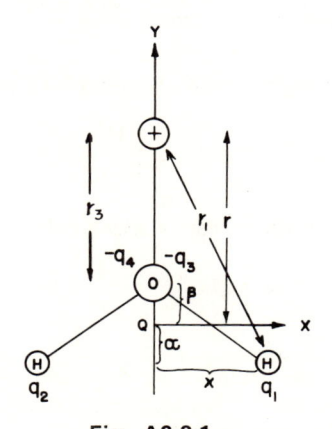

Fig. A2.3.1.

are negative charges. Further, the *magnitudes* of all the charges are equal

$$|q_1| = |q_2| = |q_3| = |q_4| = q \tag{A2.3.4}$$

and, because of symmetrical disposition of the water molecule,

$$r_2 = r_1 \quad \text{and} \quad r_4 = r_3 \tag{A2.3.5}$$

Hence, from Eqs. (A2.3.3), (A2.3.4), and (A2.3.5),

$$\psi_r = 2q\left(\frac{1}{r_1} - \frac{1}{r_3}\right) \tag{A2.3.6}$$

It is obvious (see Fig. A2.3.1) that

$$r_1^2 = (r + \alpha)^2 + x^2 \tag{A2.3.7}$$

$$= r^2\left(1 + \frac{\alpha^2 + x^2}{r^2} + \frac{2\alpha}{r}\right) \tag{A2.3.8}$$

and

$$r_3 = r - \beta \tag{A2.3.9}$$

$$= r\left(1 - \frac{\beta}{r}\right) \tag{A2.3.10}$$

Thus,

$$\frac{1}{r_1} = \frac{1}{r}\left(1 + \frac{\alpha^2 + x^2}{r^2} + \frac{2\alpha}{r}\right)^{-\frac{1}{2}} \tag{A2.3.11}$$

and

$$\frac{1}{r_3} = \frac{1}{r}\left(1 - \frac{\beta}{r}\right)^{-1} \tag{A2.3.12}$$

One can now use the binomial expansion, i.e.,

$$(1 \pm m)^{-n} = 1 \mp nm + \frac{n(n+1)}{2}m^2 \mp \cdots \tag{A2.3.13}$$

and drop off all terms higher than the *third*.[†] Thus,

$$\frac{1}{r_1} \sim \left\{\frac{1}{r} - \frac{1}{2}\frac{\alpha^2 + x^2}{r^3} - \frac{\alpha}{r^2} + \frac{3}{8}\left[\frac{(\alpha^2 + x^2)^2}{r^5} + \frac{4\alpha^2}{r^3} + \frac{4\alpha(\alpha^2 + x^2)}{r^4}\right]\right\} \tag{A2.3.14}$$

and, omitting all terms with powers r greater than 3, one has

$$\frac{1}{r_1} \approx \frac{1}{r} - \frac{\alpha}{r^2} + \frac{1}{2r^3}(2\alpha^2 - x^2) \tag{A2.3.15}$$

[†] It is at this stage that the treatment of ion–quadrupole interactions diverges from that of ion–dipole interactions (*cf.* Appendix 2.2). In the latter, the binomial expansion was terminated after the second term.

Further,

$$\frac{1}{r_3} \sim \frac{1}{r}\left(1 + \frac{\beta}{r} + \frac{\beta^2}{r^2}\right)$$
$$= \frac{1}{r} + \frac{\beta}{r^2} + \frac{\beta^2}{r^3} \qquad (A2.3.16)$$

Subtracting Eq. (A2.3.16) from Eq. (A2.3.15), one has

$$\frac{1}{r_1} - \frac{1}{r_3} = -\frac{(\alpha + \beta)}{r^2} + \frac{1}{2r^3}(2\alpha^2 - x^2 - 2\beta^2) \qquad (A2.3.17)$$

and, therefore, by substitution of (A2.3.17) in (A2.3.6),

$$\psi_r = -\frac{2q(\alpha + \beta)}{r^2} + \frac{1}{2r^3}[2(2q\alpha^2 - 2q\beta^2) - 2qx^2] \qquad (A2.3.18)$$

The first term on the right-hand side of (A2.3.18) can be rearranged as follows:
$$2q(\alpha + \beta) = 2q\alpha + 2q\beta$$
$$= \Sigma (2q)d \quad \text{where} \quad d = \alpha \text{ or } \beta \qquad (A2.3.19)$$

Thus, as a first approximation, the water molecule can be represented as a dipolar charge distribution in which there is a positive charge of $+2q$ (due to the H atoms) at a distance α from the origin on the bisector of the H—O—H angle and a charge of $-2q$ (due to the lone electron pair) at a distance $-\beta$ from the origin, it follows that

$$\Sigma (2q)d = \Sigma \text{ magnitude of each charge of dipole}$$
$$\times \text{ distance of the charge from origin} \qquad (A2.3.20)$$

The right-hand side of this expression is the *general* expression for the dipole moment μ, as is seen by considering the situation when $\alpha = \beta$, i.e., $\Sigma (2q)d = (2q)2d$, where $2d$ is the distance between the charges of the dipole, in which case one obtains the familiar expression for the dipole moment μ,

$$\mu = 2q2d \qquad (A2.3.21)$$

Thus, the first term on the right-hand side of Eq. (A2.3.18) is

$$-\frac{2q(\alpha + \beta)}{r^2} = -\frac{\mu}{r^2} \qquad (A2.3.22)$$

The second term can be interpreted as follows: Consider $(2q\alpha^2 - 2q\beta^2)$. It can be written thus

$$2q\alpha^2 - 2q\beta^2 = q\alpha^2 + q\alpha^2 + (-q)\beta^2 + (-q)\beta^2$$
$$= \Sigma qd_y^2 \quad \text{where} \quad d_y = \alpha \text{ or } \beta \qquad (A2.3.23)$$

174 CHAPTER 2

But the general definition of a quadrupole moment is the magnitude of each charge of quadrupole times the *square* of the distance of the charge from the origin. Thus, $\Sigma\, qd_y^2$ is the y component p_{yy} of the quadrupole moment for the particular coordinate system which has been chosen. Similarly, $\Sigma\, 2qx^2$ is the x component p_{xx} of the quadrupole moment. Thus,

$$2(2q\alpha^2 - 2q\beta^2) - 2qx^2 = 2p_{yy} - p_{xx} \quad (A2.3.24)$$

One can combine $2p_{yy} - p_{xx}$ into a single symbol and talk of the quadrupole moment p_W of the water molecule in the particular orientation of Fig. A2.3.1. Hence,

$$2(2q\alpha^2 - 2q\beta^2) - 2qx^2 = p_W \quad (A2.3.25)$$

and, therefore, by substituting (A2.3.22) and (A2.3.25) in (A2.3.18),

$$\psi_r = -\frac{\mu}{r^2} + \frac{p_W}{2r^3} \quad (A2.3.26)$$

The ion–quadrupole interaction energy [*cf.* Eq. (A2.3.1)] thus becomes

$$U_{I-Q} = -\frac{z_i e_0 \mu}{r^2} + \frac{z_i e_0 p_W}{2r^3} \quad (A2.3.27)$$

When a negative ion is considered, the water molecule turns around through π, and one obtains by an argument similar to that for positive ions

$$U_{I-Q} = -\frac{z_i e_0 \mu}{r^2} - \frac{z_i e_0 p_W}{2r^3} \quad (A2.3.28)$$

CHAPTER 3
ION–ION INTERACTIONS

3.1. INTRODUCTION

A model has been given for the breaking-up of an ionic crystal into free ions which stabilize themselves in solution with solvent sheaths. One central theme guided the account, the interaction of an ion with its neighboring water molecule.

But ion–solvent interactions are only part of the story relating an ion to its environment. When an ion looks out upon its surroundings, it sees not only solvent dipoles but also other ions. The mutual interaction between these ions constitutes an essential part of the picture of an electrolytic solution.

Why are ion–ion interactions of importance? Because, as will be shown, they affect the equilibrium properties of ionic solutions, and also because they interfere with the drift of ions, for instance, under an externally applied electric field (Chapter 4).

Now, the degree to which these interactions affect the properties of solutions will depend on the mean distance apart of the ions, i.e., on how densely the solution is populated with ions, because the interionic fields are distance dependent. This ionic population density will in turn depend on the nature of the electrolyte, i.e., on the extent to which the electrolyte gives rise to ions in solution.

3.2. TRUE AND POTENTIAL ELECTROLYTES

3.2.1. Ionic Crystal Are True Electrolytes

An important point to recall regarding the dissolution of an ionic crystal (Chapter 2) is that *ionic* lattices consist of ions *even before they come into contact with a solvent*. In fact, all that a polar solvent does is to use ion–dipole (or ion–quadrupole) forces to disengage the ions from the lattice sites, solvate them, and disperse them in solution.

Such ionic crystals are known as *true* electrolytes or *ionophores* (the Greek suffix *phore* means "bearer of"; thus, an ionophore is a "substance which bears ions"). When a true electrolyte is melted, its ionic lattice is dismantled and the *pure* liquid true electrolyte shows considerable ionic conduction (Chapter 2). Thus, the characteristic of a *true* electrolyte is that, in the pure liquid form, it is an ionic conductor. All salts belong to this class. Sodium chloride, therefore, is a typical true electrolyte.

3.2.2. Potential Electrolytes: Nonionic Substances Which React with the Solvent to Yield Ions

A large number of substances, e.g., organic acids, show little conductivity in the pure liquid state. Evidently, there must be some fundamental difference in structure between organic acids and inorganic salts, and this difference is responsible for the fact that one pure liquid (the true electrolyte) is an ionic conductor and the other is not.

What is this difference between, say, sodium chloride and acetic acid? Electron diffraction studies furnish an answer. They show that gaseous acetic acid consists of *separate, neutral* molecules and the bonding of the atoms inside these molecules is essentially nonionic. These neutral molecules retain their identity and separate existence when the gas condenses to give liquid acetic acid. Hence, there are hardly any ions in liquid acetic acid and, therefore, little conductivity.

Now, the first requirement of an electrolyte is that it should give rise to a conducting solution. From this point of view, it appears that acetic acid will never answer the requirements of an electrolyte; it is nonionic. When, however, acetic acid is dissolved in water, an interesting phenomenon occurs: ions are *produced*, and, therefore, the solutions conduct electricity. Thus, acetic acid, too, is a type of electrolyte; it is not a true electrolyte, but a *potential* one ("one which can, but has not yet, become"). Potential electrolytes are also called *ionogens*, i.e., "ion producers."

How does acetic acid, which does not consist of ions in the pure liquid state, generate ions when dissolved in water? In short, how do potential electrolytes work? Obviously, there must be some reaction between neutral acetic acid molecules and water, and this reaction must lead to the splitting of the acetic acid molecules into charged fragments, or ions.

A simple picture is as follows. Suppose that an acetic acid molecule collides with a water molecule and, in the process, the H of the acetic acid OH group is transferred from the oxygen atom of the OH to the oxygen atom of the H_2O. A proton has been transferred from CH_3COOH to H_2O

$$\begin{array}{c}H\\|\\H-C-C\\|\ \ \ \ \ \diagdown\\H\ \ \ \ OH\end{array}\begin{array}{c}O\\\diagup\\\\ \diagdown\end{array}+\begin{array}{c}H\\\diagup\\O\\\diagdown\\H\end{array}\longrightarrow\left[\begin{array}{c}H\\|\\H-C-C\\|\ \ \ \ \ \diagdown\\H\ \ \ \ O\end{array}\begin{array}{c}O\\\diagup\\\\\end{array}\right]^-+\left[\begin{array}{c}\\H-O\\\diagdown\\H\end{array}\begin{array}{c}H\\\diagup\\\\\end{array}\right]^+$$

The result of the proton transfer is that two ions have been produced: (1) an acetate ion and (2) a hydrated proton. Thus, potential electrolytes (organic acids and most bases) dissociate into ions by ionogenic, or ion-forming, *chemical reactions* with solvent molecules, in contrast to true electrolytes, which give rise to ionic solutions by *physical interactions* between ions present in the ionic crystal and solvent molecules (Fig. 3.1).

The mechanism of the functioning of potential electrolytes will be described in detail later (Chapter 5).

3.2.3. An Obsolete Classification: Strong and Weak Electrolytes

The classification into true and potential electrolytes is a modern one. It is based on a knowledge of the structure of the electrolyte: whether, in the pure form, it consists of an ionic lattice or neutral molecules (Fig. 3.2). It is not based on the behavior of the solute in any particular solvent.

Historically, however, the classification of electrolytes was made on the basis of their behavior in one particular solvent, i.e., water. *Weak* electrolytes were those which yielded relatively poorly conducting solutions when dissolved in *water*, and *strong* electrolytes were those which gave highly conducting solutions when dissolved in water.

The disadvantage of this classification into strong and weak electrolytes lies in the following fact: As soon as a different solvent, i.e., a nonaqueous solvent, is chosen, what was a strong electrolyte in water may behave as a weak electrolyte in the nonaqueous solvent. For example, sodium chloride behaves like a strong electrolyte (i.e., yields highly conducting solutions) in water; and acetic acid, like a weak electrolyte. In liquid ammonia, however, the conductance behavior of acetic acid is similar to that of sodium

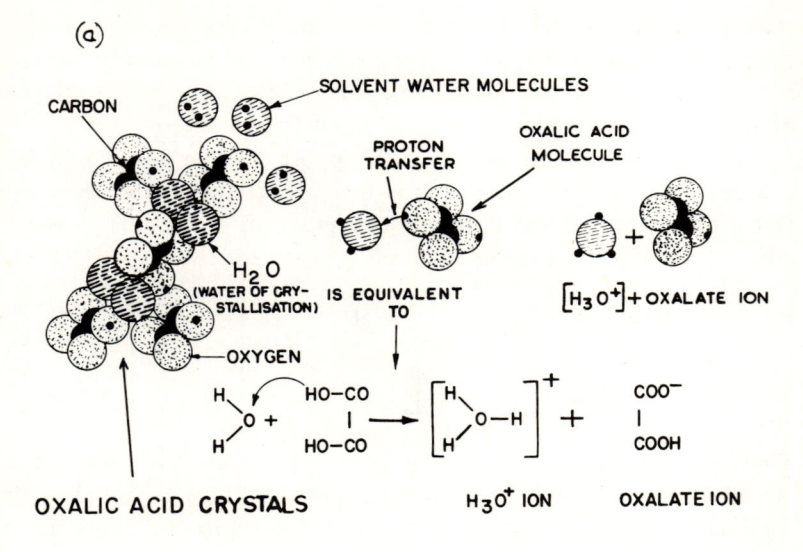

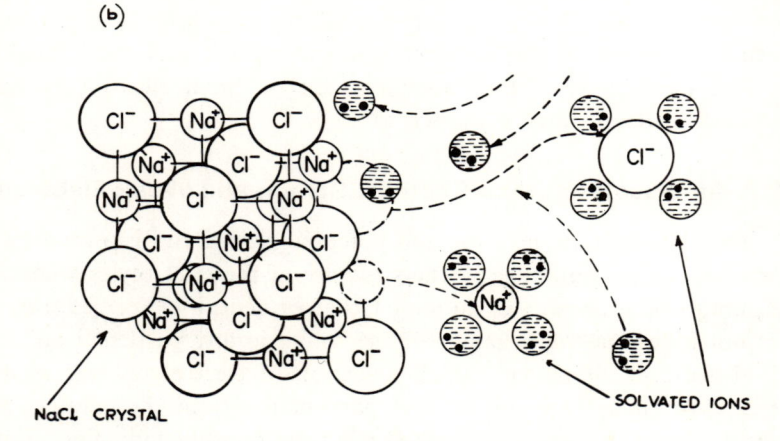

Fig. 3.1. Schematic diagram to illustrate the difference in the way potential electrolytes and true electrolytes dissolve to give ionic solutions: (a) Oxalic acid (a potential electrolyte) undergoes a proton-transfer chemical reaction with water to give rise to hydrogen ions and oxalate ions. (b) Sodium chloride (a true electrolyte) dissolves by the solvation of the Na+ and Cl+ ions in the crystal.

chloride in water, i.e., the solutions are highly conducting (Table 3.1). This is an embarrassing situation. Can one say: Acetic acid is weak in water and strong in liquid ammonia? What is wanted is a classification

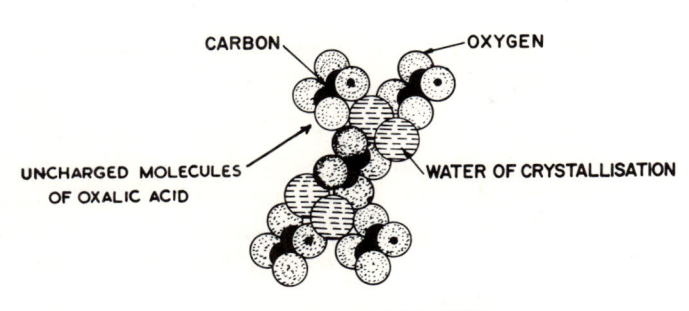

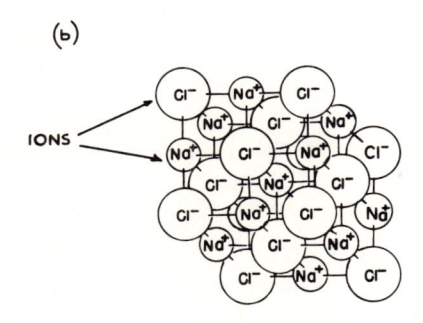

ION–ION INTERACTIONS

Fig. 3.2. Electrolytes can be classified as (a) potential electrolytes (e.g., oxalic acid) which, in the pure state, consist of uncharged molecules and (b) true electrolytes (e.g., sodium chloride) which, in the pure state, consist of ions.

TABLE 3.1

Conductance Behavior of Substances in Different Media

	Equivalent conductance	
	Water	Liquid ammonia
NaCl	106.7	284.0
Acetic acid	4.7	216.6

of electrolytes which is independent of the solvent concerned. The classification into true and potential electrolytes is such a classification. It does not depend on the solvent.

3.2.4. The Nature of the Electrolyte and the Relevance of Ion–Ion Interactions

Solutions of most potential electrolytes in *water* generally contain only small concentrations of ions, and, therefore, ion–ion interactions in these solutions are negligible; the ions are on the average too far apart. The behavior of such solutions is governed predominantly by the position of the equilibrium in the proton-transfer reaction between the potential electrolyte and water (see Chapter 5).

In contrast, true electrolytes are completely dissociated into ions when the parent salts are dissolved in water. The resulting solutions generally consist only of solvated ions and solvent molecules. The dependence of many of their properties on concentration (and, therefore, mean distance apart of the ions in the solution) is determined, therefore, by the interactions between ions. To understand these properties, one must understand ion–ion interactions.

Further Reading

1. G. Kortum and J. O'M. Bockris, *Textbook of Electrochemistry*, Vol. I, Elsevier, Amsterdam, 1951.
2. R. M. Fuoss and F. Accascina, *Electrolytic Conductance*, Interscience Publishers, Inc., New York, 1959.

3.3. THE DEBYE–HÜCKEL (OR ION-CLOUD) THEORY OF ION–ION INTERACTIONS

3.3.1. A Strategy for a Quantitative Understanding of Ion–Ion Interactions

The first task in thinking in detail about ion–ion interactions is to evolve a quantitative measure of these interactions.[†]

One approach is to follow a procedure similar to that used in the discussion of ion–solvent interactions (*cf.* Section 2.2.1). Thus, one can consider an initial state in which ion–ion interactions do not exist (are

---

[†] The question of how one obtains an experimental measure of ion–ion interactions is discussed in Section 3.4.

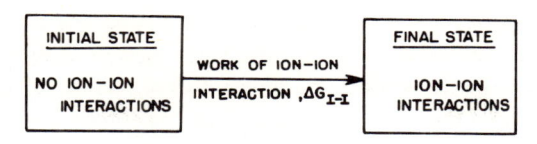

Fig. 3.3 The free energy ΔG_{I-I} of ion–ion interactions is the free-energy change in going from a hypothetical electrolytic solution, in which ion–ion interactions do not operate, to a real solution, in which these interactions do operate.

"switched off") and a final state in which the interactions are in play (are "switched on"). Then, the free-energy change in going from the initial state to the final state can be considered the free energy ΔG_{I-I}, of ion–ion interactions (Fig. 3.3).

The final state is obvious; it is ions in solution. The initial state is not so straightforward; one cannot take ions in vacuum, because then there will be ion–solvent interactions when these ions enter the solvent. The following approach is therefore adopted. One conceives of a *hypothetical* situation in which the ions are there in solution but are nevertheless not interacting. Now, if ion–ion interactions are assumed to be electrostatic in origin (a similar assumption was made with regard to ion–solvent interactions, *cf.* Section 2.2.2), then the imaginary initial state of noninteracting ions implies an assembly of *discharged* ions.

Thus, the process of going from an initial state of noninteracting ions to a final state of ion–ion interactions is equivalent to taking an assembly of discharged ions, charging them up, and setting the electrostatic charging work equal to the free energy ΔG_{I-I} of ion–ion interactions (Fig. 3.4).

One point about the above procedure should be borne in mind. Since,

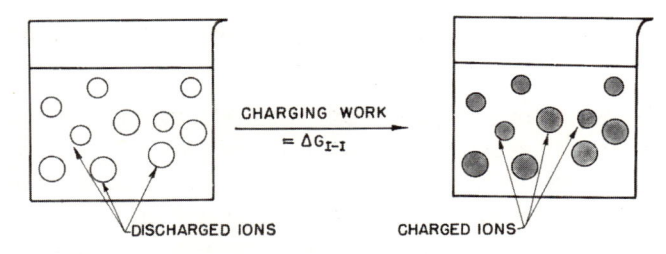

Fig. 3.4. The free energy ΔG_{I-I} of ion–ion interactions is the electrostatic work of taking an imaginary assembly of discharged ions and charging them up to obtain a solution of charged ions.

in the charging process, both the positively charged and negatively charged ionic species are charged up, one obtains a free-energy change which involves *all* the ionic species constituting the electrolyte. Generally, however, the desire is to isolate the contribution to the free energy of ion–ion interactions arising from one ionic species i only. This partial free-energy change is, by definition, the *chemical-potential change* $\Delta\mu_{i-I}$ arising from the interactions of one ionic species with the ionic assembly.

To compute this chemical-potential change $\Delta\mu_{i-I}$, rather than the free-energy change ΔG_{I-I}, one must adopt an approach similar to that used in the Born theory of solvation. One thinks of an ion of species i and imagines that this reference ion alone of all the ions in solution is in a state of zero charge (Fig. 3.5). If one computes the work of charging up the reference ion (of radius r_i) from a state of zero charge to its final charge of $z_i e_0$, then the charging work W times the Avogadro number N_A is equal to the partial molar free energy of ion–ion interactions, i.e., to the chemical potential of ion–ion interactions

$$\Delta\mu_{i-I} = N_A W \qquad (3.1)$$

Further, one can consider a charged sphere (of radius r_i and charge $z_i e_0$) as a model for an ion (*cf.* Section 2.2.2) and use the expression for the work of charging the sphere from a state of zero charge to a charge of $z_i e_0$ to represent the work W of charging an ion, i.e.,

$$W = \frac{(z_i e_0)^2}{2\varepsilon r_i} \qquad (2.16)$$

$$= \frac{z_i e_0}{2} \frac{z_i e_0}{\varepsilon r_i} \qquad (3.2)$$

But $z_i e_0/\varepsilon r_i$ is the electrostatic potential ψ at the surface of the ion, and, therefore,

$$\Delta\mu_{i-I} = N_A W = \frac{N_A z_i e_0}{2} \psi \qquad (3.3)$$

The essence of the task, therefore, in computing the chemical-potential change due to the interactions of the ionic species i with the ionic solution, is the calculation of the electrostatic potential produced at a reference ion by the rest of the ions in solution. Theory must aim at this quantity.

If one knew the *time-average spatial distribution* of the ions, then one could find out how all the other charges are distributed as a function of distance from the reference ion. At that stage, one of the fundamental laws of electrostatics could be used, namely, the law of the superposition of

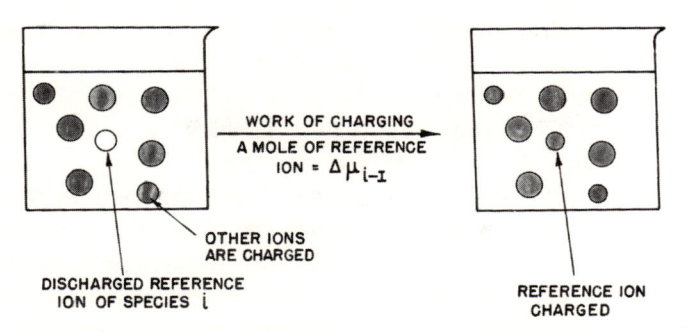

Fig. 3.5. The chemical potential $\Delta\mu_{i-I}$ arising from the interactions of an ionic species i with the electrolytic solution is equal to the Avogadro number times the electrostatic work of taking an imaginary solution in which one reference ion alone is discharged and charging this reference ion up to its normal charge.

potentials, according to which the potential at a point due to an assembly of charges is the sum of the potentials due to each of the charges in the assembly.

Thus, the problem of calculating the chemical-potential change $\Delta\mu_{i-I}$ due to the interactions between one ionic species and the assembly of all the other ions has been reduced to the following problem: On a time average, how are the ions distributed around any specified ion? If that distribution becomes known, it would be easy to calculate the electrostatic potential of the specified ion, due to the other ions and then, by Eq. (3.3), the energy of that interaction. Thus, the task is to develop a model that describes the equilibrium spatial distribution of ions inside an electrolytic solution and then to describe the model mathematically.

3.3.2. A Prelude to the Ionic-Cloud Theory

A spectacular advance in the understanding of the distribution of charges around an ion in solution was achieved in 1923 by Debye and Hückel. It was as significant in the understanding of ionic solutions as the Maxwell theory of the distribution of velocities in the understanding of gases.

Before going into the details of their theory, a moment's reflection on the magnitude of the problem would promote appreciation of their achievement.

Consider, for example, a 10^{-3} mole liter^{-1} aqueous solution of sodium chloride. There will be $10^{-6} \times 6.023 \times 10^{23}$ sodium ions per cubic centimeter of solution and the same number of chloride ions, together, of course, with the corresponding number of water molecules. Nature takes these $2 \times 6.023 \times 10^{17}$ ions cm^{-3} and arranges them so that there is a particular time-average† spatial distribution of the ions. The number of particles involved are enormous, and the situation appears far too complex for mathematical treatment.

But there exist conceptual techniques for tackling complex situations. One of them is model building. What is done is to conceive a model which contains only the *essential* features of the real situation. All the thinking and mathematical analysis is done on the (relatively simple) model, and, then, the theoretical predictions are compared with the experimental behavior of the real system. A good model simulates nature. If the model yields wrong answers, then one tries again by changing the imagined model until one arrives at a model, the theoretical predictions of which agree well with experimental observations.

The genius of Debye and Hückel lay in their formulation of a very simple but powerful model for the time-average distribution of ions in very dilute solutions of electrolytes. From this distribution, they were able to get the electrostatic potential contributed by the surrounding ions to the total electrostatic potential at the reference ion and, hence, the chemical-potential change arising from ion–ion interactions [*cf.* Eq.(3.3)]. Attention will now be focused on their approach.

The electrolytic solution consists of solvated ions and water molecules. The first step in the Debye–Hückel approach is to select arbitrarily any one ion out of the assembly and call it a *reference ion* or *central ion*. Only the reference ion is given the individuality of a discrete charge. What is done with the water molecules and the remaining ions? The water molecules are looked upon as a continuous dielectric medium. The remaining ions of the solution (i.e., all ions except the central ion) are lapsed into anonymity, their charges being "smeared out" into a continuous spatial distribution of charge (Fig. 3.6). Whenever the concentration of ions of one sign exceeds that of the opposite sign, there will arise a *net* or excess charge in the particular region under consideration. Obviously, the total charge in the atmosphere must be of opposite sign and exactly equal to the charge on the reference ion.

---

† Using an imaginary camera (with exposure time of $\sim 10^{-12}$ sec), suppose that it were possible to take snapshots of the ions in an electrolytic solution. Different snapshots would show the ions distributed differently in the space containing the solution; but the scrutiny of a large enough number of snapshots (say, $\sim 10^{12}$) would permit one to recognize a certain average distribution characterized by average positions of the ions; this is the time-average spatial distribution of the ions.

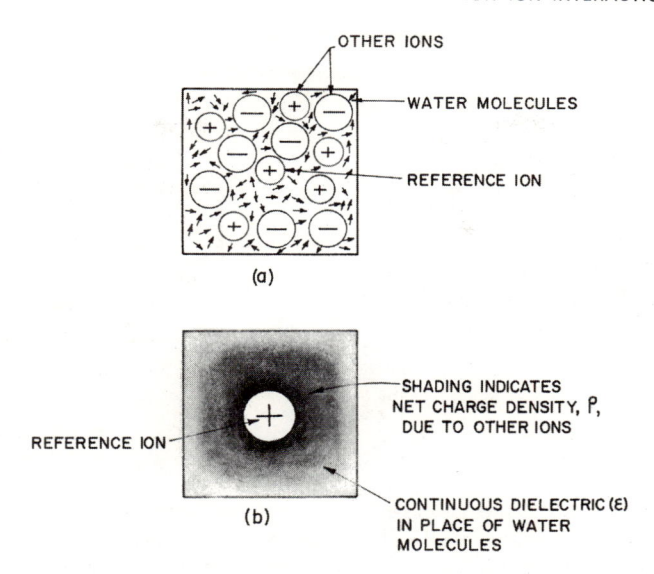

Fig. 3.6. A schematic comparison of (a) the assembly of ions and solvent molecules which constitute a real electrolytic solution and (b) the Debye–Hückel picture in which a reference ion is surrounded by net charge density ϱ due to the surrounding ions and a dielectric continuum of the same dielectric constant ε as the bulk solvent.

Thus, the electrolytic solution is considered to consist of a central ion standing alone in a continuum. Thanks to the water molecules, this continuum acquires a dielectric constant (taken to be the value for bulk water). The charges of the discrete ions which populate the environment of the central ion are thought of as smoothed out and contribute to the continuum dielectric a *net* charge density (*excess* charge per unit volume). Thus, water enters the analysis in the guise of a dielectric constant ε; and the ions, except the specific one chosen as the central ion, in the form of an excess charge density ϱ (Fig. 3.7).

Thus, the complicated problem of the time-average distribution of ions inside an electrolytic solution reduces, in the Debye–Hückel model, to the mathematically simpler problem of finding out how the excess charge density ϱ varies with distance r from the central ion.

An objection may be raised at this point. The electrolytic solution as a whole is electroneutral, i.e., the net charge density ϱ is zero. Then, why is not $\varrho = 0$ everywhere?

So as not to anticipate the detailed discussion, an intuitive answer

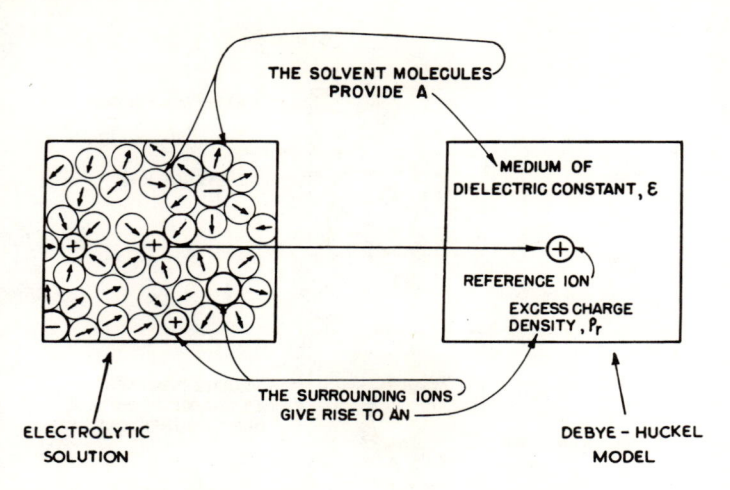

Fig. 3.7. The Debye-Hückel model is based upon selecting one ion as a reference ion, replacing the solvent molecules by a continuous medium of dielectric constant ε and the remaining ions by an excess charge density ϱ_r (the shading used in this book to represent the charge density is not indicated in this figure).

will first be given. If the central ion is, for example, positive, it will exert an attraction for negative ions; hence, there should be a greater aggregation of negative ions than of positive ions in the neighborhood of the central positive ion, i.e., $\varrho \neq 0$. An analogous situation, but with a change in sign, obtains near a central negative ion. At the same time, the thermal forces are knocking the ions about in all directions and trying to establish electroneutrality, i.e., the thermal motions try to smooth everything to $\varrho = 0$. Thus, the time average of the electrostatic forces of ordering and the thermal forces of disordering is a *local* excess of negative charge near a positive ion and an excess of positive charge near a negative ion. Of course, the excess positive charge near a negative ion compensates the excess negative charge near a positive ion, and the overall effect is electroneutrality, i.e., a ϱ of zero for the whole solution.

3.3.3. How the Charge Density near the Central Ion Is Determined by Electrostatics: Poisson's Equation

Consider an infinitesimally small volume element dv situated at a distance r from the arbitrarily selected central ion, upon which attention is to be fixed during the discussion (Fig. 3.8), and let the net charge density

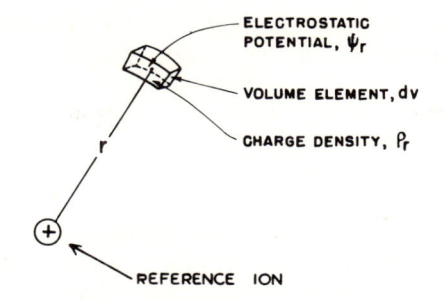

Fig. 3.8. At a distance r from the reference ion, the excess charge density and electrostatic potential, in an infinitesimal volume element dv, are ϱ_r and ψ_r, respectively.

inside the volume element be ϱ_r. Further, let the average[†] electrostatic potential in the volume element be ψ_r. The question is: What is the relation between the excess density ϱ_r in the volume element and the time-average electrostatic potential ψ_r?

One relation between ϱ_r and ψ_r is given by Poisson's equation (Appendix 3.1). There is no reason to doubt that there is spherically symmetrical distribution of positive and negative charge and, therefore, excess charge density around a given central ion. Hence, Poisson's equation can be written as

$$\frac{1}{r^2}\frac{d}{dr}\left(r^2\frac{d\psi}{dr}\right) = -\frac{4\pi}{\varepsilon}\varrho_r \qquad (3.4)$$

where ε is the dielectric constant of the medium and is taken to be that of bulk water.

3.3.4. How the Excess Charge Density near the Central Ion Is Given by a Classical Law for the Distribution of Point Charges in a Coulombic Field

The excess charge density in the volume element dv is equal to the total ion density (total number of ions per unit volume) times the charge on these ions. Let there be, *per unit volume*, n_1 ions of type 1, each bearing charge z_1e_0, n_2 of type 2 with charge z_2e_0, and n_i of type i with charge z_ie_0,

---

[†] Actually, there are discrete charges in the neighborhood of the central ion and, therefore, discontinuous variations in the potential. But, because in the Debye–Hückel model the charges are smoothed out, the potential is averaged out.

where z_i is the valency of the ion and e_0 is the electronic charge. Then, the excess charge density ϱ_r in the volume element dv is given by

$$\varrho_r = n_1 z_1 e_0 + n_2 z_2 e_0 + \cdots + n_i z_i e_0 \tag{3.5}$$

$$= \sum n_i z_i e_0 \tag{3.6}$$

To proceed further, one must link up the unknown quantities n_1, $n_2, \ldots, n_i, \ldots$ to known quantities. The link is made on the basis of the Boltzmann distribution law of classical statistical mechanics. Thus, one writes

$$n_i = n_i^0 e^{-U/(kT)} \tag{3.7}$$

where U can be described either as the change in potential energy of the i particles when their concentration in the volume element dv is changed from the bulk value n_i^0 to n_i or as the work that must be done by a hypothetical external agency against the time average of the electrical and other forces between ions in producing the above concentration change. Since the potential energy U relates to the time average of the forces between ions rather than to the actual forces for a given distribution, it is also known as the *potential of average force*.

If there are no ion–ion interactional forces, $U = 0$; then, $n_i = n_i^0$, which means that the local concentration would be equal to the bulk concentration. If the forces are attractive, then the potential change U is negative (i.e., negative work is done by the hypothetical external agency) and $n_i > n_i^0$; there is a local *accumulation* of ions in excess of their bulk concentrations. If the forces are repulsive, the potential-energy change is positive (i.e., the work done by the external agency is positive) and $n_i < n_i^0$; there is local *depletion* of ions.

In the first instance, and as a first approximation, one may ignore all types of ion–ion interactions except those deriving from simple coulombic[†] forces. Thus, short-range interactions (e.g., dispersion interactions) are excluded. This is a fundamental assumption of the Debye–Hückel theory. Thus, the potential of average force U simply becomes the *coulombic* potential energy of an ion of charge $z_i e_0$ in the volume element dv, i.e., to

---

[†] In this book, the term *coulombic* is restricted to forces (with r^{-2} dependence on distance) which are based directly on Coulomb's law. More complex forces, e.g., those which vary as r^{-4} or r^{-7}, may result as a net force from the resultant of several different coulombic interactions. Nevertheless, such more complex results of the interplay of several coulombic forces will be called *noncoulombic*.

the charge $z_i e_0$ on the ion times the electrostatic potential ψ_r in the volume element dv. That is,

$$U = z_i e_0 \psi_r \tag{3.8}$$

The Boltzmann distribution law (3.7) thus assumes the form

$$n_i = n_i^0 e^{-z_i e_0 \psi_r/(kT)} \tag{3.9}$$

Now that n_i, the concentration of the ionic species i in the volume element dv, has been related to its bulk concentration n_i^0, the expression (3.6) for the excess charge density in the volume element dv becomes

$$\varrho_r = \sum_i n_i z_i e_0 = \sum_i n_i^0 z_i e_0 e^{-z_i e_0 \psi_r/(kT)} \tag{3.10}$$

3.3.5. A Vital Step in the Debye–Hückel Theory of the Charge Distribution around Ions: Linearization of the Boltzmann Equation

At this point of the theory, Debye and Hückel made a move which was not only mathematically expedient but also turned out to be wise. They decided to carry out the analysis only for systems in which the average electrostatic potential ψ_r would be small so that

$$z_i e_0 \psi_r \ll kT \quad \text{or} \quad \frac{z_i e_0 \psi_r}{kT} \ll 1 \tag{3.11}$$

Based on this assumption, one can expand the exponential of Eq. (3.10) in a Taylor series, i.e.,

$$e^{-z_i e_0 \psi_r/(kT)} = 1 - \frac{z_i e_0 \psi_r}{kT} + \frac{1}{2}\left(\frac{z_i e_0 \psi_r}{kT}\right)^2 \cdots \tag{3.12}$$

and neglect all except the first two terms. Thus, in (3.10),

$$\varrho_r = \sum_i n_i^0 z_i e_0 \left(1 - \frac{z_i e_0 \psi_r}{kT}\right) \tag{3.13}$$

$$= \sum_i n_i^0 z_i e_0 - \sum_i \frac{n_i^0 z_i^2 e_0^2 \psi_r}{kT} \tag{3.14}$$

The first term $\sum n_i^0 z_i e_0$ gives the charge on the electrolytic solution as a whole. But this is zero because the solution as a whole must be electrically neutral. The local excess charge densities near ions cancel out because the excess positive charge density near a negative ion is compensated for by an excess negative charge density near a positive ion. Hence,

$$\sum_i n_i^0 z_i e_0 = 0 \qquad (3.15)$$

and one is left with

$$\varrho_r = -\sum_i \frac{n_i^0 z_i^2 e_0^2 \psi_r}{kT} \qquad (3.16)$$

3.3.6. The Linearized Poisson–Boltzmann Equation

The stage is now set for the calculation of the potential ψ_r and the charge density ϱ_r in terms of known parameters of the solution.

Notice that one has obtained two expressions for the charge density ϱ_r in the volume element dv at a distance r from the central ion. One has the Poisson equation [cf. Eq. (3.4)]

$$\varrho_r = -\frac{\varepsilon}{4\pi}\left[\frac{1}{r^2}\frac{d}{dr}\left(r^2 \frac{d\psi_r}{dr}\right)\right] \qquad (3.17)$$

and one has the "linearized" Boltzmann distribution

$$\varrho_r = -\sum_i \frac{n_i^0 z_i^2 e_0^2 \psi_r}{kT} \qquad (3.18)$$

where $\sum_i$ refers to the summation over all species of ions typified by i.

If one equates these two expressions one can obtain the linearized Poisson–Boltzmann (P–B) expression

$$\frac{1}{r^2}\frac{d}{dr}\left(r^2 \frac{d\psi_r}{dr}\right) = \left(\frac{4\pi}{\varepsilon kT}\sum_i n_i^0 z_i^2 e_0^2\right)\psi_r \qquad (3.19)$$

The constants in the right-hand parentheses can all be lumped together and called a new constant $\varkappa^2$, i.e.,

$$\varkappa^2 = \frac{4\pi}{\varepsilon kT}\sum_i n_i^0 z_i^2 e_0^2 \qquad (3.20)$$

At this point, the symbol $\varkappa$ has come in only to reduce the tedium of writing. It turns out later, however, that $\varkappa$ is not only a shorthand symbol; it contains information concerning several fundamental aspects of the distribution of ions around an ion in solution. In Chapter 7, it will be shown that it also contains information concerning the distribution of charges near a metal surface in contact with an ionic solution. In terms of $\varkappa$, the linearized P–B expression (3.19) is

$$\frac{1}{r^2}\frac{d}{dr}\left(r^2\frac{d\psi_r}{dr}\right) = \varkappa^2\psi_r \tag{3.21}$$

3.3.7. The Solution of the Linearized P–B Equation

The rather messy-looking linearized P–B equation (3.21) can be tidied up by a mathematical trick. Introducing a new variable μ defined by

$$\psi_r = \frac{\mu}{r} \tag{3.22}$$

one has

$$\frac{d\psi_r}{dr} = \frac{d}{dr}\frac{\mu}{r} = -\frac{\mu}{r^2} + \frac{1}{r}\frac{d\mu}{dr}$$

and, therefore,

$$\frac{1}{r^2}\frac{d}{dr}\left(r^2\frac{d\psi_r}{dr}\right) = \frac{1}{r^2}\frac{d}{dr}\left(-\mu + r\frac{d\mu}{dr}\right)$$

$$= \frac{1}{r^2}\left(-\frac{d\mu}{dr} + r\frac{d^2\mu}{dr^2} + \frac{d\mu}{dr}\right)$$

$$= \frac{1}{r}\frac{d^2\mu}{dr^2} \tag{3.23}$$

Hence, the differential equation (3.21) becomes

$$\frac{1}{r}\frac{d^2\mu}{dr^2} = \varkappa^2\frac{\mu}{r} \tag{3.24}$$

or

$$\frac{d^2\mu}{dr^2} = \varkappa^2\mu \tag{3.25}$$

To solve this simple differential equation, it is recalled that the differentiation of an exponential function results in the multiplication of that function by the constant in the component. For example,

$$\frac{d}{dr}e^{\pm\varkappa r} = \pm\varkappa e^{\pm\varkappa r}$$

and $\tag{3.26}$

$$\frac{d^2}{dr^2}e^{\pm\varkappa r} = \varkappa^2 e^{\pm\varkappa r}$$

Hence, if μ is an exponential function of r, one will obtain a differential equation of the form of Eq. (3.25). In other words, the "primitive" or "origin" of the differential equation must have had an exponential in $\varkappa r$.

Two possible exponential functions, however, would lead to the same final differential equation; one of them would have a positive exponent and the other a negative one [cf. Eq. (3.26)]. The general solution of the linearized P–B equation can, therefore, be written as

$$\mu = Ae^{-\varkappa r} + Be^{+\varkappa r} \tag{3.27}$$

where A and B are constants to be evaluated. Or, from Eq. (3.22),

$$\psi_r = A\frac{e^{-\varkappa r}}{r} + B\frac{e^{+\varkappa r}}{r} \tag{3.28}$$

The constant B is evaluated by using the boundary condition that, far enough from a central ion situated at $r = 0$, the thermal forces completely dominate the coulombic forces which decrease as r^2, and there is electroneutrality, i.e., the electrostatic potential ψ_r vanishes at distances sufficiently far from such an ion, $\psi_r \to 0$ as $r \to \infty$. This condition would be satisfied only if $B = 0$. Thus, if B had a finite value, Eq. (3.29) shows that the electrostatic potential would shoot up to infinity, i.e., $\psi_r \to \infty$ as $r \to \infty$, a physically unreasonable proposition. Hence,

$$\psi_r = A\frac{e^{-\varkappa r}}{r} \tag{3.29}$$

To evaluate the integration constant A, a hypothetical condition will be considered in which the solution is so dilute and, on the average, the ions are so far apart that there is a negligible interionic field. Further, the central ion is assumed to be a *point* charge, i.e., to have a radius negligible compared with the distances otherwise to be considered. Hence, the potential near the central ion is, in this special case, simply that due to an isolated point charge of value $z_i e_0$,

$$\psi_r = \frac{z_i e_0}{\varepsilon r} \tag{3.30}$$

At the same time, for this hypothetical solution in which the concentration tends to zero, i.e., $n_i^0 \to 0$, it is seen from Eq. (3.20) that $\varkappa \to 0$. Thus, in Eq. (3.29), $e^{-\varkappa r} \to 1$, and one has

$$\psi_r = \frac{A}{r} \tag{3.31}$$

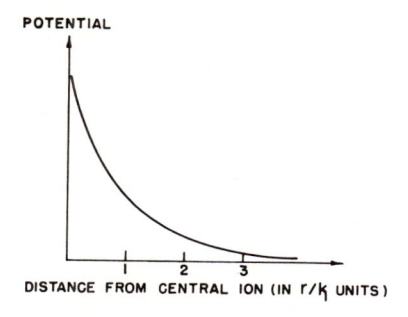

Fig. 3.9. The variation of the electrostatic potential ψ as a function of distance from the central ion expressed in units of $r/\varkappa$.

Hence, by combining Eqs. (3.30) and (3.31),

$$A = \frac{z_i e_0}{\varepsilon} \tag{3.32}$$

By introducing this expression for A into Eq. (3.29), the result is

$$\psi_r = \frac{z_i e_0}{\varepsilon} \frac{e^{-\varkappa r}}{r} \tag{3.33}$$

Here, then, is the appropriate solution of the *linearized* P–B equation (3.21). It shows how the electrostatic potential varies with distance r from an arbitrarily chosen reference ion (Fig. 3.9).

3.3.8. The Ionic Cloud around a Central Ion

In the Debye–Hückel model of a dilute electrolytic solution, a reference ion sitting at the origin of the spherical coordinate system is surrounded by the smoothed-out charge of the other ions. Further, because of the local inequalities in the concentrations of the positive and negative ions, the smoothed-out charge of one sign does not (locally) cancel out the smoothed-out charge of the opposite sign; there is a local excess charge density of one sign.

Now, as explained in Section 3.3.2, the principal objective of the Debye–Hückel theory is to calculate the time-average spatial distribution of the excess charge density around a reference ion. How is this objective attained?

The Poisson equation (3.4) relates the potential at r from the sample ion to the charge density at r, i.e.,

$$\frac{1}{r^2} \frac{d}{dr}\left(r^2 \frac{d\psi_r}{dr}\right) = -\frac{4\pi}{\varepsilon} \varrho_r \tag{3.4}$$

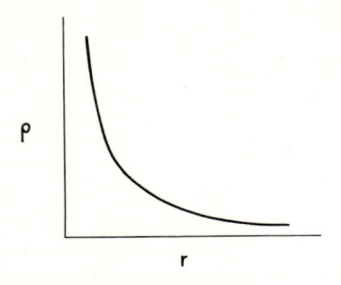

Fig. 3.10. The variation of the excess charge density ϱ as a function of distance from the central ion.

Further, one has the linearized P–B equation

$$\frac{1}{r^2}\frac{d}{dr}\left(r^2\frac{d\psi_r}{dr}\right) = \varkappa^2\psi_r \tag{3.21}$$

From these two Eqs. (3.4) and (3.21), one has the linear relation between excess charge density and potential, i.e.,

$$\varrho_r = -\frac{\varepsilon}{4\pi}\varkappa^2\psi_r \tag{3.34}$$

and by inserting the solution (3.33) for the linearized P–B equation, the result is

$$\varrho_r = -\frac{z_i e_0}{4\pi}\varkappa^2\frac{e^{-\varkappa r}}{r} \tag{3.35}$$

Here then is the desired expression for the spatial distribution of the charge density with distance r from the central ion (Fig. 3.10). Since the excess charge density results from an unequal distribution of positive and negative ions, Eq. (3.35) also describes the distribution of ions around a reference or sample ion.

To understand this distribution of ions, however, one must be sufficiently attuned to mathematical language to read the physical significance of Eq. (3.35). The physical ideas implicit in the distribution will therefore be stated in pictorial terms. One can say that the central reference ion is surrounded by a "cloud," or "atmosphere," of excess charge (Fig. 3.11). This ionic cloud extends into the solution (i.e., r increases), and the excess charge density ϱ decays with distance r in an exponential way. The excess charge residing on the ion cloud is opposite in sign to that of the

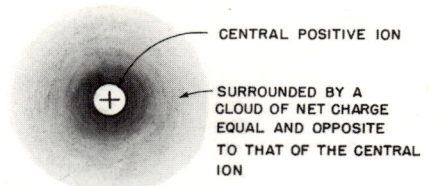

Fig. 3.11. The distribution of excess charge density around a central ion can be pictured as a cloud, or atmosphere, of net charge around the central ion.

central ion. Thus, a positively charged reference ion has a negatively charged ion atmosphere, and *vice versa* (Fig. 3.12).

Up to now, the charge *density* at a given distance has been discussed. The *total* excess charge contained in the ionic atmosphere which surrounds the central ion can, however, easily be computed. Consider a spherical shell of thickness dr at a distance r from the origin, i.e., from the center of the reference ion (Fig. 3.13). The charge dq in this thin shell is equal to the charge density ϱ_r times the volume $4\pi r^2\, dr$ of the shell, i.e.,

$$dq = \varrho_r 4\pi r^2\, dr \tag{3.36}$$

The total charge q_{cloud} contained in the ion atmosphere is obtained by summing the charges dq contained in all the infinitesimally thick spherical shells. In other words, the total excess charge surrounding the reference ion is computed by integrating dq (which is a function of the distance r from the central ion) from a lower limit corresponding to the distance from the central ion at which the cloud is taken to commence to the point where the cloud ends. Now, the ion atmosphere begins at the surface of the ion; so the lower limit depends upon the model of the ion. The first model chosen by Debye and Hückel was that of point-charge ions, in which case the lower limit is $r = 0$. The upper limit for the integration is $r \to \infty$ be-

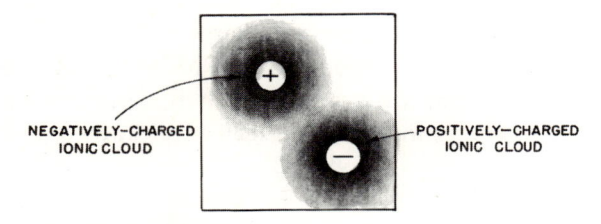

Fig. 3.12. A positively charged ion has a negatively charged ionic cloud, and *vice versa*.

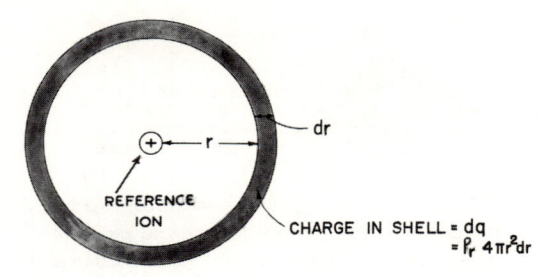

Fig. 3.13. A spherical shell, of thickness dr, at a distance r from the center of the reference ion.

cause the charge of the ionic cloud decays exponentially into the solution and becomes zero only in the limit $r \to \infty$.

Thus,

$$q_{\text{cloud}} = \int_{r=0}^{r \to \infty} dq = \int_{r=0}^{r \to \infty} \varrho_r 4\pi r^2 \, dr \tag{3.37}$$

and, by substituting for ϱ_r from Eq. (3.35), the result is

$$q_{\text{cloud}} = -\int_{r=0}^{r \to \infty} \frac{z_i e_0}{4\pi} \varkappa^2 \frac{e^{-\varkappa r}}{r} 4\pi r^2 \, dr$$

$$= -z_i e_0 \int_{r=0}^{r \to \infty} e^{-\varkappa r} (\varkappa r) \, d(\varkappa r) \tag{3.38}$$

The integration can be done by parts (Appendix 3.2), leading to the result

$$q_{\text{cloud}} = -z_i e_0 \tag{3.39}$$

which means that a central ion of charge $+z_i e_0$ is enveloped by a cloud containing a total charge of $-z_i e_0$ (Fig. 3.14). Thus, the total charge on the surrounding volume is just equal and opposite to that on the reference ion. This is of course precisely how things should be so that there can be electroneutrality for the ionic solution taken as a whole; a given ion, together with its cloud, has a zero net charge.

How is this equal and opposite charge of the ion atmosphere distributed in the space around the central ion? It is seen from Eqs. (3.35) and (3.36) that the net charge in a spherical shell of thickness dr and at a distance r from the origin is

$$dq = -z_i e_0 e^{-\varkappa r} \varkappa^2 r \, dr \tag{3.40}$$

Thus, the excess charge on a spherical shell varies with r and has a maximum

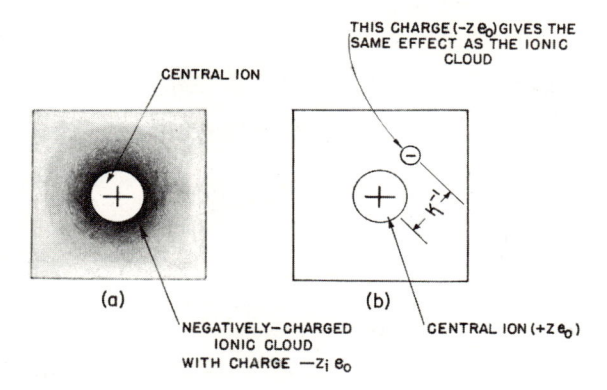

Fig. 3.14. The total charge $-z_i e_0$ on the ionic cloud is just equal and opposite to that $+z_i e_0$ on the central ion.

value for a value of r given by

$$0 = \frac{dq}{dr}$$

$$= \frac{d}{dr}[-z_i e_0 \varkappa^2 (e^{-\varkappa r} r)]$$

$$= -z_i e_0 \varkappa^2 \frac{d}{dr}(e^{-\varkappa r} r)$$

$$= -z_i e_0 \varkappa^2 (e^{-\varkappa r} - r\varkappa e^{-\varkappa r}) \tag{3.41}$$

Since $(z_i e_0 \varkappa^2)$ is finite, Eq. (3.41) can be true only when

$$0 = e^{-\varkappa r} - r\varkappa e^{-\varkappa r}$$

or

$$r = \varkappa^{-1} \tag{3.42}$$

Hence, the maximum value of the charge contained in a spherical shell (of infinitesimal thickness dr) is attained when the spherical shell is at a distance $r = \varkappa^{-1}$ from the reference ion (Fig. 3.15). For this reason (but see also Section 3.3.9), $\varkappa^{-1}$ is known as the *thickness*, or *radius*, of the ionic cloud which surrounds a reference ion. An elementary dimensional analysis [e.g., of Eq. (3.43)] will indeed reveal that $\varkappa^{-1}$ has the dimensions of length. Also, $\varkappa^{-1}$ is sometimes referred to as *the Debye–Hückel reciprocal length*.

198 CHAPTER 3

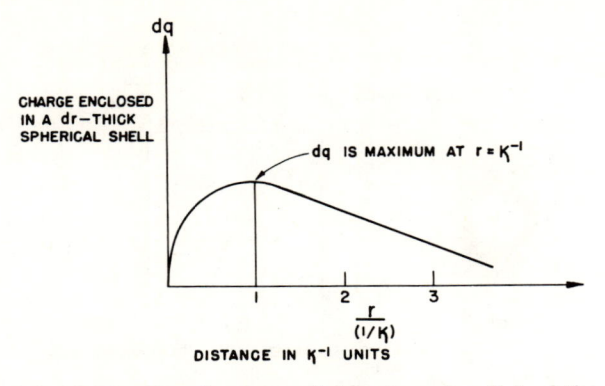

Fig. 3.15. The distance variation (in $\varkappa^{-1}$ units) of the charge dq enclosed in a dr-thick spherical shell, showing that dq is a maximum at $r = \varkappa^{-1}$.

It may be recalled that $\varkappa^{-1}$ is given [from Eq. (3.20)] by

$$\varkappa^{-1} = \left(\frac{\varepsilon kT}{4\pi} \frac{1}{\sum_i n_i^0 z_i^2 e_0^2} \right)^{\frac{1}{2}} \tag{3.43}$$

As the concentration tends toward zero, the cloud tends to spread out increasingly (Fig. 3.16). Values of the thickness of the ion atmosphere for various concentrations of the electrolyte are presented in Table 3.2.

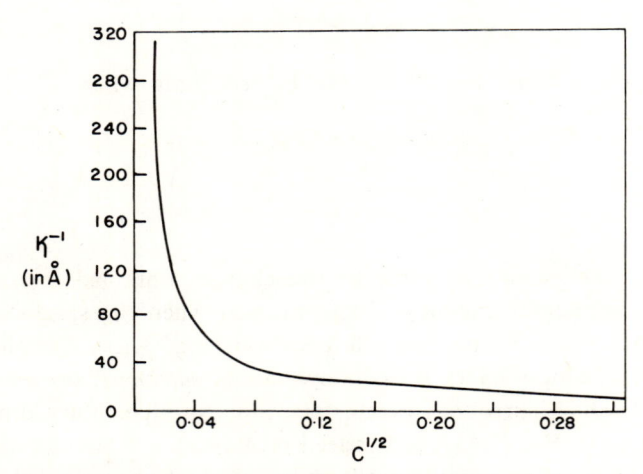

Fig. 3.16. The variation in the thickness $\varkappa^{-1}$ of the ionic cloud as a function of electrolyte concentration.

TABLE 3.2
Thickness of Ionic Atmosphere (in Angstroms) at Various Concentrations and for Various Types of Salts

C, moles liter^{-1}	Type of salt			
	1,1	1,2	2,2	1,3
10^{-4}	304	176	152	124
10^{-3}	96	55.5	48.1	39.3
10^{-2}	30.4	17.6	15.2	12.4
10^{-1}	9.6	5.5	4.8	3.9

3.3.9. How Much Does the Ionic Cloud Contribute to the Electrostatic Potential ψ_r at a Distance r from the Central Ion?

An improved feel for the effects of ionic clouds emerges from considering the following interesting problem.

Imagine, in a thought experiment, that the *charge* on the ionic cloud does not exist. There is only one charge now, that on the central ion. What is the potential at distance r from the central ion? It is simply given by the familiar formula for the potential at a distance r from a single charge, namely,

$$\psi_r = \frac{z_i e_0}{\varepsilon r} \tag{3.44}$$

Then, let the charge on the cloud be switched on. The potential ψ_r at the distance r from the central ion is no longer given by the central ion only. It is given by the law of superposition of potentials (Fig. 3.17), i.e.,

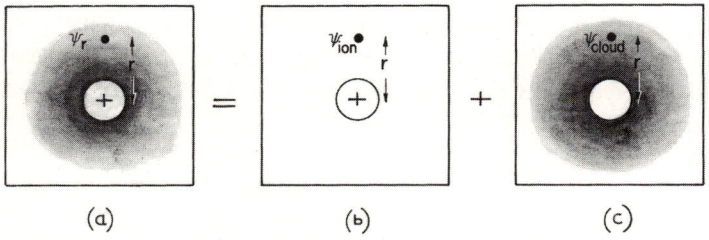

Fig. 3.17. The superposition of the potential ψ_{ion} due to the ion and that ψ_{cloud} due to the cloud yields the total potential at a distance r from the central ion.

ψ_r is the sum of the potential due to the central ion and that due to the ionic cloud

$$\psi_r = \psi_{\text{ion}} + \psi_{\text{cloud}} \qquad (3.45)$$

The contribution ψ_{cloud} can thus be easily found. One rearranges Eq. (3.45) to read

$$\psi_{\text{cloud}} = \psi_r - \psi_{\text{ion}} \qquad (3.46)$$

and substitutes for ψ_{ion} with Eq. (3.44) and for ψ_r with the Debye–Hückel expression [Eq. (3.33)]. Then,

$$\psi_{\text{cloud}} = \frac{z_i e_0}{\varepsilon} \frac{e^{-\varkappa r}}{r} - \frac{z_i e_0}{\varepsilon r}$$

$$= \frac{z_i e_0}{\varepsilon r} (e^{-\varkappa r} - 1) \qquad (3.47)$$

The value of $\varkappa$ [cf. Eq. (3.20)] is proportional to $\sum n_i^0 z_i^2 e_0^2$. In sufficiently dilute solutions, $\sum n_i^0 z_i^2 e_0^2$ can be taken as sufficiently small to make $\varkappa r \ll 1$,

$$e^{-\varkappa r} - 1 \sim 1 - \varkappa r - 1 \sim -\varkappa r \qquad (3.48)$$

and, based on this approximation,

$$\psi_{\text{cloud}} = -\frac{z_i e_0}{\varepsilon \varkappa^{-1}} \qquad (3.49)$$

By introducing the expressions (3.44) and (3.49) into the expression (3.45) for the total potential ψ_r at a distance r from the central ion, it follows that

$$\psi_r = \frac{z_i e_0}{\varepsilon r} - \frac{z_i e_0}{\varepsilon \varkappa^{-1}} \qquad (3.50)$$

The second term, which arises from the cloud, reduces the value of the potential to a value *less* than that if there were no cloud. This is consistent with the model; the cloud has a charge opposite to that on the central ion and must, therefore, alter the potential in a sense opposite to that due to the central ion.

The expression

$$\psi_{\text{cloud}} = -\frac{z_i e_0}{\varepsilon \varkappa^{-1}} \qquad (3.49)$$

leads to another, and helpful, way of looking at the quantity $\varkappa^{-1}$. It is seen that ψ_{cloud} is independent of r, and, therefore, the contribution of the

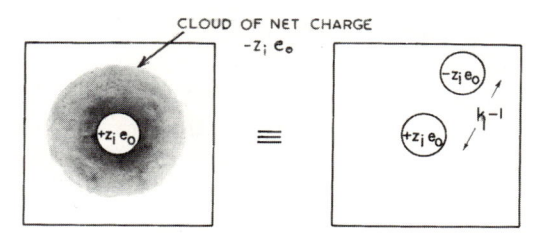

Fig. 3.18. The contribution ψ_{cloud} of the ionic cloud to the potential at the central ion is equivalent to the potential due to a single charge, equal and opposite to that of the central ion, placed at a distance $\varkappa^{-1}$ from the central ion.

cloud to the potential *at the site of the point-charge central ion* can be considered given by Eq. (3.49) above. But, if the entire charge of the ionic atmosphere [which is $-z_i e_0$ as required by electroneutrality—*cf.* Eq. (3.39)] were placed at a distance $\varkappa^{-1}$ from the central ion, then the potential produced at the reference ion would be $-z_i e_0/(\varepsilon \varkappa^{-1})$. It is seen therefore from Eq. (3.49) that the effect of the ion cloud, namely, ψ_{cloud}, is equivalent to that of a single charge, equal in magnitude but opposite in sign to that of the central ion, placed at a distance $\varkappa^{-1}$ from the reference ion (Fig. 3.18). This is an added—and more important—reason that the quantity $\varkappa^{-1}$ is termed the effective thickness or radius of the ion atmosphere surrounding a central ion (*cf.* Section 3.3.8).

3.3.10. The Ionic Cloud and the Chemical-Potential Change Arising from Ion–Ion Interactions

It will be recalled (see Section 3.3.1) that it was the potential at the surface of the reference ion which needed to be known in order to calculate the chemical-potential change $\Delta\mu_{i-I}$ arising from the interactions between a particular ionic species i and the rest of the ions of the solution, i.e., one needed to know ψ in Eq. (3.3),

$$\Delta\mu_{i-I} = \frac{N_A z_i e_0}{2} \psi \qquad (3.3)$$

It was to obtain this potential ψ that Debye and Hückel conceived their model of an ionic solution. The analysis threw up the picture of an ion being enveloped in an ionic cloud. But what is the origin of the ionic cloud? It is born of the interactions between the central ion and the ions of the environment. If there were no interactions (e.g., coulombic forces between ions), thermal forces would prevail, distribute the ions randomly ($\varrho = 0$), and wash out the ionic atmosphere. It appears therefore that the simple ionic cloud picture has not only led to success in describing the distribution of ions but also given the electrostatic potential ψ_{cloud} at the surface of a reference ion due to the interactions between this reference ion and the rest of the ions in the solution (the quantity required for reasons declared in Section 3.3.1).

Thus, the expression (3.49) for ψ_{cloud} can be substituted for ψ in Eq. (3.3) with the result that

$$\Delta\mu_{i-I} = -\frac{N_A}{2} \frac{(z_i e_0)^2}{\varepsilon \varkappa^{-1}} \tag{3.51}$$

The Debye–Hückel ionic-cloud model for the distribution of ions in an electrolytic solution has permitted the theoretical calculation of the chemical-potential change arising from ion–ion interactions. But, how is this theoretical expression to be checked, i.e., connected with a measured quantity? It is to this testing of the Debye–Hückel theory that attention will now be turned.

Further Reading

1. P. Debye and E. Hückel, *Z. Physik*, **24**, 185 (1923).
2. H. S. Harned and B. B. Owen, *The Physical Chemistry of Electrolytic Solutions*, 3rd ed., Reinhold Publishing Corp., New York, 1958.
3. R. A. Robinson and R. H. Stokes, *Electrolyte Solutions*, 2nd ed., Butterworth's Publications, Ltd., London, 1959.

3.4. ACTIVITY COEFFICIENTS AND ION–ION INTERACTIONS

3.4.1. The Evolution of the Concept of Activity Coefficient

The existence of ions in solution, of interactions between these ions, and of a chemical-potential change $\Delta\mu_{i-I}$ arising from ion–ion interactions have all been taken to be self-evident in the treatment hitherto presented here. This, however, is a modern point of view. The thinking about electrolytic solutions actually developed along a different path.

Ionic solutions were at first treated in the same way as nonelectrolytic solutions, though the latter do not contain (interacting) charged species. The starting point was the classical thermodynamic formula for the chemical potential μ_i of a nonelectrolyte solute

$$\mu_i = \mu_i^0 + RT \ln x_i \tag{3.52}$$

In this expression, x_i is the concentration[†] of the solute in mole fraction units, and μ_i^0 is its chemical potential in the standard state, i.e., when x_i assumes a standard or normalized value of unity

$$\mu_i = \mu_i^0 \quad \text{when} \quad x_i = 1 \qquad (3.53)$$

Since the solute particles in a solution of a nonelectrolyte are uncharged, they do not engage in long-range coulombic interactions. The short-range interactions arising from dipole–dipole or dispersion forces become significant only when the mean distance between the solute particles is small, i.e., when the concentration of the solute is high. Thus, one can to a good approximation say that there are no interactions between solute particles in dilute nonelectrolyte solutions. Hence, if Eq. (3.52) for the chemical potential of a solute in a nonelectrolyte solution (with noninteracting particles) is used for the chemical potential of a ionic species i in an electrolytic solution, then it is tantamount to ignoring the long-range coulombic interactions between ions. In an actual electrolytic solution, however, ion–ion interactions operate whether one ignores them or not. It is obvious, therefore, that measurements of the chemical potential μ_i of an ionic species—or, rather, measurements of any property which depends on the chemical potential—would reveal the error in Eq. (3.52), which is blind to ion–ion interactions. In other words, experiments show that, *even in dilute solutions*,

$$\mu_i - \mu_i^0 \neq RT \ln x_i$$

In this context, a frankly empirical approach was adopted by earlier workers, not yet blessed by Debye and Hückel's light. Solutions that obeyed Eq. (3.52) were characterized as *ideal* solutions since this equation applies to systems of noninteracting solute particles, i.e., ideal particles. Electrolytic solutions which do not obey the equation were said to be *nonideal*. In order to use an equation of the form of (3.52) to treat nonideal electrolytic solutions, an empirical correction factor f_i was introduced by Lewis as a modifier of the concentration term[‡]

$$\mu_i - \mu_i^0 = RT \ln x_i f_i \qquad (3.54)$$

[†] The value of x_i^0 in the case of an electrolyte derives from the number of moles of ions in species i actually present in solution. This number need not be equal to the number of moles of i expected of dissolved electrolyte; if, for instance, the electrolyte is a potential one, then only a fraction of the electrolyte may react with the solvent to form ions, i.e., the electrolyte may be incompletely dissociated.

[‡] The standard chemical potential μ_i^0 has the same significance here as in Eq. (3.52) for ideal solutions. Thus, μ_i^0 can be defined either as the chemical potential of an *ideal*

204   CHAPTER 3

It was argued that, in nonideal solutions, it was not just the analytical concentration x_i of species i, but its *effective* concentration $x_i f_i$ which determined the chemical-potential change $\mu_i - \mu_i^0$. This effective concentration $x_i f_i$ was also known as the *activity* a_i of the species i, i.e.,

$$a_i = x_i f_i \qquad (3.55)$$

and the correction factor f_i, as the *activity coefficient*. For ideal solutions, the activity coefficient is unity, and the activity a_i becomes identical to the concentration x_i, i.e.,

$$a_i = x_i \quad \text{when} \quad f_i = 1 \qquad (3.56)$$

Thus, the chemical-potential change in going from the standard state to the final state can be written as

$$\mu_i - \mu_i^0 = RT \ln x_i + RT \ln f_i \qquad (3.57)$$

Equation (3.57) summarizes the empirical or formal treatment of the behavior of electrolytic solutions. Such a treatment cannot furnish a theoretical expression for the activity coefficient f_i. It merely recognizes that expressions such as (3.52) must be modified if significant interaction forces exist between solute particles.

3.4.2. The Physical Significance of Activity Coefficients

For a hypothetical system of *ideal* (noninteracting) particles, the chemical potential has been stated to be given by

$$\mu_i \text{ (ideal)} = \mu_i^0 + RT \ln x_i \qquad (3.52)$$

For a real system of interacting particles, the chemical potential has been expressed in the form

$$\mu_i \text{ (real)} = \mu_i^0 + RT \ln x_i + RT \ln f_i \qquad (3.57)$$

Hence, to analyze the physical significance of the activity coefficient term in Eq. (3.57), it is necessary to compare this equation with Eq. (3.52). It is obvious that, when Eq. (3.52) is subtracted from Eq. (3.57), the differ-

---

solution in its standard state of $x_i = 1$ or as the chemical potential of a solution in its state of $x_i = 1$ and $f_i = 1$, i.e., $a_i = 1$. No real solution can have $f_i = 1$ when $x_i = 1$; so, the standard state pertains to the same hypothetical solution as the standard state of an ideal solution.

ence, i.e., μ_i (real) $- \mu_i$ (ideal), is the chemical-potential change $\Delta\mu_{i-I}$ arising from interactions between the solute particles (ions in the case of electrolyte solutions). That is,

$$\mu_i \text{ (real)} - \mu_i \text{ (ideal)} = \Delta\mu_{i-I} \tag{3.58}$$

and, therefore,

$$\Delta\mu_{i-I} = RT \ln f_i \tag{3.59}$$

Thus, the activity coefficient is a measure of the chemical-potential change arising from ion–ion interactions. There are several well-established methods of experimentally determining activity coefficients, and these methods are treated in adequate detail in standard treatises (*cf.* Further Reading at the end of this section).

Now, according to the Debye–Hückel theory, the chemical-potential change $\Delta\mu_{i-I}$ arising from ion–ion interactions has been shown to be given by

$$\Delta\mu_{i-I} = -\frac{N_A(z_i e_0)^2}{2\varepsilon\varkappa^{-1}} \tag{3.51}$$

Hence, by combining Eqs. (3.51) and (3.58), the result is

$$RT \ln f_i = -\frac{N_A(z_i e_0)^2}{2\varepsilon\varkappa^{-1}} \tag{3.60}$$

Thus, the Debye–Hückel ionic-cloud model for ion–ion interactions has permitted a theoretical calculation of activity coefficients resulting in Eq. (3.59).

The activity coefficient in Eq. (3.59) arises from the formula (3.57) for the chemical potential, in which the concentration of the species i is expressed in mole fraction x_i units. But one can also express the concentration in moles per liter of solution (molarity) or in moles per kilogram of solvent (molality). Thus, alternate formulas for the chemical potential of a species i in an *ideal* solution read

$$\mu_i = \mu_i^0(c) + RT \ln c_i \tag{3.61}$$

and

$$\mu_i = \mu_i^0(m) + RT \ln m_i \tag{3.62}$$

where c_i and m_i are the molarity and molality of the species i, respectively; and $\mu_i^0(c)$ and $\mu_i^0(m)$, the corresponding standard chemical potentials.

When the concentration of the ionic species in a *real* solution is expressed as a molarity c_i or a molality m_i, there are corresponding activity coefficients γ_c and γ_m and corresponding expressions for μ_i

$$\mu_i = \mu_i^0(c) + RT \ln c_i + RT \ln \gamma_c \qquad (3.63)$$

and

$$\mu_i = \mu_i^0(m) + RT \ln m_i + RT \ln \gamma_m \qquad (3.64)$$

3.4.3. The Activity Coefficient of a Single Ionic Species Cannot Be Measured

Before the activity coefficients calculated on the basis of the Debye–Hückel model can be compared with experiment, there arises a problem similar to one faced in the discussion of ion–solvent interactions (Chapter 2).

There, it was realized the heat of hydration of an individual ionic species could not be measured because such a measurement would involve the transfer of ions of only one species into a solvent instead of ions of two species with equal and opposite charges. Even if such a transfer were physically possible, it would result in a charged solution† and, therefore, an extra, undesired interaction between the ions and the electrified solution. The only way out was to transfer a neutral electrolyte (an equal number of positive and negative ions) into the solvent, but this meant that one could only measure the heat of interactions of a *salt* with the solvent and this experimental quantity could not be separated into the individual ionic heats of hydration.

Here, in the case of ion–ion interactions, the desired quantity is the activity coefficient f_i,‡ which depends through Eq. (3.57) on $\mu_i - \mu_i^0$. This means that one seeks the free-energy change of an ionic solution per mole of ions of a single species i. To measure this quantity, one would have a problem similar to that experienced with ion–solvent interactions, namely, the measurement of the change of free energy of a solution, resulting from a change in the concentration of one ionic species only.

This change in free energy associated with the addition of one ionic species only would include an undesired work term representing the electrical work of interaction between the ionic species being added and the charged solution.† To avoid free-energy changes associated with interacting

---

† The solution may not be initially charged but will become so once an ionic species is added to it.

‡ The use of the symbol γ for the activity coefficients when the concentration is expressed in molarities and molalities should be noted. When the concentration is expressed as a mole fraction, f_i has here been used. For dilute solutions, the numerical values of activity coefficients for these different systems of units are almost the same.

with a solution, it is necessary that, after changing the concentration of the ionic species, the electrolytic solution should end up uncharged and electroneutral. This aim is easily accomplished by adding an electroneutral electrolyte containing the ionic species i. Thus, the concentration of sodium ions can be altered by adding sodium chloride. The solvent, water, maintains its electroneutrality when the uncharged ionic lattice (containing two ionic species of opposite charge) is dissolved in it.

When ionic lattices, i.e., salts, are dissolved instead of individual ionic species, one eliminates the problem of ending up with charged solutions but another problem emerges. If one increases the concentration of sodium ions by adding the salt sodium chloride, one has perforce to produce a simultaneous increase of the concentration of chloride ions. This means, however, that there are two contributions to the change in free energy associated with a change in salt concentration: (1) the contribution of the positive ions, and (2) the contribution of the negative ions.

Since neither the positive nor the negative ions can be added separately, the individual contributions of the ionic species to the free energy of the system cannot be determined. Thus, the activity coefficients of individual ions, which depend by, e.g., (3.63) on the free-energy changes when the particular individual species alone is added to the solution, are inaccessible to experimental measurement. One can only measure the activity coefficient of the net electrolyte, i.e., of at least two ionic species together. It is necessary, therefore, to establish a conceptual link between the activity coefficient of an *electrolyte* in solution (that quantity accessible to experiment) and that of only one of its ionic species [*not* accessible to experiment, but calculable theoretically from (3.60)].

3.4.4. The Mean Ionic Activity Coefficient

Consider a uni-univalent electrolyte MA (e.g., NaCl). The chemical potential of the M^+ ions is [*cf.* (3.58)]

$$\mu_{M^+} = \mu_{M^+}^0 + RT \ln x_{M^+} + RT \ln f_{M^+} \qquad (3.65)$$

and the chemical potential of the A^- ions is

$$\mu_{A^-} = \mu_{A^-}^0 + RT \ln x_{A^-} + RT \ln f_{A^-} \qquad (3.66)$$

Adding the two expressions, one obtains

$$\mu_{M^+} + \mu_{A^-} = (\mu_{M^+}^0 + \mu_{A^-}^0) + RT \ln(x_{M^+} x_{A^-}) + RT \ln(f_{M^+} f_{A^-}) \qquad (3.67)$$

208 CHAPTER 3

What has been obtained here is the change in the free energy of the system due to the addition of 2 moles of ions—1 mole of M$^+$ ions and 1 mole of A$^-$ ions—which are contained in 1 mole of electroneutral salt MA.

Now, suppose that one is only interested in the *average* contribution to the free energy of the system from 1 mole of both M$^+$ and A$^-$ ions. One has to divide Eq. (3.67) by 2

$$\frac{\mu_{M^+} + \mu_{A^-}}{2} = \frac{\mu^0_{M^+} + \mu^0_{A^-}}{2} + RT \ln(x_{M^+}x_{A^-})^{\frac{1}{2}} + RT \ln(f_{M^+}f_{A^-})^{\frac{1}{2}} \quad (3.68)$$

At this stage, one can define several new quantities

$$\mu_\pm = \frac{\mu_{M^+} + \mu_{A^-}}{2} \quad (3.69)$$

$$\mu^0_\pm = \frac{\mu^0_{M^+} + \mu^0_{A^-}}{2} \quad (3.70)$$

$$x_\pm = (x_{M^+}x_{A^-})^{\frac{1}{2}} \quad (3.71)$$

and

$$f_\pm = (f_{M^+}f_{A^-})^{\frac{1}{2}} \quad (3.72)$$

What is the significance of these quantities $\mu_\pm$, $\mu_\pm^0$, $x_\pm$, and $f_\pm$? It is obvious they are all *average* quantities—the mean chemical potential $\mu_\pm$, the mean standard chemical potential $\mu_\pm^0$, the mean ionic mole fraction $x_\pm$, and the mean ionic-activity coefficient $f_\pm$. In the case of $\mu_\pm$ and $\mu_\pm^0$, the arithmetic mean (half the sum) is taken because free energies are additive; but, in the case of $x_\pm$ and $f_\pm$, the geometric mean (the square root of the product) is taken because the effects of mole fraction and activity coefficient on free energy are multiplicative.

In this notation, Eq. (3.68) for the *average* contribution of a mole of ions to the free energy of the system becomes

$$\mu_\pm = \mu_\pm^0 + RT \ln x_\pm + RT \ln f_\pm \quad (3.73)$$

since a mole of ions is produced by the dissolution of half a mole of salt. In other words, $\mu_\pm$ is half the chemical potential μ_{MA} of the salt.[†]

$$\tfrac{1}{2}\mu_{MA} = \mu_\pm = \mu_\pm^0 + RT \ln x_\pm + RT \ln f_\pm. \quad (3.74)$$

[†] The symbol μ_{MA} should not be taken to mean that *molecules* of MA exist in the solution; μ_{MA} is the observed free-energy change of the system resulting from the dissolution of a mole of electrolyte.

Thus, a clear connection has been set up between observed free-energy changes μ_{MA} consequent upon the change from a state in which the two ionic species of a salt are infinitely far apart to a state corresponding to the given concentration, and its mean ionic-activity coefficient $f_\pm$. Hence the value of $f_\pm$ is experimentally measurable. This mean ionic-activity coefficient cannot, however, be *experimentally* split into the individual ionic-activity coefficients. All that can be obtained from $f_\pm$ is the *product* of the individual ionic-activity coefficients [Eq. (3.72)]. The theoretical approach must hence be to calculate the activity coefficients f_+ and f_- for the positive and negative ions [*cf.* Eq. (3.60)] and combine them through Eq. (3.72) into a mean ionic-activity coefficient $f_\pm$ which can be compared with the experimentally derived mean ionic-activity coefficient.

3.4.5. The Conversion of Theoretical Activity-Coefficient Expressions into a Testable Form

Individual ionic-activity coefficients are experimentally inaccessible (Section 3.4.3); hence, it is necessary to relate the theoretical individual activity coefficient f_i [Eq. (3.64)] to the experimentally accessible *mean* ionic-activity coefficient $f_\pm$ so that the Debye–Hückel model can be tested. The procedure is to make use of the relation (3.72)

$$f_\pm = (f_{M+} f_{A-})^{\frac{1}{2}} \tag{3.72}$$

of which the general form for an electrolyte which dissolves to give ν_+ z_+-valent positive ions and ν_- z_--valent negative ions can be shown to be (*cf.* Appendix 3.3)

$$f_\pm = (f_+^{\nu_+} f_-^{\nu_-})^{1/\nu} \tag{3.75}$$

where f_+ and f_- are the activity coefficients of the positive and negative ions, and

$$\nu = \nu_+ + \nu_- \tag{3.76}$$

By taking logarithms of both sides of Eq. (3.75), the result is

$$\ln f_\pm = \frac{1}{\nu}(\nu_+ \ln f_+ + \nu_- \ln f_-) \tag{3.77}$$

At this stage, the Debye–Hückel expressions (3.60) for f_+ and f_- can be introduced into Eq. (3.77) to give

$$\ln f_\pm = -\frac{1}{\nu}\left[\frac{N_A e_0^2}{2\varepsilon RT} \varkappa(\nu_+ z_+^2 + \nu_- z_-^2)\right] \tag{3.78}$$

210   CHAPTER 3

Since the solution as a whole is electroneutral, v_+z_+ must be equal to v_-z_-, and, therefore,

$$v_+z_+^2 + v_-z_-^2 = v_-z_-z_+ + v_+z_+z_-$$
$$= z_+z_-(v_+ + v_-)$$
$$= z_+z_-v \qquad (3.79)$$

Using this relation in Eq. (3.78), one obtains

$$\ln f_\pm = -\frac{N_A(z_+z_-)e_0^2}{2\varepsilon RT}\varkappa \qquad (3.80)$$

Now, one can substitute for $\varkappa$ from Eq. (3.43)

$$\varkappa = \left(\frac{4\pi}{\varepsilon kT}\sum n_i^0 z_i^2 e_0^2\right)^{\frac{1}{2}} \qquad (3.43)$$

but, before this substitution is made, $\varkappa$ can be expressed in a different form. Since

$$n_i^0 = \frac{c_i N_A}{1000} \qquad (3.81)$$

where c is the concentration in moles per liter, it follows that

$$\sum n_i^0 z_i^2 e_0^2 = \frac{N_A e_0^2}{1000}\sum c_i z_i^2 \qquad (3.82)$$

Prior to the Debye–Hückel theory, $\frac{1}{2}\sum c_i z_i^2$ had been empirically introduced by Lewis as a quantity of importance in the treatment of ionic solutions. Since it quantifies the charge in an electrolytic solution, it was known as the *ionic strength* and given the symbol I

$$I = \frac{1}{2}\sum c_i z_i^2 \qquad (3.83)$$

In terms of the ionic strength I, $\varkappa$ can be written as [cf. Eqs. (3.43), (3.82), and (3.83)]

$$\varkappa = \left(\frac{8\pi N_A e_0^2}{1000\varepsilon kT}\right)^{\frac{1}{2}} I^{\frac{1}{2}} \qquad (3.84)$$

or as

$$\varkappa = BI^{\frac{1}{2}} \qquad (3.85)$$

where

$$B = \left(\frac{8\pi N_A e_0^2}{1000\varepsilon kT}\right)^{\frac{1}{2}} \qquad (3.86)$$

ION–ION INTERACTIONS

TABLE 3.3

Value of the Parameter B for Water at Various Temperatures

Temperature, °C	$10^{-8}B$
0	0.3248
10	0.3264
20	0.3282
25	0.3291
30	0.3301
35	0.3312
40	0.3323
50	0.3346
60	0.3371
80	0.3426
100	0.3488

Values of B for water at various temperatures are given in Table 3.3. On the basis of the expression (3.85) for $\varkappa$, Eq. (3.80) becomes

$$\ln f_\pm = -\frac{N_A(z_+z_-)e_0^2}{2\varepsilon RT} BI^{\frac{1}{2}} \qquad (3.87)$$

or

$$\log f_\pm = -\frac{1}{2.303} \frac{N_A e_0^2}{2\varepsilon RT} B(z_+z_-)I^{\frac{1}{2}} \qquad (3.88)$$

For greater compactness, one can define a constant A given by

$$A = \frac{1}{2.303} \frac{N_A e_0^2}{2\varepsilon RT} B \qquad (3.89)$$

and write Eq. (3.88) in the form

$$\log f_\pm = -A(z_+z_-)I^{\frac{1}{2}} \qquad (3.90)$$

For 1:1-valent electrolytes, $z_+ = z_- = 1$ and $I = c$, and, therefore,

$$\log f_\pm = -Ac^{\frac{1}{2}} \qquad (3.91)$$

Values of the constant A for water at various temperatures are given in Table 3.4.

In Eqs. (3.90) and (3.91), the theoretical mean ionic-activity coefficients are in a form directly comparable with experiment. A quantitative compari-

TABLE 3.4
Values of Constant A for Water at Various Temperatures

Temperature, °C	Values of constant A
0	0.4918
10	0.4989
20	0.5070
25	0.5115
30	0.5161
40	0.5262
50	0.5373
60	0.5494
80	0.5767
100	0.6086

son of the experimentally observed activity coefficients with those calculated with the Debye–Huckel model can now be made.

Further Reading

1. H. S. Harned and B. B. Owen, *The Physical Chemistry of Electrolytic Solution*, 3rd ed., Reinhold Publishing Corp., New York, 1958.
2. R. A. Robinson and R. H. Stokes, *Electrolytic Solutions*, Butterworth's Publications, Ltd., London, 1959.
3. G. Kortüm, *Treatise on Electrochemistry*, Elsevier, Amsterdam, 1965.

3.5. THE TRIUMPHS AND LIMITATIONS OF THE DEBYE–HÜCKEL THEORY OF ACTIVITY COEFFICIENTS

3.5.1. How Well Does the Debye–Hückel Theoretical Expression for Activity Coefficients Predict Experimental Values?

The approximate theoretical equation

$$\log f_{\pm} = -A(z_+ z_-) I^{\frac{1}{2}} \tag{3.90}$$

indicates that the logarithm of the activity coefficient must decrease linearly with the square root of the ionic strength or, in the case of 1:1-valent electrolytes,[†] with $c^{\frac{1}{2}}$. Further, the slope of the $\log f_{\pm}$ versus $I^{\frac{1}{2}}$ straight line can be unambiguously evaluated from fundamental physical constants and

---

[†] That is, $I = \frac{1}{2} \sum c_i z_i^2$. For a 1:1 electrolyte, $I = \frac{1}{2}(c_i 1^2 + c_j 1^2)$. As $c_i = c_j = c$, $I = c$.

TABLE 3.5

Experimental Values of Activity Coefficients of Various Electrolytes at Different Concentrations at 25°C

1:1 electrolyte, HCl					
Concentration, molal	0.0001	0.0002	0.0005	0.001	0.002
Mean activity coefficient	0.9891	0.9842	0.9752	0.9656	0.9521
2:1 electrolyte, $CaCl_2$					
Concentration, (moles per liter)	0.0018	0.0061	0.0095		
Mean activity coefficient	0.8588	0.7745	0.7361		
2:2 electrolyte, $CdSO_4$					
Concentration, molal	0.0005	0.001	0.005		
Mean activity coefficient	0.774	0.697	0.476		

from (z_+z_-). Finally, the slope does not depend on the particular electrolyte (i.e., whether it is NaCl or KBr, etc.) but only on its valence type, i.e., on the charges borne by the ions of the electrolyte, whether it is a 1:1-valent or 2:2-valent electrolyte, etc. These are clear-cut predictions.

Even before any detailed comparison with experiment, one can use an elementary spot check: At infinite dilution, where the interionic forces are negligible, does the theory yield the activity coefficient which one would expect from experiment, i.e., unity? At infinite dilution, c or $I \to 0$, which means that $\log f_\pm \to 0$ or $f_\pm \to 1$. The properties of an extremely dilute solution of ions should be the same as those of a solution containing nonelectrolyte particles. Thus, the Debye–Hückel theory comes out successfully from the infinite dilution test.

Further, if one takes the experimental values of the activity coefficient (Table 3.5) at extremely low electrolyte concentration and plots $\log f_\pm$ versus $I^{\frac{1}{2}}$ curves, it is seen that: (1) They are linear (cf. Fig. 3.19), and (2) they are grouped according to the valence type of the electrolyte (Fig. 3.20). Finally, when one compares the calculated and observed slopes, it becomes clear that there is excellent agreement to an error of $\pm 0.5\%$ (Table 3.6 and Fig. 3.21) between the results of experiment and the conclusions emerging from an analysis of the ionic-cloud model of the distribution of ions in an electrolyte. Since Eq. (3.90) has been found to be valid at limiting low electrolyte concentrations, it is generally referred to as the *Debye–Hückel limiting law.*

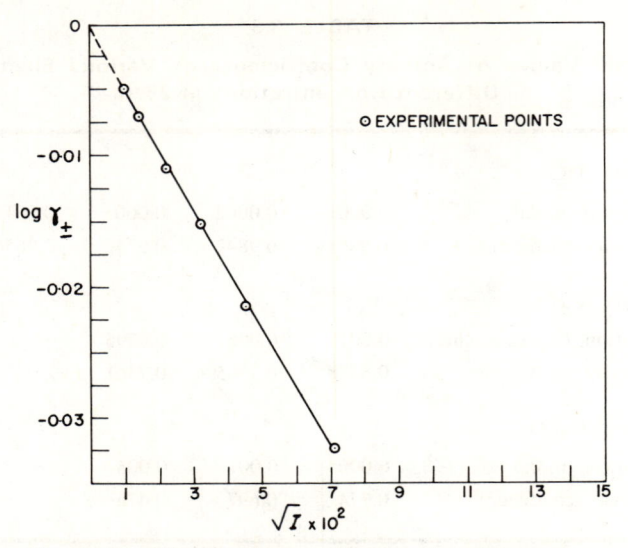

Fig. 3.19. The logarithm of the experimental mean activity coefficient of HCl varies linearly with the square root of the ionic strength.

The success of the Debye–Hückel limiting law is no mean achievement. One has only to think of the complex nature of the real system, of the presence of the solvent which has been recognized only through a dielectric

TABLE 3.6

Experimental and Calculated Values of the Slope of $\log f_{\pm} - \sqrt{I}$ for Alcohol Water Mixtures at 25°C

Solvent mole fraction water	Dielectric constant	Slope Observed	Slope Calculated
\multicolumn{4}{c}{1:1 *type of salt, Croceo tetranitro diamino cobaltiate*}
1.00	78.8	0.50	0.50
0.80	54.0	0.89	0.89
\multicolumn{4}{c}{1:2 *type of salt, Croceo sulfate*}
1.00	78.8	1.10	1.08
0.80	54.0	1.74	1.76
\multicolumn{4}{c}{3:1 *type of salt, Luteo iodate*}
1.00	78.8	1.52	1.51

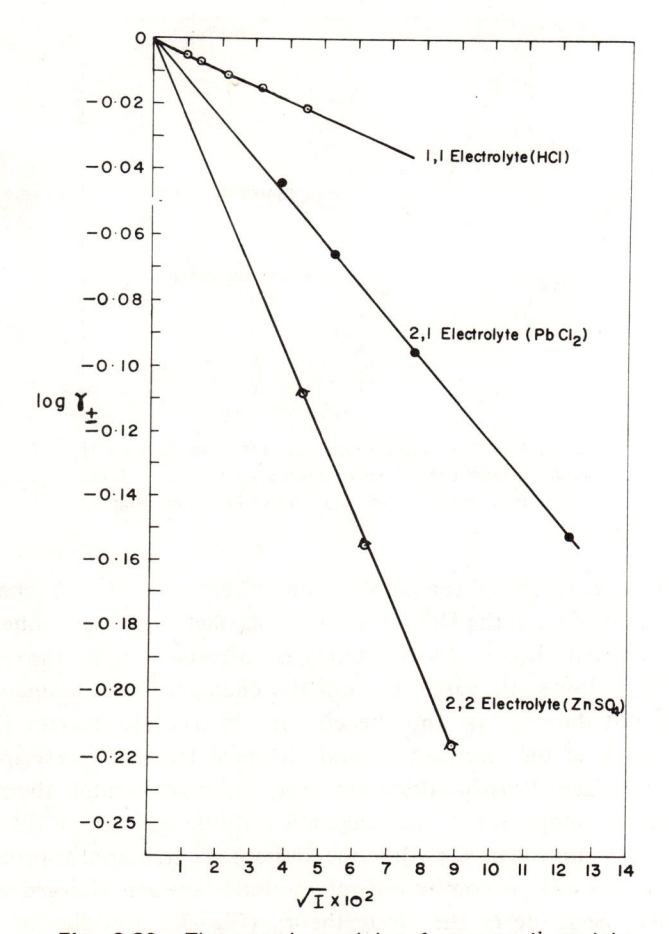

Fig. 3.20. The experimental log $f_\pm$ versus $I^{\frac{1}{2}}$ straight-line plots for different electrolytes can be grouped according to valence type.

constant, of the simplicity of the coulomb force law used, and, finally, of the fact that the ions are not point charges, to realize (*cf.* Table 3.6) that the simple ion-cloud model has been brilliantly successful—almost unexpectedly so. It has grasped the essential truth about electrolytic solutions, albeit about solutions of extreme dilution. The success of the model is so remarkable and the implications so wide (see Section 3.5.6), that the Debye–Hückel approach is to be regarded as one of the most significant pieces of theory in the ionics part of electrochemistry.

It is a theme of this book that model-oriented electrochemistry is to a

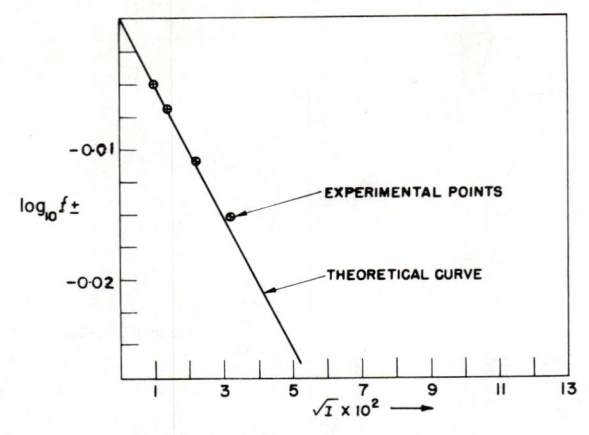

Fig. 3.21. The comparison of the experimentally observed mean activity coefficients of HCl and those that are calculated from the Debye–Hückel limiting law.

great extent the result of the application of electrostatics to chemistry. From this point of view, the Debye–Hückel approach is an excellent example of electrochemical theory. Electrostatics is introduced into the problem in the form of Poisson's equation, and the chemistry is contained in the Boltzmann distribution law and the concept of true electrolytes (Section 3.2). The union of the electrostatic and chemical modes of description to give the linearized Poisson–Boltzmann equation illustrates therefore a characteristic development of electrochemical thinking.

It is hence not surprising that the Poisson–Boltzmann approach has been used frequently in computing interactions between charged entities. Mention may be made of the Gouy theory (Fig. 3.22) of the interaction

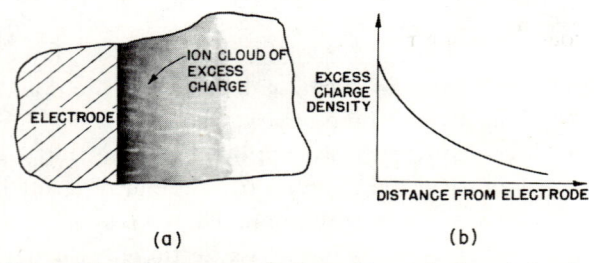

Fig. 3.22. An electrode immersed in an ionic solution is often enveloped by an ionic cloud [see Fig. 3.22 (a)] in which the excess charge density varies with distance as shown in Fig. 3.22 (b).

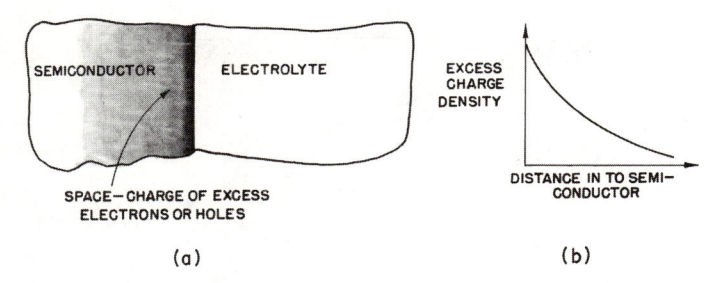

Fig. 3.23. (a) A space charge produced by excess electrons or holes often exists inside the semiconductor. (b) The space charge density varies with distance from the semiconductor–electrolyte interface.

between a charged electrode and the ions in a solution (see Section 7.4). Other examples are the distribution (Fig. 3.23) of electrons or holes inside a semiconductor and in the vicinity of the semiconductor electrolyte interface (see Section 7.7), and the distribution (Fig. 3.24) of charges near a polyelectrolyte molecule or a colloidal particle (see Section 7.8).

However, one must not overstress the triumphs of the Debye–Hückel limiting law [Eq. (3.90)]. Models are always simplifications of reality. They never treat all its complexities, and, thus, there can never be a *perfect* fit between experiment and the predictions based on a model.

What, then, are the inadequacies of the Debye–Hückel limiting law? One does not have to look far. If one examines the experimental $\log f_\pm$ versus $I^{\frac{1}{2}}$ curve, not just in the extreme dilution regions, but at higher concentrations, it turns out that the simple Debye–Hückel limiting law falters. The plot of $\log f_\pm$ versus $I^{\frac{1}{2}}$ is a curve (Fig. 3.25 and Table 3.7)

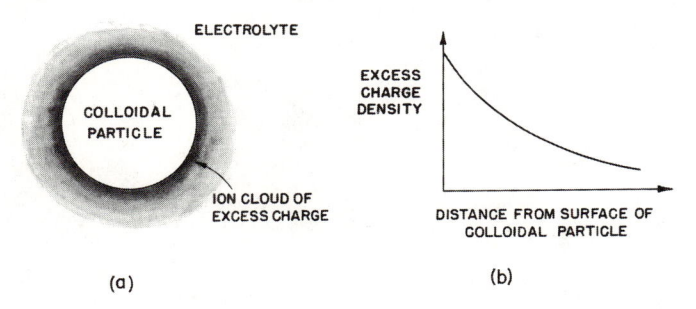

Fig. 3.24. (a) A colloidal particle is surrounded by an ionic cloud of excess charge density. (b) The excess charge density in the cloud varies with distance from the surface of the colloidal particle.

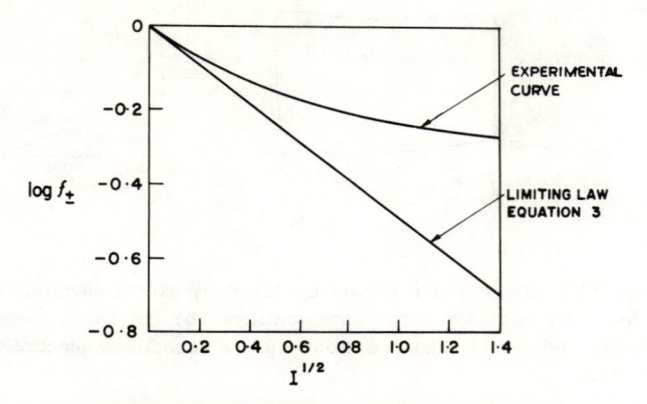

Fig. 3.25. The experimental log $f_\pm$ versus $I^{1/2}$ curve is a straight line only at extremely low concentrations.

and not a straight line as promised by Eq. (3.90). Further, the curves depend not only on valence type (e.g., 1:1 or 2:2) but also (Fig. 3.26) on the particular electrolyte (e.g., NaCl or KCl).

It appears that the Debye–Hückel law is the law for the tangent to the $\log f_\pm$ versus $I^{1/2}$ curve at very low concentrations, say, up to $0.01 N$ for 1:1 electrolytes in aqueous solutions. At higher concentrations, the model must be improved. What refinements can be made?

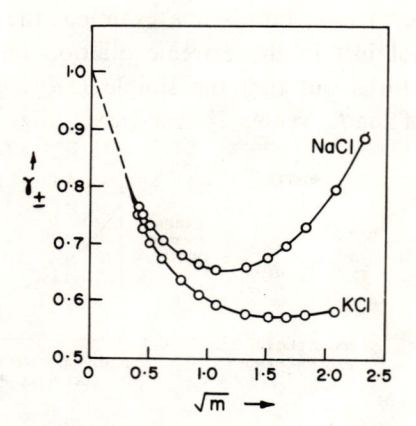

Fig. 3.26. Even though NaCl and KCl are 1 : 1 electrolytes, their activity coefficients vary in different ways with concentration directly one examines to higher concentrations.

TABLE 3.7
Comparison of Calculated [Eq. (3.90)] and Experimental Values of log $f_\pm$ for NaCl at 25°C

Concentration, molal	$-\log f_\pm$ experimental	$-\log f_\pm$ calculated
0.001	0.0155	0.0162
0.002	0.0214	0.0229
0.005	0.0327	0.0361
0.01	0.0446	0.0510
0.02	0.0599	0.0722

3.5.2. Ions Are of Finite Size, Not Point Charges

One of the general procedures for refining a model which has been successful in an extreme situation is to liberate the theory from its approximations. So one has to recall what approximations have been used to derive the Debye–Hückel limiting law. The first one that comes to mind is the point-charge approximation.† One now asks: Is it reasonable to consider ions as point charges?

It has been shown (cf. Section 3.3.8) that the mean thickness $\varkappa^{-1}$ of the ionic cloud depends on the concentration. As the concentration of a 1:1 electrolyte increases from $0.001N$ to $0.01N$ to $0.1N$, $\varkappa^{-1}$ decreases from about 100 to 30 to about 10 Å. This means that the relative dimensions of the ion cloud and of the ion change with concentration. Whereas the radius of the cloud is 100 times the radius of the ion at $0.001N$, it is only about 10 times the dimensions of an ion at $0.1N$. Obviously, under these latter circumstances, an ion cannot be considered a geometrical point charge in comparison with a dimension only 10 times its size (Fig. 3.27). The more concentrated the solution, i.e., the smaller the size $\varkappa^{-1}$ of the ion cloud (Section 3.3.8), the less valid is the point-charge approximation.

If, therefore, one wants the theory to be applicable to $0.1N$ solutions or to solutions of even higher concentration, the finite size of the ions must be introduced into the mathematical formulation.

To remove the assumption that ions can be treated as point charges,

---

† Another approximation in the Debye–Hückel model involves the use of Poisson's equation, which is based on the smearing-out of the charges into a continuously varying charge density. At high concentrations, the mean distance between charges is low, and the ions see each other as discrete point charges, not as smoothed-out charges. Thus, the use of Poisson's equation becomes less and less justified as the solution becomes more and more concentrated.

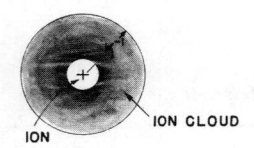

Fig. 3.27. At 0.1*N*, the thickness of the ion cloud is only 10 times the radius of the central ion.

it is necessary, at first, to recall at what stage in the derivation of the theory the assumption was invoked.

The linearized P–B equation involved neither the point-charge approximation nor any considerations of the dimensions of the ions. Hence, the basic differential equation

$$\frac{1}{r^2}\frac{\partial}{\partial r}\left(r^2\frac{\partial \psi_r}{dr}\right) = \varkappa^2 \psi_r \qquad (3.21)$$

and its general solution, i.e.,

$$\psi_r = A\frac{e^{-\varkappa r}}{r} + B\frac{e^{+\varkappa r}}{r} \qquad (3.28)$$

can be taken as the basis for the generalization of the theory for finite-sized ions.

As before (*cf.* Section 3.3.7), the integration constant B must be zero because, otherwise, one cannot satisfy the requirement of physical sense that, as $r \to \infty$, $\psi \to 0$. Hence, Eq. (3.28) reduces to

$$\psi_r = A\frac{e^{-\varkappa r}}{r} \qquad (3.29)$$

In evaluating the constant A, a procedure different from that used after (3.29) is adopted. The charge dq in any particular spherical shell (of thickness dr) situated at a distance r from the origin is, as argued earlier,

$$dq = \varrho_r 4\pi r^2 \, dr \qquad (3.36)$$

The charge density ϱ_r is obtained thus

$$\varrho_r = -\frac{\varepsilon}{4\pi}\left[\frac{1}{r^2}\frac{\partial}{\partial r}\left(r^2\frac{\partial \psi}{\partial r}\right)\right] = -\frac{\varepsilon}{4\pi}\varkappa^2 \psi_r \qquad (3.34)$$

and, inserting the expression for ψ_r from Eq. (3.29), one obtains

$$\varrho_r = -\frac{\varepsilon}{4\pi}\varkappa^2 A\frac{e^{-\varkappa r}}{r} \qquad (3.92)$$

ION–ION INTERACTIONS 221

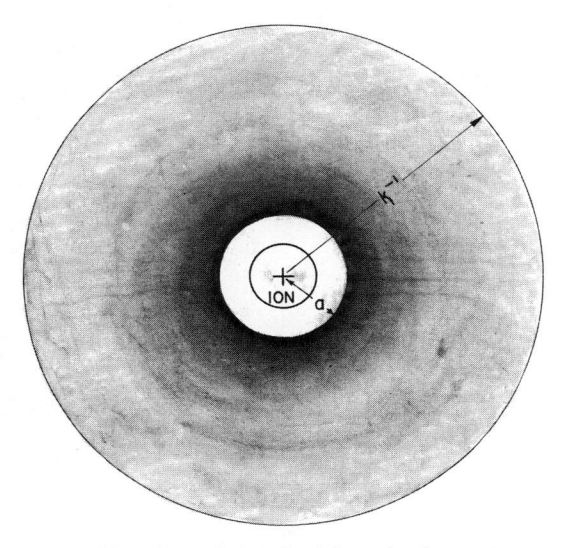

Fig. 3.28. For a finite-sized ion, the ion atmosphere starts at a distance *a* from the center of the reference ion.

Thus, by combining Eqs. (3.36) and (3.92),

$$dq = -A\varkappa^2\varepsilon(e^{-\varkappa r}r\, dr) \tag{3.93}$$

The total charge in the ion cloud q_{cloud} is, on the one hand, equal to $-z_i e_0$ [*cf.* Eq. (3.41)] as required by the electroneutrality condition and, on the other hand, the result of integrating dq. Thus,

$$q_{\text{cloud}} = -z_i e_0 = \int_?^\infty dq\, dr = -A\varkappa^2\varepsilon \int_?^\infty e^{-\varkappa r}r\, dr \tag{3.94}$$

What lower limit should be used for the integration? In the point-charge model, one used a lower limit of zero, meaning that the ion cloud commences from zero (i.e., from the surface of a zero-radius ion) and extends to infinity. But now the ions are taken to be of finite size, and a lower limit of zero is obviously wrong. The lower limit should be a distance corresponding to the distance from the ion center at which the ionic atmosphere starts (Fig. 3.28).

As a first step, one can use for the lower limit of the integration a distance parameter which is greater than zero. Then, one can go through the mathematics and later worry about the physical implications of the ion-size parameter. Let this procedure be adopted and symbol a be used for the ion-size parameter.

One has, then,

$$\int_a^\infty dq\, dr = -A\varkappa^2\varepsilon \int_a^\infty e^{-\varkappa r} r\, dr$$
$$= -A\varepsilon \int_a^\infty \varkappa r e^{-(\varkappa r)}\, d\varkappa r \qquad (3.95)$$

As before (cf. Appendix 3.2), one can integrate by parts, thus,

$$\int_a^\infty \varkappa r e^{-(\varkappa r)}\, d\varkappa r = -[\varkappa r e^{-\varkappa r}]_a^\infty + \int_a^\infty e^{-\varkappa r}\, d\varkappa r$$
$$= \varkappa a e^{-\varkappa a} - [e^{-\varkappa r}]_a^\infty \qquad (3.96)$$

Hence, inserting Eq. (3.96) in Eq. (3.95), one obtains

$$\int_a^\infty dq\, dr = -A\varepsilon e^{-\varkappa a}(1 + \varkappa a) = -z_i e_0 \qquad (3.97)$$

from which

$$A = \frac{z_i e_0}{\varepsilon} \frac{e^{\varkappa a}}{1 + \varkappa a} \qquad (3.98)$$

Using this value of A in Eq. (3.30), one obtains a new and less approximate expression for the potential ψ_r at a distance r from a finite-size central ion,

$$\psi_r = \frac{z_i e_0}{\varepsilon} \frac{e^{\varkappa a}}{1 + \varkappa a} \frac{e^{-\varkappa r}}{r} \qquad (3.99)$$

3.5.3. The Theoretical Mean Ionic-Activity Coefficient in the Case of Ionic Clouds with Finite-Sized Ions

Once again (cf. Section 3.3.9), one can use the law of superposition of potentials to obtain the ionic-atmosphere contribution ψ_{cloud} to the potential ψ_r at a distance r from the central ion. From Eq. (3.46), i.e.,

$$\psi_{\text{cloud}} = \psi_r - \psi_{\text{ion}} \qquad (3.46)$$

it follows by substitution of the expression (3.99) for ψ_r and Eq. (3.44) for ψ_{ion} that

$$\psi_{\text{cloud}} = \frac{z_i e_0}{\varepsilon r} \frac{e^{\varkappa(a-r)}}{1 + \varkappa a} - \frac{z_i e_0}{\varepsilon r}$$
$$= \frac{z_i e_0}{\varepsilon r} \left[\frac{e^{\varkappa(a-r)}}{1 + \varkappa a} - 1 \right] \qquad (3.100)$$

It will be recalled, however, that, in order to calculate the activity coefficient from the expressions

$$RT \ln f_i = \Delta\mu_{i-I} \tag{3.59}$$

and

$$\Delta\mu_{i-I} = \frac{N_A z_i e_0}{2} \psi \tag{3.3}$$

i.e., from

$$\ln f_i = \frac{N_A z_i e_0}{2RT} \psi \tag{3.101}$$

it is necessary to know ψ, which is the potential at the surface of the ion due to the surrounding ions, i.e., due to the cloud. Since, in the finite-ion-size model, the ion is taken to have a size a, it means that ψ is the value of ψ_{cloud} at $r = a$,

$$\psi = \psi_{\text{cloud}} \qquad r = a \tag{3.102}$$

The value of ψ_{cloud} at $r = a$ is got by setting $r = a$ in Eq. (3.100). Hence,

$$\psi = \psi_{\text{cloud}(r=a)} = -\frac{z_i e_0}{\varepsilon \varkappa^{-1}} \frac{1}{1 + \varkappa a} \tag{3.103}$$

By substitution of the expression (3.103) for $\psi = \psi_{\text{cloud}(r=a)}$ in Eq. (3.101), one obtains

$$\ln f_i = -\frac{N_A(z_i e_0)^2}{2\varepsilon RT \varkappa^{-1}} \frac{1}{1 + \varkappa a} \tag{3.104}$$

This individual ionic-activity coefficient can be transformed into a mean ionic-activity coefficient by the same procedure as for the Debye–Hückel limiting law (*cf.* Section 3.4.12). On going through the algebra, one finds that the expression for $\log f_\pm$ in the finite-ion-size model is

$$\log f_\pm = -\frac{A(z_+ z_-)I^{\frac{1}{2}}}{1 + \varkappa a} \tag{3.105}$$

It will be recalled, however, that the thickness $\varkappa^{-1}$ of the ionic cloud can be written as [Eq. (3.85)]

$$\varkappa = BI^{\frac{1}{2}} \tag{3.85}$$

Using this notation, one ends up with the final expression

$$\log f_\pm = -\frac{A(z_+ z_-)I^{\frac{1}{2}}}{1 + BaI^{\frac{1}{2}}} \tag{3.106}$$

If one compares Eq. (3.105) of the finite-ion-size model with Eq. (3.90) of the point-charge approximation, it is clear that the only difference between the two expressions is that the former contains a term $1/(1 + \varkappa a)$ in the denominator. Now, one of the tests of a more general version of a theory is the *correspondence principle*, i.e., the general version of a theory must reduce to the approximate version under the conditions of applicability of the latter. Does the Eq. (3.105) from the finite-ion-size model reduce to Eq. (3.90) from the point-charge model?

Rewrite Eq. (3.105) in the form

$$\log f_\pm = -A(z_+ z_-) I^{\frac{1}{2}} \frac{1}{1 + a/\varkappa^{-1}} \tag{3.107}$$

and consider the term $a/\varkappa^{-1}$. As the solution becomes increasingly dilute, the radius $\varkappa^{-1}$ of the ionic cloud becomes increasingly large compared with the ion size, and, simultaneously, $a/\varkappa^{-1}$ becomes increasingly small compared with unity, or

$$\frac{1}{1 + a/\varkappa^{-1}} \sim 1 \tag{3.108}$$

Thus, directly the solution is sufficiently dilute to make $a \ll \varkappa^{-1}$, i.e., to make the ion size insignificant in comparison with the radius of the ion atmosphere, the finite-ion-size model Eq. (3.105) reduces to the corresponding Eq. (3.90) of the point-charge model because the extra term $1/(1 + a/\varkappa^{-1})$ tends to unity

$$-\left[\frac{A(z_+ z_-) I^{\frac{1}{2}}}{1 + \varkappa a} \right]_{a \ll \varkappa^{-1}} = -A(z_+ z_-) I^{\frac{1}{2}} \tag{3.109}$$

The physical significance of $a/\varkappa^{-1} \ll 1$ is that, at very low concentrations, the ion atmosphere has such a large radius compared with that of the ion that one need not consider the ion as having a finite size a. Considering $a/\varkappa^{-1} \ll 1$ is tantamount to reverting to the point-charge model.

One can now proceed rapidly to compare this theoretical expression for $\log f_\pm$ with experiment; but what value of the ion-size parameter should be used? The time has come to worry about the precise physical meaning of the parameter a which was introduced to allow for the finite size of ions.

3.5.4. The Ion-Size Parameter a

One can at first try to speculate on what value of the ion-size parameter is appropriate. A lower limit is the sum of the *crystallographic* radii of the

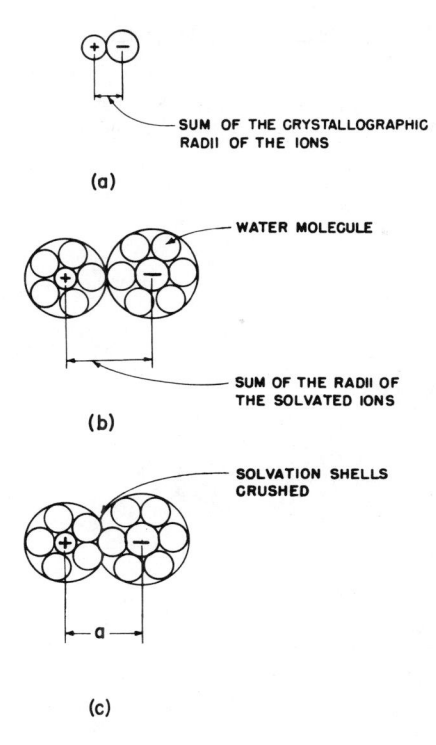

Fig. 3.29. The ion-size parameter cannot be (a) less than the sum of the crystallographic radii of the ions or (b) more than the sum of the radii of the solvated ions and is most probably (c) less than the sum of the radii of the solvated ions because the solvation shells may be crushed.

positive and negative ions present in solution; ions cannot come closer than this distance [Fig. 3.29(a)]. But, in a solution, the ions are generally solvated (*cf.* Chapter 2). So perhaps the sum of the solvated radii should be used [Fig. 3.29(b)]. However, when two solvated ions collide, is it not likely [Fig. 3.29(c)] that their hydration shells are crushed to some extent? This means that the ion-size parameter a should be greater than the sum of the crystallographic radii and perhaps less than the sum of the solvated radii. It should best be called the *mean distance of closest approach*, but, beneath the apparent wisdom of this term, there lies a measure of ignorance. For example, an attempted calculation of just how crushed together two solvated ions are would involve many difficulties.

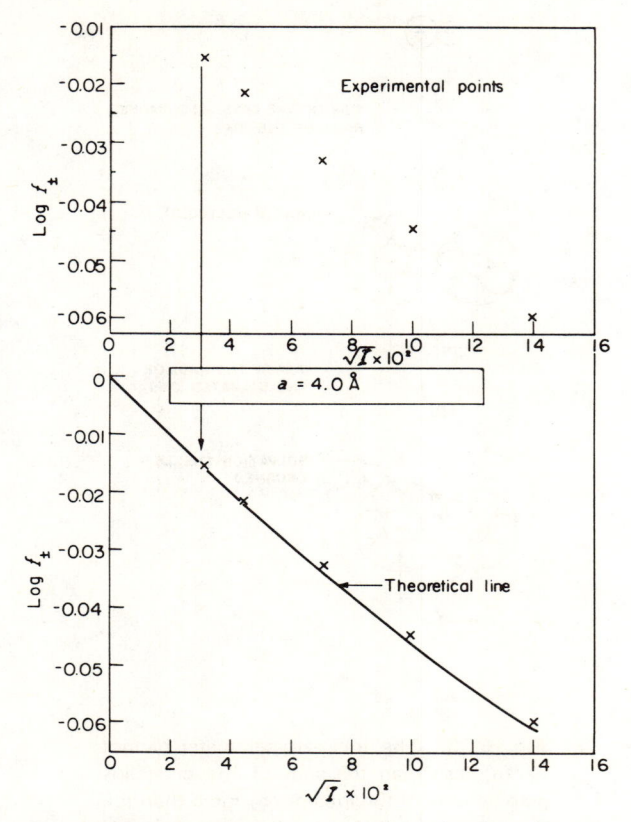

Fig. 3.30. Procedure for recovering the ion-size parameter from experiment and then using it to produce a theoretical log $f_\pm$ versus $I^{\frac{1}{2}}$ curve which can be compared with an experimental curve.

To circumvent the uncertainty in the quantitative definition of *a*, it is best to regard it as a *parameter* in Eq. (3.106), i.e., a quantity, the numerical value of which is left to be calibrated or adjusted on the basis of experiment. The procedure (Fig. 3.30) is to assume that the expression for $\log f_\pm$ [Eq. (3.106)] is correct at one concentration, then to equate this theoretical expression to the experimental value of $\log f_\pm$ corresponding to that concentration, and to solve the resulting equation for *a*. Once the ion-size parameter, or mean distance of closest approach, is thus obtained at one concentration, the value can be used for the calculation of values of the activity coefficient over a range of other and higher concentrations. Then, the situation is regarded as satisfactory if the value of *a* obtained from experiments at one concentration can be used in Eq. (3.106) to reproduce the results of experiments over a range of concentrations.

3.5.5. Comparison of the Finite-Ion-Size Model with Experiment

After taking into account the fact that ions have finite dimensions and cannot therefore be treated as point charges, the following expression has been derived for the logarithm of the activity coefficient:

$$\log f_\pm = -\frac{A(z_+ z_-) I^{\frac{1}{2}}}{1 + BaI^{\frac{1}{2}}} \qquad (3.106)$$

How does the general form of this expression compare with the Debye–Hückel limiting law as far as agreement with experiment is concerned? To see what the extra term $(1 + BaI^{\frac{1}{2}})^{-1}$ does to the shape of the $\log f_\pm$ versus $I^{\frac{1}{2}}$ curve, one can expand it in the form of a binomial series

$$\frac{1}{1+x} = (1+x)^{-1} = 1 - x + \frac{x^2}{2!} - \cdots \qquad (3.110)$$

and use only the first two terms. Thus,

$$\frac{1}{1 + BaI^{\frac{1}{2}}} = 1 - BaI^{\frac{1}{2}} \qquad (3.111)$$

and, therefore,

$$\log f_\pm \sim -A(z_+ z_-) I^{\frac{1}{2}} (1 - BaI^{\frac{1}{2}}) \qquad (3.112)$$

$$\sim -A(z_+ z_-) I^{\frac{1}{2}} + \text{constant}(I^{\frac{1}{2}})^2 \qquad (3.113)$$

This result is encouraging. It shows that the $\log f_\pm$ versus $I^{\frac{1}{2}}$ curve give values of $\log f_\pm$ higher than those given by the limiting law, the deviation increasing with concentration. In fact, the general shape of the predicted curve (Fig. 3.31) is very much on the right lines.

The values of the ion-size parameter, or closest distance of approach, which are recovered from experiment are physically reasonable for many electrolytes. They lie around 3 to 5 Å, which is greater than the sum of the crystallographic radii of the positive and negative ions and pertains more to the solvated ion (Table 3.8).

By picking on a reasonable value of the ion-size parameter a, independent of concentration, it is found that, in many cases, Eq. (3.112) gives a very good fit with experiment, often for ionic strengths up toward 0.1. For example, on the basis of $a = 4.0$ Å, Eq. (3.112) gives an almost exact

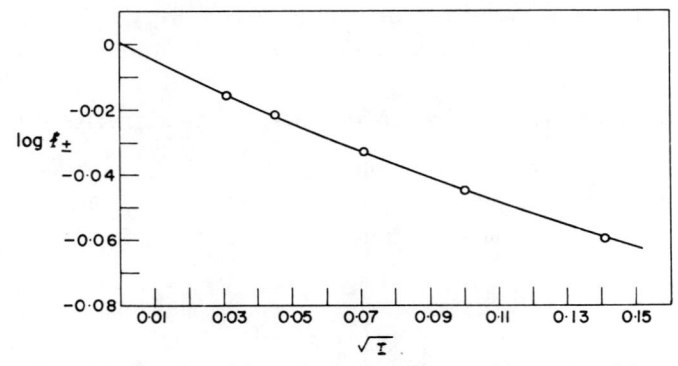

Fig. 3.31. Comparison of the experimental mean activity coefficients with theory for Eq. (3.112).

agreement up to $0.02M$ in the case of sodium chloride (Fig. 3.32 and Table 3.9).

The ion-size parameter a has done part of the job of extending the range of concentration in which the Debye–Hückel theory of ionic clouds agrees with experiment. But has it done the whole job? One must therefore start looking for discrepancies between theory and fact and for the less satisfactory features of the model.

The most obvious drawback of the finite-ion-size version of the Debye–Hückel theory lies in the fact that a is an *adjustable parameter*. When parameters which have to be taken from experiment enter a theory, they imply that the physical situation has been incompletely comprehended or is too complex to be mathematically analyzed. In contrast, the constants of the limiting law were calculated without recourse to experiment.

The best illustration of the fact that a has to be adjusted is its concentration dependence. As the concentration changes, the ion-size para-

TABLE 3.8

Values of Ion-Size Parameter for a Few Electrolytes

Salt	a, Å
HCl	4.5
HBr	5.2
LiCl	4.3
NaCl	4.0
KCl	3.6

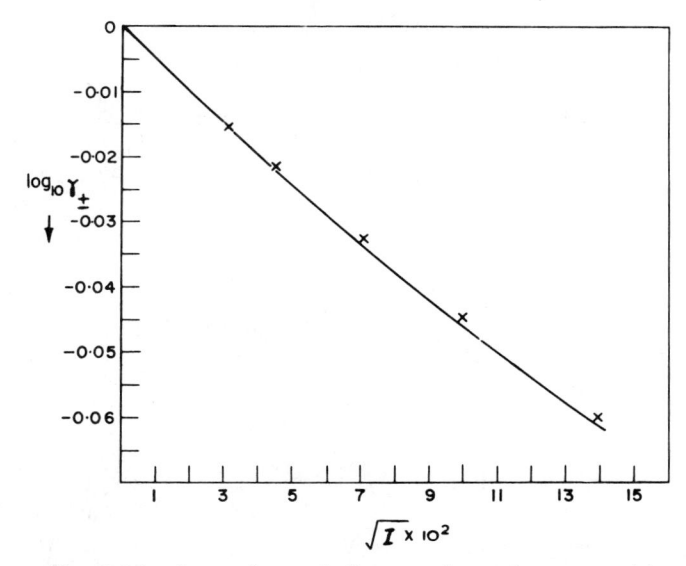

Fig. 3.32. Comparison of the experimental mean activity coefficients for sodium chloride with the theoretical log $f_{\pm}$ versus $I^{\frac{1}{2}}$ curve based on Eq. (3.112) with $a = 4.0$ Å.

meter has to be modified (Fig. 3.33). Further, for some electrolytes at higher concentrations, a has to assume quite impossible (i.e., large negative, irregular) values to fit the theory to experiment (Table 3.10).

Evidently, there are factors at work in an electrolytic solution which have not yet been reckoned with, and the ion-size parameter is being asked to include the effects of all these factors simultaneously, even though these other factors probably have little to do with the size of the ions and

TABLE 3.9

Experimental Mean Activity Coefficients and Those Calculated from Eq. (3.112) with $a = 4.0$ Å at 25°C at Various Concentrations of NaCl

Molality	Experimental mean activity coefficient $-\log f_{\pm}$	Calculated
0.001	0.0155	0.0155
0.002	0.0214	0.0216
0.005	0.0327	0.0330
0.01	0.0446	0.0451
0.02	0.0599	0.0609

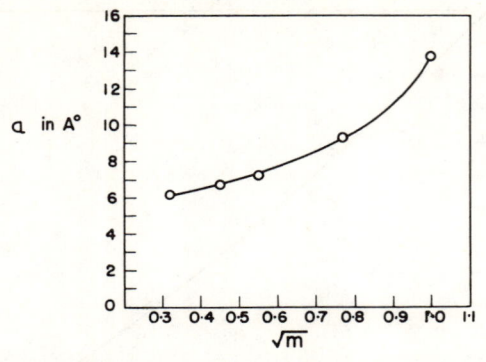

Fig. 3.33. The variation of the ion-size parameter with concentration of NaCl.

may vary with concentration. If this were so, the ion-size parameter a, calculated back from experiment, would indeed have to vary with concentration. The problem, therefore, is: What factors, forces, and interactions were neglected in the Debye–Hückel theory of ionic clouds?

3.5.6. The Debye–Hückel Theory of Ionic Solutions: An Assessment

It is appropriate at this stage to register the achievement in the theory of ionic solutions described thus far.

Starting with the point of view that ion–ion interactions are bound to operate in an electrolytic solution, the chemical-potential change $\Delta\mu_{i-I}$, in going from a hypothetical state of noninteracting ions to a state in which the ions of species i interact with the ionic solution, was considered a quantitative measure of these interactions. As a first approximation, the

TABLE 3.10
Values of Parameter a at Higher Concentrations

Concentration, molality	Value of a for HCl, Å	Concentration, molality	Value of a for LiCl, Å
1	13.8	2	41.3
1.4	24.5		
1.8	85.0	2.5	−141.9
2	−411.2		
2.5	− 27.9	3	− 26.4
3	− 14.8		

ion–ion interactions were assumed to be purely coulombic in origin. Hence, the chemical-potential change arising from the interactions of species *i* with the electrolytic solution is given by the Avogadro number times the electrostatic work W resulting from taking a discharged reference ion and charging it up in the solution to its final charge. In other words, the charging work is given by the same formula as that used in the Born theory of solvation, i.e.,

$$W = \frac{z_i e_0}{2} \psi \tag{3.3}$$

where ψ is the electrostatic potential at the surface of the reference ion, contributed by the other ions in the ionic solution. The problem, therefore, was to obtain a theoretical expression for the potential ψ. This involved an understanding of the distribution of ions around a given reference ion.

It was in tackling this apparently complicated task that appeal was made to the Debye–Hückel simplifying model for the distribution of ions in an ionic solution. This model treats only one ion—the central ion—as a discrete charge, the charge of the other ions being smoothed out to give a continuous charge density. Because of the tendency of negative charge to accumulate near a positive ion, and *vice versa*, the smoothed-out positive and negative charge densities do not cancel out; rather, their imbalance gives rise to an excess local charge density ϱ_r, which of course dies away toward zero as the distance from the central ion is increased. Thus, the calculation of the distribution of ions in an electrolytic solution reduces to the calculation of the variation of excess charge density ϱ_r with distance r from the central ion.

The excess charge density ϱ_r was taken to be given, on the one hand, by Poisson's equation of electrostatics

$$\varrho_r = -\frac{\varepsilon}{4\pi} \frac{1}{r^2} \frac{d}{dr}\left(r^2 \frac{d\psi_r}{dr}\right) \tag{3.17}$$

and, on the other, by the linearized Boltzmann distribution law

$$\varrho_r = -\sum \frac{n_i^0 z_i^2 e_0^2 \psi_r}{kT} \tag{3.18}$$

The result of equating these two expressions for the excess charge density is the fundamental partial differential equation of the Debye–Hückel model, the linearized P–B equation (*cf.* Fig. 3.34)

$$\frac{1}{r^2} \frac{d}{dr}\left(r^2 \frac{d\psi_r}{dr}\right) = \varkappa^2 \psi_r \tag{3.21}$$

232 CHAPTER 3

Poisson Equation

$$\varrho_r = -\frac{\varepsilon}{4\pi}\left[\frac{1}{r^2}\frac{d}{dr}\left(r^2\frac{d\psi_r}{dr}\right)\right]$$

Boltzmann Equation

$$\varrho_r = \sum n_i^0 z_i e_0 \exp\left(-\frac{z_i e_0 \psi_r}{kT}\right)$$

Linearized Boltzmann Equation

Electroneutrality

$$\sum n_i^0 z_i e_0 = 0$$

$$\varrho_r = \sum n_i^0 z_i e_0 - \sum \frac{n_i^0 z_i^2 e_0^2 \psi_r}{kT}$$

$$\varrho_r = -\sum \frac{n_i z_i^2 e_0^2 \psi_r}{kT}$$

Linearized Poisson–Boltzmann Equation

$$\frac{1}{r^2}\frac{d}{dr}\left(r^2\frac{d\psi_r}{dr}\right) = \left[\frac{4\pi}{\varepsilon kT}\sum n_i^0 z_i^2 e_0^2\right]\psi_r$$

$$\varkappa^2 = \frac{4\pi}{\varepsilon kT}\sum n_i^0 z_i^2 e_0^2$$

$$\frac{1}{r^2}\frac{d}{dr}\left(r^2\frac{d\psi_r}{dr}\right) = \varkappa^2 \psi_r$$

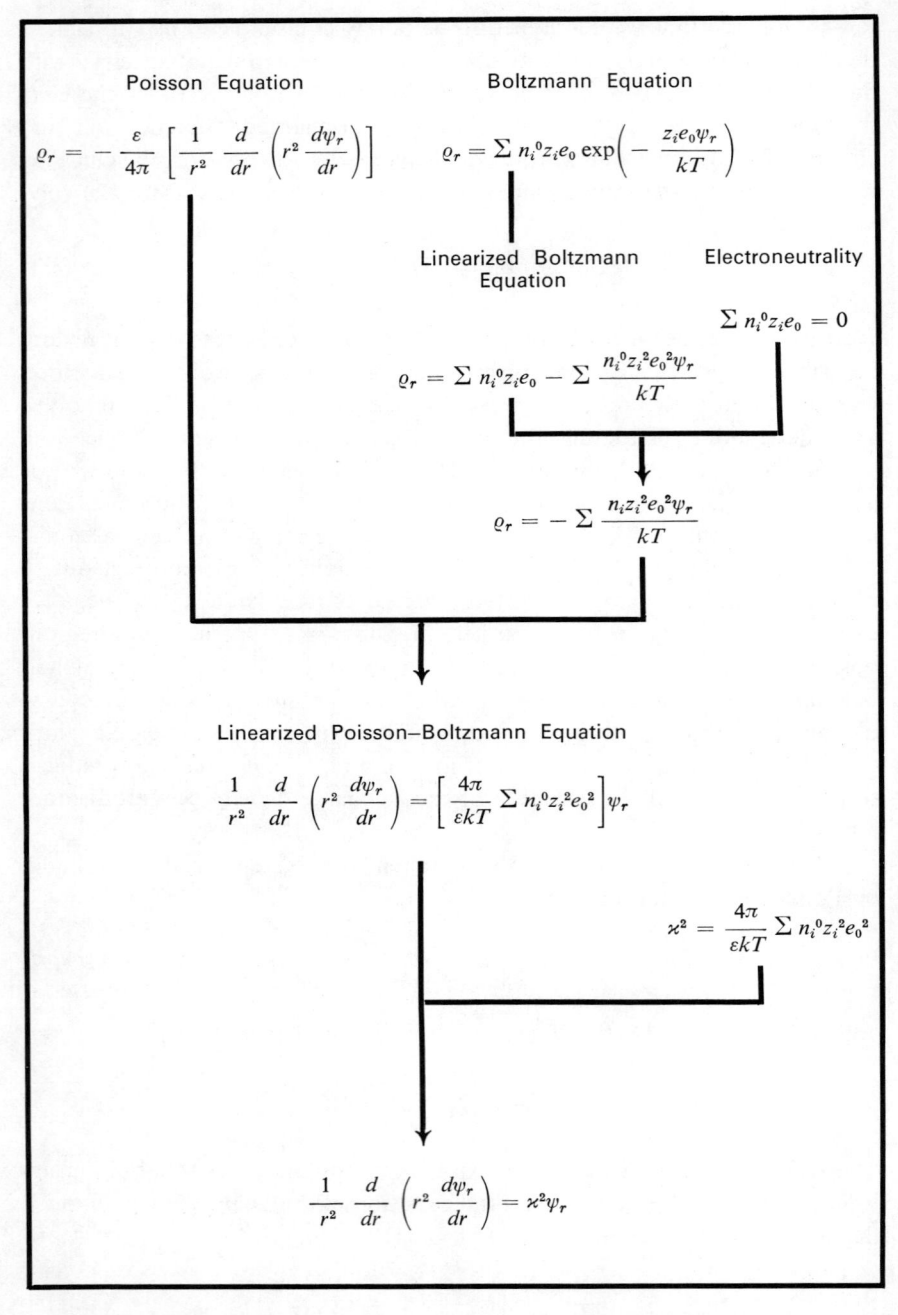

Fig. 3.34. Steps in the derivation of the linearized Poisson–Boltzmann equation.

where

$$\varkappa^2 = \frac{4\pi}{\varepsilon kT} \sum n_i^0 z_i^2 e_0^2 \tag{3.20}$$

By assuming that ions can be regarded as point charges, the solution of the linearized P–B equation turns out to be (*cf.* Fig. 3.35)

$$\psi_r = \frac{z_i e_0}{\varepsilon} \frac{e^{-\varkappa r}}{r} \tag{3.33}$$

Such a variation of potential with distance from a typical (central or reference) ion corresponded to a charge distribution which can be expressed as a function of distance r from the central ion by

$$\varrho_r = -\frac{z_i e_0}{4\pi} \varkappa^2 \frac{e^{-\varkappa r}}{r} \tag{3.35}$$

This variation of the excess charge density with distance around the central or typical ion yielded a simple physical picture. A reference positive ion can be thought of as being surrounded by a cloud of negative charge of radius $\varkappa^{-1}$. The charge density in this ionic atmosphere, or ionic cloud, decays in the manner indicated by Eq. (3.35). Thus, the interactions between a reference ion and the surrounding ions of the solution is equivalent to the interactions between the reference ion and the ionic cloud which, in the point-charge model, sets up at the central ion a potential ψ_{cloud} given by

$$\psi_{\text{cloud}} = -\frac{z_i e_0}{\varepsilon \varkappa^{-1}} \tag{3.49}$$

The magnitude of central ion–ionic-cloud interactions is given by introducing the expression for ψ_{cloud} into the expression (3.3) for the work of creating the ionic cloud, i.e., setting up the ionic interaction situation. Thus, one obtains for the energy of such interactions

$$\Delta \mu_{i-I} = -\frac{N_A (z_i e_0)^2}{2\varepsilon \varkappa^{-1}} \tag{3.51}$$

In order to test these predictions, attention was drawn to an empirical treatment of ionic solutions. For solutions of noninteracting particles, the chemical-potential change in going from a solution of unit concentration to one of concentration x_i is described by the equation

$$\mu_i - \mu_i^0 = RT \ln x_i \tag{3.52}$$

Linearized Poisson–Boltzmann Equations

$$\frac{1}{r^2}\frac{d}{dr}\left[r^2\frac{d\psi_r}{dr}\right] = \varkappa^2\psi_r$$

↓ solution

$$\psi_r = \frac{Ae^{-\varkappa r}}{r} + \frac{Be^{\varkappa r}}{r}$$

Boundary Condition I

As $r \to \infty$, $\psi_r \to 0$

$$\psi_r = \frac{Ae^{-\varkappa r}}{r}$$

Boundary Condition II

As $n_i^0 \to 0$, i.e., $\varkappa \to 0$

(1) $\psi_r = \dfrac{A}{r}$

(2) $\psi_r = \psi_{\text{due to point charge}}$

$$= \frac{z_i e_0}{\varepsilon r}$$

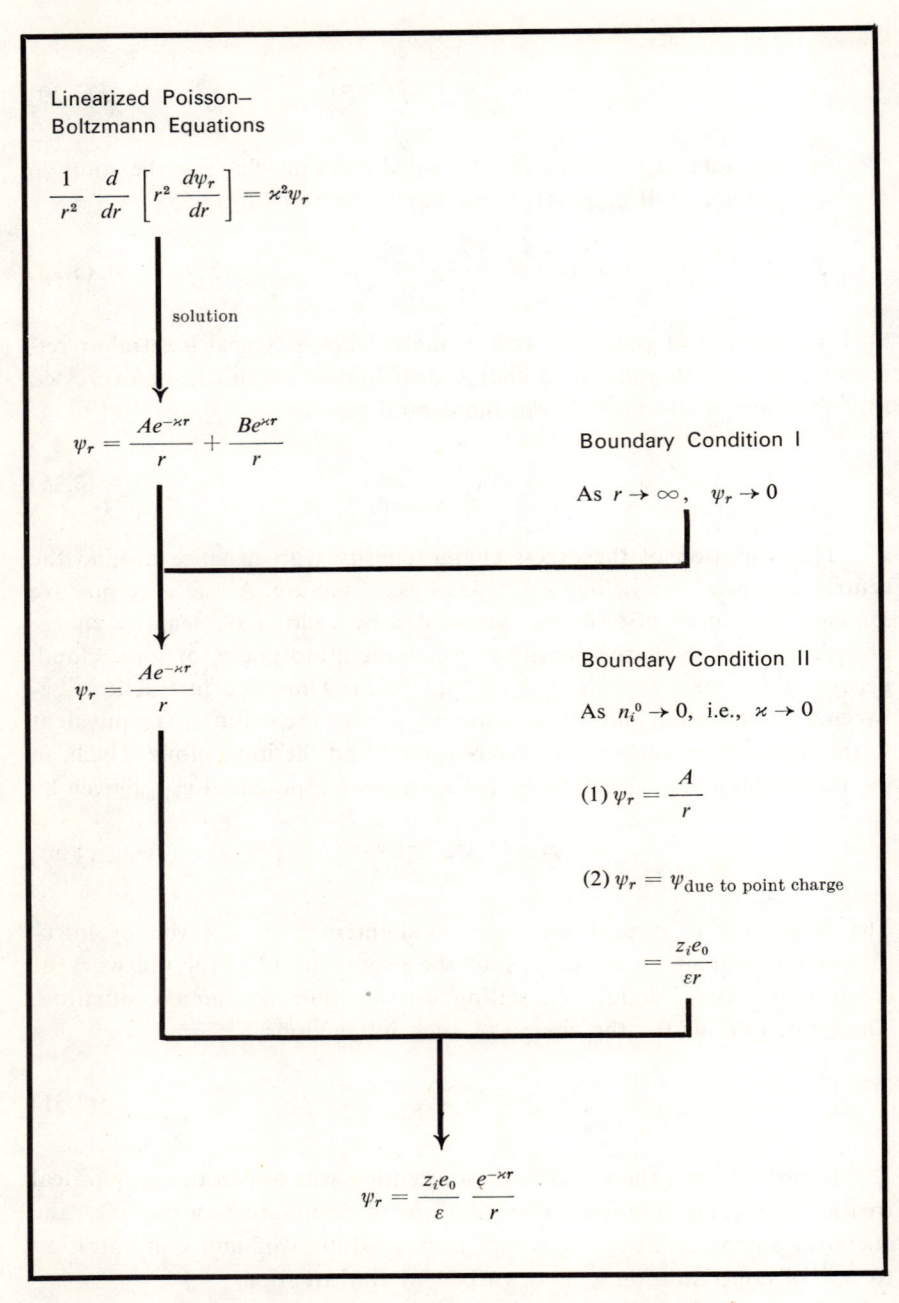

$$\psi_r = \frac{z_i e_0}{\varepsilon}\frac{e^{-\varkappa r}}{r}$$

Fig. 3.35. Steps in the solution of the linearized Poisson–Boltzmann equation for point-charge ions.

However, in the case of an electrolytic solution in which there are ion–ion interactions, it is experimentally observed that

$$\mu_i - \mu_i^0 \neq RT \ln x_i$$

If one is unaware of the nature of these interactions, one can write an empirical equation to compensate for one's ignorance

$$RT \ln f_i = (\mu_i - \mu_i^0) - RT \ln x_i \qquad (3.58)$$

and say that solutions behave ideally if the so-called activity coefficient f_i is unity, i.e., $RT \ln f_i = 0$, and, in real solutions, $f_i \neq 1$. It is clear that f_i corresponds to a coefficient to account for the behavior of ionic solutions, which differs from those in which there are no charges. Thus, f_i accounts for the interactions of the charges, so that

$$RT \ln f_i = \Delta \mu_{i-I} = -\frac{N_A (z_i e_0)^2}{2\varepsilon \varkappa^{-1}} \qquad (3.60)$$

Thus arose the Debye–Hückel expression for the experimentally inaccessible individual ionic-activity coefficient. This expression could be transformed into the Debye–Hückel limiting law for the experimentally measurable mean ionic-activity coefficient

$$\log f_\pm = -A(z_+ z_-) I^{\frac{1}{2}} \qquad (3.90)$$

which would indicate that the logarithm of the mean activity coefficient falls linearly with the square root of the ionic strength $I (= \frac{1}{2} \sum c_i z_i^2)$, which is a measure of the total number of electric charges in the solution.

The agreement of the Debye–Hückel limiting law with experiment improved with decreasing electrolyte concentration and became excellent for the limiting tangent to the $\log f_\pm$ versus $I^{\frac{1}{2}}$ curve. With increasing concentration, however, experiment deviated more and more from theory, and, at concentrations above $1N$, even showed an *increase* in $f_\pm$ with increase of concentration, whereas theory indicated a continued decrease.

An obvious improvement of the theory consisted in removing the assumption of point-charge ions and taking into account their finite size. With the use of an ion-size parameter a, the expression for the mean ionic-activity coefficient became

$$\log f_\pm = -\frac{A(z_+ z_-) I^{\frac{1}{2}}}{1 + \varkappa a} \qquad (3.105)$$

However, the value of the ion-size parameter a could not be theoretically evaluated. Hence, an experimentally calibrated value was used. With this calibrated value for a, the values of $f_\pm$ at other concentrations [calculated from (3.105)] were compared with experiment.

The finite-ion-size model yielded agreement with experiment at concentrations up toward $0.1 N$. It also introduced through the value of a, the ion-size parameter, a specificity to the electrolyte (making NaCl different from KCl), whereas the point-charge model yielded activity coefficients which depended only upon the valency type of electrolyte. Thus, while the limiting law sees only the charges on the ions, it is blind to the specific characteristics which an ionic species may have, and this defect is overcome by the finite-ion-size model.

Unfortunately, the value of a obtained from experiment by Eq. (3.105) varies with concentration (as it should not if it represented simply the collisional diameters), and, as the concentration increases beyond about $0.1 M$, a has sometimes to assume physically impossible (e.g., negative) values. Evidently, these changes demanded by experiment do not only reflect real changes in the sizes of ions, but they represent other effects neglected in the simplifying Debye–Hückel model. Hence, the basic postulates of the Debye–Hückel model must be scrutinized.

The basic postulates can be put down as follows: (1) The central ion sees the surrounding ions in the form of a smoothed-out charge density and not as discrete charges. (2) All the ions in the electrolytic solution are free to contribute to the charge density, and there is, for instance, no pairing up of positive and negative ions to form any electrically neutral couples. (3) Only long-range coulombic forces are relevant to ion–ion interactions; and short-range noncoulombic forces, such as dispersion forces, play a negligible role. (4) The solution is sufficiently dilute to make ψ_r [which depends on concentration through $\varkappa$—cf. Eq. (3.20)] small enough to warrant the linearization of the Boltzmann equation (3.10). (5) The only role of the solvent is to provide a dielectric medium for the operation of interionic forces, i.e., the removal of a number of ions from the solvent to cling more or less permanently to ions other than the central ion is neglected.

It is because it is implicitly attempting to represent all these various aspects of the real situation inside an ionic solution that the experimentally calibrated ion-size parameter varies with concentration. Of course, a certain amount of concentration variation of the ion-size parameter is understandable because the parameter depends upon the radius of solvated ions and this time-average radius might be expected to decrease with an increase of concentration. Hence, one must try to isolate that part of the changes in the ion-size parameter which does not reflect real changes in the sizes of ions but represents the impact of, for instance, ionic solvation upon activity coefficients.

This question of the influence of ion–solvent interactions (Chapter 2) upon the ion–ion interactions will be considered in Section 3.6.

3.5.7. On the Parentage of the Theory of Ion–Ion Interactions

Stress has been laid on the contribution of Debye and Hückel (1923) to the development of the theory of ion–ion interactions. It was Debye and Hückel who ushered in the electrostatic theory of ionic solutions and worked out its excellent predictions.

It is not often realized, however, that the credit due to Debye and Hückel as the parents of the theory of ionic solutions is the credit that is quite justifiably accorded to foster parents. The true parents were Milner and Gouy. These authors made important contributions very early in the growth of the theory of ion–ion interactions.

Milner's contribution (1912) was direct. He attempted to find out the virial[†] equation for a mixture of ions. However, Milner's statistical mechanical approach lacked the mathematical simplicity of the ionic-cloud model of Debye and Hückel and proved too unwieldy to yield a general solution testable by experiment. Nevertheless, the contribution of Milner was a seminal one in that, for the first time, the behavior of an ionic solution had been linked mathematically to the interionic forces.

The contribution of Gouy (1910) was indirect.[‡] Milner's treatment was not sufficiently fruitful because he did not formulate a mathematically treatable model. Gouy developed such a model in his treatment of the distribution of the excess charge density in the solution near an electrode. Whereas Milner sought to describe the interactions between series of discrete ions, it was Gouy who suggested the smoothing-out of the ionic charges into a continuous distribution of charge and took the vital step of using Poisson's equation to relate the electrostatic potential and the charge density

---

[†] *Virial* is derived from the Latin word for *force*, and the virial equation of state is a relationship between pressure, volume, and temperature of the form

$$\frac{PV}{RT} = 1 + \frac{K_2}{V} + \frac{K_3}{V^2} + \cdots$$

where $K_2, K_3, \ldots$, the virial coefficients, represent interactions between constituent particles.

[‡] Chapman, in 1913, made an independent contribution on the same lines as that of Gouy.

in the continuum. Thus, Gouy was the first to evolve the ionic-atmosphere model.

It was with an awareness of the work of Milner and Gouy that Debye and Hückel attacked the problem. Their contributions, however, were vital ones. By choosing one ion out of the ionic solution and making an analogy between this charged reference ion and the charged electrode of Gouy, by using the Gouy type of approach to obtain the variation of charge density and potential with distance from the central ion and thus to get the contribution to the potential arising from interionic forces, and, finally, by evolving a charging process to get the chemical-potential change due to ion–ion interactions, they were able to link the chemical-potential change caused by interionic forces to the experimentally measurable activity coefficient. Without these essential contributions of Debye and Hückel, a viable theory of ionic solutions would not have emerged.

Further Reading

1. G. Gouy, *J. Phys.* **9**, 457 (1910).
2. S. R. Milner, *Phil. Mag.* **6** (23): 551 (1912).
3. P. Debye and E. Hückel, *Z. Physik*, **24**, 305 (1923).
4. R. A. Robinson and R. H. Stokes, *Electrolytic Solutions*, Butterworth's Publications, Ltd., London, 1959.
5. H. S. Harned and B. B. Owen, *The Physical Chemistry of Electrolytic Solutions*, 3rd ed., Reinhold Publishing, Corps., New York, 1958.
6. H. L. Friedman, *Ionic Solution Theory*, Interscience Publishers, Inc., New York, 1962.
7. M. H. Lietzke, R. W. Stroughton, and R. M. Fuoss, *Proc. Nat. Acad. Sci., U. S.*, **59**: 39 (1968).

3.6. ION–SOLVENT INTERACTIONS AND THE ACTIVITY COEFFICIENT

3.6.1. The Effect of Water Bound to Ions on the Theory of Deviations from Ideality

The theory of behavior in ionic solutions arising from ion–ion interactions has been seen (Section 3.5) to give rise to expressions in which, as the ionic concentration increases, the activity coefficient *decreases*. In spite of the excellent numerical agreement between the predictions of the interionic attraction theory and experimental values of activity coefficients at sufficiently low concentrations (e.g., $< 3 \times 10^{-3} N$), there is a most sharp disagreement at concentrations above about $1 N$, when the activity coef-

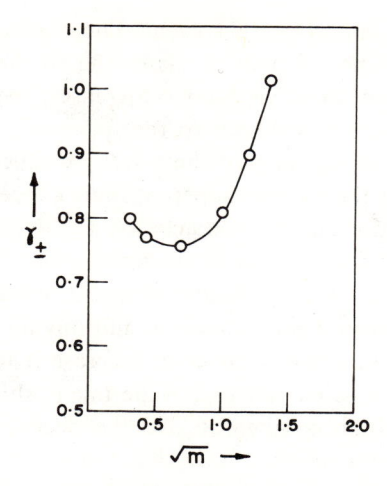

Fig. 3.36. The observed $\gamma_{\pm}$ versus $\sqrt{m}$ curve for lithium chloride showing a minimum.

ficient begins to *increase* back toward the values it had in limitingly dilute solutions. In fact, at sufficiently high concentrations (one might have argued, when the ionic interactions are greatest), the activity coefficient, instead of continuing to decrease, begins to exceed the value of unity characteristic of the reference state of noninteraction, i.e., of infinite dilution (Fig. 3.36).

A qualitative picture can at once be given for these apparently anomalous happenings. In Chapter 2, it has been argued that ions exist in solution in various states of interaction with solvent particles. There is a consequence which must therefore follow for the effectiveness of some of these water

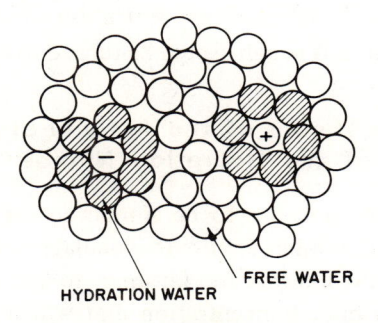

Fig. 3.37. The distinction between free water and hydration water which is locked up in the solvent sheaths of ions.

molecules in counting as part of the solvent. Those which are tightly bound to certain ions cannot be effective in dissolving further ions added (Fig. 3.37). As the concentration of electrolyte increases, therefore, the amount of *effective* solvent decreases. In this way, the apparently anomalous increase in the activity coefficient occurs. For the activity coefficient is in effect that factor which multiplies the simple, apparent ionic concentration and makes it the *effective* concentration, i.e., the activity. If the hydration of the ions reduces the amount of free solvent from that present for a given stoichiometric concentration, then the effective concentration increases and the activity coefficient must increase so that its multiplying effect on the simple stoichiometric concentration is such as to increase it to take into account the reduction of the effective solvent. Experiment shows that sometimes these increases more than compensate for the decrease due to interionic forces, and it is thus not unreasonable that the activity coefficient should rise above unity.

Some glimmering of the quantitative side of this can be seen by taking the number of waters in the primary hydration sheath of the ions as those which are "no longer effective solvent particles." For NaCl, for example, Table 2.20 indicates that this number is about 7. If the salt concentration is, e.g., $10^{-2}N$, the number of moles of water per liter withdrawn from effect as free solvent would be 0.07. As the number of moles of water per liter is $1000/18 = 55.5$, the number of moles of free water is 55.43 and the effects arising from such a small change are not observable. But now consider a $1N$ solution of NaCl. The water withdrawn is 7 moles liter^{-1}, and the change in the number of moles of free water is from 55.5 in the infinitely dilute situation to 48.5, a significant change. At $5N$ NaCl, more than half of the water in the solution is associated with the ions, and a sharp increase of activity coefficient—somewhat of a doubling, in fact—would be expected to express the increase in effective concentration of the ions.

To what extent can this rough sketch be turned to a quantitative model?

3.6.2. Quantitative Theory of the Activity of an Electrolyte as a Function of the Hydration Number

The basic thought here has to be similar to that which lay beneath the theory of electrostatic interactions: to calculate the work done in going from a state in which the ions are too far apart to feel any interionic attraction, to the state at a finite concentration c at which part of the ions behavior is due to this. This work was then [Eq. (3.59)] placed equal to $RT \ln f$, where f is the activity coefficient, which was thereby calculated.

When one realizes that one has to explain a reversal of the direction in which the activity coefficient varies with concentration by taking into account the removal of some of the solvent from effective partaking in the ionic solution's activity, the philosophy of the calculation of what effect this has on the activity coefficient becomes clear. One must calculate the work done in the changes caused by solvent removal and add this to the work done in building up the ionic atmosphere. What, then, is the contribution due to these water-removing processes? [Note that ions are hydrated at all times in which they are in the solution. One is not going to calculate the heat of hydration; that was done in Chapter 2. Here, the task is to calculate the work done as a consequence of the fact that, when water molecules enter the solvation sheath, they are, so to write, *hors de combat* as far as the solvent is concerned. It is to be an $RT \ln(c_1/c_2)$ type of calculation.]

Let it be assumed, as a device for the calculation, that the interionic attraction is switched off. (Reasons for employing this artifice will duly be given.) Thus, there are two kinds of work which must be taken into account.

1. The $RT \ln(c_1/c_2)$ kind of work done when the change of concentration of the free solvent *water*, caused by the introduction into the solution of ions of a certain concentration, takes place.
2. The $RT \ln(c_1/c_2)$ kind of work done when the corresponding change of concentration of the *ions*, due to the removal of the water to their sheaths, occurs.

The work done for process 1 is easy to calculate. Before the ions have been added, the concentration of the water is unaffected by anything; it is the concentration of pure water, and its activity, the activity of a pure substance, can be regarded as unity. After the ions are there, the activity of the water is, say, a_w. Then, the work when the activity of the water goes from 1 to a_w is $RT \ln(a_w/1)$.

However, one wishes to know the change of activity caused to the *electrolyte* by this change of activity of the water. Further, the calculation must be reduced to that for 1 mole of electrolyte. Let the sum of the moles of water in the primary sheath per liter of solution for both ions of the imagined 1:1 electrolyte be n_h (for 1 molar solutions, this is the hydration number). Then, if there are n moles of electrolyte in the water, the change in free energy due to the removal of the water to the ions' sheaths is $-(n_h/n)RT \ln a_w$ per mole of electrolyte.

One now comes to the work 2, see above, and realizes why the calculation is best made as a thought process in which the interionic attraction

242 CHAPTER 3

is shut off while this calculation of work is done. For one wants to be able to use the ideal-solution (no interaction) equation for the work done, $RT \ln(c_1/c_2)$, and not $RT \ln(a_1/a_2)$. Thus, to have to use the latter expression would be awkward; it needs a knowledge of the activities themselves, and, that, one is trying to calculate.

Now, the free-energy change due to the change of concentration of the *ions*—consequent upon the removal of the effective solvent molecule—is

$$RT \ln \frac{x_{\text{after water removal from free to solvated state}}}{x_{\text{before water removal from free to solvated state}}}$$

where x is the mole fraction of the electrolyte in the solution.

Before the water is removed,

$$x_{\text{before}} = \frac{n}{n_w + n} \qquad (3.114)$$

where n is the number of moles of electrolyte present in n_w of water. Then, after the water is removed to the sheaths,

$$x_{\text{after}} = \frac{n}{n_w - n_h + n} \qquad (3.115)$$

The change in free energy is hence:

$$RT \ln \frac{n_w + n}{n_w + n - n_h}$$

Hence, the total free-energy change in the solution, calculated per mole of the electrolyte present, is

$$-\frac{n_h}{n} RT \ln a_w + RT \ln \frac{n_w + n}{n_w + n - n_h}$$

Now, one has to switch back on the coulombic interactions. If the expression for the work done in building up an ionic atmosphere [e.g., Eq. (3.106)] were still valid in the region of relatively high concentrations in which the effect of change of concentration is occurring, then,[†]

$$RT \log f_{\pm(\text{exp})} = -\frac{A\sqrt{c}}{1 + Ba\sqrt{c}} - 2.303 RT \frac{n_h}{n} \log a_w + 2.303 RT \log \frac{n_w + n}{n_w + n - n_h} \qquad (3.116)$$

[†] Here $\sqrt{c}$ has been written instead of the $I^{\frac{1}{2}}$ of Eq. (3.106). For 1:1 electrolytes, c and I are identical.

One sees at once that there is a possibility of a change of direction of the change of $\log f_\pm$ with increase of concentration in the solution. If the last term predominates, $RT \ln f_\pm$ may *in*crease with concentration.

But the situation here does have a fairly large shadow on it because of the use of the expression (3.106) in $\sqrt{c}$. It will be seen (Section 3.9) that, at concentrations as high as $1N$, there are some fundamental difficulties for the ionic cloud model on which this $\sqrt{c}$ expression of (3.106) was based (the ionic atmosphere can no longer be considered a continuum of smoothed-out charge). It is clear that, when the necessary mathematics can be done, there will be an improvement on the $\sqrt{c}$ expression, and one will hope to get it more correct than it now is. Because of this shadow, a comparison of (3.116) with experiment to test the validity of the model for removing solvent molecules to the ions' sheath should be done a little with tongue in cheek.

3.6.3. The Water Removal Theory of Activity Coefficients and Its Apparent Consistency with Experiment at High Electrolytic Concentrations

If one examines the ion–solvent terms in Eq. (3.116), one sees that, since $a_W \leq 1$ and in general $n_h > n$ (more than one hydration water per ion), both the terms are positive. Hence, one can conclude that the Debye–Hückel treatment, which ignores the withdrawal of solvent from solution, gives values of activity coefficients which are smaller than those which take into account these effects. Further, the difference arises from the ion–solvent terms, i.e.,

$$-2.303 RT \frac{n_h}{n} \log a_W + 2.303 RT \log \frac{n_W + n}{n_W + n - n_h}$$

As the electrolyte concentration increases, a_W decreases and n_h increases; hence both ion–solvent terms increase the value of $\log f$. Further, the numerical evaluation shows that the above ion–solvent term can equal and become larger than the Debye–Hückel ($\sqrt{c}$) coulombic term. This means that the $\log f_\pm$ versus $I^{\frac{1}{2}}$ curve can pass through a minimum and then start rising, which is precisely what is observed (*cf.* Fig. 3.36, where an activity coefficient is plotted against the corresponding molality).

On the other hand, with increasing dilution, $n_W + n \gg n_h$ or $n_W + n - n_h \sim n_W + n$ and $a_W \to 1$, and, hence, the terms vanish, which indicates that ion–solvent interactions (which are of short range) are significant for the theory of activity coefficients only in concentrated so-

TABLE 3.11
Water Activities in Sodium Chloride Solutions

M	a_W	M	a_W
0.1	0.99665	3	0.8932
0.5	0.98355	3.4	0.8769
1	0.96686	3.8	0.8600
1.4	0.9532	4.2	0.8428
1.8	0.9389	4.6	0.8250
2.2	0.9242	5	0.8068
2.6	0.9089	5.4	0.7883

lutions. At extreme dilutions, only the ion–ion long-range coulombic interactions are of importance.

In order to test Eq. (3.116), which is a quantitative statement of the influence of ionic hydration on activity coefficients, it is necessary to know the quantity n_h and the activity a_W of water, it being assumed that an experimentally calibrated value of the ion-size parameter is available. The activity of water can be obtained from independent experiments (Table 3.11). The quantity n_h can be used as a parameter. If Eq. (3.116) is tested as a two-parameter equation (n_h and a being the two parameters, Table 3.12) it is found that theory is in excellent accord with experiment. For instance, in the case of NaCl, the calculated activity coefficient agrees with the experimental value for solutions as concentrated as $5M$ (Fig. 3.38).

TABLE 3.12
Values of n_h of a One Molar Solution and a

Salt	n_h	a, Å
HCl	8.0	4.47
HBr	8.6	5.18
NaCl	3.5	3.97
NaBr	4.2	4.24
KCl	1.9	3.63
$MgCl_2$	13.7	5.02
$MgBr_2$	17.0	5.46

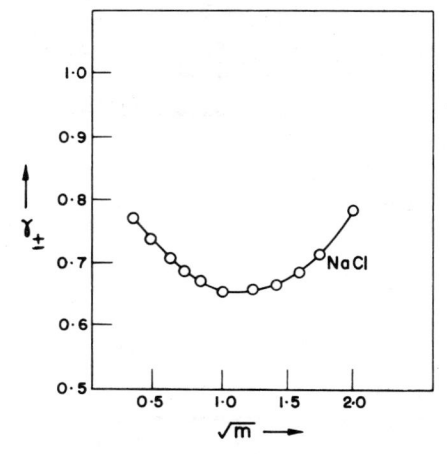

Fig. 3.38. Comparison of the activity coefficients of NaCl calculated from Eq. (3.116) with $a = 3.97$ and $n_h = 3.5$, with the experimentally observed activity coefficients for NaCl (the full line, theoretical curve; open circles, experimental points).

Since the quantity n_h is the number of moles of water used up in solvating $n = n_+ + n_-$ moles of ions, it can be split up into two terms: $(n_h)^+$ moles required to hydrate n_+ moles of cations, and $(n_h)^-$ moles required to hydrate n_- moles of anions. It follows that $[(n_h)^+/n_+]$ and $[(n_h)^-/n_-]$ are the hydration numbers (see Chapter 2) of the positive and negative ions and $[(n_h)^+/n_+] + [(n_h)^-/n_-]$ is the hydration number of the electrolyte.

It has been found that, in the case of several electrolytes, the values of the hydration numbers obtained by fitting the theory [Eq. (3.116)] to experiment are in reasonable agreement with hydration numbers determined by an independent method (Table 3.13). Alternatively, one can say that, when independently obtained hydration numbers are substituted in Eq. (3.116), the resulting values of $\log f_\pm$ show fair agreement with experiment.

In conclusion, therefore, it may be said that the treatment of the influence of ion–solvent interactions on ion–ion interactions has extended the range of concentration of an ionic solution which is accessible to theory. Whereas the finite-ion-size version of the Debye–Hückel theory did not permit theory to deal with solutions in a range of concentrations corresponding to those of real life, Eq. (3.116) advances theory into the range of practical concentrations. Apart from this numerical agreement with ex-

TABLE 3.13
Nearest Integer Hydration Number of Electrolytes from Eq. (3.139) and the Most Probable Value from Independent Experiments (cf. Table 2.20)

Salt	Hydration number from Eq. (3.116) (nearest integer)	Hydration number from other experimental methods
LiCl	7	7 ± 1
LiBr	8	7 ± 1
NaCl	4	6 ± 2
KCl	2	5 ± 2
KI	3	4 ± 2

periment, Eq. (3.116) unites two basic aspects of the situation inside an electrolytic solution, namely ion–solvent interactions and ion–ion interactions.

Further Reading

1. N. Bjerrum, *Z. Anorg. Allgem. Chem.*, **109**: 175 (1920).
2. R. H. Stokes and R. A. Robinson, *J. Amer. Chem. Soc.*, **70**: 1870 (1948).
3. E. Glueckauf, *Trans. Faraday Soc.*, **51**: 1235 (1955).
4. R. H. Stokes and R. A. Robinson, *Trans. Faraday Soc.*, **53**: 301 (1957).
5. R. A. Robinson and R. H. Stokes, *Electrolyte Solutions*, 2nd ed., Butterworth's Publications, Ltd., London, 1959.
6. H. S. Frank, "Solvent Models and the Interpretation of Ionization and Solvation Phenomena," in: B. E. Conway and R. G. Barradas, eds., John Wiley & Sons, Inc., New York, 1966.

3.7. THE SO-CALLED "RIGOROUS" SOLUTIONS OF THE POISSON–BOLTZMANN EQUATION

One approach to the understanding of the discrepancies between the experimental values of the activity coefficient and the predictions of the Debye–Hückel model has just been described (Section 3.6); it involved a consideration of the influence of solvation.

An alternative approach is based on the view that the failure of the Debye–Hückel theory at high concentrations stems from the fact that the development of the theory involved the linearization of the Boltzmann equation (*cf.* Section 3.3.5). If such a view is taken, there is an obvious solution to the problem: instead of linearizing the Boltzmann equation, one can take the higher terms. Thus, one obtains the unlinearized P–B equation

$$\frac{1}{r^2}\frac{d}{dr}\left(r^2\frac{d\psi_r}{dr}\right) = -\frac{4\pi}{\varepsilon}\varrho_r = -\frac{4\pi}{\varepsilon}\sum z_i e_0 n_i^0 e^{-z_i e_0 \psi_r/kT} \quad (3.117)$$

In the special case of a symmetrical electrolyte ($z_+ = -z_- = z$) with equal concentrations of positive and negative ions, i.e., $n_+^0 = n_-^0 = n^0$, one gets

$$\sum z_i e_0 n_i^0 e^{-z_i e_0 \psi_r/kT} = n^0 z_+ e_0 e^{-z_+ e_0 \psi_r/kT} - n^0 z_- e_0 e^{+z_- e_0 \psi_r/kT}$$
$$= n^0 z e_0 (e^{-z e_0 \psi_r/kT} - e^{+z e_0 \psi_r/kT}) \quad (3.118)$$

But

$$e^{+x} - e^{-x} = 2\sinh x$$

and, therefore,

$$\varrho_r = -2n^0 z e_0 \sinh \frac{z e_0 \psi_r}{kT} \quad (3.119)$$

or

$$\frac{1}{r^2}\frac{d}{dr}\left(r^2\frac{d\psi}{dr}\right) = \frac{8\pi z e_0 n^0}{\varepsilon}\sinh\frac{z e_0 \psi}{kT} \quad (3.120)$$

By utilizing a computer program, one could obtain from (3.120) so-called "rigorous" solutions.

Before proceeding further, however, it is appropriate to stress a logical inconsistency in working with the unlinearized P–B equation (3.117). The unlinearized Boltzmann equation (3.10) implies a *nonlinear* relationship between charge density and potential. In contrast, the *linearized* Boltzmann equation (3.16) implies a *linear* relationship of ϱ_r to ψ_r.

Now, a *linear* charge density–potential relation is consistent with the law of superposition of potentials, which states that the electrostatic potential at a point due to an assembly of charges is the sum of the potentials due to the individual charges. Thus, when one uses an unlinearized P–B equation, one is assuming the validity of the law of superposition of potentials in the Poisson equation and its invalidity in the Boltzmann equation. This is a basic logical inconsistency which must reveal itself in the predictions which emerge from the so-called rigorous solutions. This is indeed the case, as will be shown below.

Recall that, after obtaining the contribution of the ionic atmosphere to the potential at the central ion, the coulombic interaction between the central ions and the cloud was calculated by an imaginary charging process, generally known as the *Guntelberg charging process* in recognition of its originator.

In the Guntelberg charging process, the central ion *i* is assumed to be in a hypothetical condition of zero charge. The rest of the ions, fully charged, are in the positions that they would hypothetically have were the central ion charged to its normal value $z_i e_0$; i.e., the other ions constitute an ionic atmosphere enveloping the central ion (Fig. 3.39). The ionic cloud sets up a potential $\psi_{\text{cloud}} = -(z_i e_0/\varepsilon \varkappa^{-1})$ at the site of the central ion. Now, the charge of the central ion is built up (Fig. 3.39) from zero to its final value $z_i e_0$, and the work done in this process is calculated by the usual formula for the electrostatic work of charging a sphere (see Section 3.3.1) i.e.,

$$W = \frac{z_i e_0}{2} \psi \quad (3.3)$$

Since, during the charging, only ions of the *i*th type are considered, the Guntelberg charging process gives that part of the *chemical potential* due to electrostatic interactions.

Now, the Guntelberg charging process was suggested several years after Debye and Hückel made their theoretical calculation of the activity coefficient. These authors carried out another charging process, the *Debye charging process*. All the ions are assumed to be in their equilibrium, or

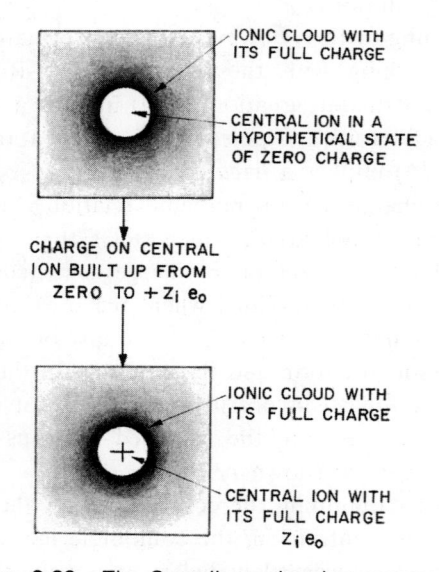

Fig. 3.39. The Guntelberg charging process.

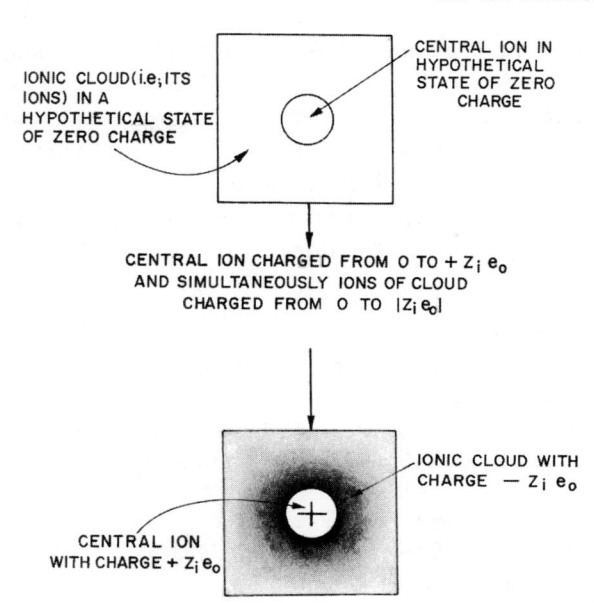

Fig. 3.40. The Debye charging process.

time-average, positions in the ionic atmosphere (Fig. 3.40), but the central ion and the cloud ions are *all* considered in a hypothetical condition of zero charge. All the ions of the assembly are then *simultaneously* brought to their final values of charge by an imaginary charging process in which there are small additions of charges to each. Since ions of all types (not only of the *i*th type) are considered, the work done in this charging process yields the *free-energy* change arising from the electrostatic interactions in solution. Differentiation of the free energy with respect to the number of moles of the *i*th species gives the chemical potential

$$\frac{\partial \Delta G_{I-I}}{\partial n_i} = \Delta \mu_{i-I} \tag{3.121}$$

From the rigorous solutions of the unlinearized P–B equation, one gets the cloud contribution to the electrostatic potential at the central ion; and, when this value of the electrostatic potential is used in the two charging processes to get the chemical-potential change $\Delta \mu_{i-I}$ arising from ion–ion interactions, it is found that the Guntelberg and Debye charging processes give discordant results. As shown by Onsager, this discrepancy is not due to the invalidity of either of the two charging processes: it is a symptom of the logical inconsistency intrinsic in the unlinearized P–B equation.

This discussion of rigorous solutions has thus brought out an important point: The agreement between the chemical-potential change $\Delta\mu_{i-I}$ calculated by the Debye and Guntelberg charging processes serves as a fundamental test of any theory of the electrostatic potential set up at a central ion by the remaining ions of an ionic solution.

Further Reading

1. P. Debye and E. Hückel, *Z. Physik*, **24**: 305 (1923).
2. H. Müller, *Z. Physik*, **28**: 324 (1927).
3. T. H. Gronwall, V. K. La Mer, and K. Sandved, *Z. Physik*, **29**: 358 (1929).
4. R. H. Fowler, *Statistical Mechanics*, Cambridge University Press, New York (1929).
5. L. Onsager, *Chem. Rev.*, **13**: 73 (1933).
6. J. G. Kirkwood, *J. Chem. Phys.*, **2**: 351 (1934).
7. J. G. Kirkwood, *J. Chem. Phys.*, **2**: 767 (1934).
8. J. G. Kirkwood, *Chem. Rev.*, **19**: 275 (1936).
9. W. G. McMillan, Jr., and J. E. Mayer, *J. Chem. Phys.*, **13**: 276 (1945).
10. J. E. Mayer, *J. Chem. Phys.*, **18**: 1426 (1950).
11. H. Falkenhagen and G. Kelbg, *Ann. Physik* VI, **11**: 60 (1952).
12. H. Falkenhagen and G. Kelbg, *Z. Elektrochem.*, **56**: 834 (1952).
13. H. Falkenhagen, M. Leist, and H. Kelbg, *Ann. Physik* VI, **11**: 51 (1952).
14. J. C. Poirier, *J. Chem. Phys.*, **21**: 965 and 972 (1953).
15. J. G. Kirkwood and J. C. Poirier, *J. Phys. Chem.*, **58**: 591 (1954).
16. H. S. Frank, *Discussions Faraday Soc.*, **24**: 66 (1957).
17. E. A. Guggenheim, *Trans. Faraday Soc.*, **55**: 1714 (1959).
18. E. A. Guggenheim, *Trans. Faraday Soc.*, **56**: 1152 (1960).
19. F. H. Stillinger, Jr., J. G. Kirkwood, and P. J. Wojtowicz, *J. Chem. Phys.*, **32**: 1837 (1960).
20. F. H. Stillinger, Jr., and J. G. Kirkwood, *J. Chem. Phys.*, **33**: 1282 (1960).
21. F. H. Stillinger, Jr., *J. Chem. Phys.*, **35**: 1581 (1961).
22. E. A. Guggenheim, *Trans. Faraday Soc.*, **58**: 86 (1962).
23. H. L. Friedman, *Ionic Solution Theory*, Interscience Publishers, Inc., New York, 1962.
24. J. C. Poirier, "Current Status of the Statistical Mechanical Theory of Ionic Solutions," in: B. E. Conway and R. G. Barradas, eds., *Chemical Physics of Ionic Solutions*, John Wiley & Sons, Inc., New York, 1966.

3.8. TEMPORARY ION ASSOCIATION IN AN ELECTROLYTIC SOLUTION: FORMATION OF PAIRS, TRIPLETS, ETC.

3.8.1. Positive and Negative Ions Can Stick Together: Ion-Pair Formation

The Debye–Hückel model assumed the ions to be in *almost* random thermal notions and therefore in *almost* random positions. The slight deviation from randomness was pictured as giving rise to an ionic cloud around a given ion, a positive ion (of charge $+ze_0$) being surrounded by a cloud of excess negative charge $(-ze_0)$. However, the possibility was not considered that some negative ions in the cloud would get sufficiently close to the central positive ion in the course of their quasi-random solution movements so that their thermal translational energy would not be sufficient for them to continue their independent movements in the solution. Bjerrum suggested that a pair of oppositely charged ions may get trapped in each other's coulombic field. An *ion pair* may be formed.

The ions of the pair together form an ionic dipole on which the net charge is zero. Within the ionic cloud, the locations of such uncharged ion pairs are completely random, since, being uncharged, they are not acted upon by the coulombic field of the central ion. Further, on the average, a certain fraction of the ions in the electrolytic solution will be stuck together in the form of ion pairs. This fraction must now be evaluated.

3.8.2. The Probability of Finding Oppositely Charged Ions near Each Other

Consider a spherical shell of thickness dr and of radius r from a reference positive ion (Fig. 3.41). The probability P_r that a negative ion is in the spherical shell is proportional, firstly, to the ratio of the volume $4\pi r^2\, dr$ of the shell to the total volume V of the solution; secondly, to the total number N_- of negative ions present; and, thirdly, to the Boltzmann factor $\exp(-U/kT)$, where U is the potential energy of a negative ion at a distance r from a cation, i.e.,

$$P_r = 4\pi r^2\, dr\, \frac{N_-}{V}\, e^{-U/kT} \qquad (3.122)$$

Since N_-/V is the concentration n_-^0 of negative ions in the solution and

$$U = \frac{-z_- z_+ e_0^2}{\varepsilon r} \qquad (3.123)$$

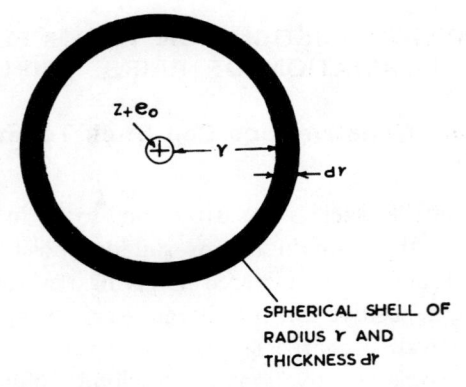

Fig. 3.41. The probability P_r of finding an ion of charge z_-e_0 in a dr-thick spherical shell of radius r around a reference ion of charge z_+e_0.

it is clear that

$$P_r = (4\pi n_-^0)r^2 e^{z_-z_+e_0^2/\varepsilon r kT}\, dr \tag{3.124}$$

or, writing

$$\lambda = \frac{z_-z_+e_0^2}{\varepsilon kT} \tag{3.125}$$

one has

$$P_r = (4\pi n_-^0)e^{\lambda/r}r^2\, dr \tag{3.126}$$

A similar equation is valid for the probability of finding a positive ion in a dr-thick shell at a radius r from a reference negative ion. Hence, in general, one may write for the probability of finding an i type of ion in a dr-thick spherical shell at a radius r from a reference ion k of opposite charge

$$P_r = (4\pi n_i^0)e^{\lambda/r}r^2\, dr \tag{3.127}$$

where

$$\lambda = \frac{z_iz_ke_0^2}{\varepsilon kT} \tag{3.128}$$

This probability of finding an ion of one type of charge near an ion of the opposite charge varies in an interesting way with distance (Fig. 3.42). For small values of r, the function P_r is dominated by $e^{\lambda/r}$ rather than by r^2, and, under these conditions, *P_r increases with decreasing r*; for large values of r, $e^{\lambda/r} \to 1$ and *P_r increases with increasing r* because the volume $4\pi r^2\, dr$ of the spherical shell increases as r^2. It follows from these considerations

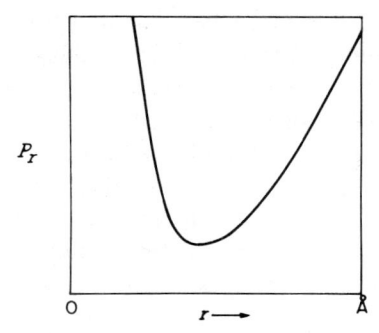

Fig. 3.42. The probability P_r of finding an ion of one type of charge as a function of distance.

that P_r goes through a minimum for a particular, critical value of r. This conclusion may also be reached by computing the number of ions in a series of shells, each of an arbitrarily selected thickness of 0.1 Å (Table 3.14).

3.8.3. The Fraction of Ion Pairs, According to Bjerrum

If one integrates P_r between a lower and an upper limit, one gets the probability P_r of finding a negative ion within a distance from the reference positive ion, defined by the limits. Now, for two oppositely charged ions to stick together to form an ion pair, it is necessary that they should be close enough for the coulombic attractive energy to overcome the thermal energy which scatters them apart. Let this "close-enough" distance be q. Then, one can say that an ion pair will form when the distance r between a

TABLE 3.14

Number of Ions in Spherical Shells at Various Distances

r, Å	Number of ions in shell $\times 10^{22}$	
	Of opposite charge	Of like charge
2	$1.77n_i$	$0.001n_j$
2.5	$1.37n_i$	$0.005n_j$
3	$1.22n_i$	$0.01n_j$
3.57	$1.18n_i$	$0.02n_j$
4	$1.20n_i$	$0.03n_j$
5	$1.31n_i$	$0.08n_j$

positive and negative ion becomes less than q. Thus, the probability of ion-pair formation is given by the integral of P_r between a lower limit of a, the closest distance of approach of ions, and an upper limit of q.

Now, the probability of any particular event is the number of times that the particular event is expected to be observed divided by the total number of observations. Hence, the probability of ion-pair formation is the number of ions of species i which are associated into ion pairs divided by the total number of i ions, i.e., the probability of ion-pair formation is the fraction θ of ions which are associated into ion pairs. Thus,

$$\theta = \int_a^q P_r \, dr = \int_a^q 4\pi n_i^0 e^{\lambda/r} r^2 \, dr \tag{3.129}$$

It is seen from Figure 3.43 that the integral in Eq. (3.129) is the area under the curve between the limits $r = a$ and $r = q$. But it is obvious that, as r increases past the minimum, the integral becomes greater than unity. Since, however, θ is a fraction, this means that the integral diverges.

In this context, Bjerrum took the arbitrary step of cutting off the integral at the value of $r = q$ corresponding to the minimum of the P_r versus r curve. This minimum can easily be shown (Appendix 3.4) to occur at

$$q = \frac{z_- z_+ e_0^2}{2\varepsilon kT} = \frac{\lambda}{2} \tag{3.130}$$

Bjerrum justified this step by arguing that it is only short-range coulombic interactions that lead to ion-pair formation and, further, when a pair of oppositely charged ions are situated at a distance apart of $r > q$, it is more appropriate to consider them free ions.

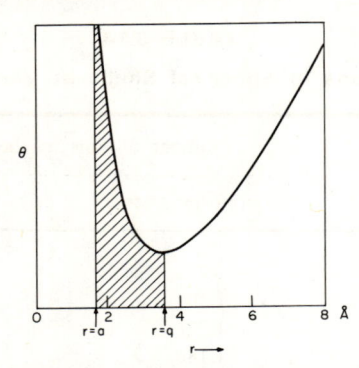

Fig. 3.43. The integral in Eq. (3.129) is the area under the curve between the limits $r = a$ and $r = q$.

ION–ION INTERACTIONS 255

Bjerrum concluded, therefore, that ion-pair formation occurs when an ion of one type of charge, e.g., a negative ion, enters a sphere of radius q drawn around a reference ion of the opposite charge, e.g., a positive ion. But it is the ion-size parameter which defines the closest distance of approach of a pair of ions. The Bjerrum hypothesis can therefore be stated as follows: If $a < q$, then ion-pair formation can occur; if $a > q$ the ions remain free (Fig. 3.44).

Now that the upper limit of the integral in Eq. (3.129) has been taken to be $q = \lambda/2$, the fraction of ion pairs is given by carrying out the integration. It is

$$\theta = 4\pi n_i^0 \int_a^{q=\lambda/2} e^{\lambda/r} r^2 \, dr \tag{3.131}$$

For mathematical convenience, a new variable y is defined as

$$y = \frac{\lambda}{r} = \frac{2q}{r} \tag{3.132}$$

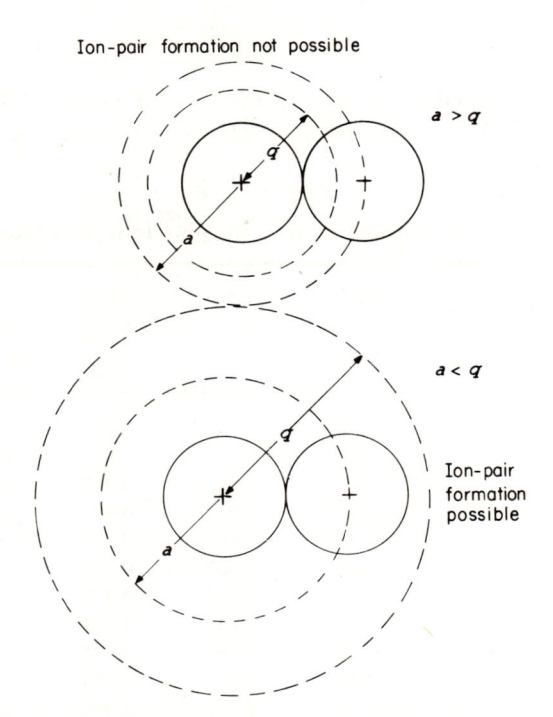

Fig. 3.44. (a) Ion-pair formation occurs if $a \leq q$; (b) ion-pair formation does not occur if $a > q$.

TABLE 3.15
Value of the Integral $\int_2^b e^y y^{-4} \, dy$

b	$\int_2^b e^y y^{-4} \, dy$	b	$\int_2^b e^y y^{-4} \, dy$	b	$\int_2^b e^y y^{-4} \, dy$
2.0	0	3	0.326	10	4.63
2.1	0.0440	3.5	0.442	15	93.0
2.2	0.0843	4	0.550		
2.4	0.156	5	0.771		
2.6	0.218	6	1.041		
2.8	0.274				

Hence, in terms of the new variable y, Eq. (3.131) becomes (*cf.* Appendix 3.5)

$$\theta = 4\pi n_i^0 \left(\frac{z_+ z_- e_0^2}{\varepsilon kT} \right)^3 \int_2^b e^y y^{-4} \, dy \tag{3.133}$$

where

$$b = \frac{\lambda}{a} = \frac{z_+ z_- e_0^2}{\varepsilon a kT} = \frac{2q}{a} \tag{3.134}$$

Bjerrum has tabulated the integral $\int_2^b e^y y^{-4} \, dy$ for various values of b (Table 3.15). This means that, by reading off the value of the corresponding

TABLE 3.16
Fraction of Association, θ, of Univalent Ions in Water at 18°C

$a \times 10^8$ cm:	2.82	2.35	1.76	1.01	0.70	0.47
q/a:	2.5	3	4	7	10	15
$c^\dagger$, moles liter^{-1}						
0.0001	—	—	—	—	0.001	0.027
0.0002	—	—	—	—	0.002	0.049
0.0005	—	—	—	0.002	0.006	0.106
0.001	—	0.001	0.001	0.004	0.011	0.177
0.002	0.002	0.002	0.003	0.007	0.021	0.274
0.005	0.002	0.004	0.007	0.016	0.048	0.418
0.01	0.005	0.008	0.012	0.030	0.030	0.529
0.02	0.008	0.013	0.022	0.053	0.137	0.632
0.05	0.017	0.028	0.046	0.105	0.240	0.741
0.1	0.029	0.048	0.072	0.163	0.336	0.804
0.2	0.048	0.079	0.121	0.240	0.437	0.854

$\dagger$ c in moles liter^{-1} = $1000 n_i^0/N$.

integral and substituting for the various other terms in (3.133), the degree of association of an electrolyte may be computed if the ion sizes, the dielectric constant, and the concentrations are known (Table 3.16).

3.8.4. The Ion-Association Constant K_A of Bjerrum

The quantity θ yields a clear idea of the fraction of ions which are associated in ion pairs in a particular electrolytic solution, at a given concentration. It would, however, be advantageous if each electrolyte, e.g., NaCl, BaSO$_4$, and La(NO$_3$)$_3$, were assigned a particular number which would reveal, without going through the calculation of θ, the extent to which the ions of that electrolyte associate in ion pairs. The quantitative measure chosen to represent the tendency for ion-pair formation was guided by historical considerations.

Arrhenius in 1887 had suggested that many properties of electrolytes could be explained by a *dissociation* hypothesis: The neutral molecules AB of the electrolyte dissociate to form ions A$^+$ and B$^-$, and this dissociation is governed by an equilibrium

$$AB \rightleftharpoons A^+ + B^- \tag{3.135}$$

Applying the law of mass action to this equilibrium, one can define a dissociation constant

$$K = \frac{a_{A^+} a_{B^-}}{a_{AB}} \tag{3.136}$$

By analogy,[†] one can define an *association* constant K_A for ion-pair formation. Thus, one can consider an equilibrium between free ions (the positive M$^+$ ions and the negative A$^-$ ions) and the associated ion pairs (symbolized IP)

$$M^+ + A^- \rightleftharpoons IP \tag{3.137}$$

The equilibrium sanctions the use of the law of mass action

$$K_A = \frac{a_{IP}}{a_{M^+} a_{A^-}} \tag{3.138}$$

[†] The analogy must not be carried too far because it is only a formal analogy. Arrhenius's hypothesis can now be seen to be valid for ionogens (i.e., potential electrolytes), in which case the neutral ionogenic molecules (e.g., acetic acid) consist of aggregates of atoms held together by covalent bonds. What is under discussion here is ion association, or ion-pair formation, of ionophores (i.e., true electrolytes). In these ion pairs, the positive and negative ions retain their identity as ions and are held together by electrostatic attraction.

where the a's are the activities of the relevant species. From (3.138), it is seen that K_A is the reciprocal of the ion pair's dissociation constant.

Since θ is the fraction of ions in the form of ion pairs, θc is the *concentration* of ion pairs, and $(1 - \theta)c$ is the concentration of free ions. If the activity coefficients of the positive and negative free ions are f_+ and f_-, respectively, and that of the ion pairs is f_{IP}, one can write

$$K_A = \frac{\theta c f_{IP}}{(1-\theta)cf_+(1-\theta)cf_-}$$

$$= \frac{\theta}{(1-\theta)^2} \frac{1}{c} \frac{f_{IP}}{f_+ f_-} \qquad (3.139)$$

or, using the definition of the mean ionic-activity coefficient [*cf.* Eq. (3.72)],

$$K_A = \frac{\theta}{(1-\theta)^2} \frac{1}{c} \frac{f_{IP}}{f_\pm^2} \qquad (3.140)$$

Some simplifications can now be introduced. The ion-pair activity coefficient f_{IP} is assumed to be unity because deviations of activity coefficients from unity are ascribed in the Debye–Hückel theory to electrostatic interactions. But ion pairs are not involved in such interactions owing to their zero charge, and, hence, they behave ideally like uncharged particles, i.e., $f_{IP} = 1$.

Further, in very dilute solutions: (1) The ions rarely come close enough together (i.e., to within a distance q) to form ion pairs, and one can consider $\theta \ll 1$ or $1 - \theta \sim 1$; (2) activity coefficients tend to unity, i.e., f_i or $f_\pm \to 1$.

Hence, under these conditions of very dilute solutions, Eq. (3.140) becomes

$$K_A \sim \frac{\theta}{c} \qquad (3.141)$$

and, substituting for θ from Eq. (3.133), one has

$$K_A = \frac{4\pi n_i^0}{c} \left(\frac{z_+ z_- e_0^2}{\varepsilon kT} \right)^3 \int_2^b e^y y^{-4} \, dy \qquad (3.142)$$

But

$$n_i^0 = \frac{cN_A}{1000} \qquad (3.143)$$

and, therefore,

$$K_A = \frac{4\pi N_A}{1000} \left(\frac{z_+ z_- e_0^2}{\varepsilon kT} \right)^3 \int_2^b e^y y^{-4} \, dy \qquad (3.144)$$

TABLE 3.17

Ion Association Constant K_A: Extent to Which Ion-Pair Formation Occurs

Salt	Solvent	Temperature, °C	ε	K
KBr	Acetic acid	30	6.20	9.09×10^6
KBr	Ammonia	−34	22	5.29×10^2
CsCl	Ethanol	25	24.30	1.51×10^2
KI	Acetone	25	20.70	1.25×10^2
KI	Pyridine	25	12.0	4.76×10^3

The value of the association constant provides an indication of whether ion-pair formation is significant. The higher the value of K_A, the more extensive is the ion-pair formation (Table 3.17).

What are the factors which increase K_A and therefore increase the degree of ion-pair formation? From Eq. (3.144), it can be seen that the factors which increase K_A are (1) low dielectric constant ε; (2) small ionic radii, which lead to a small value of a and hence [cf. Eq. (3.134)] to a large value of the upper limit b of the integral in Eq. (3.144); and (3) large z_+ and z_-.

These ideas based on Bjerrum's picture of ion-pair formation have received considerable experimental support. Thus, in Fig. 3.45, the association constant is seen to increase markedly with decrease of dielectric con-

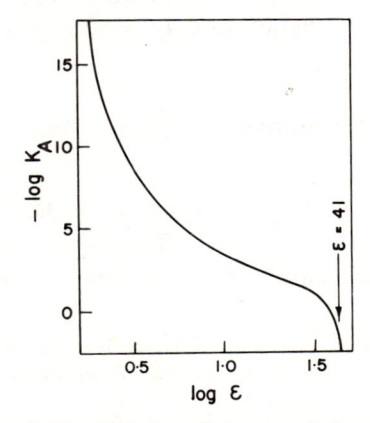

Fig. 3.45. Variation of the association constant K_A with dielectric constant for 1:1 salts.

260 CHAPTER 3

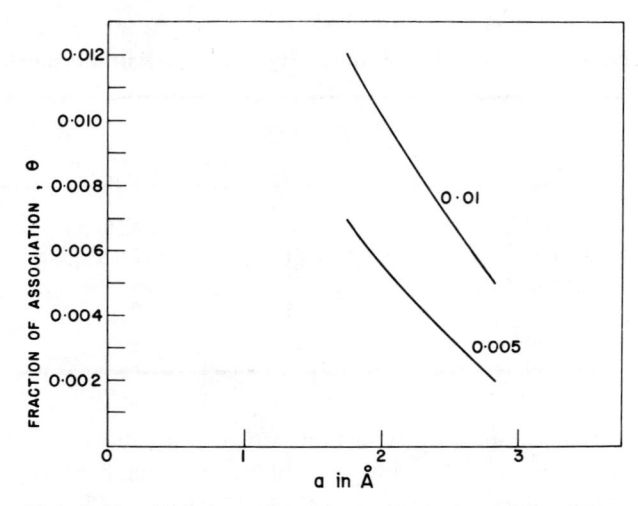

Fig. 3.46. Variation of θ, the fraction of associated ions, with a, the closest distance of approach (i.e., the ion-size parameter).

stant.[†] The dependence of ion-pair formation on the distance of closest approach is seen in Fig. 3.46.

When numerical calculations are carried out with these equations, the essential conclusion which emerges is that, in aqueous media, ion association in pairs scarcely occurs for 1:1-valent electrolytes but can be of importance for 2:2-valent electrolytes. The reason is that K_A depends on z_+z_- through Eq. (3.142). In nonaqueous solutions, most of which have dielectric constants much less than that of water ($\varepsilon = 80$), ion association is extremely important.

3.8.5. Activity Coefficients, Bjerrum's Ion Pairs, and Debye's Free Ions

What direct role do the ion pairs have in the Debye–Hückel electrostatic theory of activity coefficients? The answer simply is: None. Since ion pairs carry no net charge,[‡] they are ineligible for membership in the

---

[†] But the critical dielectric constant above which there is no more ion-pair formation (as indicated by Fig. 3.45) is really a result of the arbitrary cutting off of ion-pair formation at the distance q [see (3.8.6)].

[‡] Remember that the equations for the Bjerrum theory, as presented here, are correct only for electrolytes yielding ions of the same valency z, i.e., only for symmetrical 1:1- or 2:2-valent electrolytes.

ion cloud where the essential qualification is *charge*. Hence, ion pairs are dismissed from a direct consideration in the Debye–Hückel theory.

This does not mean that the Debye–Hückel theory gives the right answer when there is ion-pair formation. The extent of ion-pair formation decides the value of the concentration to be used in the ionic-cloud model. By removing a fraction θ of the total number of ions, only a fraction $1 - \theta$ of the ions remain for the Debye–Hückel treatment which interests itself only in the *free* charges. Thus, the Debye–Hückel expression for the activity coefficient [Eq. (3.106)] is valid for the free ions with two important modifications: (1) Instead of there being a concentration c of ions, there is only $(1 - \theta)c$; the remainder θc is not reckoned with owing to association. (2) The closest distance of approach of *free* ions is q and not a. These modifications yield

$$\log f_\pm = - \frac{A(z_+ z_-)\sqrt{(1-\theta)c}}{1 + Bq\sqrt{(1-\theta)c}} \qquad (3.145)$$

This calculated mean activity coefficient is related to the measured mean activity coefficient of the electrolyte $(f_\pm)_{obs}$ by the relation (for the derivation, see Appendix 3.6)

$$(f_\pm)_{obs} = (1 - \theta) f_\pm \qquad (3.146)$$

or

$$\log (f_\pm)_{obs} = \log f_\pm + \log (1 - \theta)$$

$$= - \frac{A(z_+ z_-)\sqrt{(1-\theta)c}}{1 + Bq\sqrt{(1-\theta)c}} + \log (1 - \theta) \qquad (3.147)$$

This equation indicates how the activity coefficient depends on the extent of ion association. In fact, this equation constitutes the bridge between the treatment of solutions of true electrolytes and solutions of potential electrolytes. More will be said on this matter in the chapter on protons in solution (Chapter 5), part of which deals with potential electrolytes.

3.8.6. The Fuoss Approach to Ion-Pair Formation

Despite a considerable agreement with experiment, there are several unsatisfactory features of the Bjerrum picture of ion-pair formation.

The first and most important defect of the Bjerrum picture is that it identifies, as ion pairs, ions which are not in physical contact; the pair is counted as an ion pair as long as $r < q$.

A second defect is the arbitrary way in which the probability integral [cf. Eq. (3.129)] is terminated at q, i.e., the distance r at which P_r is a minimum. The physical reasons supporting this choice of a condition for the maximum distance between ion centers at which pairing can occur are not clear. In practice, however, the value of the ion-association constant K_A is not very sensitive to the actual numerical value chosen for the upper limit of the integral.

It is these considerations that led Fuoss to present an alternative picture of ion pairs and a derivation of the ion-association constant.

An ion pair is defined in a straighforward manner. For the period of time (irrespective of its magnitude) that two oppositely charged ions are in contact and, therefore, at a distance apart of $r = a$, the two ions function as neutral dipole and can be defined as an ion pair.

To get an idea of the fraction of ion pairs in a solution, the following thought experiment is a useful device. Let the motion of all the ions in a solution be frozen and the number of oppositely charged pairs of ions *in contact* be counted. If this thought experiment is repeated many times, then one can determine N_{IP}, the average number of ion pairs. The fraction of ion pairs is then obtained by dividing N_{IP} by N_i, the average number of ions.

The calculation of the fraction $\theta = N_{IP}/N_i$ is done as follows. Suppose Z positive ions and an equal number of negative ions exist in a volume V of solution. Let there be Z_{IP} ion pairs; then there will be $Z_{FI} = Z - Z_{IP}$ free ions of each species. Now, suppose one adds δZ positive ions and a similar number of anions. Since some of these will form ion pairs and some will remain free,

$$\delta Z = \delta Z_{FI} + \delta Z_{IP} \qquad (3.148)$$

The number δZ_{FI} of added negative ions that remain free is proportional, firstly, to the number δZ of negative ions added to the solution and, secondly, to the free volume $V - v_+ Z$ not occupied by positive ions of volume v_+

$$\delta Z_{FI} = (V - v_+ Z)\, \delta Z \qquad (3.149)$$

The number δZ_{IP} of negative ions that form pairs with positive ions is proportional, firstly, to the number δZ of negative ions added; secondly, to the volume $v_+ Z_{FI}$ occupied by free positive ions; and, finally, to the Boltzmann factor $e^{-U/kT}$, where U is the potential energy of a negative ion in contact with a positive ion[†]

$$\delta Z_{\text{IP}} = 2v_+ Z_{\text{FI}} e^{-U/kT} \, \delta Z \tag{3.150}$$

By dividing (3.150) by (3.149), the result is

$$\frac{\delta Z_{\text{IP}}}{\delta Z_{\text{FI}}} = 2v_+ \frac{Z_{\text{FI}} e^{-U/kT} \, \delta Z}{(V - v_+ Z) \, \delta Z} \tag{3.151}$$

or

$$\delta Z_{\text{IP}} = 2v_+ \frac{e^{-U/kT}}{(V - v_+ Z)} Z_{\text{FI}} \, \delta Z_{\text{FI}} \tag{3.152}$$

If dilute solutions are considered, one can neglect the total volume $v_+ Z$ occupied by positive ions in comparison with the volume V of the solution, and, therefore,

$$\delta Z_{\text{IP}} = \frac{2v_+ e^{-U/kT}}{V} Z_{\text{FI}} \, \delta Z_{\text{FI}} \tag{3.153}$$

Now suppose that one adds increments δZ of positive and negative ions until a total of N positive and N negative ions are present in the volume V; this process is equivalent to integrating Eq. (3.153). The result is

$$N_{\text{IP}} = \frac{N_{\text{FI}}^2}{V} v_+ e^{-U/kT} \tag{3.154}$$

or

$$\frac{N_{\text{IP}}}{V} = \left(\frac{N_{\text{FI}}}{V}\right)^2 v_+ e^{-U/kT} \tag{3.155}$$

As a model of an ion pair, one can consider the positive ions to be charged spheres of radius a and the negative ions point charges; this gives a contact distance of a. Then, the volume of the cations is given by

$$v_+ = \tfrac{4}{3} \pi a^3 \tag{3.156}$$

With regard to U, the potential energy of a negative ion in contact with a positive ion, the following argument is adopted. It has been shown that the potential ψ_r at a distance r from a central positive ion [cf. Eq. (3.99)] is

$$\psi_r = \frac{z_+ e_0}{\varepsilon r} \frac{1}{1 + \varkappa a} e^{\varkappa(a-r)} \tag{3.157}$$

[†] The factor of 2 comes in because, when one adds δZ negative ions, one must also add δZ positive ions, which also form ion pairs.

264 CHAPTER 3

Hence, at the surface of the positive ion of radius $r = a$, the potential is

$$\psi_{r=a} = \frac{z_+ e_0}{\varepsilon a} \frac{1}{1 + \varkappa a} \tag{3.158}$$

It follows therefore that

$$U = -z_- e_0 \psi_{r=a} = -\frac{z_+ z_- e_0^2}{\varepsilon a} \frac{1}{1 + \varkappa a} \tag{3.159}$$

or

$$\frac{U}{kT} = -\frac{z_+ z_- e_0^2}{\varepsilon a k T} \frac{1}{1 + \varkappa a} = -\frac{b}{1 + \varkappa a} \tag{3.160}$$

where, as stated earlier,

$$b = \frac{z_+ z_- e_0^2}{\varepsilon a k T} \tag{3.134}$$

By substituting for U/kT and v_+ from (3.160) and (3.156) in (3.155), the result is

$$\frac{N_{\text{IP}}}{V} = \left(\frac{N_{\text{FI}}}{V}\right)^2 \frac{4}{3} \pi a^3 \, e^{b/(1+\varkappa a)} \tag{3.161}$$

But

$$\frac{N_{\text{IP}}}{V} = n_{\text{IP}} = \frac{c_{\text{IP}} N_A}{1000} = \frac{\theta c N_A}{1000} \tag{3.162}$$

where c_{IP} and c are the concentration of ion pairs and electrolyte, respectively, and

$$\frac{N_{\text{FI}}}{V} = n_{\text{FI}} = \frac{(1 - \theta) c N_A}{1000} \tag{3.163}$$

where c_{FI} is the concentration of free ions. Hence, from (3.161), (3.162), and (3.163).

$$\theta c = \frac{(1 - \theta)^2 c^2 N_A}{1000} \left(\frac{4}{3} \pi a^3\right) e^{b/(1+\varkappa a)} \tag{3.164}$$

If the solution is considered dilute, $1 - \theta \sim 1$, and, since $\varkappa^{-1} \to \infty$ (i.e., $\varkappa \to 0$),

$$e^{b/(1+\varkappa a)} \sim e^b \tag{3.165}$$

in which case, Eq. (3.164) becomes

$$\frac{\theta}{c} = \frac{4\pi N_A}{3000} a^3 e^b \tag{3.166}$$

But, according to (3.141),

$$K_A = \frac{\theta}{c} \tag{3.141}$$

Hence, according to the Fuoss approach, the ion-association constant is given by

$$K_A = \frac{4\pi N_A}{3000} a^3 e^b \tag{3.167}$$

in contrast to the following expression (3.144) from the Bjerrum approach,

$$K_A = \frac{4\pi N_A}{1000} \left(\frac{z_+ z_- e_0^2}{\varepsilon kT}\right)^3 \int_2^b e^y y^{-4} \, dy \tag{3.144}$$

Now, the Bjerrum theory was tested in solutions with dielectric constants such that $b = z_+ z_- e_0^2 / \varepsilon akT = 2q/a$ was significantly larger than 2. Under these conditions, the following approximation to the integral in Bjerrum's equation can be made

$$\left(\frac{z_+ z_- e_0^2}{\varepsilon kT}\right)^3 \int_2^b e^y y^{-4} \, dy \sim a^3 \frac{e^b}{b} \tag{3.168}$$

in which case, the ion association constant of Bjerrum reduces to

$$K_A = \frac{4\pi N_A a^3}{1000} \frac{e^b}{b} \tag{3.169}$$

As the dielectric constant of the solution is changed, $b = z_+ z_- e_0^2 / \varepsilon akT$ changes, but this change is overshadowed by the e^b term. In other words, the experimental results do not permit a distinction between the Fuoss dependence of K_A on e^b [cf. Eq. (3.167)] and the Bjerrum dependence of K_A on e^b/b [cf. Eq. (3.169)]. However, the Fuoss approach is to be preferred because it is on a simpler and less arbitrary conceptual basis.

3.8.7. From Ion Pairs to Triple Ions to Clusters of Ions...

The coulombic attractive forces given by $z_+ z_- e_0^2 / \varepsilon r^2$ are large when the dielectric constant is small. When nonaqueous solvents of low dielectric constant are used, the values of dielectric constant are small. In such solutions of electrolytes, therefore, it has already been stated that ion-pair formation is favored.

Suppose that the electrostatic forces are sufficiently strong; then, it may well happen that the ion-pair "dipoles" may attract ions and *triple ions* are formed thus

$$M^+ + (A^-M^+)_{\text{ion pair}} = [M^+A^-M^+]_{\text{triple ion}} \quad (3.170)$$

or

$$A^- + (M^+A^-)_{\text{ion pair}} = [A^-M^+A^-]_{\text{triple ion}} \quad (3.171)$$

From uncharged ion pairs, charged triple ions have been formed. These charged triple ions play a role in determining activity coefficients. Triple-ion formation has been suggested in solvents for which $\varepsilon < 15$. The question of triple-ion formation can be treated on the same lines as has been done for ion-pair formation.

Further decrease of dielectric constant below a value of about 10 may make possible the formation of still larger clusters of four, five, or more ions. In fact, there is some evidence for the clustering of ions into groups containing four ions in solvents of low dielectric constant.

Further Reading

1. S. Arrhenius, *Z. Phys. Chem.*, **1**, 631 (1887).
2. N. Bjerrum, *Kgl. Danske Videnskab. Selskab.*, **4**: 26 (1906).
3. N. Bjerrum, *Kgl. Danske Videnskab. Selskab.*, **7**: 9 (1926).
4. R. M. Fuoss and C. A. Kraus, *J. Am. Chem. Soc.*, **55**: 476 (1933).
5. R. M. Fuoss and C. A. Kraus, *J. Am. Chem. Soc.*, **55**, 1019 (1933).
6. R. M. Fuoss, *Z. Physik*, **35**: 59 (1934).
7. R. M. Fuoss, *Trans. Faraday Soc.*, **30**: 967 (1934).
8. R. M. Fuoss, *Chem. Rev.*, **17**: 27 (1935).
9. J. G. Kirkwood, *J. Chem. Phys.*, **18**: 380 (1950).
10. R. M. Fuoss and L. Onsager, *Proc. Natl. Acad. Sci. U.S.*, **41**: 274, 1010 (1955).
11. J. T. Denison and J. B. Ramsey, *J. Am. Chem. Soc.*, **77**: 2615 (1955).
12. W. R. Gilkerson, *J. Chem. Phys.*, **25**: 1199 (1956).
13. R. M. Fuoss and L. Onsager, *J. Phys. Chem.*, **61**: 668 (1957).
14. H. S. Harned and B. B. Owen, *The Physical Chemistry of Electrolytic Solution*, 3rd ed., Reinhold Publishing Corp., New York, 1957.
15. R. M. Fuoss, *J. Am. Chem. Soc.*, **80**: 5059 (1958).
16. R. A. Robinson and R. H. Stokes, *Electrolyte Solutions*, 2nd ed., Butterworth's Publications, Ltd., London, 1959.
17. R. M. Fuoss and F. Accascina, *Electrolytic Conductance*, Interscience Publishers, Inc., New York, 1959.
18. C. B. Monk, *Electrolytic Dissociation*, Academic Press, London, 1961.
19. C. W. Davies, *Ion Association*, Butterworth's Publications, Ltd., London, 1962.
20. S. Levine and D. K. Rozenthal, "The Interaction Energy of an Ion Pair in an Aqueous Medium," in: B. E. Conway and R. G. Barradas, eds., *Chemical Physics of Ionic Solutions*, John Wiley & Sons, Inc., New York, 1966.

21. J. E. Prue, "Ion Association and Solvation," in: B. E. Conway and R. G. Barradas, eds., *Chemical Physics of Ionic Solutions*, John Wiley & Sons, Inc., New York, 1966.
22. G. H. Nancollas, "Thermodynamic and Kinetic Aspects of Ion Association in Solutions of Electrolytes," in: B. E. Conway and R. G. Barradas, eds., *Chemical Physics of Ionic Solutions*, John Wiley & Sons, Inc., New York, 1966.
23. R. M. Fuoss, *Proc. Nat. Acad. Sci.*, **57**, 1550 (1967).

3.9. THE QUASI-LATTICE APPROACH TO CONCENTRATED ELECTROLYTIC SOLUTIONS

3.9.1. At What Concentration Does the Ionic-Cloud Model Break Down?

A powerful stimulus for a somewhat different model for concentrated solutions has been an examination of the concentration dependence of the validity of the Debye–Hückel ionic-cloud model. A simple question is posed: According to the ionic-cloud picture, what is the number of ions responsible for any given fraction of the influence of the ionic cloud on the central ion? In 10^{-8} mole liter^{-1} solutions of 1:1-valent electrolytes, calculations show that 474 ions produce about 50% of the effect of the ionic cloud on the central ion. To achieve this effect, the 474 ions must be smeared around the central ion over a spherical region of a radius of 21,000 Å. Under these circumstances, it is quite legitimate to argue that the central ion sees a smoothed-out cloud of charge around it and one can use the Poisson equation with its implication of a continuous charge distribution.

However, as the electrolyte concentration is increased, the situation ceases to be so satisfactory. Thus, in 10^{-2} mole liter^{-1} solutions, only one ion produces the 50% of the effect of the ionic atmosphere on the central ion. To do this, the ion must be smeared around the central ion over a spherical volume with a radius of only 25 Å *and* yet yield a spherically symmetric continuous charge distribution. This is an impossible demand. Hence, under these circumstances, the central ion experiences a *discrete* charge, not a *smoothed-out* cloud of charge.

It has been concluded, therefore, that any specified fraction of the ion cloud contains a number of ions which decreases as the concentration increases. In other words, the cloud gets more and more *coarse grained*; smoothness decreases, discreteness increases. The increase of coarse grainedness leads to large fluctuations with time of the electrostatic potential ψ. Hence, the model of a continuous, smoothed-out charge density—the ionic cloud—must break down at some concentration less than 10^{-2} mole liter^{-1}. But at what concentration?

The question at issue is how coarse grained the cloud is. The measure of coarse grainedness shall be taken to be the dimension of the ion cloud compared to the mean distance between ions. This decision has the following rationale: The Deybe–Hückel ionic-cloud model implies that neighboring ions in dilute solutions are so far apart that the short-range noncoulombic interaction of a neighboring ion on the central ion is less important than the long-range coulombic interactions which set up the ionic cloud. Now, the long-range interaction between an ion and the ionic cloud can be taken as equivalent to the interaction between the central ion and the electrical image (i.e., an ion of opposite charge) situated at a distance apart of $\varkappa^{-1}$. Thus, the relative values of $\varkappa^{-1}$ and l, the mean distance between ions, provides an indication of the relative importance of long-range and short-range nearest-neighbor interactions.

The mean distance l between ions is obtained thus: If c is the concentration of the electrolyte in moles per liter, then $2cN_A$ ions† are contained in 1000 cm³. The volume occupied by one ion is $1000/2N_A c$, and, therefore, the average distance l apart of any two ions is

$$l = \left(\frac{1000}{2N_A c}\right)^{\frac{1}{3}} \tag{3.172}$$

If $\varkappa^{-1} > l$, then long-range interactions outweigh the nearest-neighbor interactions, the ion cloud is fine grained, and the Debye–Hückel theory is in its element. If $\varkappa^{-1} < l$, the cloud cannot be fine grained for the average nearest ion is not within it. It can be shown that for 1:1-valent electrolytes dissolved in water at 25 °C

$$\varkappa = \left(\frac{8\pi N_A c z_i^2 e_0^2}{1000\, kT}\right)^{\frac{1}{2}} c^{\frac{1}{2}} \quad \text{or} \quad \varkappa^{-1} = 3.043 c^{-\frac{1}{2}} \text{ [Å]} \tag{3.173}$$

and

$$l = \left(\frac{1000}{2N_A c}\right)^{\frac{1}{3}} = 9.398 c^{-\frac{1}{3}} \text{ [Å]} \tag{3.174}$$

Figure 3.47 has a plot of $\varkappa^{-1}$ and l versus $c^{\frac{1}{2}}$. The two curves intersect at $c = c_{\text{lim}} = 0.001$ mole liter^{-1}. At concentrations $c < c_{\text{lim}}$, $\varkappa^{-1} > l$, i.e., there is fine grainedness of the ion cloud and the Debye–Hückel approach is

---

† The factor 2 occurs because both positive and negative ions are counted.

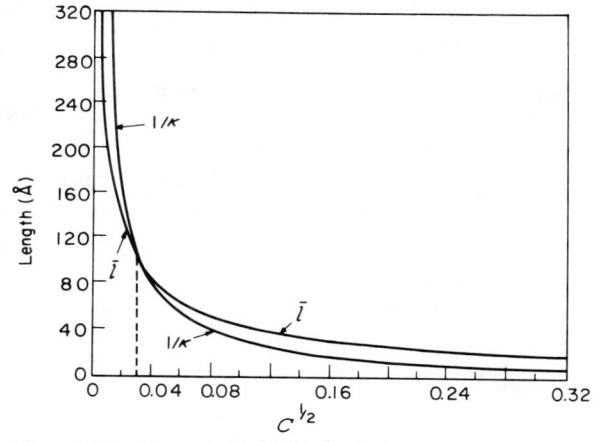

Fig. 3.47. The mean distance, l between ions and the thickness of the ionic cloud, $\varkappa^{-1}$, as a function of $c^{\frac{1}{2}}$.

fundamentally valid; at concentrations $c > c_{\text{lim}}$, $\varkappa^{-1} < l$ and the ion cloud is too coarse grained to permit the type of smoothing of charge, etc., to allow the application of the ionic-cloud model.

The concentration of 0.001 mole liter^{-1} can therefore be taken as a limit for the validity of the simple ionic-cloud picture.[†]

3.9.2. The Case for a Cube-Root Law for the Dependence of the Activity Coefficient on Electrolyte Concentration

For concentrations above c_{lim}, the mean distance apart of ions, i.e., l, would appear to be a fundamental quantity in a revised theory. But l is obtained by taking the $2N_A c$ ions and *imagining* them to be arranged in a periodic manner to give a volume of 1000 cm³. In such a regular arrangement, or "lattice," the "lattice constant" is the mean distance apart of the ions. Thus, l is a "lattice parameter," and the theoretical approach to solutions more concentrated than c_{lim} should be based on a "lattice model."

[†] Of course, the calculation of the concentration at which coarse grainedness begins, itself, depends through $\varkappa^{-1}$ on the use of the linearized P–B equations. When $\varkappa^{-1} < l$, the theory used for the calculation of the condition for coarse grainedness is not valid. Hence, the above calculation of the concentration at which coarse grainedness sets in is in fact on the borderline of validity and can only be considered to provide a rough answer. However, in partial conformity with this answer are the data for HCl, which show that the limiting law does begin to break down at 3 to 5 × 10⁻³N.

270 CHAPTER 3

Since $l \propto c^{-\frac{1}{3}}$, it would be expected that, if the parameter l (characterizing a hypothetical lattice in solution) were of any importance, there should be a dependence of the properties of an electrolytic solution (e.g., $\log f_{\pm}$) on $c^{\frac{1}{3}}$. Such an observation was indeed made before the advent of the Debye–Hückel ionic-cloud theory. When $\log f_{\pm}$ is plotted against $c^{\frac{1}{3}}$, long linear regions can be noted (see Fig. 3.48) at concentrations above about $c_0 = 0.001$ mole liter^{-1}. This strengthens the argument for thinking in terms of a lattice parameter, rather than an ionic cloud, for all but very dilute solutions.

Some years before the Debye–Hückel theory, Ghosh had suggested that, when an ionic crystal such as sodium chloride is dissolved in water, it exists in the form of an expanded lattice. He obtained a dependence of $\log f_{\pm}$ on $c^{\frac{1}{3}}$ by using the idea of lattice energy (*cf.* Chapter 2). Thermal motions, however, would surely disrupt and wash out the long-range order so much a characteristic of a lattice. In fact, it was the realization of this point that stimulated Debye and Hückel to develop the ionic-cloud model according to which the excess charge density decays monotonically with distance. Sufficiently far from the central ion, the excess charge density is virtually equal to zero, i.e., indicates equal probabilities of finding negative and positive ions. However, with increasing electrolyte concentration, the ionic cloud becomes more and more coarse grained, which indicates short-range order and a quasi-lattice structure. Under these conditions of electrolyte concentration, a distance characteristic of the short-range (lattice-like) order, rather than $\varkappa^{-1}$, becomes important in calculating interactions in solution. This distance would be related to l, the mean distance apart of the ions.

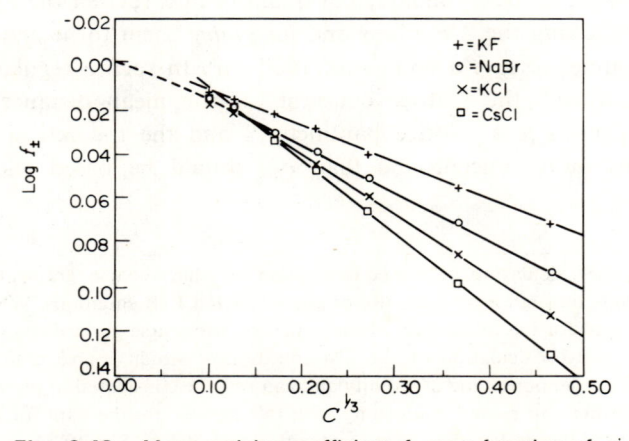

Fig. 3.48. Mean activity coefficient $f_{\pm}$ as a function of $c^{\frac{1}{3}}$.

These matters relevant to the possibility of a quasi lattice in concentrated aqueous solutions are closely connected to structural models for pure liquid electrolytes (Chapter 6). Thus, one view of a molten electrolyte is of the regular lattice of the parent ionic crystal from which about 10 to 30% of the ions are removed to give vacancies, the remaining ions being placed in rapid motion. This pure liquid electrolyte would still show appreciable vestiges (over local regions) of the lattice structure of the solid crystal. Around any particular ion, the opposite charge would not be smeared out into a density which falls away monotonically with radial distance away from the ion. On the other hand, the charge density would show periodic, damped maxima. These correspond to the probability of finding alternate positive and negative ions, as in a lattice. The damping represents the decrease in the ordering of ions—hence, the term *quasi lattice*.

3.9.3. The Beginnings of a Quasi-Lattice Theory for Concentrated Electrolytic Solutions

The basic characteristics which a quasi-lattice theory must possess in order satisfactorily to account for the behavior of concentrated electrolytic solutions are now considered.

The contribution of interactions to the chemical potential of an ionic species must be proportional to the average interaction energy for a pair of ions, i.e., to $z_i e_0 z_j e_0 / \varepsilon l$. Why is this an average interaction energy? Because l is the average ion–ion distance. The proportionality constant must express the "latticeness" of the electrolytic solution and the "quasiness" of the lattice.

The latticeness needs the use of a factor which describes the arrangement of ions in its geometrical aspect. For ionic crystals, the geometry of the particular arrangement of ions is taken into account in calculations of lattice energy (*cf.* Chapter 2) by the Madelung number. Hence, a Madelung-*like* number M^* can be inserted into the proportionality constant as an ion-arrangement factor. One has, through the Madelung-like number, some view of a greatly expanded and thermally smeared-out lattice. Over local regions, there is an alternation of positive and negative ions, but thermal forces wash out any long-range order. The exact value of the Madelung-like number depends, for instance, on the average locations of remote ions. Since periodicity is not simply analyzable in the liquid, the values of the Madelung number will differ from those in the crystalline state.

To bring out the quasiness of the lattice, it is necessary, firstly, to insert an inverse function of the temperature and, secondly, to include terms for the short-range interactions which produce the ordering. The quasiness factor will represent the disintegration of the lattice (i.e., the decrease in the degree of order) as the temperature increases. By analogy with the Debye–Hückel dependence of interaction energy on $T^{-\frac{1}{2}}$, the proportionality constant may be considered to include a function of $T^{-\frac{1}{2}}$. The functional relationship, yet unknown, is intended to represent short-range interactions which produce local order.

By such physical arguments, Frank and Thompson have arrived at a dimensionally acceptable *guess* in an expression for the activity coefficient. Their expression has the form

$$- \log f_{\pm} = \alpha - \beta c^{\frac{1}{2}} + \gamma c^{\frac{2}{3}} \tag{3.175}$$

in accordance with experiment.

It is important to emphasize that what has been accomplished is an essay in interpretation rather than a theory. The matters discussed here lie at the frontiers of research into the structure of concentrate solution. The position is that no rigorous quasi-lattice theory exists today. The problem has been attacked by several authors but has so far resisted unravelment. The attempts do not survive simple criteria, for example, that the Debye and Guntelberg charging processes must yield the same expression for the logarithm of the activity coefficient (Section 3.7).

The essential problem remains. It is the mathematical description of the time-varying, thermally smeared partial order of ions in the electrolytic solution, under conditions in which coarse-grainedness makes the ionic cloud model invalid.

Further Reading

1. J. C. Ghosh, *J. Chem. Soc.*, **113**: 449, 707 (1918).
2. L. Onsager, *Chem. Rev.*, **13**: 73 (1933).
3. H. S. Frank, *J. Chem. Phys.*, **13**: 478 (1945).
4. R. H. Fowler and E. A. Guggenheim, *Statistical Thermodynamics*, Cambridge University Press, New York, 1952.
5. R. A. Robinson and R. H. Stokes, *Electrolyte Solutions*, Butterworths Scientific Publications Ltd., London, 1955.
6. H. S. Frank and P. T. Thompson, in: W. Hamer, ed., *The Structure of Electrolyte Solutions*, John Wiley & Sons, New York, 1960.

7. H. S. Frank, "Solvent Models and the Interpretation of Ionization and Solvation Phenomena," in: B. E. Conway and R. G. Barradas, eds. *Chemical Physics of Ionic Solution,* John Wiley & Sons, Inc., New York, 1966.
8. S. R. Cohen, *Trans. Faraday Soc.,* **63** (9): 2225 (1967).

3.10. THE STUDY OF THE CONSTITUTION OF ELECTROLYTIC SOLUTIONS

3.10.1. The Temporary and Permanent Association of Ions

The effect of the temporary association of ions on the mean activity coefficient has been considered in Section 3.8. The association of positive and negative ions in ion pairs removed them from consideration in the ionic cloud, and the Debye–Hückel activity coefficient pertains only to the free ions.

In the case of ion pairs, the forces are coulombic (electrostatic) in origin. The ion pairs are formed when the thermal-dissociating forces are exceeded by the coulombic-associating forces. But are the thermal forces permanently exceeded?

For the following reasons, it is unlikely that ion pairs are permanent. Pair formation involves solvated ions, and, hence, the center-to-center distance of the two ions is about 5 to 10 Å. Coulombic forces are long range in character. The electrostatic interaction between a pair of oppositely charged ions may therefore be sufficient to overcome thermal forces temporarily, but, at some later instant, one ion of the pair may be persuaded by a momentum-imparting collision to break away from its partner. The association of ions in ion pairs is therefore generally a temporary affair. Ion pairs have a short lifetime.

Such electrostatic ion-pair formation does not exhaust the theoretical possibilities or the experimental realities. When a cation and an anion come closer than about 5 to 10 Å, short-range chemical forces between the ions are likely to predominate, particularly in the absence of intervening solvent molecules. In such cases, where cations and anions are adjacent and not separated by neutral solvent molecules, there is the possibility of forming bonds stronger and more lasting than electrostatic interactions. Such "chemical" interaction is distinguished from the coulombic variety by stating that a *complex*—in contrast to an ion pair—is formed.

The formation of complexes would also lead to a deviation of the Debye–Hückel activity coefficient from the observed value. How, then, to distinguish between ion-pair formation and complex formation? This question increases the need for the study of the constitution of electrolytic solutions.

3.10.2. Electromagnetic Radiation, a Tool for the Study of Electrolytic Solutions

The understanding of the concentration dependence of activity coefficients required the postulate of the concepts of ion-pair formation and complex formation. Certain structural questions, however, could not be answered unequivocally by these considerations alone. For instance, it was not possible to decide whether pure coulombic or chemical forces were involved in the process of ion association, i.e., whether the associated entities were ion pairs or complexes. The approach has been to postulate one of these types of association, then to work out the consequences of such an association on the value of the activity coefficient, and finally to compare the observed and calculated values. Proceeding on this basis, it is inevitable that the postulate will always stand in need of confirmation because the path from postulate to fact is indirect.

It is fortunately possible to gain direct information on the constitution and concentrations of the various species in an electrolytic solution by "seeing" the various species. These methods of seeing are based upon shining "light"[†] into the electrolytic solution and studying the light which is transmitted, scattered, or refracted. The light emerging from the electrolytic solution, in the ways just mentioned, is altered or modulated as a result of the *interaction* between the free ionic or associated species and the incident light.

There are many types of interactions between electromagnetic radiation and matter. The species as a whole can change its rotational state; the bonds (if any) within the species can bend, stretch, or twist and thus alter its vibrational state; the electrons in the species may undergo transitions between various energy states; and, finally, the atomic nuclei of the species can absorb energy from the incident radiation by making transitions between the different orientations to an externally imposed magnetic field.

All these responses of a species to the stimulus of the incident electromagnetic radiation involve energy exchange. This energy exchange must, according to the quantum laws, occur only in finite jumps of energy. Hence, the light which has been modulated by these interactions contains information about the *energy* that has been exchanged between the incident light and the species present in the electrolytic solution. If the free, unassociated ions and the associated aggregates (ion pairs, triplets, complex ions, etc.)

[†] The word *light* is used here to represent not only the visible range of the electromagnetic spectrum but all the other ranges for which analytical methods have been developed.

interact differently with the incident radiation—and they must at some frequency (each species has a characteristic frequency)—, then information concerning the structures will be observed in the radiation emerging from the electrolytic solution.

These aspects of the interaction of radiation and matter are the concern of *absorption spectroscopy* in the most general sense of the term. It is customary, however, to use different terms depending on the wavelength of the incident radiation used and the particular types of interaction involved. Thus, one speaks of various kinds of spectroscopy—visible and ultraviolet (UV) absorption spectroscopy, Raman and infrared (IR) spectroscopy, nuclear magnetic resonance (NMR) spectroscopy, and electron-spin resonance (ESR) spectroscopy. A brief description of the principles of these techniques and their application to the study of electrolytic solutions will now be given.

3.10.3. Visible and Ultraviolet Absorption Spectroscopy

Atoms, neutral molecules, and ions (simple and associated) can exist in several possible *electronic* states; this is the basis of visible and UV absorption spectroscopy. Transitions between these energy states occur by the absorption of discrete energy quanta ΔE which are related to the frequency ν of the light absorbed by the well-known relation

$$\Delta E = h\nu$$

where h is Planck's constant. When light is passed through the electrolytic solution, there is absorption at the characteristic frequencies corresponding to the electronic transitions of the species present in the solution. Owing to the absorption, the intensity of transmitted light of the absorption frequency is less than that of the incident light; and the fall-off in intensity at a particular wavelength is given by an exponential relation known as the Beer–Lambert law

$$I = I^0 e^{-\varepsilon cl}$$

where I^0 and I are the incident and transmitted intensities, ε is a characteristic of the absorbing species and is known as *molar absorptivity*, l is the length of the material (e.g., electrolytic solution) through which the light passes, and c is the concentration of the absorbing molecules.

A historic use of the Beer–Lambert law was by Bjerrum, who studied the absorption spectra of dilute copper sulphate solutions and found that the molar absorptivity was independent of the concentration. Bjerrum concluded that the only species present in dilute copper sulphate solutions are free, unassociated copper and sulphate ions and not, as was thought at the time, undissociated copper sulphate molecules which dissociate into ions to an extent which depends on the concentration. For, if any undissociated molecules were present, then the molar absorptivity of the copper sulphate solution would have been dependent on the concentration.

In recent years, it has been found that the molar absorptivity of *concentrated* copper sulphate solutions does show a slight concentration dependence. This concentration dependence has been attributed to ion-pair formation occurring through the operation of coulombic forces between the copper and sulphate ions. This is perhaps ironic because Bjerrum's concept of ion pairs is being used to contradict Bjerrum's conclusion that there are only free ions in copper sulphate solutions. Nevertheless, there is a fundamental difference between the erroneous idea that a copper sulphate crystal dissolves to give copper sulphate *molecules*, which then dissociate into free ions, and the modern point of view that the ions of an ionic crystal pass into solution as free solvated ions which, under certain conditions, associate into ion pairs.

The method of visible and UV absorption spectroscopy is at its best when the absorption spectra of the free ions and the associated ions are quite different and known. When the associated ions cannot be chemically isolated and their spectra studied, the type of absorption by the associated ions has to be attributed to electronic transitions known from other well-studied systems. For example, there can be an electron transfer to the ion from its immediate environment (charge-transfer spectra), i.e., from the entities which are associated with the ion, or transitions between new electronic levels produced in the ion under the influence of the electrostatic field of the species associated with the ion (crystal-field splitting). Thus, there is an influence of the environment on the absorption characteristics of a species, and this influence reduces the unambiguity with which spectra are characteristic of species rather than of their environment. Herein lies what may be considered a disadvantage of the technique of visible and UV absorption spectroscopy.

3.10.4. Raman Spectroscopy

Visible and UV absorption spectroscopy are based on studying that part of the incident light which is transmitted (after absorption) through an electrolytic solution in the same direction as the original beam. There is, however, a certain amount of light which is scattered in other directions.

The scattered light in any direction consists mainly of light of the same frequency as the incident radiation (in which case the phenomenon is known as *Rayleigh scattering*) and gives no information concerning the intramolecular structure of the molecules responsible for the scattering. In addition, however, it is observed that the scattered beam includes light, the frequency of which is shifted from the incident frequency by an amount $\Delta \nu$ either added or subtracted from the incident frequency; this phenomenon is known as the *Raman effect* and the type of scattering, as *Raman scattering*. The Raman shift $\pm \Delta \nu$ is due to energy exchanged $\pm \Delta E$ between the incident radiation and the scattering species

$$\Delta E = h \Delta \nu$$

A simple view of the origin of the Raman effect is as follows: Rayleigh scattering is produced because the electric field of the incident light induces a dipole moment in the scattering species, and, since the incident field is oscillating, the induced dipole moment also varies periodically. Such an oscillating dipole acts as an antenna and radiates light of the same frequency as the incident light.

It has already been stated (*cf*. 2.5.4) that it is the deformation polarizability α_deform which determines the magnitude of dipole moment induced by a particular field. If this polarizability changes from its time-average value, then the induced-dipole antenna will be radiating at a new frequency that is different from the incident frequency; in other words, there is a Raman shift. Now the changes in polarizability of the scattering species can be correlated with their rotations and vibrations and also their symmetry characteristics. Hence, the Raman shifts are characteristic of the rotational and vibrational energy levels of the scattering species and provide direct information about these levels. Though the presence of electrostatic bonding produces second-order perturbations in the Raman lines, it is the species with covalent bonds to which Raman spectroscopy is sensitive and not ion pairs.

Another feature which makes Raman spectra useful is the fact that the integrated intensity of the Raman line is in general proportional to the concentration of the scattering species.

The problem with Raman spectroscopy is the low intensity of the Raman lines, which permits an easy detection of species only in concentrations of at least 10%. Thus, at low concentrations of the scattering species, the method presents great difficulties in separating out the *signal* from the *noise*, i.e., identifying the Raman lines from the ever-present general background intensity. Fortunately, the availability of lasers, which are intense sources of monochromatic light, is stimulating further applications of this powerful technique which is noted for the lack of ambiguity with which it can report on the species in an electrolytic solution. Devices which distinguish a low-intensity signal from noise, recently available, also help.

3.10.5. Infrared Spectroscopy

Infrared spectroscopy resembles Raman spectroscopy in that it provides information on the vibrational and rotational energy levels of a species, but it differs from the latter technique in that it is based on studying the light transmitted *through* a medium after absorption, and not that *scattered* by it.

The techniques of Raman and IR spectroscopy are generally considered complementary in the gas and solid phases because some of the species under study may reveal themselves in only one of the techniques. Nevertheless, it must be stressed that Raman scattering is not affected by the aqueous medium, whereas strong absorption in the infrared shown by *water* proves to be a troublesome interfering factor in the study of aqueous solutions by the IR method.

3.10.6. Nuclear Magnetic Resonance Spectroscopy

The nuclei of atoms can be likened in some respects to elementary magnets. In a strong magnetic field, the different orientations that the elementary magnets assume correspond to different energies. Thus, transitions of the nuclear magnets between these different energy levels correspond to different frequencies of radiation in the short-wave, radio-frequency range. Hence, if an electrolytic solution is placed in a strong magnetic field and an oscillating electromagnetic field is applied, the nuclear magnets exchange energy (exhibit resonant absorption) when the incident frequency equals that for the transitions of nuclei between various levels.

Were this NMR to depend only on the nuclei of the species present in the solution, the technique would be without point for the identification of species in a solution. But the nuclei sense the applied field as modified by the environment of the nuclei. The modification is almost exclusively due to the nuclei and electrons in the neighborhood of the sensing nucleus, i.e., due to the adjacent atoms and bonds. Thus, NMR studies can be used to provide information on the type of association between an ion and its environmental particles, e.g., on ion–solvent interactions or ion association.

Further Reading

1. J. E. Hinton and E. S. Amis, *Chem. Rev.*, **67**: 367 (1967).
2. T. T. Wall and D. F. Hornig, *J. Chem. Phys.*, **47**: 784 (1967).
3. N. A. Matwiyoff, P. E. Darby, and W. G. Movius, *Inorg. Chem.*, **7**: 2173 (1968).
4. E. R. Malinowski and P. S. Knapp, *J. Chem. Phys.*, **48**: 4989 (1968).

3.11. A PERSPECTIVE VIEW ON THE THEORY OF ION–ION INTERACTIONS

A host of new ideas have been described in the present chapter. It is time to lay them out in perspective.

The development was triggered by the intuitive idea that the presence of free ions in an electrolytic solution must surely lead to some phenomenon caused by the energy of interionic attraction. Hence, the analysis of ion–ion interactions are as important to the understanding of ionic solutions as are ion–solvent interactions.

The original Debye–Hückel theory was a brilliant first step in the theoretical treatment of ion–ion interactions. It was worked out by assuming a very simple idealized model: (1) point-charge ions; (2) only coulombic forces between ions; (3) ions smoothed out into a charged continuum; (4) a small enough electrostatic potential near an ion, to permit the linearization of the Boltzmann equation.

Many beautiful results fell out of the theory. For instance, it became possible to present a mathematical description of the distribution of excess charge density as a function of distance from any ion chosen as a reference ion. This distribution function also yielded a physical picture of ions in an electrolytic solution surrounded by clouds of charge, the opposite charges predominating. The interactions between an ion and the rest of the electrolytic solution thus became equivalent to the interaction between an ion and its ionic cloud. This interaction between the reference (or sample) ion and the ionic cloud could be quantitatively treated in a simple way. All one had to do was to calculate the potential ψ_{cloud}, set up at the surface of the reference ion by the ionic cloud and use this ψ_{cloud} in the expression for charging up the reference ion. Thus was obtained the chemical-potential change arising from the interactions between an ionic species and the ionic solution.

The linkup between theoretical calculations and experimental measurements was made on the basis that the chemical-potential change arising from ion–ion interactions was set equal to the empirical correction factor

$RT \ln f_\pm$ which had been introduced to account for the fact that ionic solutions do not behave like so-called ideal solutions of noninteracting solute particles. In this manner, the Debye–Hückel ionic-cloud model could be made to yield the limiting law—an expression for the concentration variation of the logarithm of the activity coefficient $f_\pm$. The experimental $\log f_\pm$ versus $I^{\frac{1}{2}}$ curves corresponded more and more closely to the limiting law as the solutions were made more and more dilute. But the limiting law argued for a constant slope at all concentrations, whereas the experimental curves showed a slope which decreased with concentration. In fact, the curves passed through a minimum and even acquired a positive slope. The theory had to be improved.

The first step in the development of the theory for concentrated solutions was to give the ions a finite size rather than to consider them point charges.

An ion-size parameter a was defined, and the beginning of the decrease of slope of the theoretical $\log f_\pm$ versus $I^{\frac{1}{2}}$ curve could be explained. It is important to realize that the introduction of an ion-size parameter was equivalent to an incorporation into the model of a short-range repulsive force which would forbid the ions to approach closer than a particular distance a. The parameter a, being different for different electrolytes of the same valence type, introduced specific differences between these electrolytes and gave to each of them a different $\log f_\pm$ versus $I^{\frac{1}{2}}$ curve, as is experimentally observed. These specific differences between two electrolytes of the same valence type were unforeseen by the point-charge version of the theory, which argued that all electrolytes of the same valence type would have the same $\log f_\pm$ versus $I^{\frac{1}{2}}$ curve.

The actual values of the ion-size parameter, which had to be used to make theory fit experiment, were larger than the sum of the crystallographic radii of naked ions. Hence, the value of the ion-size parameter obviously reflected the attachment of a sheath of water molecules to an ion, i.e., the primary hydration sheath. To the extent, therefore, that the simple version of the Debye–Hückel theory relegated the solvent to the role of a ubiquitous dielectric and gave it a nonessential role, one had to introduce the influence of solvation upon ion–ion interactions through the sizes of ions.

The effect of the sizes of ions was only a subsidiary influence of the phenomenon of solvation. Its main effect arose in a different way. As a solution became more and more concentrated, a greater and greater fraction of the solvent became trapped in the solvent sheaths of ions, became *hors de combat*, which resulted in an increase of the effective concentration.

It was this influence of solvation on the activity coefficient which was of greater significance than the impact of solvation on the sizes of ions. When the effect of solvation on the activity coefficient was added onto the Debye–Hückel activity coefficient, one obtained an expression for $\log f_\pm$ which made the $\log f_\pm$ versus $I^{\frac{1}{2}}$ curve pass through a minimum and then turn upward in accordance with experiment.

At this stage of the presentation, a fundamentally different approach was tried. Taking the view that the potential ψ_r may not be small enough to justify the linearization of the Boltzmann equation, the obvious solution was to work with an *un*linearized Boltzmann equation. It turned out, however, that attempts to use the unlinearized P–B equation have hitherto led to a logical inconsistency most clearly seen in the fact that the Güntelberg and the Debye procedures for arriving at the chemical potential of ion–ion interactions lead to different results even though the two procedures ought to be equivalent.

The whole theory, till that stage, had been worked out on the basis that every ion from the parent crystal remained free in solution. But ions tend not only to aggregate around each other in clouds but also to stick together or associate to form neutral pairs. The Debye–Hückel theory was held to remain valid for free ions, but the formation of uncharged ion pairs reduced the concentration of free ions. The ion-association model led to the concept of a minimum distance q, closer than which two ions approach only at the risk of sticking together as an ion pair. By using a corrected value for ionic concentration and the Debye–Hückel model only for the free ions, one obtained an expression for the activity coefficient which was valid when the solvation corrections were not important.

As the ideas of finite ion size, solvation effects, and ion association were invoked, the model of an ionic solution slowly approached reality. The experimental $\log f_\pm$ versus $I^{\frac{1}{2}}$ curves were quasi-empirically "reproduced" by the theory even at a very high concentration, e.g., $5M$.

The gathering triumphs and advances had been accompanied, of course, by less satisfactory aspects.

Thus, the finite-ion-size model could not generate the value of the ion-size parameter from theoretical arguments alone. The *effective* size of solvated ions in collision was too complex a many-body problem for accurate calculation. The ion size had to be incorporated as an adjustable parameter. In contrast, the Debye–Hückel limiting law represented an "absolute" theory with no semiempirical parameters.

Likewise, the influence of solvation on the activity coefficient, i.e., the effect of taking water molecules from their positions in the water network into the solvent sheaths around ions, was difficult to calculate. There were difficulties in counting the number of water molecules "removed" from the bulk water. The primary water molecules must, of course, be counted, but how much of the secondary solvation sheath? It turned out that, by using the primary hydration numbers alone, quite good consistency with experiment is observed. But these hydration numbers too are usually obtained by calibration from experiment (*cf.* Section 2.4). Finally, in the Stokes–Robinson treatment, it is assumed that the Debye–Hückel activity coefficient must be modified by a solvation correction, i.e., it is assumed that the basis of the Debye–Hückel model is valid in concentrated solutions.

When, however, one makes a fundamental analysis of the Debye–Hückel model, it becomes clear that, even at concentrations as low as about 10^{-3} mole liter^{-1}, the thickness $\varkappa^{-1}$ of the ionic atmosphere is less than the mean distance *l* between ions. In other words, the image charge representing the ionic cloud is closer to the central ion than is the average nearest-neighbor ion. This is ridiculous, for the ionic cloud consists of ions. Thus, there is a fundamental invalidity of the Debye–Hückel model above any but the lowest practical concentrations.

A clue to an alternative approach for higher concentrations lies in observed experimental results that indicate a linear relationship between $\log f_\pm$ and $c^{\frac{1}{2}}$. This points to the necessity of developing a quasi-lattice model for concentrated solutions—a feat which has not yet been achieved.

Appendix 3.1. Poisson's Equation for a Spherically Symmetrical Charge Distribution

A starting point for the derivation of Poisson's equation is Gauss's law which can be stated as follows: Consider a sphere of radius *r*. The electric field X_r due to charge in the sphere will be normal to the surface of the sphere and equal everywhere on this surface. The total field over the surface of the sphere, i.e., the surface integral of the normal component of the field, will be equal to X_r times the area of the surface, i.e., $X_r 4\pi r^2$. According to Gauss's law, this surface integral of the normal component of the field is equal to $4\pi/\varepsilon$ times the total charge contained in the sphere. If ϱ_r is the charge density at a distance *r* from the center of the sphere, then $4\pi r^2 \varrho_r \, dr$ is the charge contained in a *dr*-thick shell distance *r* from the center and $\int_0^r 4\pi r^2 \varrho_r \, dr$ is the total charge contained in the sphere. Thus, Gauss's law states that

$$X_r 4\pi r^2 = \frac{4\pi}{\varepsilon} \int_0^r 4\pi r^2 \varrho_r \, dr$$

ION–ION INTERACTIONS

i.e.,

$$r^2 X_r = \frac{4\pi}{\varepsilon} \int_0^r r^2 \varrho_r \, dr \tag{A3.1.1}$$

Now, according to the definition of electrostatic potential ψ_r, at a point r,

$$\psi_r = -\int_\infty^r X_r \, dr \tag{2.4}$$

or

$$X_r = -\frac{d\psi_r}{dr} \tag{A3.1.2}$$

Substituting (A3.1.2) in (A3.1.1), one gets

$$r^2 \frac{d\psi_r}{dr} = -\frac{4\pi}{\varepsilon} \int_0^r r^2 \varrho_r \, dr \tag{A3.1.3}$$

Differentiating both sides with respect to r,

$$\frac{d}{dr}\left(r^2 \frac{d\psi_r}{dr}\right) = -\frac{4\pi}{\varepsilon} r^2 \varrho_r$$

i.e.,

$$\frac{1}{r^2} \frac{d}{dr}\left(r^2 \frac{d\psi_r}{dr}\right) = -\frac{4\pi}{\varepsilon} \varrho_r \tag{A3.1.4}$$

which is Poisson's equation for a spherically-symmetrical charge distribution.

Appendix 3.2. Evaluation of the Integral $\int_{r=0}^{r\to\infty} e^{-(\varkappa r)}(\varkappa r)\, d(\varkappa r)$

According to the rule for integration by parts,

$$\int v \, du = uv - \int u \, dv$$

Thus,

$$\int_{r=0}^{r\to\infty} (\varkappa r) e^{-\varkappa r} \, d(\varkappa r) = \left[-\varkappa r e^{-\varkappa r} - \int \frac{e^{-\varkappa r}}{(-1)} d(\varkappa r) \right]_{r=0}^{r\to\infty}$$

$$= \left[-\varkappa r e^{-\varkappa r} - e^{-\varkappa r} \right]_{r=0}^{r\to\infty}$$

$$= \left[-e^{-\varkappa r}(\varkappa r + 1) \right]_{r=0}^{r\to\infty}$$

$$= +1 \tag{A3.2.1}$$

Appendix 3.3. Derivation of the Result $f_\pm = (f_+^{\nu_+} f_-^{\nu_-})^{1/\nu}$

Suppose that, on dissolution, 1 mole of salt gives rise to ν_+ moles of positive ions and ν_- moles of negative ions. Then, instead of Eqs. (3.65) and (3.66), one has

$$\nu_+ \mu_+ = \nu_+ \mu_+^0 + \nu_+ RT \ln x_+ + \nu_+ RT \ln f_+ \qquad (A.3.3.1)$$

and

$$\nu_- \mu_- = \nu_- \mu_-^0 + \nu_- RT \ln x_- + \nu_- RT \ln f_- \qquad (A3.3.2)$$

Upon adding the two expressions (A3.3.1) and (A3.3.2) and dividing by $\nu = \nu_+ + \nu_-$ to get the average contribution to the free-energy change of the system per mole of both positive and negative ions, the result is

$$\mu_\pm = \frac{\nu_+ \mu_+ + \nu_- \mu_-}{\nu} = \frac{\nu_+ \mu_+^0 + \nu_- \mu_-^0}{\nu} + RT \ln(x_+ x_-)^{1/\nu} + RT \ln(f_+ f_-)^{1/\nu} \qquad (A3.3.3)$$

which may be written as

$$\mu_\pm = \mu_\pm^0 + RT \ln x_\pm + RT \ln f_\pm \qquad (A3.3.4)$$

where

$$\mu_\pm^0 = \frac{\nu_+ \mu_+^0 + \nu_- \mu_-^0}{\nu} \qquad (A3.3.5)$$

$$x_\pm = (x_+^{\nu_+} x_-^{\nu_-})^{1/\nu} \qquad (A3.3.6)$$

and

$$f_\pm = (f_+^{\nu_+} f_-^{\nu_-})^{1/\nu} \qquad (A3.3.7)$$

Appendix 3.4. To Show That the Minimum in the P_r versus r Curve Occurs at $r = \lambda/2$

From

$$P_r = 4\pi n_i^0 e^{\lambda/r} r^2 \qquad (A3.4.1)$$

one finds the minimum in the P_r versus r curve by differentiating the expression for P_r with respect to r and setting the result equal to zero. That is,

$$\frac{dP_r}{dr} = 4\pi n_i e^{\lambda/r} 2r - 4\pi n_i^0 r^2 e^{\lambda/r} \frac{\lambda}{r^2} = 0 \qquad (A3.4.2)$$

i.e.,

$$2r_{\min} - \lambda = 0$$

or

$$r_{\min} = q = \frac{\lambda}{2} \qquad (A3.4.3)$$

Appendix 3.5. Transformation from the Variable r to the Variable $y = \lambda/r$

From $y = \lambda/r$, it is obvious that

$$r^2 = \frac{\lambda^2}{y^2} \tag{A3.5.1}$$

$$dr = -\frac{\lambda}{y^2} dy \tag{A3.5.2}$$

$$e^{\lambda/r} = e^y \tag{A3.5.3}$$

Further, when $r = q = \lambda/2$, it is clear that

$$y = 2 \tag{A3.5.4}$$

and, when $r = a$, it follows that

$$y = \frac{\lambda}{a} = b \tag{A3.5.5}$$

By introducing the substitutions (A3.5.1), (A3.5.2), (A3.5.3), (A3.5.4), and (A3.5.5) into Eq. (3.13.1), the result is

$$0 = 4\pi n_i^0 \lambda^3 \int_{y=2}^{y=b} e^y y^{-4} \, dy$$

or

$$0 = 4\pi n_i^0 \left(\frac{z_+ z_- e_0^2}{\varepsilon \varkappa T} \right)^3 \int_{y=2}^{y=b} e^y y^{-4} \, dy \tag{A3.5.6}$$

Appendix 3.6. Relation between Calculated and Observed Activity Coefficients

Consider a solution consisting of 1 kg of solvent in which is dissolved m moles of a 1:1 valent salt.

Ignoring the question of ion-pair formation, there are m moles of positive ions and m moles of negative ions in the solution. Hence, the total free energy, G, of the solution is given by

$$G = \left(\frac{1000}{M_{H_2O}} \right) \mu_{H_2O} + m\mu_+ + m\mu_- \tag{A3.6.1}$$

where M_{H_2O} is the molecular weight of the water and μ_{H_2O}, μ_+, and μ_- are the chemical potentials of water, positive ions, and negative ions, respectively.

Now, if a fraction θ of the positive and negative ions form ion pairs designated $(+\ -)$, then the free energy of the solution assuming ion-pair formation is

$$G = \left(\frac{1000}{M_{H_2O}}\right)\mu_{H_2O} + (1-\theta)m\mu_+' + (1-\theta)m\mu_-' + \theta m\mu_{+-}' \quad (A3.6.2)$$

where the primed chemical potentials are based on ion association.

Since, however, the free energy of the solution must be independent of whether ion association is considered or not, Eqs. (A3.6.1) and (A3.6.2) can be equated. Hence

$$\mu_+ + \mu_- = (1-\theta)[\mu_+' + \mu_-'] + \theta\mu_{+-}' \quad (A3.6.3)$$

But, ions are in equilibrium with ion pairs; therefore

$$\mu_{+-} = \mu_+' + \mu_-' \quad (A3.6.4)$$

Combining Eqs. (A3.6.4) and (A3.6.3), one has

$$\mu_+ + \mu_- = \mu_+' + \mu_-' \quad (A3.6.5)$$

or

$$\mu_+^0 + RT \ln \gamma_+ m + \mu_-^0 + RT \ln \gamma_- m = \mu_+^{0'} + RT \ln \gamma_+'(1-\theta)m + \mu_-^{0'}$$
$$+ RT \ln \ln \gamma_-'(1-\theta)m \quad (A3.6.6)$$

It is clear, however, that $\mu_+^0 = \mu_+^{0'}$ and $\mu_-^0 = \mu_-^{0'}$ since, whether ion association is considered or not, the standard chemical potentials of the positive and negative ions are the chemical potentials corresponding to ideal solutions of unit molality of these ions. Hence

$$\gamma_+\gamma_- = (1-\theta)^2\gamma_+'\gamma_-' \quad (A3.6.7)$$

or

$$\gamma_\pm = (1-\theta)\gamma_\pm' \quad (A3.6.8)$$

But, $\gamma_\pm$ is the observed or stoichiometric activity coefficient, and, therefore,

$$(\gamma_\pm)_{obs} = (1-\theta)\gamma_\pm' \quad (A3.6.9)$$

Similarly,

$$(f_\pm)_{obs} = (1-\theta)f_\pm \quad (3.146)$$

CHAPTER 4
ION TRANSPORT IN SOLUTIONS

4.1. INTRODUCTION

The interaction of an ion in solution with its environment of solvent molecules and other ions has been the subject of the previous two chapters. Now, attention will be focused on the motion of ions through their environment. The treatment will restrict itself to solutions of true electrolytes.

There are two aspects to these ionic motions. Firstly, there is the *individual* aspect. This concerns the dynamical behavior of ions as individuals—the trajectories they trace out in the electrolyte, and the speeds with which they dart around. These ionic movements are basically random in direction and speed. Secondly, ionic motions have a *group* aspect which is of particular significance when more ions move in certain directions than in others and thus produce a drift, or flux,[†] of ions. This drift has important consequences because an ion has a mass and bears a charge. Consequently, the flux of ions in a preferred direction results in the transport of matter and a flow of charge.

If the directional drift of ions did not occur, the interfaces between the electrodes and electrolyte of an electrochemical system would run out

---

[†] The word *flux* occurs very frequently in the treatment of transport phenomena. The flux of any species i is the number of moles of that species crossing a unit area of a reference plane in 1 sec; hence, *flux is the rate of transport.*

288 CHAPTER 4

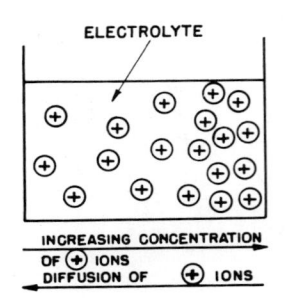

Fig. 4.1. The diffusion of positive ions resulting from a concentration gradient of these ions in an electrolytic solution. The directions of increasing ionic concentration and of ionic diffusion are shown below the diagram.

of ions to fuel the charge-transfer reactions which occur at such interfaces. Hence, the movements and drift of ions is of vital significance to the continued functioning of an electrochemical system.

A flux of ions can come about in three ways. If there is a difference in the concentration of ions in different regions of the electrolyte, the resulting concentration gradient produces a flow of ions. This phenomenon is termed *diffusion* (Fig. 4.1). If there are differences in electrostatic potential at various points in the electrolyte, then the resulting electric field produces a flow of charge in the direction of the field. This is termed *migration* or *conduction* (Fig. 4.2). Finally, if a difference of pressure or density or temperature exists in various parts of the electrolyte, then the liquid begins to move as a whole or parts of it move relative to other parts. This is *hydrodynamic* flow.

It is intended to restrict the present discussion to the transport processes of diffusion and conduction and their interconnection. (The laws of hydrodynamic flow will not be described mainly because they are not particular to the flow of electrolytes; they are characteristic of the flow of all gases and liquids, i.e., of fluids). The initial treatment of diffusion and conduction will be in phenomenological terms; then, the molecular happenings underlying these transport processes will be explored.

Now, in looking at ion–solvent and ion–ion interactions, it has been possible to present the phenomenological or nonstructural treatment in the framework of equilibrium thermodynamics, which excludes time and therefore fluxes, from its analyses.

Such a straightforward application of thermodynamics cannot be made,

ION TRANSPORT IN SOLUTIONS 289

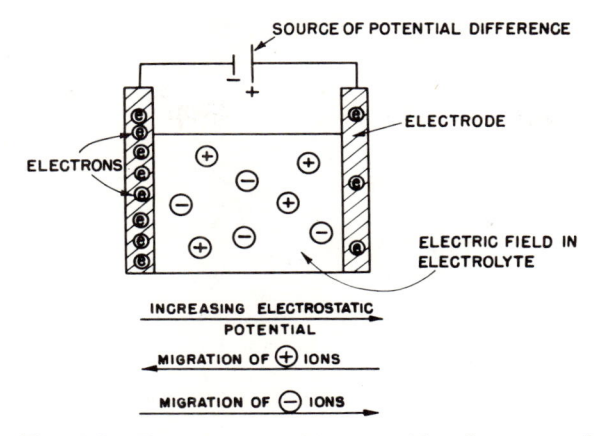

Fig. 4.2. The migration of ions resulting from a gradient of electrostatic potential (i.e., an electric field) in an electrolyte. The electric field is produced by the application of a potential difference between two electrodes immersed in the electrolyte. The directions of increasing electrostatic potentials and of ionic migration are shown below the diagram.

however, in the consideration of transport processes. The drift of ions occurs precisely because the system is not at equilibrium; rather, the system is seeking to attain equilibrium. In other words, the system undergoes change (there cannot be transport without temporal change!) because the free energy is not uniform and a minimum everywhere. It is the existence of such gradients of free energy which set up the process of ionic drift and make the system strive for the attainment of equilibrium by the dissipation of free energy. An appropriate framework for the phenomenological or gross view of ionic drift may be that of nonequilibrium thermodynamics, a subject concerned with the rates at which systems near equilibrium move toward it.

4.2. IONIC DRIFT UNDER A CHEMICAL-POTENTIAL GRADIENT: DIFFUSION

4.2.1. The Driving Force for Diffusion

It has been remarked in the previous section that diffusion occurs when a concentration gradient exists. The theoretical basis of this observation will now be examined.

Consider that, in an electrolytic solution, the concentration of an ionic

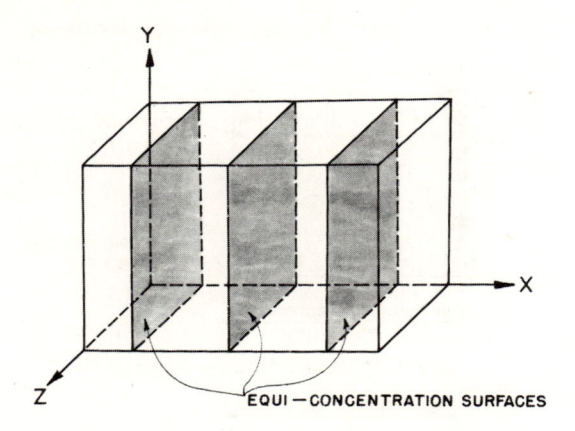

Fig. 4.3. A schematic representation of a slab of electrolytic solution in which the concentration of a species i is constant on the shaded equiconcentration surfaces parallel to the yz-plane.

species i varies in the x direction but is constant in the y and z directions. If desired, one can map equiconcentration surfaces (they will be parallel to the yz plane) (Fig. 4.3).

The situation pictured in Fig. 4.3 can also be considered in terms of the partial molar free energy, or chemical potential, of the particular species i. This is achieved through the use of the defining equation for the chemical potential [Eq. (3.56)]

$$\mu_i = \mu_i^0 + RT \ln c_i$$

(The use of concentration, rather than activity, implies the solution is assumed to behave ideally.) Since c_i is a function of x, the chemical potential also is a function of x. Thus, the chemical potential varies along the x coordinate, and, if desired, equi-μ surfaces can be drawn. Once again, these surfaces will be parallel to the yz plane.

Now, if one transfers a mole of the species i from an initial concentration c_I at x_I to a final concentration c_F at x_F, then the change in free energy, or chemical potential, of the system is (Fig. 4.4):

$$\Delta \mu = \mu_F - \mu_I = RT \ln \frac{c_F}{c_I} \tag{4.1}$$

But the change in free energy is equal to the net work done *on* the system in an isothermal, constant-pressure reversible process (*cf.* Appendix 2.1).

ION TRANSPORT IN SOLUTIONS 291

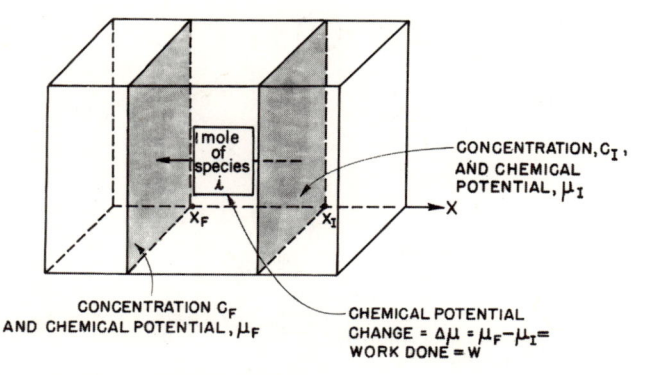

Fig. 4.4. A schematic representation of the work W done in transporting a mole of species i from an equiconcentration surface where its concentration and chemical potential are c_I and μ_I to a surface where its concentration and chemical potential are c_F and μ_F.

Thus, the work done to transport a mole of species i from x_I to x_F is

$$W = \Delta\mu \qquad (4.2)$$

Think of the analogous situation in mechanics. The work done to lift a mass from an initial height x_I to a final height x_F is equal to the difference in gravitational potential (energies), ΔU, at the two positions (Fig. 4.5):

$$W = \Delta U \qquad (4.3)$$

One goes further and says that this work has to be done because a gravitational force F_G acts on the body and that[†]

$$W = -F_G(x_I - x_F) = -F_G \Delta x = \Delta U \qquad (4.4)$$

In other words, the gravitational force can be defined thus

$$F_G = -\frac{\Delta U}{\Delta x} \qquad (4.5)$$

[†] The minus sign arises from the following argument: The displacement $x_I - x_F$ of the mass is *upward* and the force F_G acts *downward*; hence, the product of the displacement and force vectors is negative. If a minus sign is not introduced, the work done W will turn out to be negative. But it is desirable to have W as a positive quantity because of the convention that work done *on* a system is taken to be positive, hence, a minus sign must be inserted (see also Section 2.2.3).

Fig. 4.5. Schematic diagram to illustrate that the mechanical work W done in lifting a mass from an initial height x_I to a final height x_F is $W = \Delta U = -F_G(x_I - x_F)$.

The potential energy, however, may not vary linearly with distance, and, thus, the ratio $\Delta U/\Delta x$ may not be a constant. So it is better to consider infinitesimal changes in energy and distance and write

$$F_G = -\frac{dU}{dx} \qquad (4.6)$$

Thus, the gravitational force is given by the *gradient* of the gravitational potential energy, and the region of space in which it operates is said to be *a gravitational field*.

A similar situation exists in electrostatics. The electrostatic work done in moving a unit charge from x to $x + dx$ defines the difference $d\psi$ in electrostatic potential between the two points (Fig. 4.6)

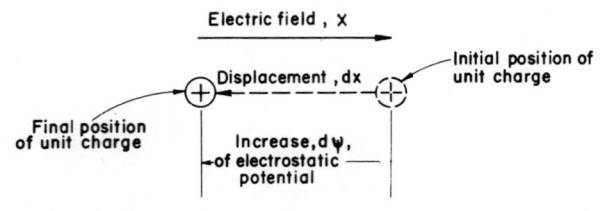

$$W = d\psi \qquad (4.7)$$

Fig. 4.6. Schematic diagram to illustrate the electrostatic work $W = d\psi = -X\,dx$ done in moving a unit positive charge through a distance dx against an electric field X.

Further, the electrostatic work is the product of the electric field, or force per unit charge X and the distance dx[†]

$$-X\,dx = d\psi \tag{4.8}$$

or

$$X = -\frac{d\psi}{dx} \tag{4.9}$$

The electric force per unit charge is therefore given by the negative of the gradient of the electrostatic potentials, and the region of space in which the force operates is known as the *electric field*.

Since the negative of the gradient of gravitational potential energy defines the gravitational force and the negative of the gradient of electrostatic potential defines the electric force, one would expect that the negative of the gradient of the chemical potential would act formally like a force. Further, just as the gravitational force results in the motion of a mass and the electric force, in the motion of a charge, the chemical-potential gradient results in the net motion, or transfer, of the species i from a region of high chemical potential to a region of low chemical potential. This net flow of the species i down the chemical-potential gradient is diffusion, and, therefore, the gradient of chemical potential may be looked upon[‡] as the diffusional force F_D. Thus, one can write

$$F_D = -\frac{d\mu_i}{dx} \tag{4.10}$$

by analogy with the gravitational and electric forces [Eqs. (4.6) and (4.9)] and consider that the diffusional force produces a diffusional flux J, the number of moles of species i crossing per second per unit area of a plane normal to the flow direction (Table 4.1).

4.2.2. The "Deduction" of an Empirical Law: Fick's First Law of Steady-State Diffusion

Qualitatively speaking, the macroscopic description of the transport process of diffusion is simple. The gradient of chemical potential resulting

---

[†] The origin of the minus sign has been explained in Section 2.2.3.

[‡] It will be shown later on (*cf.* Section 4.2.6) that, for the phenomenon of diffusion to occur, all that is necessary is an inequality of the number of diffusing particles in different regions; there is, in fact, no directed force on the individual particles. Thus, $-d\mu/dx$ is only a *pseudo*force like the centrifugal force; it is formally equivalent to a force.

294 CHAPTER 4

TABLE 4.1

Certain Forces in the Phenomenological Treatment of Transport

Force	Acting on	Results in
$-\dfrac{dU}{dx}$	Mass	Movement of mass
$-\dfrac{d\psi}{dx}$	Charge	Movement of charge (current)
$-\dfrac{d\mu}{dx}$	Species i	Movement of species i, i.e., diffusional flux of species i

from a nonuniform concentration is equivalent to a driving force for diffusion and produces a diffusion flux (Fig. 4.7). But what is the quantitative cause-and-effect relation between the driving force $d\mu_i/dx$ and the flux J? This question must now be considered.

Suppose that when diffusion is occurring, the driving force F_D and the flux J reach values which do not change with time. The system can be

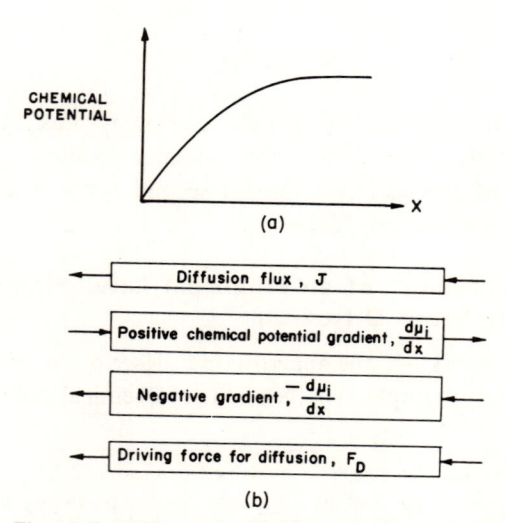

Fig. 4.7. Schematic diagram to show (a) the distance variation of the chemical potential of a species i; and (b) the relative directions of the diffusion flux, driving force for diffusion, etc.

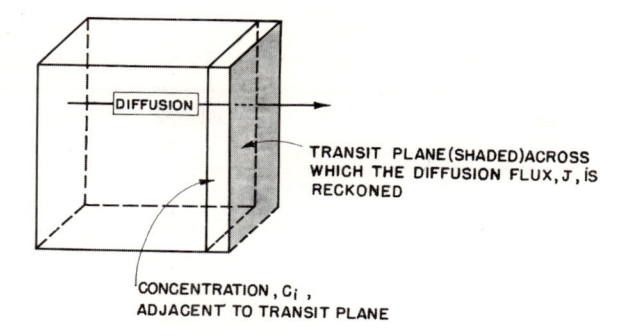

Fig. 4.8. Diagram for the derivation of the linear relation between the diffusion flux J_i and the concentration gradient dc_i/dx.

said to have attained a *steady state*. Then, the as-yet-unknown relation between the diffusion flux J and the diffusional force F_D can be represented quite generally by a power series

$$J = A + BF_D + CF_D^2 + DF_D^3 + \cdots \quad (4.11)$$

where A, B, C, etc., are constants. If, however, F_D is less than unity and sufficiently small,[†] the terms containing the powers (of F_D) greater than unity can be neglected.

Thus, one is left with

$$J = A + BF_D \quad (4.12)$$

But the constant A must be equal to zero; otherwise, it will mean that one would have the impossible situation of having diffusion even though there is no driving force for diffusion.

Hence, the assumption of a sufficiently small driving force leads to the result

$$J = BF_D \quad (4.13)$$

i.e., the flux is linearly related to the driving force. Now, the value of $F_D = 0$ (zero driving force) corresponds to an equilibrium situation; therefore, the assumption of a small value of F_D required to ensure the linear relation

---

[†] Caution should be exercised in applying the criterion. The value of F_D which will give rise to unity will depend on the units chosen to express F_D. Thus, the extent to which F_D is less than unity will depend on the units, but one can always restrict F_D to an appropriately small value.

TABLE 4.2
Diffusion Coefficient D of Ions in Aqueous Solutions

Ion	Diffusion coefficient, cm² sec⁻¹
Li⁺	1.028×10^{-5}
Na⁺	1.334×10^{-5}
K⁺	1.569×10^{-5}
Cl⁻	2.032×10^{-5}
Br⁻	2.080×10^{-5}

(4.13) between flux and force is tantamount to saying that the system is *near equilibrium, but not at equilibrium.*

The driving force on 1 mole of ions has been stated to be $-d\mu_i/dx$ [Eq. (4.10)]. If, therefore, the concentration of the diffusing species adjacent to the transit plane (Fig. 4.8), across which the flux is reckoned, is c_i moles per unit volume, the driving force F_D at this plane is $-c_i(d\mu_i/dx)$. Thus, from relation (4.13), one obtains

$$J_i = -Bc_i \frac{d\mu_i}{dx} \tag{4.14}$$

Writing

$$\mu = \mu^0 + RT \ln c_i$$

which is tantamount to assuming ideal behavior, Eq. (4.14) becomes

$$J_i = -Bc_i \frac{RT}{c_i} \frac{dc_i}{dx} = -BRT \frac{dc_i}{dx} \tag{4.15}$$

Thus, *the steady-state diffusion flux has been theoretically shown to be proportional to the gradient of concentration.* That such a proportionality existed has been known *empirically* since 1855 through the statement of Fick's first law of steady-state diffusion, which reads

$$J_i = -D \frac{dc_i}{dx} \tag{4.16}$$

where D is termed the *diffusion coefficient* (Table 4.2).

4.2.3. On the Diffusion Coefficient D

It is important to stress that, in the empirical Fick's first law, the concentration c is expressed in moles per cubic centimeter, and not in moles per liter. The flux is expressed in moles of diffusing material crossing

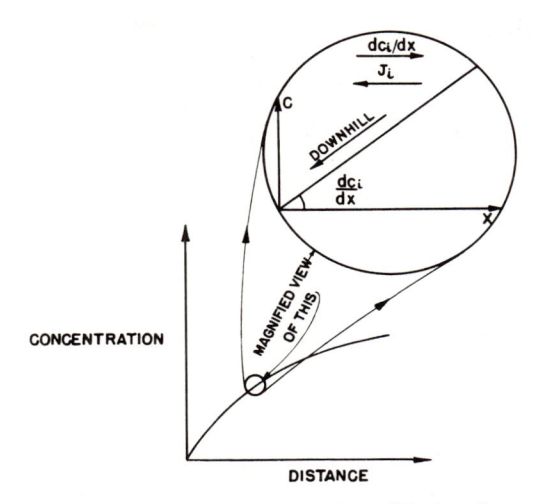

Fig. 4.9. Diagram to show that diffusion flow J_i is in a direction opposite to the direction of positive concentration gradient dc_i/dx. Matter flows downhill, i.e., down the concentration gradient.

unit area of a transit plane per unit of time, i.e., in moles per square centimeter per second, and, therefore, the diffusion coefficient D has the dimensions of centimeters squared per second. The negative sign is usually inserted in the right-hand side of the empirical Fick's law for the following reason: The flux J_i and the concentration gradient dc_i/dx are vectors, or quantities which have both magnitude and direction. But the vector J is in an opposite sense to the vector representing a positive gradient dc_i/dx. Matter flows downhill (Fig. 4.9). Hence, if J_i is taken as positive, dc_i/dx must be negative, and, if there is no negative sign in Fick's first law, the diffusion coefficient will appear as a negative quantity—perhaps an undesirable state of affairs. Hence, to make D come out a positive quantity, a negative sign is added to the right-hand side of the equation which states the empirical law of Fick.

Equating the coefficients of dc_i/dx in the phenomenological equation (4.15) with that in Fick's law [Eq. (4.16)], it is seen that

$$BRT = D \qquad (4.17)$$

Now, is the diffusion coefficient a concentration-independent constant? A naïve answer would run thus: B is a constant, and, therefore, it *appears* that D also is a constant. But the expression (4.17) was obtained only because an ideal solution was considered, and activity coefficients f_i were ignored in Eq. (3.61). Activity coefficients, however, are concentration dependent. So, if the solution does not behave ideally, one has, starting from Eq. (4.14), and using (3.63),

$$\begin{aligned} J_i &= -Bc_i \frac{d\mu_i}{dx} \\ &= -Bc_i \frac{d}{dx}(\mu_i^0 + RT \ln f_i c_i) \\ &= -Bc_i \frac{RT}{f_i c_i} \frac{d}{dx} f_i c_i \\ &= -BRT \frac{dc_i}{dx} - \frac{BRTc_i}{f_i} \frac{df_i}{dx} \\ &= -BRT \frac{dc_i}{dx} - \frac{BRTc_i}{f_i} \frac{df_i}{dc_i} \frac{dc_i}{dx} \\ &= -BRT \frac{dc_i}{dx} \left(1 + \frac{c_i}{f_i} \frac{df_i}{dc_i}\right) \\ &= -BRT \frac{dc_i}{dx} \left(1 + \frac{d \ln f_i}{d \ln c_i}\right) \end{aligned} \qquad (4.18)$$

and, therefore,

$$D = BRT\left(1 + \frac{d \ln f_i}{d \ln c_i}\right) \qquad (4.19)$$

Rigorously speaking, therefore, the diffusion coefficient is not a constant (Table 4.3). If, however, the variation of the activity coefficient is not

TABLE 4.3

Variation of the Diffusion Coefficient D with Concentration

Electrolyte	\multicolumn{4}{c}{Diffusion coefficient D in units of 10^{-5} cm² sec^{-1} at concentration (in molarity)}
	0.05	0.1	0.2	0.5
HCl	3.07	3.05	3.06	3.18
LiCl	1.28	1.27	1.27	1.28
NaCl	1.51	1.48	1.48	1.47

significant *over the concentration difference which produces diffusion*, then $(c_i/f_i)(\partial f_i/\partial c_i) \ll 1$, and, for all practical purposes, D is a constant.[†] *This effective constancy of D with concentration will be assumed in most of the discussions presented here.*

The treatment so far has been phenomenological, and, therefore, the dependence of the diffusion coefficient on factors such as temperature and type of ion can be theoretically understood only by an atomistic analysis. Thus, the quantity D can be understood in a fundamental way only by probing into the ionic movements, the results of which show up in the macroscopic world as the phenomenon of diffusion. What are these ionic movements, and how do they produce diffusion? The answering of these two questions will constitute the next topic.

4.2.4. Ionic Movements: A Case of the Random Walk

Long before the movements of ions in solution were analyzed, the kinetic theory of gases was developed and it involved a consideration of the movements of gas molecules. The *overall* pattern of ionic movements is quite similar to that of gas molecules, and, therefore, the latter will be recalled first.

Imagine a hypothetical situation in which all the gas molecules, except one, are at rest. The moving molecule will, according to Newton's first law of motion, travel with a uniform velocity until it collides with a stationary molecule. During the collision, there is a transfer of momentum (mass m times velocity v). So, the moving molecule loses some speed in the colllision, but the stationary molecule is set in motion. Now both molecules are moving, and they will undergo further collisions. The number of collisions will increase with time, and, soon, all the molecules of the gas will be continually moving, colliding, and changing their directions of motion and their velocities—a scene of hectic activity.

It would be of interest to have an idea of the path of such a gas molecule in the course of time. One might think that the detailed paths of all the particles could be predicted by applying Newton's laws to the motions of molecules. The problem, however, is obviously too complex for a practical solution. To use the laws of motion requires a knowledge of the position and velocity of each particle, and, even in 1 mole, there are 6.023×10^{23} (the Avogadro number) particles.

[†] For example, in diffusion between solutions which have a large concentration difference such as 0.1 to 0.01 mole liter^{-1}, a rough calculation suggests that the activity-coefficient correction is of the order of a few per cent.

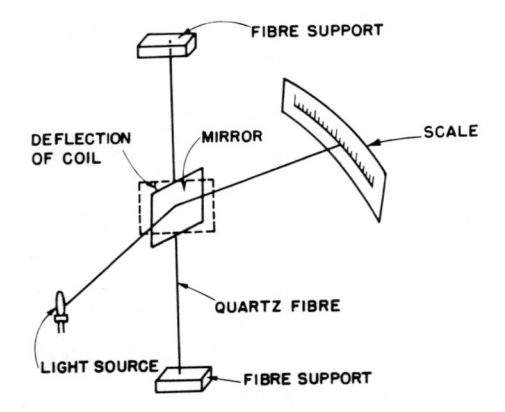

Fig. 4.10. Schematic representation of the essential parts of a mirror galvanometer, used for detecting Brownian motion.

One can however try another approach. Is the ceaseless jostling of molecules manifested in any gross (macroscopic) phenomenon? Consider a frictionless piston in mechanical equilibrium with a mass of gas enclosed in a cylinder. Owing to its weight, the piston exerts a force on the gas. But what force balances the piston's weight? One says that the gas exerts a pressure (force per unit area) on the piston owing to the continual buffeting which the piston receives from the gas molecules. Despite this fact, the bombardment by the gas molecules does not produce any *visible* motion of the piston. Evidently, the mass of the piston is so large compared to that of the gas molecules that the movements of the piston are too small to be detected.

Now, let the mass of the piston be reduced. Then, the jiggling of the extremely light piston as a result of being struck by gas molecules should make itself apparent to an observer. This is what happens if one tries to make a mirror galvanometer more and more sensitive. The essential part of this instrument is a thin quartz fiber which supports a light coil of wire seated in a magnetic field (Fig. 4.10). The deflections of the coil are made visible by fixing a mirror onto the quartz fiber and bouncing a beam of light off the mirror onto a scale. To increase the sensitivity of the instrument, one tries lighter coils, lighter mirrors, and thinner fibers. There comes a stage, however, when the "kicks" which the fiber-mirror-coil assembly receives from the air molecules are sufficient to make the assembly jiggle about. The reflected light beam then jumps about on the scale (Fig. 4.11). The excursions of the spot about a mean position on the scale represent

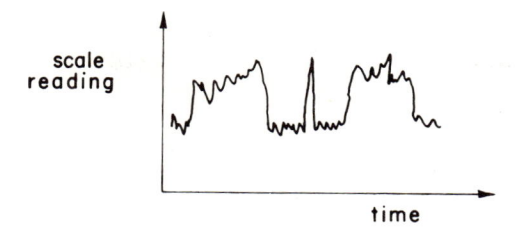

Fig. 4.11. The time variation of the reading on the scale of a mirror galvanometer.

noise. (It is as if each collision produced a sound, in which case, the irregular bombardment of the mirror assembly would result in a nonstop noise.) Signals (coil deflections) which are of this same order of magnitude obviously cannot be separated from the noise.

Instead of mirrors, one could equally consider pistons of a microscopic size, large enough to be seen with the aid of a microscope but small enough to display motions due to collisions with molecules. Such small "pistons" are present in nature. A colloidal particle in a liquid medium behaves as such a piston if it is observed in a microscope. It shows a haphazard, zig-zag motion as shown in Fig. 4.12. The irregular path of the particle must be a slow-motion version of the *random-walk* motion of the molecules in the liquid.

One has therefore a picture of the solvated ions (in an electrolytic solution) in ceaseless motion, perpetually colliding, changing direction, hopping hither and thither from site to site. This is the qualitative picture of ionic *movements*.

4.2.5. The Mean Square Distance Traveled in a Time t by a Random-Walking Particle

The movements executed by an ion in solution are a three-dimensional affair because the ion has a three-dimensional space available for roaming

Fig. 4.12. The haphazard zig-zag motion of a colloidal particle.

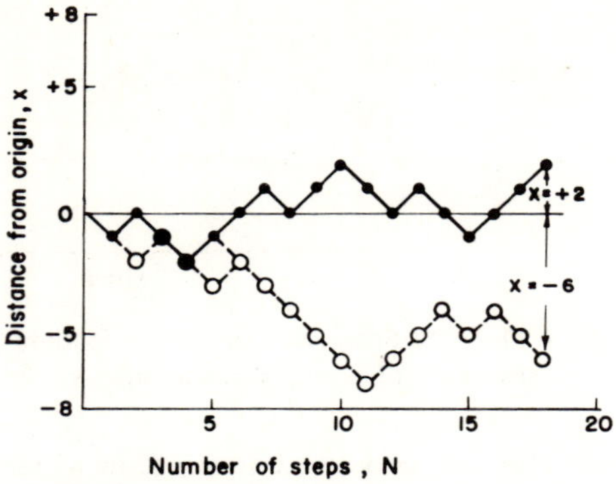

Fig. 4.13. The distance x (from the origin) traversed by the drunken sailor in two tries, each of $N = 18$ steps.

around. So one must really call the movements a "random flight" because one *flies* in three dimensions and *walks* in two dimensions. This fine difference, however, will be ignored and the term *random walk* will be retained.

The aim now is to seek a quantitative description of ionic random-walk movements. There are many exotic ways of stating the random-walk problem. It is said, for example, that a drunken sailor emerges from a bar. He intends to get back to his ship, but he is in no state to control the direction in which he takes a step. In other words, the direction of each step is completely random, all directions being equally likely.

The question is: On the average, how far does the drunken sailor progress in a time t?

For the sake of simplicity, the special case of one-dimensional random walk will be considered. The sailor starts off from $x = 0$ on the x axis. He tosses a coin: Heads!—he moves forward in the positive x direction. Tails!—he moves backward. Since, for an "honest" coin, heads are as likely as tails, the sailor is equally as likely to take a forward step as a backward step. Of course, each step is decided on a fresh toss and is uninfluenced by the results of the previous tosses. After allowing him N steps, the distance x from the origin is noted[†] (Fig. 4.13). Then, the sailor is brought back to the bar ($x = 0$) and started off on another try of N steps.

The process of starting the sailor off from $x = 0$, allowing him N

[†] It *will* be zero only if an equal number $N/2$ of heads and tails turns up. This will not happen every time. So, after each small number of tosses, the sailor is not certain to be back where he started (i.e., in the bar).

steps, and measuring the distance x traversed by the sailor is repeated many times, and it is found that the distances traversed from the origin are $x(1)$, $x(2)$, $x(3)$, ..., $x(i)$, where $x(i)$ is the distance from the origin traversed in the ith try. The *average distance* $\langle x \rangle$, from the origin is

$$\langle x \rangle = \frac{\sum_i x(i)}{\sum i} = \frac{\text{Sum of distances from origin}}{\text{Number of tries}} \qquad (4.20)$$

Since the distance x traversed by the sailor in N steps is as likely to be in the plus-x direction as in the minus-x direction, it is obvious from the canceling out of the positive and negative values of x that, for a sufficiently large number of tries, the *mean* progress from the origin is given by

$$\langle x \rangle = 0 \qquad (4.21)$$

It is obviously not very fruitful to compute the mean distance $\langle x \rangle$ traversed by the sailor in N steps. To avoid such an unenlightening result, which arises because $x(i)$ can take either positive or negative values, it is best to consider the square of $x(i)$, which is always a positive quantity whether $x(i)$ itself is negative or positive. Hence, if $x(1)$, $x(2)$, $x(3)$, ..., $x(i)$ are all squared and the mean of these quantities is taken, then one can obtain the *mean square distance* $\langle x^2 \rangle$, i.e.,

$$\langle x^2 \rangle = \frac{\sum_i x(i)^2}{\sum i} = \frac{\text{Sum of distances } x \text{ squared}}{\text{Number of tries}} \qquad (4.22)$$

Since the square of $x(i)$, i.e., $[x(i)]^2$, is always a positive quantity, the mean square distance traversed by the sailor is always a positive nonzero quantity (Table 4.4). Further, it can easily be shown (Appendix 4.1) that the magnitude of $\langle x^2 \rangle$ is proportional to N, the number of steps, and, since N itself increases linearly with time, it follows that the mean square distance traversed by the random-walking sailor is proportional to time

$$\langle x^2 \rangle \propto t \qquad (4.23)$$

It is to be noted that it is the *mean square distance*—and not the *mean distance*—that is proportional to time. If the mean distance were proportional to time, then it means that the drunken sailor (or the ion) is proceeding at a uniform velocity. This is not the case because the mean distance $\langle x \rangle$ traveled is zero. The only type of progress which the ion is making from the origin is such that the mean *square* distance is proportional to time. This is the characteristic of random walk.

TABLE 4.4
One-Dimensional Random Walk of the Sailor[†]

Trial	N_F[‡]	N_B[*]	x[**]	x^2
1	11	19	− 8	64
2	16	14	+ 2	4
3	17	13	+ 4	16
4	15	15	0	0
5	17	13	+ 4	16
6	16	14	+ 2	4
7	19	11	+ 8	64
8	18	12	+ 6	36
9	15	15	0	0
10	13	17	− 4	14
11	11	19	− 8	64
12	17	13	+ 4	16
13	17	13	+ 4	16
14	12	18	− 6	36
15	20	10	+10	100
16	23	7	+16	256
17	11	19	− 8	64
18	16	14	+ 2	4
19	17	13	+ 4	16
20	14	16	− 2	4
21	16	14	+ 2	4
22	12	18	− 6	36
23	15	15	0	0
24	10	20	−10	100
25	7	23	−16	256
26	15	15	0	0
			$\langle x \rangle = 0$	$\langle x^2 \rangle = 47.67$

[†] Total number of steps in each trial = N = 30.
[‡] Number of forward steps = N_F.
[*] Number of backwards steps = N_B.
[**] Distance from the origin = x.

4.2.6. Random-Walking Ions and Diffusion: The Einstein–Smoluchowski Equation

Consider a situation in an electrolytic solution where the concentration of the ionic species of interest is constant in the yz plane but varies in the x direction. To analyze the diffusion of ions, imagine *unit area* of a reference plane normal to the x direction. This reference plane will be termed the

ION TRANSPORT IN SOLUTIONS 305

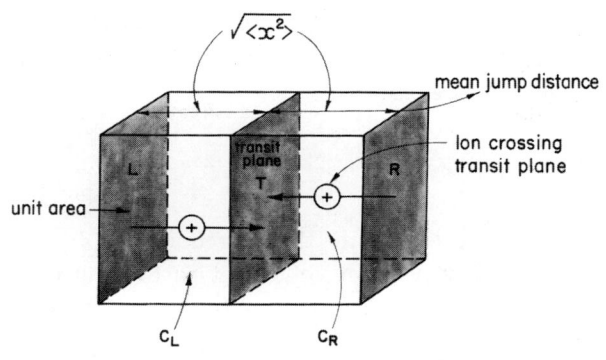

Fig. 4.14. Schematic diagram for the derivation of the Einstein–Smoluchowski relation, showing the transit plane T in between and at a distance $\sqrt{\langle x^2 \rangle}$ from the left L and right R planes. The concentrations in the left and right compartments are c_L and c_R, respectively.

transit plane T (Fig. 4.14). There is a random walk of ions across this plane both from left to right and from right to left. On either side of the transit plane, one can imagine two planes L and R which are parallel to the transit plane and situated at a distance $\sqrt{\langle x^2 \rangle}$ from it. In other words, the region under consideration has been divided into left and right compartments in which the concentrations of ions are different and designated by c_L and c_R, respectively.

In a time of t sec, a random-walking ion covers a mean square distance of $\langle x^2 \rangle$, or a mean distance of $\sqrt{\langle x^2 \rangle}$. Thus, by choosing the plane L to be at a distance $\sqrt{\langle x^2 \rangle}$ from the transit plane, one has ensured that all the ions in the left compartment will cross the transit plane in a time t *provided* they are moving in the left→right direction.

Now, the number of moles of ions in the left compartment is equal to the volume $\sqrt{\langle x^2 \rangle}$ of this compartment times the concentration c_L of ions. It follows that the number of moles of ions which make the $L \rightarrow T$ crossing in t sec is $\sqrt{\langle x^2 \rangle} c_L$ times the fraction of ions making left to right movements. Since the ions are random walking, right to left movements are as likely as left to right movements, i.e., only half the ions in the left compartment are moving away toward the right compartment. Thus, in t the number of moles of ions making the $L \rightarrow T$ crossing is $\frac{1}{2}\sqrt{\langle x^2 \rangle} c_L$, and, therefore, the number of moles of ions making the $L \rightarrow T$ crossing in 1 sec is $\frac{1}{2}(\sqrt{\langle x^2 \rangle}/t)c_L$. Similarly, the number of moles of ions making the $R \rightarrow T$ crossing in 1 sec is $\frac{1}{2}(\sqrt{\langle x^2 \rangle}/t)c_R$.

Hence, the diffusion flux of ions across the transit plane, i.e., the net number of moles of ions crossing unit area of the transit plane per second from left to right is given by

$$J = \frac{1}{2} \frac{\sqrt{\langle x^2 \rangle}}{t} (c_L - c_R) \tag{4.24}$$

This equation reveals that all that is required to have diffusion is *a difference in the numbers* per unit volume of particles in two regions. The important point is that no special diffusive force acts *on the particles* in the direction of the flux.

If no forces are pushing particles in the direction of the flow, then what about the driving force for diffusion, i.e., the gradient of chemical potential (*cf.* Section 4.2.1)? The latter is only *formally equivalent* to a force in a macroscopic treatment; it is a sort of pseudoforce like a centrifugal force. The chemical-potential gradient is not a true force which acts on the individual diffusing particles and, from this point of view, is quite unlike, for example, the coulombic force which acts on individual charges.

Now, the concentration gradient dc/dx in the left-to-right direction can be written

$$\frac{dc}{dx} = \frac{c_R - c_L}{\sqrt{\langle x^2 \rangle}} = -\frac{c_L - c_R}{\sqrt{\langle x^2 \rangle}}$$

or

$$c_L - c_R = -\sqrt{\langle x^2 \rangle}\, \frac{dc}{dx} \tag{4.25}$$

This result for $c_L - c_R$ can be substituted in Eq. (4.24) to give

$$J = -\frac{1}{2} \frac{\langle x^2 \rangle}{t} \frac{dc}{dx} \tag{4.26}$$

and, by equating the coefficients of this equation with that of Fick's first law [Eq. (4.16)], one has

$$\frac{\langle x^2 \rangle}{2t} = D$$

or

$$\langle x^2 \rangle = 2Dt \tag{4.27}$$

This is the Einstein–Smoluchowski equation; it provides a bridge between the microscopic view of random-walking ions and the coefficient D of the macroscopic Fick's law.

The coefficient 2 is intimately connected with the approximate nature of the derivation, i.e., one-dimensional random walk with ion's being permitted to jump forward and backward. More rigorous argument may yield other values of the numerical coefficient, e.g., 6.

The characteristic of *random* walk in the Einstein–Smoluchowski equation is the appearance of the mean *square* distance (i.e., square centimeters), and, since this mean square distance is proportional to time (seconds), the proportionality constant D in Eq. (4.27) must have the dimensions of centimeters squared per second. It must not be taken to mean that every ion which starts off on a random walk travels in a time t a mean square distance $\langle x^2 \rangle$ given by the Einstein–Smoluchowski relation (4.27). If a certain number of ions are, in a thought experiment, suddenly introduced on the yz plane at $x = 0$, then, in t sec, some ions would progress a distance x_1; others, x_2; still others, x_3; etc. The Einstein–Smoluchowski relation only says that [*cf.* Eqs. (4.22) and (4.27)]

$$\frac{x_1^2 + x_2^2 + \cdots + x_n^2}{n} = \langle x^2 \rangle = 2Dt \tag{4.28}$$

But how many ions travel a distance x_1; how many, x_2; etc.? In other words, how are the ions *spatially distributed* after a time t, and how does the spatial distribution vary with time? This spatial distribution of ions shall be analyzed, but only after presenting a phenomenological treatment of *non*-steady-state diffusion.

4.2.7. The Gross View of Non-Steady-State Diffusion

What has been done so far is to consider steady-state diffusion in which neither the flux nor the concentration of diffusing particles in various regions changes with time. In other words, the whole transport process is time independent. But, what happens if a concentration gradient is suddenly produced in an electrolyte initially in a time-invariant equilibrium condition? Diffusion starts of course, but it will not immediately reach a steady state which does not change with time. For example, the distance variation of concentration, which is zero at equilibrium, will not instantaneously hit the final steady-state pattern. How does the concentration vary with time?

Consider a parallelepiped (Fig. 4.15) of unit area and length dx. Ions are diffusing *in* through the left face of the parallelepiped and *out* through the right face. Let the concentration of the diffusing ions be a continuous function of x. If c is the concentration of ions at the left face, the concen-

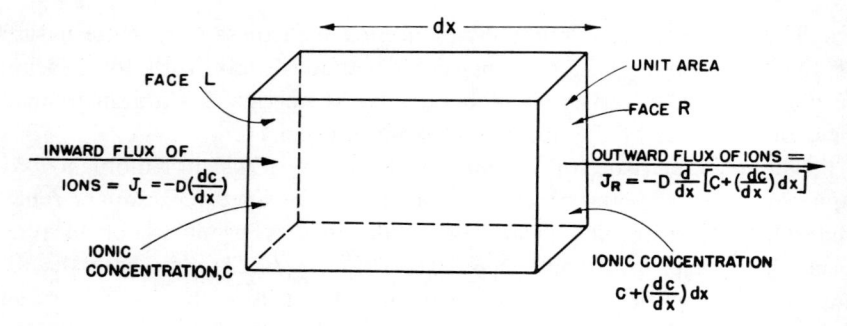

Fig. 4.15. The parallelepiped of electrolyte used in the derivation of Fick's second law.

tration at the right face is

$$c + \frac{dc}{dx} dx$$

Fick's law [Eq. (4.16)] is used to express the flux into and out of the parallelepiped. Thus the flux into the left face, J_L, is

$$J_L = -D \frac{dc}{dx} \tag{4.29}$$

and the flux out of the right face is

$$J_R = -D \frac{d}{dx} \left(c + \frac{dc}{dx} dx \right)$$

$$= -D \frac{dc}{dx} - D \frac{d^2c}{dx^2} dx \tag{4.30}$$

The net *out*flow of material from the parallelepiped of volume dx is

$$J_L - J_R = D \frac{d^2c}{dx^2} dx \tag{4.31}$$

Hence, the net outflow of ions *per unit volume* per unit time is $D (d^2c/dx^2)$. But this net outflow of ions per unit volume per unit time from the parallepiped is in fact the sought-for variation of concentration with time, i.e., dc/dt. One obtains partial differentials because the concentration depends both on time and distance, but the subscripts x and t are generally omitted because it is, for example, obvious that the time variation is in a fixed region of space, i.e., constant x. Hence,

$$\frac{\partial c}{\partial t} = D \frac{\partial^2 c}{\partial x^2} \tag{4.32}$$

This partial differential equation is known as *Fick's second law*. It is the basis for the treatment of most time-dependent diffusion problems in electrochemistry.

That Fick's second law is in the form of a differential equation implies that it describes what is common to *all* diffusion problems and it has "squeezed out" what is characteristic to any *particular* diffusion problem.[†] Thus, one always has to calculate the precise functional relationship

$$c = f(x, t)$$

for a particular situation.

The process of calculating the functional relationship consists in solving the partial differential equation, which is Fick's second law, i.e., Eq. (4.32).

4.2.8. An Often Used Device for Solving Electrochemical Diffusion Problems: The Laplace Transformation

Partial differential equations, such as Fick's second law (in which the concentration is a function of both time and space) are generally more difficult to solve than *total* differential equations, where the dependent variable is a function of only one independent variable. An example of a *total* differential equation (of second order[‡]) is the linearized Poisson–Boltzmann equation

$$\frac{1}{r^2} \frac{d}{dr} \left(r^2 \frac{d\psi}{dr} \right) = \varkappa^2 \psi$$

It has been shown (Section 3.3.7) that the solution of this equation (with ψ dependent on r only) was easily accomplished.

One may conclude, therefore, that the solution of Fick's second law (a partial differential equation) would proceed smoothly if some mathematical device could be utilized to convert it into the form of a total differential equation. The Laplace transformation method is often used as such a device.

Since the method is based on the *operation*[§] of Laplace transformation, a brief digression[*] on the nature of this operation is given before using it to solve the partial differential equation involved in non-steady-state electrochemical diffusion problems, namely, Fick's second law.

[†] This point is dealt with at greater length in Section 4.2.9.
[‡] The *order* of a differential equation is the order of its highest derivatives, which in the example quoted is a second-order derivative, $d^2\psi/dr^2$.
[§] A mathematical operation is a rule for converting one function into another.
[*] Many excellent treatments of the Laplace transformation technique are available, e.g., J. C. Jaeger, *Introduction to Applied Mathematics,* John Wiley & Sons, Inc., New York, 1952.

Consider a function y of the variable z, i.e., $y = f(z)$, represented by the plot of y against z. The familiar *operation* of differentiation performed on the function y consists in finding the *slope* of the curve representing $y = f(z)$ for various values of z, i.e., the differentiation operation consists in evaluating dy/dz. The operation of integration consists in finding the *area* under the curve, i.e., it consists in evaluating $\int_{z_1}^{z_2} f(z)\, dz$.

The operation of Laplace transformation performed on the function $y = f(z)$ consists of two steps:

1. Multiplying $y = f(z)$ by e^{-pz}, where p is a positive quantity which is independent of z
2. Integrating the resulting product ye^{-pz} $[= f(z)e^{-pz}]$ with respect to z between the limits $z = 0$ and $z = \infty$

In short, the Laplace transform $y = f(z)$ is

$$\int_0^\infty e^{-pz} y\, dz \quad \text{or} \quad \int_0^\infty e^{-pz} f(z)\, dz$$

Just as one often symbolizes the result of the differentiation of y by y', the result of the operation of Laplace transformation performed on y is often represented by a symbol $\bar{y}$. Thus,

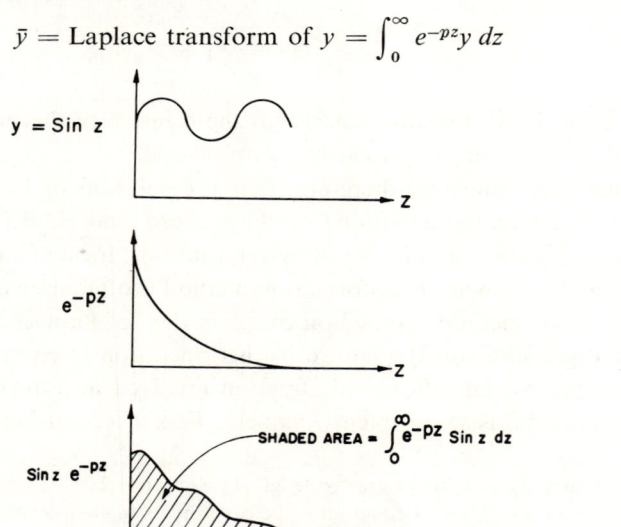

$$\bar{y} = \text{Laplace transform of } y = \int_0^\infty e^{-pz} y\, dz \qquad (4.33)$$

Fig. 4.16. Steps in the operation of the Laplace transformation of (a) the function of $y = \sin z$, showing (b) e^{-pz} and (c) the product $e^{-pz} \sin z$ integrated between the limits 0 and ∞.

ION TRANSPORT IN SOLUTIONS 311

TABLE 4.5
Laplace Transforms

Function	Transform
$f(t)$	$\bar{f} = \int_0^\infty e^{-pt} f(t)\, dt$
1	$\dfrac{1}{p}$
λ (a constant)	$\dfrac{\lambda}{p}$
$\dfrac{1}{\sqrt{\pi t}}$	$\dfrac{1}{\sqrt{p}}$
$2\sqrt{\dfrac{t}{\pi}}$	$p^{-\tfrac{3}{2}}$
$e^{\omega t}$	$\dfrac{1}{p-\omega}$
$\dfrac{1}{\omega}\sin \omega t$	$\dfrac{1}{p^2+\omega^2}$
$\dfrac{1}{\sqrt{\pi t}} \exp\!\left(-\dfrac{k^2}{4t}\right)$	$\dfrac{1}{\sqrt{p}} e^{-k\sqrt{p}}\ [k \geq 0]$
$2\sqrt{\dfrac{t}{\pi}} \exp\!\left(-\dfrac{k^2}{4t}\right) - k\,\text{erfc}\!\left(\dfrac{k}{2\sqrt{t}}\right)$	$p^{-3/2} e^{-k\sqrt{p}}\ [k \geq 0]$
$\cos \omega t$	$\dfrac{p}{p^2+\omega^2}$

What happens during Laplace transformation can be easily visualized by choosing a function, say, $y = \sin z$, and representing the operation in a figure (Fig. 4.16). It can be seen[†] that the operation consists in finding the area under the curve, ye^{-pz}, between the limits $z = 0$ and $z = \infty$.

The Laplace transforms of some functions encountered in diffusion problems are collected in Table 4.5.

[†] From Fig. 4.16, it can also be seen that, apart from having to make the integral converge, the exact value of p is not of significance for, in any case, p disappears after the operation of inverse transformation (cf. Section 4.2.11).

4.2.9. Laplace Transformation Converts the Partial Differential Equation Which Is Fick's Second Law into a Total Differential Equation

It will now be shown that, by using the operation of Laplace transformation, Fick's second law—a partial differential equation—is converted into a total differential equation which can be readily solved.

Since whatever operation is carried out on the left-hand side of an equation must be repeated on the right-hand side, both sides of Fick's second law will be subject to the operation of Laplace transformation (*cf.* Eq. (4.33)]

$$\int_0^\infty e^{-pt} \frac{\partial c}{\partial t} dt = \int_0^\infty e^{-pt} D \frac{\partial^2 c}{\partial x^2} dt \qquad (4.34)$$

which, by using the symbol for a Laplace-transformed function, can be written

$$\overline{\frac{\partial c}{\partial t}} = D \overline{\frac{\partial^2 c}{\partial x^2}} \qquad (4.35)$$

To proceed further, one must evaluate the integrals of Eq. (4.34). Consider the Laplace transform

$$\overline{\frac{\partial c}{\partial t}} = \int_0^\infty e^{-pt} \frac{\partial c}{\partial t} dt \qquad (4.36)$$

The integral can be evaluated by the rule for integration by parts as follows

$$\underbrace{\int_0^\infty e^{-pt}}_{u} \underbrace{\frac{\partial c}{\partial t} dt}_{dv} = \underbrace{e^{-pt}}_{u} \underbrace{\int_0^\infty \partial c}_{v} - \int_0^\infty \underbrace{\left[\int_0^\infty \partial c\right]}_{v} \underbrace{de^{-pt}}_{du}$$

$$= \left[e^{-pt} c \right]_0^\infty + p \int_0^\infty e^{-pt} c \, dt \qquad (4.37)$$

Since $\int_0^\infty e^{-pt} c \, dt$ is in fact the Laplace transform of c [*cf.* the defining equation (4.33)] and, for conciseness, is represented by the symbol $\bar{c}$ and since e^{-pt} is zero when $t \to \infty$ and unity when $t = 0$, Eq. (4.36) reduces to

$$\overline{\frac{\partial c}{\partial t}} = \int_0^\infty e^{-pt} \frac{\partial c}{\partial t} dt = -c[t=0] + p\bar{c} \qquad (4.38)$$

where $c[t=0]$ is the value of the concentration c at $t = 0$.

ION TRANSPORT IN SOLUTIONS 313

Next, one must evaluate the integral on the right-hand side of Eq. (4.34), i.e.,

$$D\overline{\frac{\partial^2 c}{\partial x^2}} = \int_0^\infty e^{-pt} D \frac{\partial^2 c}{\partial x^2}\, dt \tag{4.39}$$

Since the integration is with respect to the variable t and the differentiation is with respect to x, their order can be interchanged. Further, one can move the constant D outside the integral sign. Hence, one can write

$$D\overline{\frac{\partial^2 c}{\partial x^2}} = D\frac{\partial^2}{\partial x^2} \int_0^\infty e^{-pt} c\, dt \tag{4.40}$$

Once again, it is clear from Eq. (4.33) that $\int_0^\infty e^{-pt} c\, dt$ is the Laplace transform of c, i.e., $\bar{c}$ and, therefore,

$$D\overline{\frac{\partial^2 c}{\partial x^2}} = D\frac{\partial^2}{\partial x^2} \bar{c} \tag{4.41}$$

From Eqs. (4.32), (4.38), and (4.41), it follows that, after Laplace transformation, Fick's second law takes the form

$$p\bar{c} - c[t=0] = D\frac{d^2 \bar{c}}{dx^2} \tag{4.42}$$

This, however, is a *total* differential equation because it contains only the variable x. Thus, by using the operation of Laplace transformation, Fick's second law has been converted into a more easily solvable total differential equation involving $\bar{c}$, the Laplace transform of the concentration.

4.2.10. The Initial and Boundary Conditions for the Diffusion Process Stimulated by a Constant Current (or Flux)

A differential equation can be arrived at by differentiating an original equation, or *primitive*, as it is called. In the case of Fick's second law, the primitive is the equation which gives the precise nature of the functional dependence of concentration on space and time; i.e., the primitive is an elaboration on

$$c = f(x, t)$$

Since, in the process of differentiation, constants are eliminated and since three differentiations (two with respect to x and one with respect to time) are necessary to arrive at Fick's second law, three constants have been eliminated in the process of going from the precise concentration dependence which characterizes a particular problem to the general relation between the time—and space—derivatives of concentration which describes any nonstationary diffusion situation.

The three characteristics, or *conditions*, as they are called, of a particular diffusion process cannot be rediscovered by mathematical argument applied to the differential equation. To get at the three conditions, one has to resort to a physical understanding of the diffusion process. Only then can one proceed with the solution of the (now) total differential equation (4.42) and get the precise functional relationship between concentration, distance, and time.

Instead of attempting a general discussion of the three conditions characterizing a particular diffusion problem, it is best to treat a typical electrochemical diffusion problem.

Consider that, in an electrochemical system, a *constant* current is switched on at a time which is arbitrarily designated $t = 0$ (Fig. 4.17). The current is due to charge-transfer reactions at the electrode–solution interfaces, and these reactions consume a species. Since the concentration of this species at the interface falls below the bulk concentration, a concentration gradient for the species is set up and it diffuses toward the

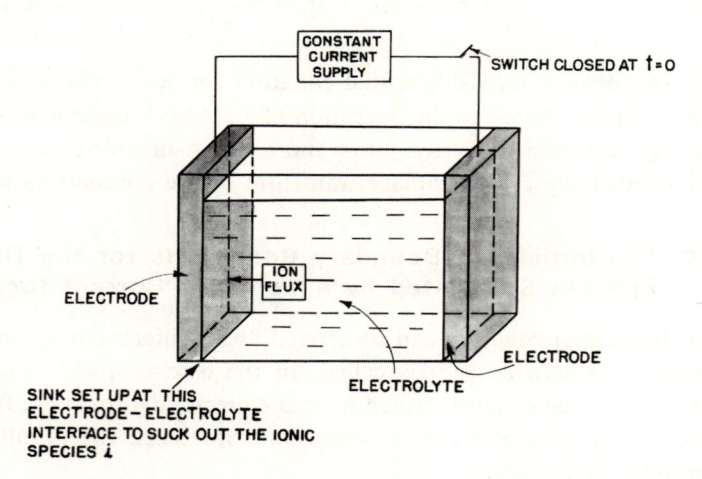

Fig. 4.17. Schematic representation of an electrochemical system connected to a constant current supply which is switched on at $t = 0$. The current promotes a charge-transfer reaction at the electrode-electrolyte interfaces, which results in the diffusion flux of the species i toward the interface.

interface. Thus, the externally controlled current sets up[†] within the solution a diffusion flux.

The diffusion is described by Fick's second law

$$\frac{\partial c}{\partial t} = D \frac{\partial^2 c}{\partial x^2} \qquad (4.32)$$

or, after Laplace transformation, by

$$p\bar{c} - c[t=0] = D \frac{d^2\bar{c}}{dx^2} \qquad (4.42)$$

To analyze the diffusion problem, one must solve the differential equation, i.e., describe how the concentration of the diffusing species varies with distance x from the electrode and with the time which has elapsed since the constant current is switched on. But, first, one must think out the three characteristics, or conditions, of the diffusion process described above.

The nature of one of the conditions becomes clear from the term $c[t=0]$ in the Laplace-transformed version [cf. Eq. (4.42)] of Fick's second law. The term $c[t=0]$ refers to the concentration before the commencement of diffusion, i.e., it describes the *initial condition* of the electrolytic solution in which diffusion is made to occur by the passage of a constant current. Since, before the constant current is switched on and diffusion starts, one has an unperturbed system, the concentration c of the species which subsequently diffuses must be the same throughout the system and equal to the bulk concentration c^0. Thus, the initial condition of the electrolytic solution is

$$c[t=0] = c^0 \qquad (4.43)$$

The other two conditions pertain to the situation after the diffusion begins, e.g., after the diffusion-causing current is switched on. Since these two conditions often pertain to what is happening to the boundaries of the system (in which diffusion is occurring), they are usually known as *boundary conditions*.

[†] When the externally imposed current sets up charge-transfer reactions which provoke the diffusion of ions, there is a very simple relation between the current density and the diffusion flux. The diffusion flux is a mole flux (number of moles crossing 1 cm² in 1 sec), and the current density is a charge flux (cf. Table 4.1.) Hence, the current density i, or charge flux, is equal to the charge zF per mole of ions (z is the valence of the diffusing ion; and F, the Faraday) times the diffusion flux J, i.e., $i = zFJ$.

316   CHAPTER 4

The first boundary condition is the expression of an obvious point, namely, that, *very far* from the boundary at which the diffusion source or sink is set up, the concentration of the diffusing species is unperturbed and remains the same as in the initial condition

$$c\,[x \to \infty] = c\,[t = 0] = c^0 \tag{4.44}$$

Thus, the concentration of the diffusing species has the same value c^0 at any x at $t = 0$ or for any $t > 0$ at $x \to \infty$. This is true for almost all electrochemical diffusion problems in which one switches on (at $t = 0$) the appropriate current or potential difference across the interface and thus sets up interfacial charge-transfer reactions which, by consuming or producing a species, provoke a diffusion flux of that species.

What, however, is characteristic of one particular electrochemical diffusion process and distinguishes it from all others is the nature of the diffusion flux which is started off at $t = 0$. Thus, the essential characteristic of the diffusion problem under discussion is the switching-on of the *constant* current, which means that the diffusing species is consumed at a constant rate at the interface and the species diffuses across the interface at a constant rate. In other words, the flux of the diffusing species at the $x = 0$ boundary of the solution is a constant.

It is convenient from many points of view to assume that the constant value of the flux is unity, i.e., 1 mole of the diffusing species crossing 1 cm² of the electrode–solution interface per second. This unit flux corresponds to a constant current density of 1 amp cm^{-2}. This normalization of the flux scarcely affects the generality of the treatment because it will later be seen

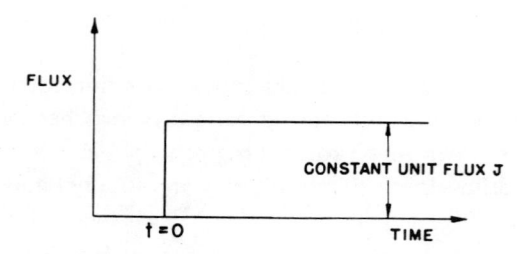

Fig. 4.18. When a constant unit flux of 1 mole cm^{-2} sec^{-1} is switched on at $t = 0$, the variation of flux with time resembles a step. (It is only an ideal switch which makes the current and, therefore, the flux instantaneously rise from zero to its constant value; this problem of technique is ignored in the diagram).

that the concentration response to an arbitrary flux can easily be obtained from the concentration response to a unit flux.

If one looks at the time variation of current or the flux across the solution boundary, it is seen that, for $t < 0$, $J = 0$ and, for $t > 0$, there is a constant flux $J = 1$ (Fig. 4.18) corresponding to the constant current switched on at $t = 0$. In other words, the time variation of the flux is like a step; that is why the flux produced in this setup is often known as a *step function* (of time).

At any instant of time, the constant flux across the boundary is related to the concentration gradient there through Fick's first law, i.e.,

$$J_{x=0} = 1 = -D\left(\frac{\partial c}{\partial x}\right)_{x=0} \tag{4.45}$$

The above initial and boundary conditions can be summarized thus

$$c[t=0] = c^0 \tag{4.43}$$

$$c[x \to \infty] = c^0 \tag{4.44}$$

$$\left(\frac{\partial c}{\partial x}\right)_{x=0} = -\frac{1}{D} \tag{4.45}$$

The three conditions just listed describe the special features of the constant (unit) flux diffusion problem. They will now be used to solve Fick's second law.

4.2.11. The Concentration Response to a Constant Flux Switched on at $t = 0$

It has been shown (Section 4.2.9) that, after Laplace transformation, Fick's second law takes the form

$$p\bar{c} - c[t=0] = D\frac{d^2\bar{c}}{dx^2} \tag{4.42}$$

The solution of an equation of this type is facilitated if the second term is zero. This objective can be attained by introducing a new variable c_1 defined thus

$$c_1 = c^0 - c \tag{4.46}$$

The variable c_1 can be recognized as the departure $c^0 - c$ of the concentration from its initial value c^0. In other words, c_1 represents the *perturbation* from the initial concentration (Fig. 4.19).

The partial differential equation [Eq. (4.32)] and the initial and bound-

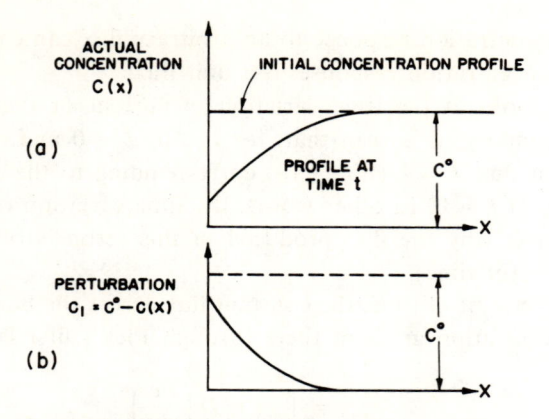

Fig. 4.19. Schematic representation of (a) the variation of concentration with distance x from the electrode at $t = 0$ and $t = t$ and (b) the variation of the perturbation $c_1 = c^0 - c_x$ in concentrations.

ary conditions have now to be restated in terms of the new variable c_1. This is easily done by using Eq. (4.46) in Eqs. (4.32), (4.43), (4.44), and (4.45). One obtains

$$\frac{\partial c_1}{\partial t} = D \frac{\partial^2 c_1}{\partial x^2} \tag{4.47}$$

$$c_1[t = 0] = 0 \tag{4.48}$$

$$c_1[x \to \infty] = 0 \tag{4.49}$$

$$\left(\frac{\partial c_1}{\partial x}\right)_{x=0} = -\frac{1}{D} \tag{4.50}$$

After Laplace transformation of Eq. (4.47), the differential equation becomes

$$p\bar{c}_1 - c_1[t = 0] = D \frac{d^2 \bar{c}_1}{dx^2} \tag{4.51}$$

Since, however, $c_1[t = 0] = 0$ [*cf.* Eq. (4.48)], it is clear that

$$\frac{d^2 \bar{c}_1}{dx^2} = \frac{p}{D} \bar{c}_1 \tag{4.52}$$

This equation is identical in form to the linearized P–B equation [*cf.* Eq.

(3.21)] and, therefore, must have the same general solution, i.e.,

$$\bar{c}_1 = Ae^{-(p/D)^{\frac{1}{2}}x} + Be^{(p/D)^{\frac{1}{2}}x} \tag{4.53}$$

where A and B are the arbitrary integration constants to be evaluated by the use of the boundary conditions. If the Laplace transformation method had not been used, the solution of (4.47) would not have been so simple.

The constant B must be zero by virtue of the following argument. From the boundary condition $c_1 [x \to \infty] = 0$, i.e., Eq. (4.49), it is clear that, after Laplace transformation,

$$\bar{c}_1 [x \to \infty] = 0 \tag{4.54}$$

Hence, as $x \to \infty$, $\bar{c}_1 \to 0$, but this will be true only if $B = 0$ as, otherwise, $\bar{c}_1$ in Eq. (4.53) will go to infinity instead of zero.

One is therefore left with

$$\bar{c}_1 = Ae^{-(p/D)^{\frac{1}{2}}x} \tag{4.55}$$

Differentiating this equation with respect to x, one obtains

$$\frac{d\bar{c}_1}{dx} = -\sqrt{\frac{p}{D}} Ae^{-(p/D)^{\frac{1}{2}}x} \tag{4.56}$$

which, at $x = 0$, leads to

$$\left(\frac{d\bar{c}_1}{dx}\right)_{x=0} = -\sqrt{\frac{p}{D}} A \tag{4.57}$$

Another expression for $(d\bar{c}_1/dx)_{x=0}$ can be obtained by applying the operation of Laplace transformation to the constant flux boundary condition (4.50). Laplace transformation on the left-hand side of the boundary condition leads to $(d\bar{c}_1/dx)_{x=0}$; and the same operation performed on the right-hand side, to $-1/Dp$ (cf. Appendix 4.2). Thus, from the boundary condition (4.50), one gets

$$\left(\frac{d\bar{c}_1}{dx}\right)_{x=0} = -\frac{1}{Dp} \tag{4.58}$$

Hence, from Eqs. (4.57) and (4.58), it is found that

$$A = \frac{1}{p^{\frac{3}{2}}D^{\frac{1}{2}}} \tag{4.59}$$

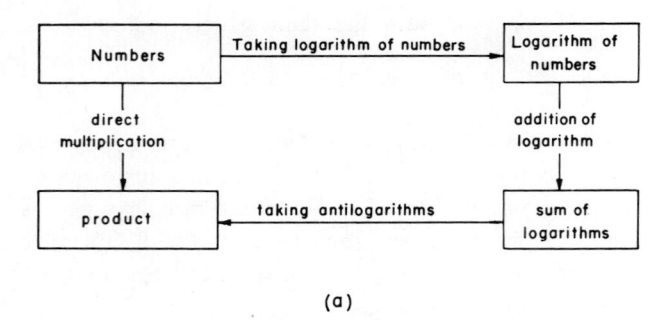

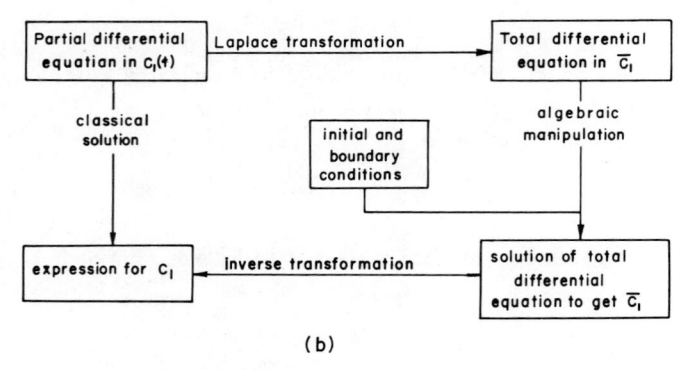

Fig. 4.20. Comparison of the use of (a) logarithms and (b) Laplace transformation.

Upon inserting this expression for A into Eq. (4.55), it follows that

$$\bar{c}_1 = \frac{1}{D^{\frac{1}{2}}p^{\frac{3}{2}}} e^{-(p/D)^{\frac{1}{2}}x} \tag{4.60}$$

The ultimate aim, however, is not to get an expression for the $\bar{c}_1$, the Laplace transform of c_1, but to get an expression for c_1 (or c) as a function of distance x and time t. The expression $\bar{c}_1$ has been obtained by a Laplace transformation of c_1, hence, to go from $\bar{c}_1$ to c_1, one must do an inverse transformation. The situation is analogous to using logarithms to facilitate the working-out of a problem (Fig. 4.20). In order to get c_1 from $\bar{c}_1$, one asks the question: What function c_1 would, under Laplace transformation, give the Laplace transform $\bar{c}_1$ of Eq. (4.60)? In other words, one has to find c_1 in the equation

$$\int_0^\infty e^{-pt} c_1 \, dt = \bar{c}_1 = \frac{1}{D^{\frac{1}{2}}p^{\frac{3}{2}}} e^{-(p/D)^{\frac{1}{2}}x} \tag{4.61}$$

ION TRANSPORT IN SOLUTIONS   321

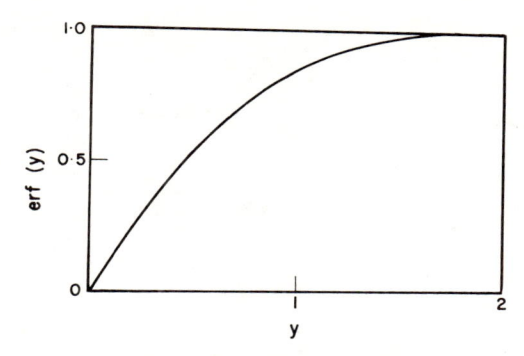

Fig. 4.21. Variation of the error function erf (y) with argument y.

A mathematician would find the function c_1, i.e., do the inverse transformation, by making use of the theory of functions of variables which are complex. Since, however, there are extensive tables of functions y and their transforms $\bar{y}$, it is only necessary to look up the tables in the column of Laplace transforms (Table 4.5). It is seen that, corresponding to the transform,

$$p^{-\frac{3}{2}} \exp\left(-\frac{x}{\sqrt{D}} p^{\frac{1}{2}}\right)$$

is the function

$$2\pi^{-\frac{1}{2}} t^{\frac{1}{2}} \exp\left(-\frac{x^2}{4Dt}\right) - xD^{-\frac{1}{2}} \operatorname{erfc}\left(\frac{x^2}{4Dt}\right)^{\frac{1}{2}}$$

where erfc is the *error function complement* defined thus

$$\operatorname{erfc}(y) = 1 - \operatorname{erf}(y) \tag{4.62}$$

erf (y) being the *error function* given by (*cf.* Fig. 4.21)

$$\operatorname{erf}(y) = \frac{2}{\sqrt{\pi}} \int_0^y e^{-u^2} du \tag{4.63}$$

Hence, the expression for the concentration perturbation c_1 in Eq. (4.61) must be

$$c_1 = \frac{1}{D^{\frac{1}{2}}} \left[\frac{2t^{\frac{1}{2}}}{\pi^{\frac{1}{2}}} \exp\left(-\frac{x^2}{4Dt}\right) - xD^{-\frac{1}{2}} \operatorname{erfc}\left(\frac{x^2}{4Dt}\right)^{\frac{1}{2}} \right] \tag{4.64}$$

If one is interested in the true concentration c, rather than the deviation c_1,

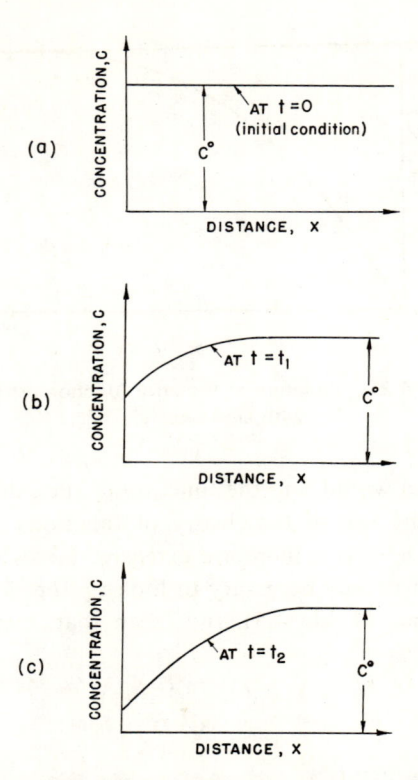

Fig. 4.22. Graphical representation of the variation of the concentration c with distance x from the electrode or diffusion sink. (a) The initial condition at $t = 0$; (b) and (c) the conditions at t_1 and t_2, where $t_2 > t_1 > t = 0$. Note that, at $t > 0$, $(dc/dx)_{x=0}$ is a constant, as it should be in the constant flux-diffusion problem.

in the concentration from the initial value c^0, one must use the defining equation for c_1

$$c_1 = c^0 - c \qquad (4.46)$$

The result is

$$c = c^0 - \frac{1}{D^{\frac{1}{2}}} \left[\frac{2t^{\frac{1}{2}}}{\pi^{\frac{1}{2}}} \exp\left(-\frac{x^2}{4Dt}\right) - \frac{x}{D^{\frac{1}{2}}} \operatorname{erfc}\left(\frac{x^2}{4Dt}\right)^{\frac{1}{2}} \right] \qquad (4.65)$$

This, then, is the fundamental equation showing how the concentration of the diffusing species varies with distance x from the electrode–solution interface and with the time t which has elapsed since a constant unit flux was switched on. In other words, Eq. (4.65) describes the diffusional response of an electrolytic solution to the stimulus of a flux which is in the form of a step function of time. The nature of the response is best appreciated by seeing how the concentration profile of the species diffusing varies as a function of time [Fig. 4.22(a), (b), and (c)]. Equation (4.65) is also of fundamental importance in describing the response of an electrochemical system to a current density which varies as a step function, i.e., to a constant current density switched on at $t = 0$. It will be seen later (Chapter 8) that this characteristic concentration response is the basis of an important electroanalytical technique.

4.2.12. How the Solution of the Constant-Flux Diffusion Problem Leads On to the Solution of Other Problems

The space and time variation of the concentration in response to the switching-on of a constant flux has been analyzed. Suppose, however, that, instead of a constant flux, one switches on a sinusoidally varying flux.[†] What is the resultant space and time variation of the concentration of the diffusing species?

One approach to this question is to set up the new diffusion problem with the initial and boundary conditions characteristic of the sinusoidally varying flux and to obtain a solution. There is, however, a simpler approach. Using the property of Laplace transforms, one can use the solution (4.65) of the constant flux-diffusion problem to generate solutions for other problems.

An electronic device is connected to an electrochemical system so that it switches on a current (Fig. 4.23) which is made to vary with time in a controllable way. The current provokes charge-transfer reactions which lead to a diffusion flux of the species involved in the charge-transfer reactions. This diffusion flux varies with time in the same way as the current. The time variation of the flux will be represented thus

$$J = g(t) \tag{4.66}$$

The imposition of this time-varying flux stimulates the electrolytic solution to respond with a space and time variation of the concentration c or the perturbation in concentration, c_1. The response depends on the

---

[†] If one feels that *current* is a more familiar word than flux, one can substitute the word current because these diffusion fluxes are often, but not always, provoked by controlling the current across an electrode–solution interface.

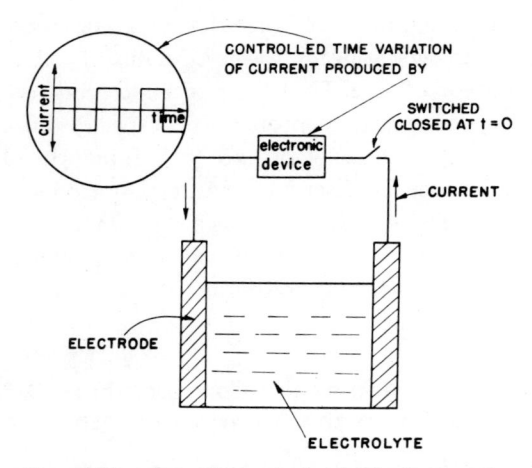

Fig. 4.23. Use of an electronic device to vary the current passing through an electrochemical system in a controlled way.

stimulus, and the mathematical relationship between the cause $J(t)$ and the effect c_1 can be represented (Fig. 4.24) quite generally thus[†]

$$\bar{c}_1 = y\bar{J} \qquad (4.67)$$

where $\bar{c}_1$ and $\bar{J}$ are the Laplace transforms of the perturbation in concentration and the flux, and y is the to-be-determined function which links the cause and effect and is characteristic of the system.

The relationship (4.67) has been defined for a flux which has an arbitrary variation with time; hence, it must also be true for the constant unit flux described in Section 4.2.10. The Laplace transform of this constant unit flux $J = 1$ is $1/p$ according to Appendix 4.2; and the Laplace transform of the concentration response to the constant unit flux is given by Eq. (4.60), i.e.,

$$\bar{c}_1 = D^{-\frac{1}{2}} p^{-\frac{1}{2}} \exp\left(-\sqrt{\frac{p}{D}}\,x\right) \qquad (4.60)$$

Hence, by substituting $1/p$ for $\bar{J}$ and $D^{-\frac{1}{2}} p^{-\frac{1}{2}} e^{-(p/D)^{\frac{1}{2}}x}$ for $\bar{c}_1$ in Eq. (4.67),

---

[†] It will be seen further on that one uses a relationship between the *Laplace transforms* of the concentration perturbation and the flux rather than the quantities c_1 and J, themselves, because the treatment in Laplace transforms is not only elegant but fruitful.

ION TRANSPORT IN SOLUTIONS 325

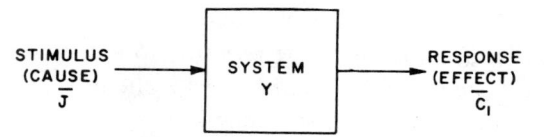

Fig. 4.24. Schematic representation of the response $\bar{c}_1$ of a system Y to a stimulus $\bar{J}$.

one has

$$y = D^{-1/2}p^{-1/2}\exp\left(-\sqrt{\frac{p}{D}}\,x\right) \qquad (4.68)$$

On introducing this expression for y into the general relationship (4.67), the result is

$$\bar{c}_1 = \left[D^{-1/2}p^{-1/2}\exp\left(-\sqrt{\frac{p}{D}}\,x\right)\right]\bar{J} \qquad (4.69)$$

This is an important result: Through the evaluation of y, it contains the concentration response to a constant unit flux switched on at $t = 0$. But, in addition, it shows how to get the concentration response to a flux $J(t)$ which is varying in a known way. All one has to do is to take the Laplace transform $\bar{J}$ of this flux $J(t)$ switched on at time $t = 0$, substitute this $\bar{J}$ in Eq. (4.69), and get $\bar{c}_1$. If one inverse-transforms the resulting expression for $\bar{c}_1$, one will obtain c_1, the perturbation in concentration as a function of x and t.

Consider a few examples. Suppose that, instead of switching on a constant unit flux at $t = 0$ (cf. Section 4.2.10), one imposes a flux which is a constant but now has a magnitude of λ moles cm^{-2} sec^{-1}, i.e., $J = \lambda$. Since the transform of a constant is $1/p$ times the constant (cf. Appendix 4.2), one obtains

$$\bar{J} = \frac{\lambda}{p} \qquad (4.70)$$

which, when introduced into Eq. (4.69), gives

$$\bar{c}_1 = \frac{\lambda}{D^{1/2}p^{3/2}}\,e^{-(x/D^{1/2})p^{1/2}} \qquad (4.71)$$

The inverse transform of the right-hand side of (4.71) is identical to that for the unit step function [cf. Eq. (4.60)] except that it is multiplied by λ. That is,

$$c_1 = \frac{\lambda}{D^{1/2}}\left[\frac{2t^{1/2}}{\pi^{1/2}}\exp\left(-\frac{x^2}{4Dt}\right) - \frac{x}{D^{1/2}}\operatorname{erfc}\left(\frac{x^2}{4Dt}\right)^{1/2}\right] \qquad (4.72)$$

In other words, the concentration response of the system to a $J = \lambda$ flux is a magnified-λ-times version of the response to a constant unit flux.

One can also understand what happens if the flux, instead of sucking ions out of the system, acts as a source and pumps in ions. This condition can be brought about by changing the direction of the constant current going through the interface and thus changing the direction of the charge-transfer reactions so that the diffusing species is produced rather than consumed. Thus, diffusion from the interface into the solution occurs. Because the direction of the flux vector is reversed, one has

$$J = -\lambda \quad \text{or} \quad \bar{J} = -\frac{\lambda}{p} \tag{4.73}$$

and, thus,

$$c_1 = -\frac{\lambda}{D^{\frac{1}{2}}} \left[\frac{2t^{\frac{1}{2}}}{\pi^{\frac{1}{2}}} \exp\left(-\frac{x^2}{4Dt}\right) - \frac{x}{D^{\frac{1}{2}}} \operatorname{erfc}\left(\frac{x^2}{4Dt}\right)^{\frac{1}{2}} \right] \tag{4.74}$$

or, in view of Eq. (4.46),

$$c = c^0 + \frac{\lambda}{D^{\frac{1}{2}}} \left[\frac{2t^{\frac{1}{2}}}{\pi^{\frac{1}{2}}} \exp\left(-\frac{x^2}{4Dt}\right) - \frac{x}{D^{\frac{1}{2}}} \operatorname{erfc}\left(\frac{x^2}{4Dt}\right)^{\frac{1}{2}} \right] \tag{4.75}$$

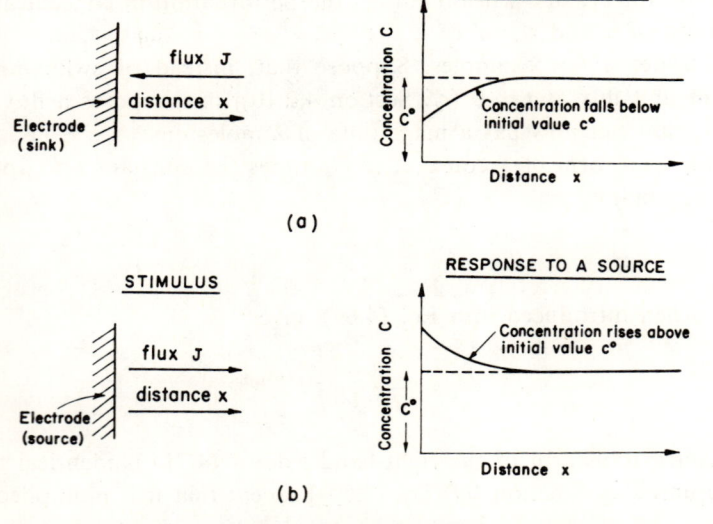

Fig. 4.25. The difference in the concentration response to the stimulus of (a) a sink and (b) a source at the electrode–electrolyte interface.

ION TRANSPORT IN SOLUTIONS 327

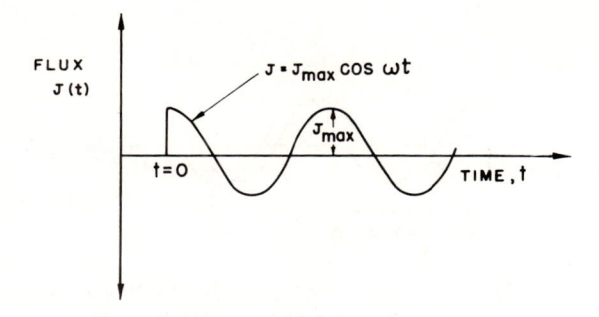

Fig. 4.26. A sinusoidally varying flux at an electrode–electrolyte interface, produced by passing a sinusoidally varying flux through the electrochemical system.

Notice the plus sign; it indicates that the concentration c rises above the initial value c^0 (Fig. 4.25).

Consider now a more interesting type of stimulus involving a periodically varying flux (Fig. 4.26). After representing the imposed flux by a cosine function

$$J = J_{max} \cos \omega t \qquad (4.76)$$

its Laplace transform is (*cf.* Table 4.5)

$$\bar{J} = J_{max} \frac{p}{p^2 + \omega^2} \qquad (4.77)$$

When this is combined with Eq. (4.69), one gets

$$\bar{c}_1 = \frac{p}{p^2 + \omega^2} \left[\frac{1}{D^{\frac{1}{2}} p^{\frac{1}{2}}} e^{-(x/D^{\frac{1}{2}})p^{\frac{1}{2}}} \right] J_{max} \qquad (4.78)$$

To simplify matters, the response of the system shall only be considered at the boundary, i.e., at $x = 0$. Hence, one can set $x = 0$ in Eq. (4.78), in which case,

$$\bar{c}_1[x = 0] = \frac{p}{p^2 + \omega^2} \frac{1}{D^{\frac{1}{2}} p^{\frac{1}{2}}} J_{max} \qquad (4.79)$$

The inverse transform reads

$$c_1[x = 0] = \frac{J_{max}}{(D\omega)^{\frac{1}{2}}} \cos\left(\omega t - \frac{\pi}{4}\right) \qquad (4.80)$$

which shows that, corresponding to a periodically varying flux (or current),

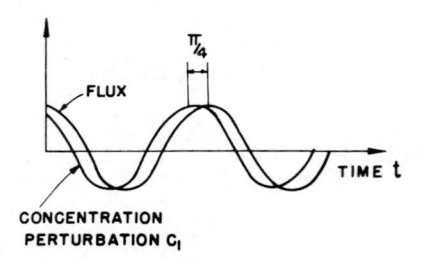

Fig. 4.27. When a sinusoidally varying diffusion flux is produced in an electrochemical system by passing a sinusoidally varying current through it, the perturbation in the concentration of the diffusing species also varies sinusoidally with time, but with a phase difference of $\pi/4$.

the concentration perturbation also varies periodically, but there is a $\pi/4$ phase difference between the flux and the concentration response (Fig. 4.27).

This is an extremely important result because an alternating flux can be produced by an alternating current density at the electrode–electrolyte interface, and, in the case of fast charge-transfer reactions, the concentration at the boundary is related to the potential difference across the interface. Thus, the current density and the potential difference both vary periodically with time, and it turns out that the phase relationship between them provides information on the rate of the charge-transfer reaction.

4.2.13. Diffusion Resulting from an Instantaneous Current Pulse

There is another important diffusion problem, the solution of which can be generated from the concentration response to a constant current (or a flux).

Consider that, in an electrochemical system, there is a plane electrode at the boundary of the electrolyte. Now, suppose that, with the aid of an electronic pulse generator, an extremely short time current pulse is sent through the system (Fig. 4.28). The current is directed so as to dissolve the metal of the electrode; hence, the effect of the pulse is to produce a burst of metal dissolution in which a layer of metal ions is piled up at the interface (Fig. 4.29).

Because the concentration of metal ions at the interface is far in excess of that in the bulk of the solution, diffusion into the solution begins. Since the source of the diffusing ions is an ion layer parallel to the plane electrode,

ION TRANSPORT IN SOLUTIONS

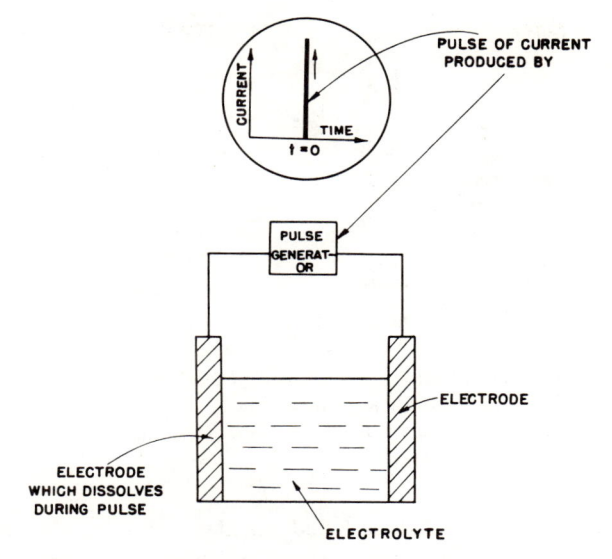

Fig. 4.28. The use of an electronic pulse generator to send an extremely short time current pulse through an electrochemical system so that there is dissolution at one electrode during the pulse.

it is known as a *plane source*; and, since the diffusing ions are produced in an *instantaneous* pulse, a fuller description of the source is contained in the term *instantaneous plane source*.

Now, as the ions from the instantaneous plane source diffuse into the solution, their concentration at various distances will change with time. The problem is to calculate the distance and time variation of this concentration.

The starting point for this calculation is the general relation between the Laplace transforms of the concentration perturbation c_1 and the time-varying flux $J(t)$

$$\bar{c}_1 = D^{-\frac{1}{2}} p^{-\frac{1}{2}} \exp\left(-\sqrt{\frac{p}{D}}\, x\right) \bar{J} \tag{4.69}$$

One has to substitute for $\bar{J}$ the Laplace transform of a flux which is an instantaneous pulse. This is done with the help of the following interesting observation.

If one takes any quantity which varies with time as a step, then the differential of that quantity with respect to time varies with time as an instantaneous pulse (Fig. 4.30). In other words, the time derivative of a step function is an instantaneous pulse. Suppose, therefore, one considers

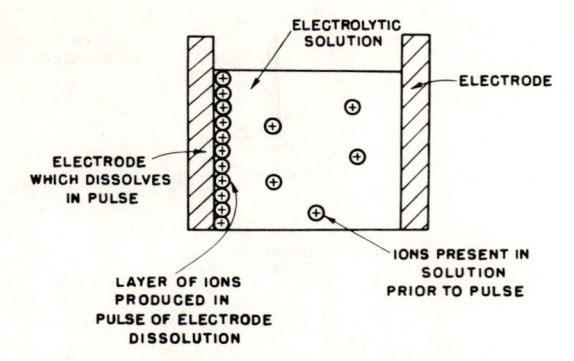

Fig. 4.29. The burst of electrode dissolution during the current pulse produces a layer of ions adjacent to the dissolving electrode (negative ions are not shown in the diagram).

a constant flux (or current) switched on at $t = 0$ (i.e., the flux is a step function of time and will be designated J_{step}), then the time derivative of that constant flux is a pulse of flux (or current) at $t = 0$ referred to by the symbol J_{pulse}, i.e.,

$$J_{\text{pulse}} = \frac{d}{dt} J_{\text{step}} \qquad (4.81)$$

If, now, one takes Laplace transforms of both sides and uses Eq. (4.38) to evaluate the right-hand side, one has

$$\bar{J}_{\text{pulse}} = p\bar{J}_{\text{step}} - J_{\text{step}}\,[t = 0] \qquad (4.82)$$

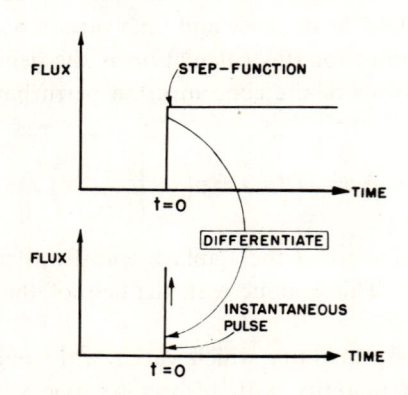

Fig. 4.30. The time derivative of a flux, which is a step function of time, is an instantaneous pulse of flux.

ION TRANSPORT IN SOLUTIONS   331

But, at $t = 0$, the magnitude of a flux which is a step function of time is zero. Hence,

$$\bar{J}_{\text{pulse}} = p\bar{J}_{\text{step}} \tag{4.83}$$

If the pumping of diffusing particles *into* the system by the step-function flux consists in switching on, at $t = 0$, a flux of $-\lambda$ moles cm^{-2} sec^{-1}, then

$$J_{\text{step}} = -\lambda \tag{4.84}$$

and

$$\bar{J}_{\text{step}} = -\frac{\lambda}{p} \tag{4.85}$$

Using this relation in Eq. (4.83), one has

$$\bar{J}_{\text{pulse}} = -\lambda \tag{4.86}$$

and, by substitution in Eq. (4.69),

$$\bar{c}_1 = -\lambda D^{-\frac{1}{2}} p^{-\frac{1}{2}} \exp\left(-\sqrt{\frac{p}{D}}\, x\right) \tag{4.87}$$

By inverse transformation (*cf.* Table 4.5),

$$c_1 = -\lambda(\pi D t)^{-\frac{1}{2}} \exp\left(-\frac{x^2}{4Dt}\right) \tag{4.88}$$

or by referring to the actual concentration c instead of the perturbation c_1 in concentration, the result is

$$c = c^0 + \lambda(\pi D t)^{-\frac{1}{2}} \exp\left(-\frac{x^2}{4Dt}\right) \tag{4.89}$$

If, prior to the current pulse, there is a zero concentration of the species which is produced by metal dissolution, i.e.,

$$c\,[t = 0] = c^0 = 0 \tag{4.90}$$

then Eq. (4.89) reduces to

$$c = \frac{\lambda}{(\pi D t)^{\frac{1}{2}}}\, e^{-x^2/4Dt} \tag{4.91}$$

It can be seen from Eq. (4.86) that λ is the Laplace transform of the pulse of flux. But a Laplace transform is an integral with respect to time. Hence, λ, which is a flux (of moles per square centimeter per second) in the constant-flux problem (*cf.* Section 4.2.12), is in fact the total concentration (moles

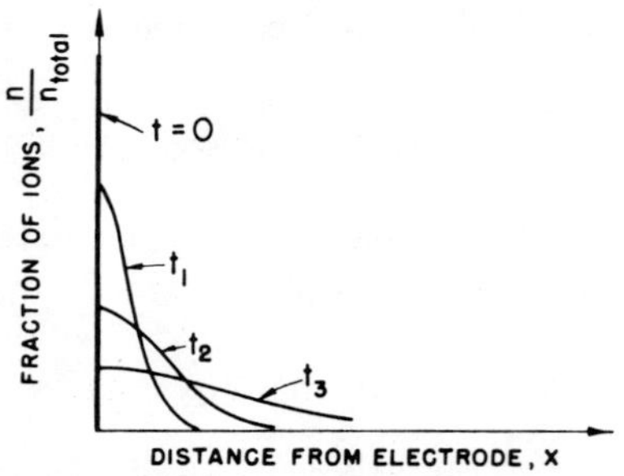

332 CHAPTER 4

Fig. 4.31. Plots of the fraction n/n_{total} of ions (produced in the pulse of electrode dissolution) against the distance x from the electrode. At $t = 0$, all the ions are on the $x = 0$ plane, and, at $t > 0$, they are distributed in the solution as a result of dissolution and diffusion. In the diagram, $t_3 > t_2 > t_1 > 0$, and the distribution curve becomes flatter and flatter.

per square centimeter) of the diffusing ions produced on the $x = 0$ plane in the burst of metal dissolution. If, instead of dealing with concentrations, one deals with *numbers* of ions, the result is

$$n = \frac{n_{total}}{(\pi Dt)^{\frac{1}{2}}} e^{-x^2/4Dt} \tag{4.92}$$

where n is the number of ions at a distance x and a time t, and n_{total} is the number of ions set up on the $x = 0$ plane at $t = 0$, i.e., n_{total} is the total number of diffusing ions.

This is the solution to the instantaneous-plane source problem. When n/n_{total} is plotted against x for various times, one obtains curves (Fig. 4.31) which show how the ions injected into the $x = 0$ plane at $t = 0$ (e.g., ions produced at the electrode in an impulse of metal dissolution) are *distributed* in space at various times. At any particular time t, a semi-bell-shaped distribution curve is obtained which shows that the ions are mainly clustered near the $x = 0$ plane, but there is a "spread." With increasing time, the spread of ions increases. This is the result of diffusion, and, after an infinitely long time, there are equal numbers of ions at any distance.

4.2.14. What Fraction of Ions Travels the Mean Square Distance $\langle x^2 \rangle$ in the Einstein–Smoluchowski Equation?

In the previous section, an experiment was described in which a pulse of current dissolves out of the electrode a certain number n_{total} of ions,

ION TRANSPORT IN SOLUTIONS   333

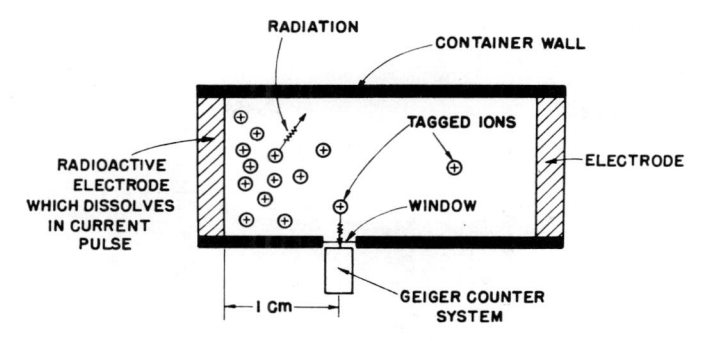

Fig. 4.32. Schematic of experiment to note the time interval between the pulse of electrode dissolution (at $t = 0$) and the arrival of radioactive ions at the window where they are registered in the Geiger-counter system.

which then start diffusing into the solution. Now suppose that the electrode material is made radioactive so that the ions produced by dissolution are detectable by a counter (Fig. 4.32). The counter head is then placed near a window in the cell at a distance of 1 cm from the dissolving electrode, so that, as soon as the tagged ions pass the window, they are registered by the counter. How long after the current pulse at $t = 0$ does the counter note the arrival of the ions?

It is experimentally observed that the counter begins to register within a few *seconds* of the termination of the instantaneous current pulse. Suppose, however, that one attempted a theoretical calculation based on the Einstein–Smoluchowski equation (4.27), i.e.,

$$\langle x^2 \rangle = 2Dt \tag{4.27}$$

using, for the diffusion coefficient of ions, the experimental value of 10^{-5} cm² sec^{-1}. Then, the estimate for the radioactive ions to reach the counter is

$$t = \frac{1}{2 \times 10^{-5} \times 60} \text{ min}$$

or

$$t \sim 10^3 \text{ min}$$

This is several orders of magnitude larger than is indicated by experience.

The dilemma may be resolved as follows. If $\langle x^2 \rangle$ in the Einstein–Smoluchowski relation pertains to the mean square distance traversed by a *majority* of the radioactive particles and if Geiger counters can—as is

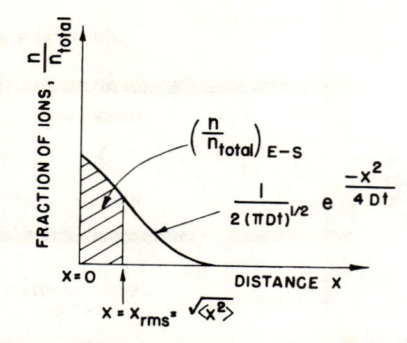

Fig. 4.33. The Einstein–Smoluchowski equation, $\langle x^2 \rangle = 2Dt$, pertains to the fraction $(n/n_{\text{total}})_{\text{ES}}$, of ions between $x = 0$ and $x = x_{\text{rms}} = \sqrt{\langle x^2 \rangle}$. This fraction is found by integrating the distribution curve between $x = 0$ and $x = x_{\text{rms}}$.

the case—detect a very small number of particles, then one can qualitatively see that there is no contradiction between the observed time and that estimated from Eq. (4.27). The time of 10^3 min estimated by the Einstein–Smoluchowski equation is far too large because it pertains to a number of radioactive ions far greater than the number needed to register in the counting apparatus. The way in which the diffusing particles spread out with time, i.e., the distribution curve for the diffusing species (Fig. 4.31) shows that, even after very short times, *some* particles have diffused to very large distances, and these are the particles registered by the counter in a time far less than that predicted by the Einstein–Smoluchowski equation.

The qualitative argument just presented can now be quantified. The central question is: To what *fraction* n/n_{total} of the ions (released at the instantaneous-plane source) does $\langle x^2 \rangle = 2Dt$ apply?[†] This question can be answered easily by integrating the n versus x distribution curve (Fig. 4.33) between the lower limit $x = 0$ (the location of the plane source) and the value of x corresponding to the square root of $\langle x^2 \rangle$. This upper limit of $\sqrt{\langle x^2 \rangle}$—the *root-mean-square distance*—is, for conciseness, represented by the symbol x_{rms}, i.e.,

$$x_{\text{rms}} = \sqrt{\langle x^2 \rangle} \qquad (4.93)$$

Thus, the Einstein–Smoluchowski fraction $(n/n_{\text{total}})_{\text{ES}}$ is given by [*cf.* Eq. (4.92)]

[†] This fraction will be termed the *Einstein–Smoluchowski fraction*.

$$\left(\frac{n}{n_{\text{total}}}\right)_{\text{ES}} = \int_0^{x_{\text{rms}}} \frac{1}{(\pi Dt)^{\frac{1}{2}}} e^{-x^2/4Dt} \, dx$$

$$= \frac{\sqrt{2}}{\pi^{\frac{1}{2}}(2Dt)^{\frac{1}{2}}} \int_0^{x_{\text{rms}}} e^{-x^2/2(2Dt)} \, dx \quad (4.94)$$

According to the Einstein–Smoluchowski relation,

$$2Dt = \langle x^2 \rangle$$

hence,

$$(2Dt)^{\frac{1}{2}} = \sqrt{\langle x^2 \rangle} = x_{\text{rms}} \quad (4.95)$$

By using this relation, Eq. (4.94) becomes

$$\left(\frac{n}{n_{\text{total}}}\right)_{\text{ES}} = \frac{\sqrt{2}}{\pi^{\frac{1}{2}} x_{\text{rms}}} \int_0^{x_{\text{rms}}} e^{-\frac{1}{2}(x/x_{\text{rms}})^2} \, dx \quad (4.96)$$

To facilitate the integration, substitute

$$\frac{x}{x_{\text{rms}}} = \sqrt{2}\, u \quad (4.97)$$

in which case, several relations follow

$$\left(\frac{x}{x_{\text{rms}}}\right)^2 = 2u^2 \quad (4.98)$$

$$dx = \sqrt{2}\, x_{\text{rms}} \, du \quad (4.99)$$

$$u = \frac{1}{\sqrt{2}} \quad \text{when} \quad x = x_{\text{rms}} \quad (4.100)$$

and

$$u = 0 \quad \text{when} \quad x = 0 \quad (4.101)$$

With the use of these relations, Eq. (4.96) becomes

$$\left(\frac{n}{n_{\text{total}}}\right)_{\text{ES}} = \frac{2}{\pi^{\frac{1}{2}}} \int_0^{1/\sqrt{2}} e^{-u^2} \, du \quad (4.102)$$

The integral on the right-hand side is the error function of $1/\sqrt{2}$ [*cf.* Eq. (4.63)].

TABLE 4.6

The Value of the Integral $\frac{2}{\sqrt{\pi}} \int_0^y e^{-u^2} du$

y	
0.00	0.00000
0.01	0.01128
0.02	0.02256
0.10	0.11246
0.20	0.22270
0.30	0.32863
0.40	0.42839
0.50	0.52050
0.60	0.60386
0.70	0.67780
0.80	0.74210
0.90	0.79691
1.00	0.84270
2.00	1.00000

Now, values of the error function have been tabulated in detail (Table 4.6). The value of the error function of $1/\sqrt{2}$, i.e.,

$$\frac{2}{\pi^{\frac{1}{2}}} \int_0^{1/\sqrt{2}} e^{-u^2} du$$

is 0.68. Hence,

$$\left(\frac{n}{n_{\text{total}}}\right)_{\text{ES}} = 0.68 \tag{4.103}$$

and, therefore, about two-thirds (68%) of the diffusing species are within the region from $x = 0$ to $x = x_{\text{rms}} = \sqrt{2Dt}$. This means however that the remaining fraction, namely, one-third, have crossed beyond this distance. Of course, the radioactive ions which are sensed by the counter almost immediately after the pulse of metal dissolution belong to the one-third group (Fig. 4.34).

In the above experiment, diffusion toward the $x \to -\infty$ direction is prevented by the presence of a physical boundary (i.e., the electrode). If no such boundary exists and diffusion in both the $+x$ and $-x$ directions is possible, then the 68% of the particles will distribute themselves in the region from $x = -x_{\text{rms}} = -\sqrt{2Dt}$ to $x = +x_{\text{rms}} = +\sqrt{2Dt}$. From

ION TRANSPORT IN SOLUTIONS 337

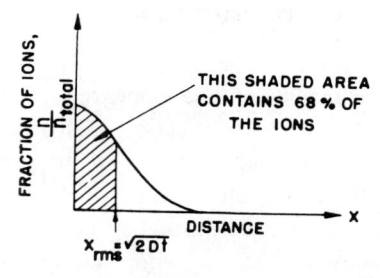

Fig. 4.34. When diffusion occurs from an instantaneous plane source (set up, e.g., by a pulse of electrode dissolution), then 68% of the ions produced in the pulse lie between $x = 0$ and $x = x_{rms}$, after the time t.

symmetry considerations, one would expect 34% to be within $x = 0$ and $x = +x_{rms}$ and an equal amount on the other side (Fig. 4.35).

From the above discussion, the advantages and limitations of using the Einstein–Smoluchowski relation become clear. If one is considering phenomena involving a few particles, then one can be misled by making Einstein–Smoluchowski calculations. If, however, one wants to know about the diffusion of a sizable fraction of the total number of particles, then the relation provides easily obtained, although rough, answers without having to go through the labor of obtaining the exact solution for the diffusion problem (see, e.g., Section 4.6.8).

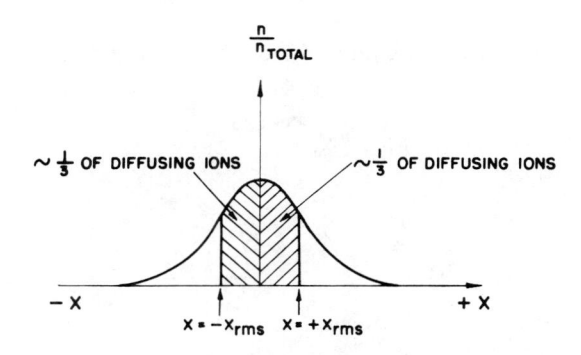

Fig. 4.35. If it were possible for diffusion to occur in the $+x$ and $-x$ directions from an instantaneous plane source at $x = 0$, then one-third of the diffusing species would lie between $x = 0$ and $x = +x_{rms}$ and a similar number between $x = 0$ and $x = -x_{rms}$.

4.2.15. How Can the Diffusion Coefficient Be Related to Molecular Quantities?

The diffusion coefficient D has appeared in both the macroscopic (Section 4.2.2) and the atomistic (Section 4.2.6) views of diffusion. But, how does the diffusion coefficient depend on the structure of the medium and the interatomic forces which operate? To answer this question, one should have a deeper understanding of this coefficient than that provided by the empirical first law of Fick, in which D appeared simply as the proportionality constant relating the flux J and the concentration gradient dc/dx. Even the random-walk interpretation of the diffusion coefficient as embodied in the Einstein–Smoluchowski equation (4.27) is not fundamental enough because it is based on the mean square distance traversed by the ion after N steps taken in a time t and does not probe into the laws governing each step taken by the random-walking ion.

This search for the atomistic basis of the diffusion coefficient will commence from the picture of random-walking ions (*cf.* Sections 4.2.4 to 4.2.6). It will be recalled that a net diffusive transport of ions occurs, in spite of the completely random zig-zag dance of individual ions, because of unequal numbers of ions in different regions.

Consider one of these random-walking ions. It can be proved (see Appendix 4.1) that the mean square distance $\langle x^2 \rangle$ traveled by an ion depends on the number N of jumps the ion takes and the mean jump distance l in the following manner (*cf.* Fig. 4.36):

$$\langle x^2 \rangle = Nl^2 \tag{4.104}$$

It has further been shown (Section 4.2.6) that, in the case of a one-dimen-

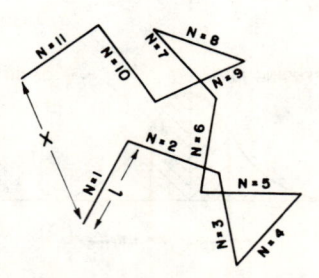

Fig. 4.36. Schematic representation of $N = 11$ steps (each of mean length l) in the random walk of an ion. After 11 steps, the ion is at a distance x from the starting point.

sional random walk, $\langle x^2 \rangle$ depends on time according to the Einstein-Smoluchowski equation

$$\langle x^2 \rangle = 2Dt \qquad (4.27)$$

Hence, by combining Eqs. (4.104) and (4.27), one has the equation

$$Nl^2 = 2Dt \qquad (4.105)$$

which relates the number of jumps and the time. If, now, only one jump of the ion is considered, i.e., $N = 1$, Eq. (4.105) reduces to

$$D = \frac{1}{2} \frac{l^2}{\tau} \qquad (4.106)$$

where τ is the *mean jump time* to cover the mean jump distance l. This mean jump time is the number of seconds per jump,[†] and, therefore, $1/\tau$ is the jump frequency, i.e., the number of jumps per second. Putting

$$k = \frac{1}{\tau} \qquad (4.107)$$

one can write Eq. (4.106) thus

$$D = \tfrac{1}{2} l^2 k \qquad (4.108)$$

Equation (4.108) shows that the diffusion depends on how far, on the average, an ion jumps and how frequently these jumps occur.

4.2.16. The Mean Jump Distance *l*, a Structural Question

To go further than Eq. (4.108), one has to examine the factors which govern the mean jump distance l and the jump frequency k. For this, the picture of the liquid (in which diffusion is occurring) as a structureless continuum is inadequate. In reality, the liquid has a structure—ions and molecules in definite arrangements at any one instant of time. This arrangement in a liquid (unlike that in a solid) is local in extent, transitory in time, and mobile in space. The details of the structure are not necessary to continue the present discussion. What counts is that ions zig-zag in a random walk and, for any particular jump, the ion has to jump *out of* one site in

---

[†] This mean jump time will include any waiting time between two successive jumps.

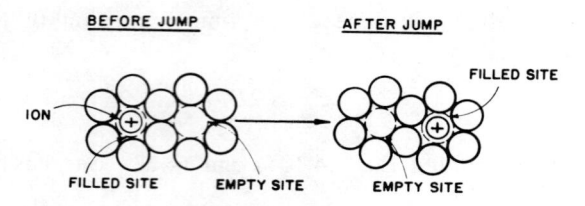

Fig. 4.37. Diagram to illustrate one step in the random walk of an ion.

the liquid structure *into* another site (Fig. 4.37). This jumping process can be symbolically represented thus

$$\boxed{\text{Ion}} + \boxed{} \rightarrow \boxed{} + \boxed{\text{Ion}} \qquad (4.109)$$

where $\boxed{\text{Ion}}$ is a site occupied by an ion and $\boxed{}$ is an empty acceptor site waiting to receive a jumping ion.

The mean jump distance l is seen to be the mean distance between sites, and its numerical value depends upon the details of the structure of the liquid, i.e., upon the instantaneous and local atomic arrangement.

4.2.17. The Jump Frequency, a Rate-Process Question

The process of diffusion always occurs at a finite rate; it is a *rate process*.

Now, chemical reactions, e.g., three atom reactions of the type

$$AB + C \rightarrow A + BC \qquad (4.110)$$

are also rate processes. Further, a three-atom reaction can be formally described as the jump of the particle B from a site in A to a site in C (Fig. 4.38). With this description, it can be seen that the notation (4.109) used

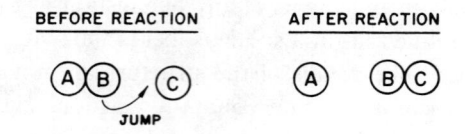

Fig. 4.38. A three-atom chemical rection $AB + C \rightarrow A + BC$ viewed as the jump of atom B from atom A to atom C.

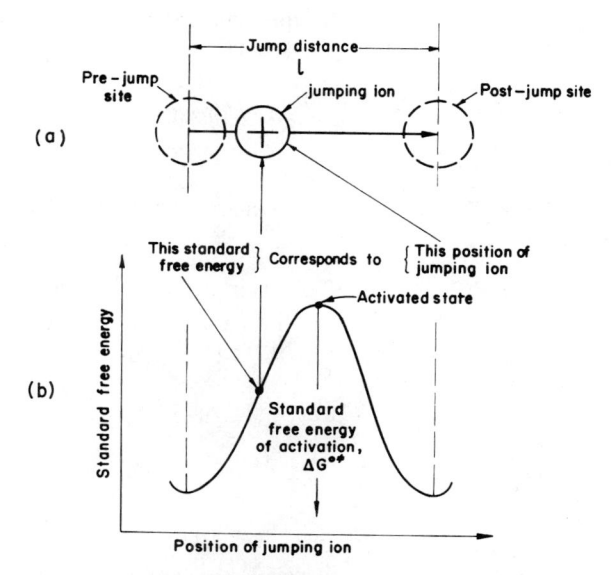

Fig. 4.39. (a) The jump of an ion from a prejump site to a postjump site through a jump distance l. (b) Corresponding to each position of the jumping ion, the system (sites + jumping ion) has a standard free energy. Thus, the standard free-energy changes corresponding to the ionic jump can be represented by the passage of a point (representing the standard free energy of the system) over the barrier by the acquisition of the standard free energy of activation.

to represent the jump of an ion has in fact established an analogy between the two rate processes, i.e., diffusion and chemical reaction. Thus, the basic theory of rate processes (Appendix 4.3) should be applicable to both the processes of diffusion and chemical reactions.

The basis of this theory is that the potential energy (and standard free energy) of the system of particles involved in the rate processes varies as the particles move to accomplish the process. Very often, the movements crucial to the process are those of a single particle, as is the case with the diffusive jump of an ion from site to site. If the free energy of the system is plotted as a function of the position of the crucial particle, e.g., the jumping ion, then the standard free energy of the system has to attain a critical value (Fig. 4.39)—the activation free energy ΔG^0 —for the process to be accomplished. One says that the system has to cross an *energy barrier* for the rate process to occur. The number of times per second, k, that the rate process occurs, i.e., the jump frequency in the case of diffusion, can be shown to be given by[†] (*cf.* Appendix 4.3)

$$\vec{k} = \frac{kT}{h} e^{-\Delta G^{0\ddagger}/RT} \tag{4.111}$$

4.2.18. The Rate-Process Expression for the Diffusion Coefficient

To obtain the diffusion coefficient in terms of atomistic quantities, one has to insert the expression for the jump frequency (4.111) into that for the diffusion coefficient [Eq. (4.108)]. The result is

$$D = \frac{1}{2} l^2 \vec{k}$$

$$= \frac{1}{2} l^2 \frac{kT}{h} e^{-\Delta G^{0\ddagger}/RT} \tag{4.112}$$

The numerical coefficient $\frac{1}{2}$ has entered here only because the Einstein–Smoluchowski equation $\langle x^2 \rangle = 2Dt$ for *one*-dimensional random walk was considered. In general, it is related to the probability of the ion's jumping in various directions, not just forward and backward. For convenience, therefore, the coefficient will be taken to be unity, in which case

$$D \sim l^2 \vec{k} \tag{4.113}$$

$$\sim l^2 \frac{kT}{h} e^{-\Delta G^{0\ddagger}/RT} \tag{4.114}$$

4.2.19. Diffusion: An Overall View

An electrochemical system runs on the basis of charge-transfer reactions at the electrode–electrolyte interfaces. These reactions involve ions or molecules which are constituents of the electrolyte. Thus, the transport of particles to or away from the interface becomes an essential condition for the continued electrochemical transformation of reactants at the interface.

One of the basic mechanisms of ionic transport is diffusion. This type of transport occurs when the concentration of the diffusing species is different in different parts of the electrolyte. What makes diffusion occur?

[†] The $\vec{k}$ on the left-hand side is the jump frequency; the k in the term kT/h is the Boltzmann constant.

This question can be answered on a macroscopic level and on a microscopic level.

The macroscopic view is based on the fact that, when the concentration varies with distance, so does the chemical potential. But a nonuniformity of chemical potential implies that the free energy is not the same everywhere and, therefore, the system is not at equilibrium. How can the system attain equilibrium? By equalizing the chemical potential, i.e., by transferring the species from the high-concentration regions to the low-concentration regions. Thus, the negative gradient of the chemical potential, $-d\mu/dx$, behaves *like* a driving force (*cf.* Section 4.2.1) which produces a net flow, or flux, of the species.

When the driving force is small, it may be taken to be linearly related to the flux. On this basis, an equation can be derived for the rate of steady-state diffusion, which is identical in form to the empirical first law of Fick,

$$J = -D\frac{dc}{dx}$$

The microscopic view of diffusion starts with the movements of individual ions. Ions dart about haphazardly, executing a random walk. By an analysis of one-dimensional random walk, a simple law can be derived (*cf.* Section 4.2.6) for the mean square distance $\langle x^2 \rangle$ traversed by an ion in a time t. This is the Einstein–Smoluchowski equation

$$\langle x^2 \rangle = 2Dt \qquad (4.27)$$

It also turns out that the random walk of individual ions is able to give rise to a flux, or flow, on the level of the group. Diffusion is simply the result of there being more random-walking particles in one region than in another (*cf.* Section 4.2.6). The gradient of chemical potential is therefore only a pseudo force which can be regarded as operating on a society of ions but not on individual ions.

The first law of Fick tells one how the concentration gradient is related to the flux under *steady-state* conditions; it says nothing about how the system goes from nonequilibrium to steady state when a diffusion source or sink is set up inside or at the boundary of the system. Thus, it says nothing about how the concentration changes with time at different distances from the source or sink. In other words, Fick's first law is inapplicable to non-steady-state diffusion. For this, one has to go to Fick's second law

$$\frac{\partial c}{\partial t} = D\frac{\partial^2 c}{\partial x^2} \qquad (4.32)$$

which relates the time and space variations of the concentration during diffusion.

Fick's second law is a partial differential equation. Thus, it describes the general characteristics of all diffusion problems but not the details of any one particular diffusion process. Hence, the second law must be solved with the aid of the initial and boundary conditions which characterize the particular problem.

The solution of Fick's second law is facilitated by the use of Laplace transforms which convert the partial differential equation into an easily integrable total differential equation. By utilizing Laplace transforms, the concentration of diffusing species as a function of time and of distance from the diffusion sink when a constant normalized current, or flux, is switched on at $t = 0$ was shown to be

$$c = c^0 - \frac{1}{D^{\frac{1}{2}}}\left[\frac{2t^{\frac{1}{2}}}{\pi^{\frac{1}{2}}}\exp\left(-\frac{x^2}{4Dt}\right) - \frac{x}{D^{\frac{1}{2}}}\operatorname{erfc}\left(\frac{x^2}{4Dt}\right)^{\frac{1}{2}}\right] \quad (4.65)$$

With the solution of this problem (in which the flux varies as a unit-step function with time), one can easily generate the solution of other problems in which the current, or flux, varies with time in other ways, e.g., as a periodic function or as a single pulse.

When the current, or flux, is a single impulse, an instantaneous-plane source for diffusion is set up and the concentration variation is given by

$$c(x,t) = \frac{\lambda}{(\pi Dt)^{\frac{1}{2}}} e^{-x^2/4Dt} \quad (4.91)$$

in the presence of a boundary. From this expression, it turns out that, in a time t, only a certain fraction ($\frac{2}{3}$) of the particles travel the distance given by the Einstein–Smoluchowski equation. Actually, the spatial distribution of the particles is given by a semi-bell-shaped distribution curve.

The final step involves the relation of the diffusion coefficient to the structure of the medium and the forces operating there. It is all a matter of the mean distance l through which ions jump during the course of their random walk and of the mean jump frequency k. The latter can be expressed in terms of the theory of rate processes, so that one ends up with an expression for the rate of diffusion which is in principle derivable from the local structure of the medium.

This, then, is an elementary picture of diffusion. The next task is to consider the phenomenon of conduction, i.e., the migration of ions in an electric field.

Further Reading

1. M. Smoluchowski, *Ann. Phys.* (*Paris*), **25**: 205 (1908).
2. A. Einstein, *Investigations on the Theory of Brownian Movement*, Methuen & Co., Ltd., London, 1926.
3. H. S. Carslaw, *Introduction to the Theory of Fourier Series and Integrals*, Macmillan & Co., Ltd., London, 1930.
4. S. Glasstone, K. J. Laidler, and H. Eyring, *The Theory of Rate Processes*, McGraw-Hill Book Company, New York, 1941.
5. S. Chandrasekhar, "Noise and Stochastic Processes," *Rev. Mod. Phys.*, **15**: 1 (1943).
6. J. C. Jaeger, *An Introduction to the Laplace Transformation*, John Wiley & Sons, Inc., New York, 1951.
7. H. S. Carslaw and J. C. Jaeger, *Conduction of Heat in Solids*, Clarendon Press, Oxford, 1959.
8. Ling Yang, and M. T. Simnad, "Measurement of Diffusivity in Liquid Systems," in: *Physicochemical Measurements at High Temperatures*, J. O'M. Bockris. J. L. White, and J. W. Tomlinson, eds., Butterworth's Publications, Ltd., London, 1959.
9. W. Jost, *Diffusion in Solids, Liquids, Gases*, Academic Press, Inc., New York, 1960.
10. P. Delahay, Chapter 5 in: P. Delahay and C. W. Tobias, eds., *Advances in Electrochemistry and Electrochemical Engineering*, Vol. 1, Interscience Publishers, Inc., New York, 1961.
11. R. P. Feynman, *Lectures on Physics*, Vol. 1, Addison-Wesley Publishing Co. Reading, Mass., 1964.

4.3. IONIC DRIFT UNDER AN ELECTRIC FIELD: CONDUCTION

4.3.1. The Creation of an Electric Field in an Electrolyte

Consider that two plane-parallel electrodes are introduced into an electrolytic solution so as to cover the ends walls of the rectangular insulating container (Fig. 4.40). With the aid of an external source, let a potential difference be applied across the electrodes. How does this applied potential difference affect the ions in the solution?

The potential in the solution has to vary from the value at one electrode, ψ_I, to that at the other electrode, ψ_II. The major portion of this potential drop $\psi_\mathrm{I} - \psi_\mathrm{II}$ occurs across the two electrode–solution interfaces (see Chapter 7), i.e., if the potentials on the solution side of the two interfaces are ψ_I' and ψ_II', then the interfacial potential differences are $\psi_\mathrm{I} - \psi_\mathrm{I}'$ and $\psi_\mathrm{II}' - \psi_\mathrm{I}$ (Fig. 4.41). The remaining potential drop $\psi_\mathrm{I}' - \psi_\mathrm{II}'$, occurs

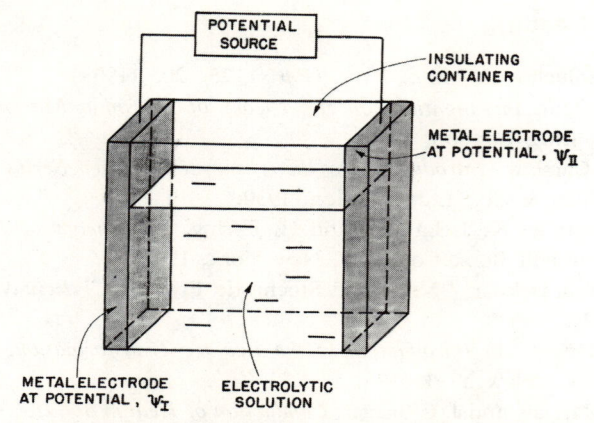

Fig. 4.40. An electrochemical system consisting of two plane-parallel electrodes immersed in an electrolytic solution is connected to a source of potential difference.

in the electrolytic solution. The electrolytic solution is therefore a region of space in which the potential at a point is a function of the distance of that point from the electrodes.

Let the test ion in the solution be at the point x_1, where the potential is ψ_1 (Fig. 4.42). This potential is by definition the work done to bring a unit of positive charge from infinity up to the particular point. (In the

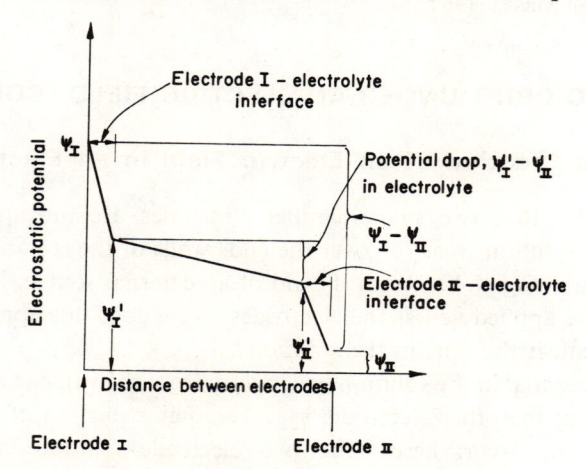

Fig. 4.41. Diagram to illustrate how the total potential difference $\psi_\mathrm{I} - \psi_\mathrm{II}$ is distributed in the region between the two electrodes.

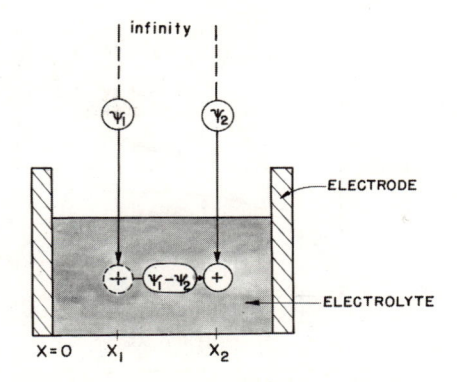

Fig. 4.42. The work done to transport a unit of positive charge from x_1 to x_2 in the solution is equal to the difference $\psi_1 - \psi_2$ in electrostatic potential at the two points.

course of this journey of the test charge from infinity to the particular point, it may have to cross phase boundaries, for example, the electrolyte–air boundary, and thereby do extra work (see Chapter 7). Such surface work terms cancel out however in discussions of the potential differences between two points in the same medium.) If another point x_2 is chosen on the normal from x to the electrodes, then the potential at x_2 is different from that at x_1 because of the variation of the potential along the distance coordinate between the electrodes. Let the potential at x_2 be ψ_2. Then, $\psi_1 - \psi_2$, the potential difference between the two points, is the work done to take a unit of charge from x_1 to x_2.

When this work $\psi_1 - \psi_2$ is divided by the distance over which the test charge is transferred, i.e., $x_1 - x_2$, one obtains the *force per unit charge*, or the electric field X:

$$X = -\frac{\psi_1 - \psi_2}{x_1 - x_2} \tag{4.115}$$

where the minus sign indicates that the force acts on a positive charge in a direction opposite to the direction of the positive gradient of potential. In the particular case under discussion, i.e., parallel electrodes covering the end walls of a rectangular container, the potential drop in the electrolyte is linear (as in the case of a parallel-plate condenser), and one can write

$$X = -\frac{\text{Potential difference across the solution}}{\text{Distance across solution}} \tag{4.116}$$

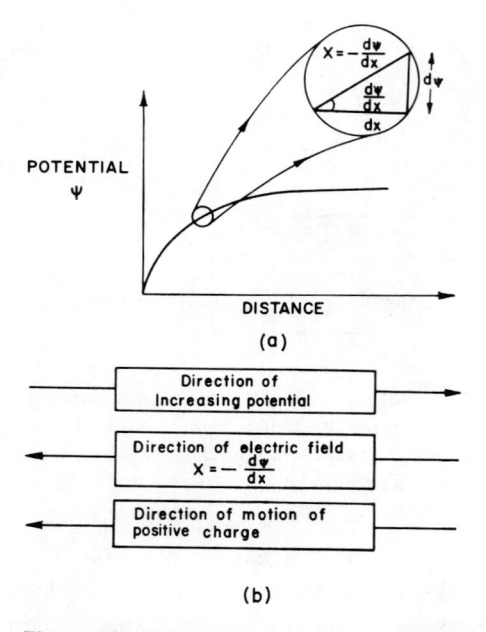

Fig. 4.43. (a) In the case of a nonlinear potential variation in the solution, the electric field at a point is the negative gradient of the electrostatic potential at that point. (b) The relative directions of increasing potential, field, and motion of a positive charge.

In general, however, it is best to be in a position to treat nonlinear potential drops. This is done by writing (Fig. 4.43)

$$X = -\frac{d\psi}{dx} \qquad (4.117)$$

where now the electric field may be a function of x.

The imposition of a potential difference between two electrodes thus makes an electrolytic solution the scene of operation of an electric field, i.e., an electric force, acting upon the charges present. This field can be mapped by drawing equipotential surfaces (all points associated with the same potential lie on the same surface). The potential map yields a geometrical representation of the field. In the case of plane-parallel electrodes extending up to the walls of a rectangular cell, the equipotential surfaces are parallel to the electrodes (Fig. 4.44).

ION TRANSPORT IN SOLUTIONS 349

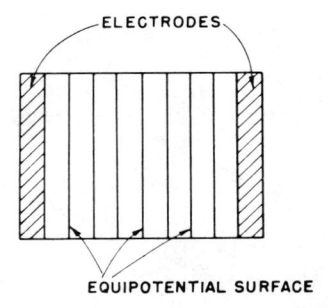

Fig. 4.44. A geometric representation of the electric field in an electrochemical system in which plane-parallel electrodes are immersed in an electrolyte so that they extend up to the walls of rectangular insulating container. The equipotential surfaces are parallel to the electrodes.

4.3.2. How Do Ions Respond to the Electric Field?

In the absence of an electric field, the ions are in ceaseless random motion. This random walk of ions has been shown to have an important characteristic: The mean distance traversed by the ions as a whole is zero because, while some are displaced in one direction, an equal number are displaced in the opposite direction. From a phenomenological view, therefore, the random walk of ions can be ignored because it does not lead to

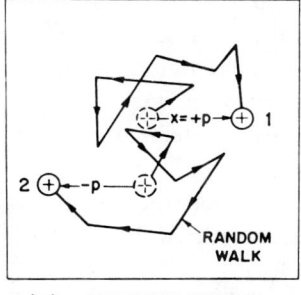

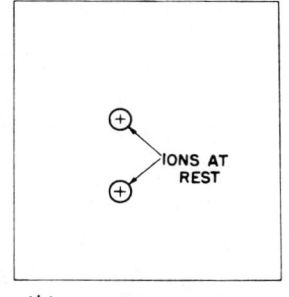

Fig. 4.45. (a) Schematic representation of the random walk of two ions showing that ion 1 is displaced a distance $x = +p$ and ion 2, a distance $x = -p$ and, hence, the mean distance traversed is zero. (b) Since the mean distance traversed by the ions is zero, there is no net flux of the ions and they can legitimately be considered, from a macroscopic point of view, at rest.

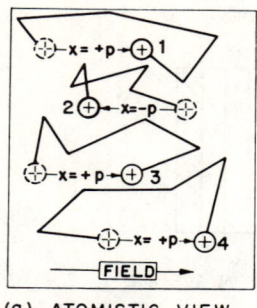

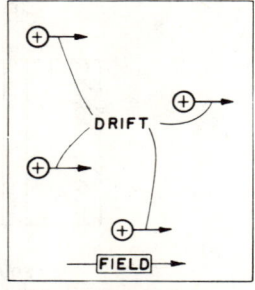

Fig. 4.46. (a) Schematic representation of the movements of four ions which random walk in the presence of a field. Their displacements are $+p$, $-p$, $+p$, and $+p$, i.e., the mean displacement is finite. (b) From a macroscopic point of view, one can ignore the random walk and consider that each ion drifts in the direction of the field.

any net transport matter (so long as there is no difference in concentration in various parts of the solution so that net diffusion down the concentration gradient occurs). The net result is as if the ions were at rest (Fig. 4.45).

Under the influence of an electric field, however, the net result of the zig-zag jumping of ions is not zero. Ions feel the electric field, i.e., they experience a force directing them toward the electrode which is charged oppositely to the charge on the ion. This directed force is equal to the charge on the ion, $z_i e_0$, times the field at the point where the ion is situated. The driving force of the electric field produces in all ions of a particular species a *velocity component* in the direction p of the potential gradient. Thus, the establishment of a potential difference between the electrodes produces a *drift*, or flux, of ions (Fig. 4.46). This drift is the *migration* (or *conduction*) of ions in response to an electric field.

As in diffusion, the relationship between the steady-state flux J of ions and the driving force of the electric field will be represented by the expression

$$J = A + BX + CX^2 + \cdots \tag{4.118}$$

For small fields, the terms higher than BX will tend to zero. Further, the constant A must be zero because the flux of ions must vanish when the field is zero. Hence, for small fields, the flux of ions is proportional to the field (*cf.* Section 4.2.2)

$$J = BX \tag{4.119}$$

4.3.3. The Tendency for a Conflict between Electroneutrality and Conduction

When a potential gradient, i.e., electric field, exists in an electrolytic solution, the positive ions drift toward the negative electrode; and the negative ions, in the opposite direction. What is the effect of this ionic drift on the state of charge of an electrolytic solution?

Prior to the application of an external field, there is a time-average electroneutrality in the electrolyte over a distance large compared with $\varkappa^{-1}$ (*cf.* Section 3.3.8), i.e., the net charge in any macroscopic volume of solution is zero because the total charge due to the positive ions is equal to the total charge due to the negative ions. Owing to the electric field, however, ionic drift tends to produce a spatial separation of charge. Positive ions will try to segregate near the negatively charged electrode; and negative ions, near the positively charged electrode.

This tendency for gross charge separation has an important implication; electroneutrality tends to be upset. Further, the separated charge causing the lack of electroneutrality tends to set up its own field, which would run counter to the externally applied field. If the two fields were to become equal in magnitude, the net field in the solution would become zero. (Thus, the driving force on an ion would vanish and ion migration stop.)

It appears from this argument that an electrolytic solution would sustain only a *transient* migration of ions, and, then, the tendency to conform to the principle of electroneutrality would result in a halt in the drift of ions after a short time. A persistent flow of charge, an electric *current*, appears to be impossible. In practice, however, an electrolytic solution can act as a *conductor* of electricity and is able to pass a current, i.e., maintain a flow of ions. Is there a paradox here?

4.3.4. The Resolution of the Electroneutrality-versus-Conduction Dilemma: Electron-Transfer Reactions

The solution to the dilemma just posed can be found by comparing an electrolytic solution with a metallic conductor. In a metallic conductor, there is a lattice of positive ions which hold their equilibrium positions during the conduction process. In addition, there are the free conduction electrons which assume responsibility for the transport of charge. Contact is made to and from the metallic conductor by means of other metallic conductors (Fig. 4.47a). Hence, electrons act as charge carriers *throughout the entire circuit*.

In the case of an electrolytic conductor, however, it is necessary to

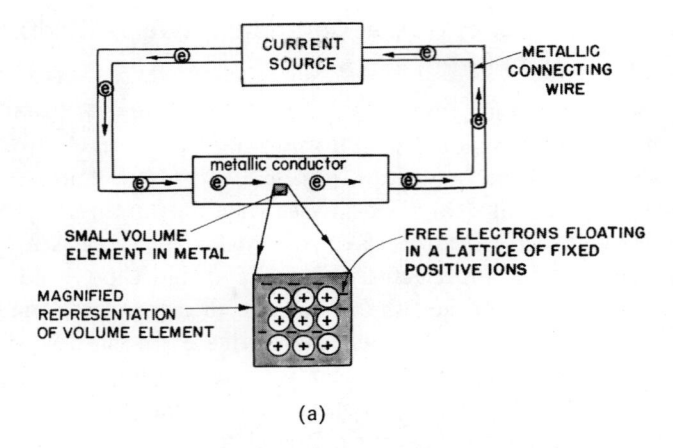

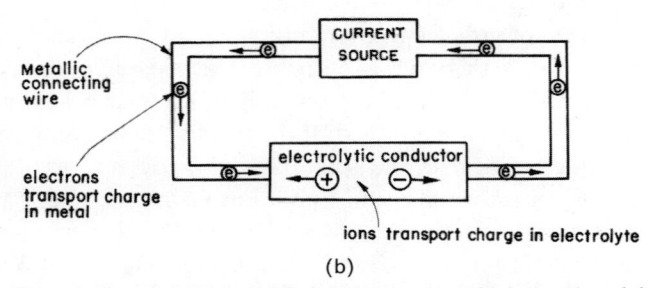

Fig. 4.47. Comparison of electric circuits which consist of (a) a metallic conductor only, and (b) an electrolytic conductor as well as a metallic one.

make electrical contact to and from the electrolyte by metallic conductors (wires). Thus, here one has the interesting situation in which electrons transport charge in the external circuit and ions carry the charge in the electrolytic solution (Fig. 4.47b). Obviously, one can maintain a steady flow of charge (current) in the entire circuit only if there is a change of charge carrier at the electrode–electrolyte interface. In other words, for a current to flow in the circuit, ions have to hand over or take electrons from the electrodes.

Such electron transfers between ions and electrodes result in chemical changes (changes in the valence or oxidation state of the ions), i.e., in electrodic reactions. When ions receive electrons from the electrode, they are said to be "electronated," or *to undergo reduction;* when ions donate electrons to the electrodes, they are said to be "deelectronated," or *to undergo oxidation.*

The occurrence of a reaction at each electrode is tantamount (oxidation = deelectronation; reduction = electronation) to removal of equal amounts of positive and negative charge from the solution. Hence, when electron-transfer reactions occur at the electrodes, ionic drift does *not* lead to segregation of charges and the building-up of an electroneutrality field (opposite to the applied field). Thus, the flow of charge can continue, i.e., the solution conducts. It is an ionic conductor.

4.3.5. The Quantitative Link between Electron Flow in the Electrodes and Ion Flow in the Electrolyte: Faraday's Law

Charge transfer is the essence of an electrodic reaction. It constitutes the bridge between the current I_e of electrons in the electrode part of the electrical circuit and the current I_i of ions in the electrolytic part of the circuit (Fig. 4.48). When a steady-state current is passing through the circuit, there must be a continuity in the currents at the electrode–electrolyte interface, i.e.,

$$I_e = I_i \qquad (4.120)$$

(This is in fact an example of Kirchhoff's law, which says that the *algebraic* sum of the currents at any junction must be zero.) Further, if one multiplies both sides of Eq. (4.120) by the time t, one obtains

$$I_e t = I_i t \qquad (4.121)$$

which means[†] that the quantity Q_e of electricity carried by the electrons is equal to that carried by the ions

$$Q_e = Q_i \qquad (4.122)$$

Let the quantity of electricity due to electron flow be the charge borne by an Avogadro number of electrons, i.e., $Q_e = N_A e_0 = F$. If the charge on each ion participating in the electrodic reaction is $z_i e_0$, it is easily seen that the *number* of ions required to preserve equality of currents [Eq. (4.120)] and equality of charge transported across the interface in time t [Eq. (4.122)] is

$$\frac{Q}{z_i e_0} = \frac{N_A e_0}{z_i e_0} = \frac{N_A}{z_i} \text{ [ions]} = \frac{1}{z_i} \text{ [mole of ions]}$$
$$= 1 \text{ g-eq} \qquad (4.123)$$

[†] The product of the current and time is the quantity of electricity.

354 CHAPTER 4

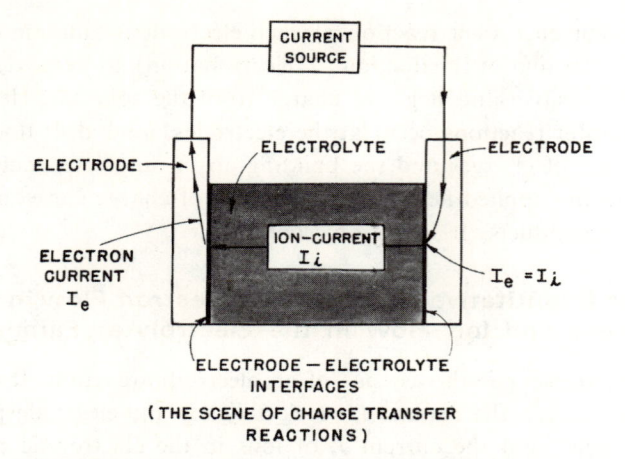

Fig. 4.48. Diagram for the derivation of Faraday's Laws. The electron current I_e in the metallic part of the circuit must be equal to the ion current I_i in the electrolytic part of the circuit.

Thus, the requirement of steady-state continuity of current at the interface leads to the following law: The passage of 1 faraday (F) of charge results in the electrodic reaction of one equivalent ($1/z_i$ moles) of ions each of charge $z_i e_0$. This is Faraday's law.† Conversely, if $1/z_i$ moles of ions undergo charge transfer, then 1 F of electricity passes through the circuit, or zF Faradays per mole of ions transformed.

4.3.6. The Proportionality Constant Relating the Electric Field and the Current Density: The Specific Conductivity

In the case of small fields, the steady-state flux of ions can be considered proportional to the driving force of an electric field (*cf.* Section 4.3.2), i.e.,

$$J = BX \qquad (4.119)$$

The quantity J is the number of *moles* of ions crossing a unit area per second. When J is multiplied by the charge borne by 1 mole of ions, zF, one obtains the current density i, or *charge flux*, i.e., the quantity of charge crossing unit area per second. Hence,

$$i = JzF = zFBX \qquad (4.124)$$

---

† Alternatively, Faraday's law states that, if a current of I amp passes for a time t sec, then It/zF moles of reactants in the electrodic reaction are produced or consumed.

TABLE 4.7
Representative Values of Specific Conductivity

Substance	Type of conductor	Specific conductivity, ohm^{-1} cm^{-1}	t °C
Copper	Metallic	5.8×10^5	20
Lead	Metallic	4.9×10^5	0
Iron	Metallic	1.1×10^5	0
4 M H$_2$SO$_4$	Electrolytic	7.5×10^{-1}	18
0.1 M KCl	Electrolytic	1.3×10^{-2}	25
Xylene	Nonelectrolyte	1×10^{-19}	25
Water	Nonelectrolyte	4×10^{-8}	18

The constant zFB can be set equal to a new constant σ, which is known as the *specific conductivity* (Table 4.7). The relation between the current density i and the electric field X is therefore

$$X = \frac{1}{\sigma} i \qquad (4.125)$$

The electric field is very simply related (Fig. 4.49) to the potential difference across the electrolyte, $(\psi_\text{I}' - \psi_\text{II}')$ [see Eq. (4.116)],

$$X = \frac{\Delta \psi}{l} \qquad (4.126)$$

where l is the distance across the electrolyte. Further, the total current I is equal to the area A of the electrodes times the current density i

$$I = iA \qquad (4.127)$$

Substituting these relations [Eqs. (4.126) and (4.127)] in the field–current-density relation [Eq. (4.125)], one has

$$\frac{\Delta \psi}{l} = \frac{1}{\sigma} \frac{I}{A}$$

or

$$\Delta \psi = \frac{l}{\sigma A} I \qquad (4.128)$$

The constants σ, l, and A determine the resistance R of the solution

$$R = \frac{l}{\sigma A} \qquad (4.129)$$

356 CHAPTER 4

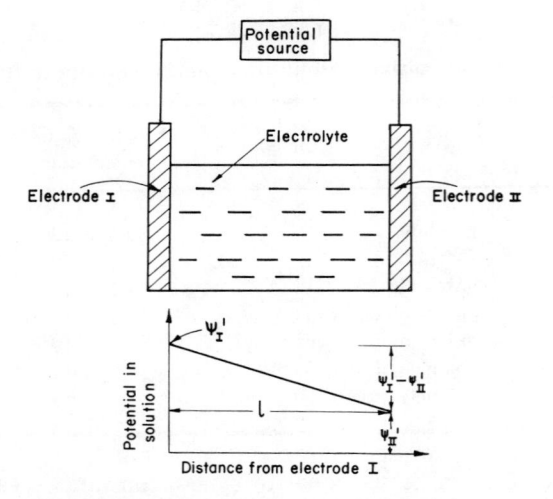

Fig. 4.49. Schematic representation of the variation of the potential in the electrolytic conductor of length *l*.

and, therefore, one has the equation

$$\Delta\psi = RI \qquad (4.130)$$

which re-expresses in the conventional Ohm's law form the assumption of (4.119) concerning flux and driving force.

Thus, an electrolytic conductor obeys Ohm's law for small fields, and, *under steady-state conditions*, it can be represented in an electrical circuit (in which there is only a dc source) by a resistor. (An analog must obey the same equation as the system it represents or simulates.)

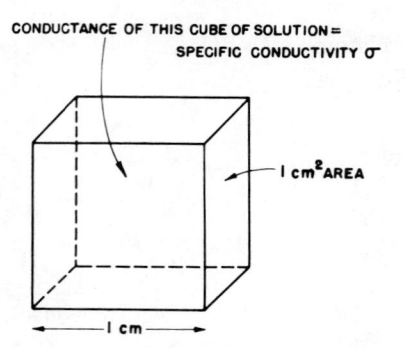

Fig. 4.50. Diagram to illustrate the meaning of the specific conductivity of an electrolyte.

As in the case of a resistor, the dc resistance of an electrolytic cell increases with the length of the conductor (distance between the electrodes) and decreases with the area [cf. Eq. (4.129)]. It can also be seen, by rearranging this equation into the form

$$\sigma = \frac{1}{R} \frac{l}{A} \qquad (4.131)$$

that the specific conductivity σ is the conductance $1/R$ of a cube of electrolytic solution 1 cm long and 1 cm² in area (Fig. 4.50).

4.3.7. Molar Conductivity and Equivalent Conductivity

In the case of metallic conductors, once the specific conductivity is defined, the *macroscopic* description of the conductor is complete. In the case of electrolytic conductors, further characterization is imperative because not only can the *concentration* of charge carriers vary but also the *charge* per charge carrier.

Thus, even though two electrolytic conductors have the same geometry, they need not necessarily have the same specific conductivity (Fig. 4.51 and Table 4.8); the number of charge carriers in that normalized geometry may be different, in which case their fluxes under an applied electric field will be different. Since *the specific conductivity of an electrolytic solution varies as the concentration*, one can write

$$\sigma = f(c) \qquad (4.132)$$

where c is the concentration of the solution in gram-moles of solute dissolved

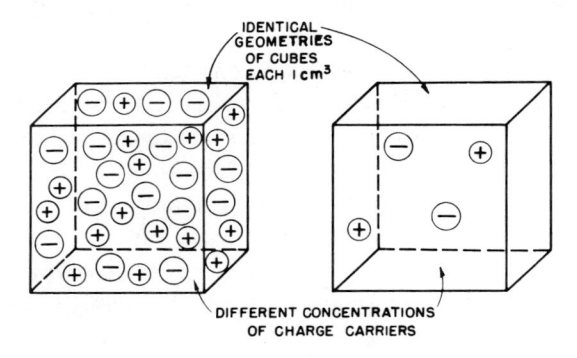

Fig. 4.51. A schematic explanation of the variation of the specific conductivity with electrolyte concentration.

TABLE 4.8

Specific Conductivity of KCl Solutions

KCl, g kg^{-1} of solution	Specific conductivities		
	0 °C	18 °C	25 °C
1.0	0.065144	0.097790	0.11187
0.1	0.0071344	0.0111612	0.012896
0.01	0.00077326	0.00121992	0.001427

in 1 cm³ of solution.[†] The specific conductivities of two solutions can be compared only if they contain the same concentration of ions. The conclusion is that, in order to compare the conductances of electrolytic conductors, one has to normalize (set the variable quantities equal to unity) not only the *geometry* but also the *concentration* of ions.

The normalization of the geometry (taking electrodes of 1 cm² in area and 1 cm apart) defines the specific conductivity; the additional normalization of the concentration (taking 1 mole of ions) defines a new quantity, the molar conductivity (Table 4.9),

$$\Lambda_m = \frac{\sigma}{c} = \sigma V \qquad (4.133)$$

where V is the volume of solution containing 1 g-mole of solute (Fig. 4.52). Defined thus, it can be seen that the molar conductivity is the specific conductivity of a solution times the volume of that solution in which is dissolved 1 g mole of solute; the molar conductivity is a kind of conductivity per particle.

One can usefully compare the molar conductivities of two electrolytic solutions only if the charges borne by the charge carriers in the two solutions are the same. If there are singly charged ions in one electrolyte (e.g., NaCl) and doubly charged ions in the other (e.g., BaSO$_4$), then the two solutions will contain different amounts of charge even though the same quantity of the two electrolytes is dissolved. In such a case, the specific conductivities of the two solutions can be compared only if they contain *equivalent* amounts of charge. This can be arranged by taking 1 *mole of charge* in each case, i.e.,

---

[†] As in the case of diffusion fluxes, the concentrations used in the definition of conduction currents (or fluxes) and conductances are not in the usual moles per liter but in moles per cubic centimeter.

TABLE 4.9
Molar Conductivities of Electrolytes

Electrolyte	Molar conductivity, at 0.01 mole liter^{-1}
KCl	141.3
NaCl	118.5
MgCl$_2$	229.2
Na$_2$SO$_4$	224.8

1 mole of ions divided by z, or 1 g-eq of the substance. Thus, the *equivalent conductivity* Λ of a solution is the specific conductivity of a solution times the volume V of that solution containing 1 g-eq of solute dissolved in it (Fig. 4.53). Hence, the equivalent conductivity is given by[†]

$$\Lambda = \frac{\sigma}{cz} \qquad (4.134)$$

where cz is the number of gram-equivalents per cubic centimeter of solution.

There is a simple relation between the molar and equivalent conductivities. It is [*cf.* Eqs. (4.133) and (4.134)]

$$\Lambda_m = z\Lambda \qquad (4.135)$$

The equivalent conductivities of some electrolytes are shown in Table 4.10.

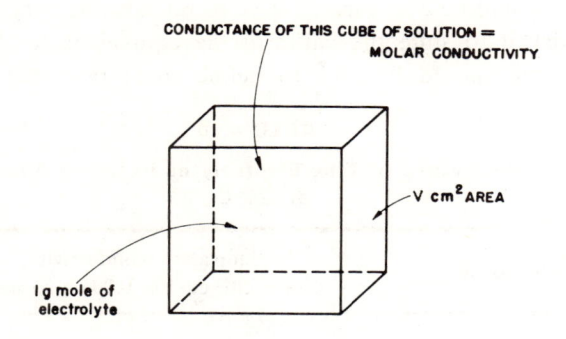

Fig. 4.52. Diagram to illustrate the meaning of the molar conductivity of an electrolyte.

[†] Since $1/z$ mole of ions is 1 g-eq, c moles is cz g-eq.

360 CHAPTER 4

CONDUCTANCE OF THIS CUBE OF SOLUTION = EQUIVALENT CONDUCTIVITY

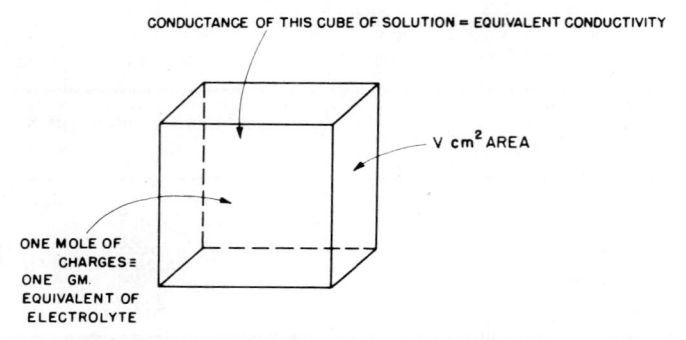

V cm² AREA

ONE MOLE OF CHARGES ≡ ONE GM. EQUIVALENT OF ELECTROLYTE

Fig. 4.53. Diagram to illustrate the meaning of the equivalent conductivity of an electrolyte.

The equivalent conductivity Λ is in the region ($\pm 25\%$) of 100 ohm^{-1} cm^{-1} for most electrolytes.

4.3.8. The Equivalent Conductivity Varies with Concentration

At first sight, the title of this section may appear surprising. The equivalent conductivity has been defined by normalizing the geometry of the system *and* the charge of the ions; why, then, should it vary with concentration? Experiment, however, gives an unexpected answer. The equivalent conductivity does vary significantly with the concentration of ions (Table 4.11). The direction of the variation may also surprise some, for the equivalent conductivity *increases* as the ionic concentration decreases (Fig. 4.54).

Now, it would be awkward to have to refer to the concentration every time one wished to state the value of the equivalent conductivity of an electrolyte. One should be able to define some reference value for the

TABLE 4.10

Equivalent Conductivities of True Electrolytes in Dilute Aqueous Solutions at 25°C

Electrolyte	Equivalent conductivity, ohm^{-1} cm^2 eq^{-1} at 0.005 g-eq liter^{-1}
KCl	143.55
NaOH	240.00
AgNO$_3$	127.20
½ BaCl$_2$	128.02
½ NiSO$_4$	93.20

TABLE 4.11
Equivalent Conductance Varies with Concentration

Concentration, eq liter^{-1} (KCl solutions)	Λ, ohm^{-1} cm^2 eq^{-1}
0.001	146.9
0.005	143.5
0.01	141.2
0.02	138.2
0.05	133.3
0.10	128.9

equivalent conductivity. Here, the facts of the experimental variation of equivalent conductivity with concentration comes to one's aid; as the electrolytic solution is made more dilute, the equivalent conductivity approaches a limiting value. This limiting value of equivalent conductivity should form an excellent basis for the comparison of the conducting powers of different electrolytes for it is the only value in which the effects of ionic concentration are removed. The limiting value shall be called the *equivalent conductivity at infinite dilution*, referred to by the symbol Λ^0 (Table 4.12).

Now, it may be argued: At infinite dilution, there are no ions of the solute; then, how can the solution conduct? The procedure for determining the equivalent conductivity of an electrolyte at infinite dilution will clarify this problem. One takes solutions of a substance in various concentrations, determines the σ, and then normalizes each to the equivalent conductivity of particular solutions. If these values of Λ are then plotted against the

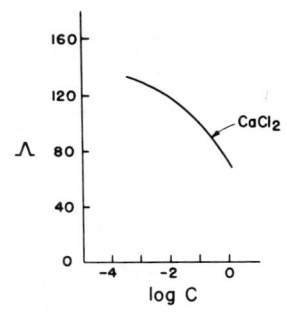

Fig. 4.54. The observed variation of the equivalent conductivity of CaCl$_2$ with concentration.

TABLE 4.12

Equivalent Conductivities at Infinite Dilution, Λ^0, of Electrolytes in Aqueous Solution at 25°C

Electrolyte	Λ^0, ohm^{-1} cm^2 eq^{-1}
HCl	426.0
NaOH	247.9
NaCl	126.4
KCl	149.8
K$_4$Fe(CN)$_6$	183.9
CaCl$_2$	135.7

reciprocals of the concentration and this Λ versus log c curve is *extrapolated*, it approaches a limiting value (Fig. 4.55). It is this *extrapolated* value at zero concentration which is known as the *equivalent conductivity at infinite dilution*.

Anticipating the atomistic treatment of conduction which follows, it may be mentioned that, at very low ionic concentrations, the ions are too far apart to exert appreciable interionic forces on each other. Only under these conditions does one obtain the pristine version of equivalent conductivity, i.e., values unperturbed by ion–ion interactions, which have been shown in Chapter 3 to be concentration dependent. The state of infinite dilution, therefore, is not only the reference state for the study of equilibrium properties (*cf.* Chapter 3.3); it is also the reference state for the study of the nonequilibrium (irreversible) process, which is called *ionic conduction*, or *migration* (*cf.* Section 4.1).

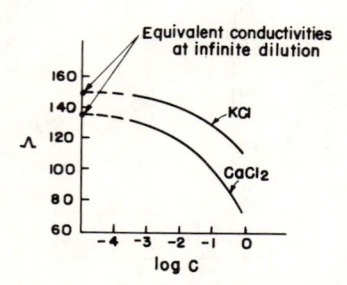

Fig. 4.55. The equivalent conductivity of an electrolyte at infinite dilution is obtained by extrapolating the Λ versus log c curve to zero concentration.

4.3.9. How the Equivalent Conductivity Changes with Concentration: Kohlrausch's Law

The experimental relationship between the equivalent conductivity and the concentration of an electrolytic solution is best brought out by plotting Λ against $c^{\frac{1}{2}}$. When this is done (Fig. 4.56), it can be seen that, up to concentrations of about $0.01N$, there is a linear relationship between Λ and $c^{\frac{1}{2}}$; thus,

$$\Lambda = \Lambda^0 - Ac^{\frac{1}{2}} \qquad (4.136)$$

where the intercept is the equivalent conductivity at infinite dilution, Λ^0, and the slope of the straight line is a positive constant A.

This *empirical* relationship between the equivalent conductivity and the square root of concentration is a law named after Kohlrausch. His extremely careful measurements of the conductance of electrolytic solutions can be considered to have played a leading role in the initiation of ionics, the physical chemistry of ionic solutions. However, Kohlrausch's law [Eq. (4.136)] had to remain nearly 40 years without a theoretical basis.

The justification of Kohlrausch's law on theoretical grounds cannot be obtained within the framework of a macroscopic description of conduction. It requires an intimate view of ions in motion. A clue to the type of theory required emerges from the empirical findings by Kohlrausch: (1) the $c^{\frac{1}{2}}$ dependence, and (2) the intercepts Λ^0 and slopes A of the Λ versus $c^{\frac{1}{2}}$ curve depend not so much on the particular electrolyte (whether it is KCl or NaCl) as on the type of electrolyte (whether it is a 1:1 or 2:2 electrolyte) (Fig. 4.57). All this is reminiscent of the dependence of the activity coefficient on $c^{\frac{1}{2}}$ (Chapter 3), to explain which, the subtleties of

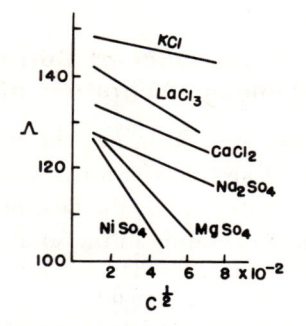

Fig. 4.56. The experimental basis for Kohlrausch's law, Λ versus $c^{\frac{1}{2}}$ plots, consist of straight lines.

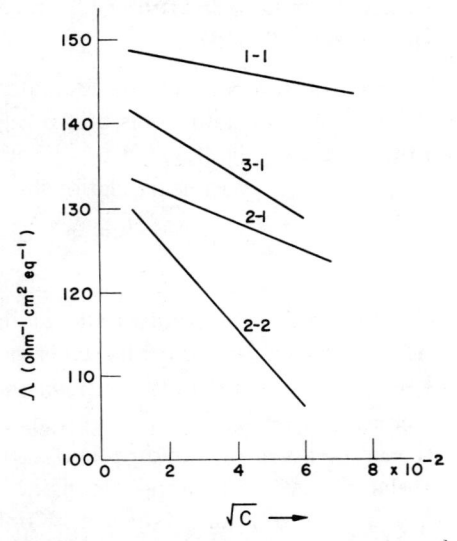

Fig. 4.57. The experimental Λ versus $c^{\frac{1}{2}}$ plots depend largely on the type of electrolyte.

ion–ion interactions had to be explored. Such interactions between positive and negative ions would determine to what extent they would influence each other when they move, and this would in turn bring about a fall of conductivity.

Kohlrausch's law will therefore be left now with only the sanction of experiment. Its incorporation into a theoretical scheme will be postponed until the section on the atomistic view of conduction is reached (see Section 4.6.12).

4.3.10. The Vectorial Character of Current: Kohlrausch's Law of the Independent Migration of Ions

The driving force for ionic drift, i.e., the electric field X, not only has a particular magnitude, it also acts in a particular direction. It is a *vector*. Since the ionic current density i, i.e., the flow of electric charge, is proportional to the electric field operating in the solution [see Eq. (4.125)],

$$i = \sigma X \qquad (4.125)$$

the ionic current density must also be a vector. Vectorial quantities are often designated by arrows placed over the quantities (unless their directed

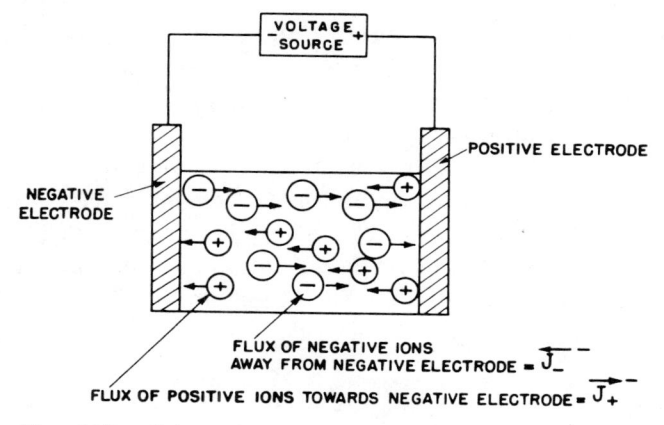

Fig. 4.58. Schematic representation of the direction of the drifts (and fluxes) of positive and negative ions acted on by an electric field.

character is obvious). Hence, Eq. (4.125) can be written thus

$$\vec{i} = \sigma \vec{X} \tag{4.137}$$

How is this current density constituted? What are its components? What is the structure of this ionic current density?

The imposition of an electric field on the electrolyte (Fig. 4.58) makes the positive ions drift toward the negative electrode and the negative ions drift in the opposite direction. The flux of positive ions, $\vec{J}_+$, gives rise to a positive-ion current density i_+; and the flux of negative ions in the opposite direction, $\overleftarrow{J}_-$, results in a negative-ion current density i_-. By *convention* the direction of current flow is taken to be either the direction in which positive charge flows or the direction opposite to that in which negative charge flows. Hence, the positive-ion flux $\vec{J}_+$ corresponds to a current *toward* the negative electrode, $\vec{i}_+$, and the negative-ion flux $\overleftarrow{J}_-$ also corresponds to a current $\vec{i}_-$ *in the same direction* as that due to the positive ions.

It can be concluded therefore that the total current density $\vec{i}$ is made up of two contributions, one due to a flux of positive ions and the other due to a flux of negative ions. Further, *assuming* for the moment that the drift of positive ions toward the negative electrode does not interfere with the drift of negative ions in the opposite direction, it follows that the component current densities are additive, i.e.,

$$\vec{i} = \vec{i}_+ + \vec{i}_- \tag{4.138}$$

But *do ions migrate independently*? Is the drift of the positive ions in one direction uninfluenced by the drift of the negative ions in the opposite direction? This is so if, and only if, the force fields of the ions do not overlap significantly, i.e., if there is negligible interaction or coupling between the ions. Coulombic ion–ion interactions usually establish such coupling. The only conditions under which the *absence* of ion–ion interactions can be assumed occur when the ions are infinitely far apart. Strictly speaking, therefore, ions migrate independently only at infinite dilution. Under these conditions, one can proceed from Eq. (4.138) to write

$$\frac{\vec{i}}{\vec{X}} = \frac{\vec{i}_+}{\vec{X}} + \frac{\vec{i}_-}{\vec{X}} \tag{4.139}$$

or [from (4.125)],

$$\sigma = \sigma_+ + \sigma_- \tag{4.140}$$

whence

$$\Lambda^0 = \lambda_+^0 + \lambda_-^0 \tag{4.141}$$

This is Kohlrausch's law of the independent migration of ions: The equivalent conductivity (*at infinite dilution*) of an electrolytic solution is the sum of the equivalent conductivities (at infinite dilution) of the ions constituting the electrolyte (Table 4.13).

At appreciable concentrations, the ions can be regarded as coupled or interacting with each other (*cf.* the ion-atmosphere model of Chapter 3). This results in the drift of positive ions toward the negative electrode, which hinders the drift of negative ions toward the positive electrode, i.e., the interionic interaction results in the positive ions' equivalent conductivity reducing the magnitude of the negative ions' equivalent conductivity to

TABLE 4.13

Equivalent Conductances of Individual Ions at Infinite Dilution ohm^{-1} cm^2 at 25°C

Cation	λ_+^0	Anion	λ_-^0
H$^+$	349.82	OH$^-$	198.5
K$^+$	73.52	Br$^-$	78.4
Na$^+$	50.11	I$^-$	76.8
Li$^+$	38.69	Cl$^-$	76.34
½ Ba^{++}	63.64	CH$_3$CO$_2^-$	40.9

below the infinite dilution value, and *vice versa*. To make quantitative estimates of these effects, however, one must make calculations of the influence of ionic-cloud effects on the phenomenon of conduction, a task which will be taken up further on.

Further Reading

1. S. Glasstone, *An Introduction to Electrochemistry*, D. van Nostrand Co., Inc., Princeton, N.J., 1949.
2. R. M. Fuoss and F. Accasina, *Electrolytic Conductance*, Interscience Publishers, Inc., New York, 1959.

4.4. THE SIMPLE ATOMISTIC PICTURE OF IONIC MIGRATION

4.4.1. Ionic Movements under the Influence of an Applied Electric Field

In seeking an atomic view of the process of conduction, one approach is to begin with the picture of ionic movements as described in the treatment of diffusion (Section 4.2.4) and then to consider how these movements are perturbed by an electric field. In the treatment of ionic movements, it was stated that the ions in solution perform a random walk in which all possible directions are equally likely for any particular step. The analysis of such a random walk indicated that the mean displacement of ions is zero (Section 4.2.4), diffusion being the result of the statistical bias in the movement of ions, due to inequalities in their numbers in different regions.

When, however, the ions are situated in an electric field, their movements are affected by the fact that they are charged. Hence, the imposition of an electric field singles out one direction in space (the direction parallel to the field) for preferential ionic movement. Positively charged particles will prefer to move toward the negative electrode; and negatively charged particles, in the opposite direction. The walk is no longer quite random. The ions drift.

Another way of looking at ionic drift is to consider the fate of any particular ion under the field. The electric force field would impart to it an acceleration according to Newton's second law. Were the ion completely isolated (e.g., in vacuum), it would accelerate indefinitely until it collided with the electrode. In an electrolytic solution, however, the ion very soon collides with some other ion or solvent molecule which crosses its path. This collision introduces a discontinuity in its speed and direction. The motion of the ion is not smooth; it is as if the medium offers resistance to the motion of the ion. Thus, the ion stops and starts and zig-zags. But the applied electric field imparts to the ion a direction (that of the oppositely charged electrode), and the ion works its way, though erratically, in the direction of this electrode. The ion *drifts* in a *preferred* direction.

4.4.2. What Is the Average Value of the Drift Velocity?

Any particular ion starts off after a collision with a velocity which may be in any direction; this is the randomness in its walk. This initial velocity can be ignored precisely because it can take place in any direction and, therefore, does not contribute to the drift (preferred motion) of the ion. But the ion is all the time under the influence of the applied-force field.[†] This force imparts a component to the velocity of the ion, an *extra* velocity component in the same direction as the force vector $\vec{F}$. It is this additional velocity component due to the force $\vec{F}$ which is called the *drift velocity* v_d. What is its average value?

From Newton's second law, it is known that the force divided by the mass of the particle is equal to the acceleration. Thus,

$$\frac{\vec{F}}{m} = \frac{dv}{dt} \tag{4.142}$$

Now, the time between collisions is a random quantity. Sometimes, the collisions may occur in rapid succession; at others, there may be fairly long intervals. It is possible however to talk of a mean time between collisions, τ. In Section 4.2.5, it has been shown that the number of collisions (steps) is proportional to the time. If N collisions occur in a time t, then the average time between collisions is t/N. Hence,

$$\tau = \frac{t}{N} \tag{4.143}$$

The average value of that component of the velocity of an ion picked up from the externally applied force is simply the product of the acceleration due to this force and the average time between collisions. Hence, the drift velocity v_d is given by

$$v_d = \frac{dv}{dt} \tau$$

$$= \frac{\vec{F}}{m} \tau \tag{4.144}$$

[†] The argument is developed in general for any force, not necessarily an electric force.

This is an important relation. It opens up many vistas. For example, through the mean time τ, one can relate the drift velocity to the details of ionic jumps between sites, as was done in the case of diffusion (Section 4.2.15).

Further, the relation (4.144) shows that the drift velocity is *proportional* to the driving force of the electric field. The flux of ions will be shown (Section 4.4.4) to be simply related to the drift velocity[†] in the following way

$$\text{Flux} = \text{Concentration of ions} \times \text{Drift velocity} \quad (4.145)$$

Thus, if the $\vec{F}$ in Eq. (4.144) is the electric force which stimulates conduction, then this equation is the molecular basis of the fundamental relation used in the macroscopic view of conduction, i.e.,

$$\text{Flux} \propto \text{Electric field} \quad (4.119)$$

In fact, the derivation of the basic relation (4.144) reveals the conditions under which the proportionality between drift velocity (or flux) and electric field breaks down. Thus, it was essential to the derivation that, in a collision, an ion does not preserve any part of its extra velocity component arising from the force field. If it did, then the actual drift velocity would be greater than that calculated by Eq. (4.144) because there would be a cumulative carry-over of the extra velocity from collision to collision. In other words, every collision must wipe out all traces of the force-derived extra velocity, and the ion must start afresh to acquire the additional velocity. This condition can be satisfied only if the drift velocity and, therefore, the field are small.

4.4.3. The Mobility of Ions

It has been shown that, when random-walking ions are subjected to a directed force $\vec{F}$, they acquire a nonrandom, directed component of velocity—the drift velocity v_d. This drift velocity is in the direction of the force $\vec{F}$ and is proportional to it

$$v_d = \frac{\tau}{m}\vec{F} \quad (4.144)$$

[†] The dimensions of flux are in moles per square centimeter per second, and they are equal to the product of the dimensions of concentration expressed in moles per cubic centimeter and velocity expressed in centimeters per second.

Since the proportionality constant τ/m is of considerable importance in discussions of ionic transport, it is useful to refer to it with a special name. It is called *the absolute mobility* because it is an index of how mobile the ions are. The absolute mobility, referred to by the symbol $\bar{u}_{\mathrm{abs}}$, is a measure of the drift velocity v_d acquired by an ionic species when it is subjected to a force $\vec{F}$, i.e.,

$$\bar{u}_{\mathrm{abs}} = \frac{\tau}{m} = \frac{v_d}{\vec{F}} \qquad (4.146)$$

which means that the absolute mobility is the drift velocity developed under unit applied force ($\vec{F} = 1$ dyne) and has its dimensions in centimeters per second per dyne.

For example, one might have an electric field X of 0.05 V cm^{-1} in the electrolyte solution and observe a drift velocity of 2×10^{-5} cm sec^{-1}. The electric force $\vec{F}$ operating on the ion is equal to the electric force per unit charge, i.e., the electric field X, times the charge $z_i e_0$ on each ion

$$\vec{F} = z_i e_0 X \frac{1}{300}$$

$$= 4.8 \times 10^{-10} \times 0.05 \times \frac{1}{300} \text{ dynes} \qquad (4.147)$$

for univalent ions. (The factor 1/300 comes in because the electronic charge is in electrostatic units (esu) and the potential involved in the field is not in electrostatic units of potential, but in volts. Hence, the potential also must be expressed in electrostatic units, the conversion factor being 1/300, i.e., 300 volts are equivalent to 1 esu of potential.) Hence, the absolute mobility is

$$\bar{u}_{\mathrm{abs}} = \frac{2 \times 10^{-5} \times 300}{4.8 \times 10^{-10} \times 0.05}$$

$$\simeq 10^8 \text{ cm sec}^{-1} \text{ dyne}^{-1}$$

In electrochemical literature, however, mobilities of ions are not usually expressed in the absolute form defined in Eq. (4.146). Instead, they have been defined as the drift velocities under the force exerted by *unit electric field* (1 V cm^{-1}) on the charge of the ion and will be referred to here as *conventional (electrochemical) mobilities* with the symbol u_{conv}. Thus,

$$u_{\mathrm{conv}} = \frac{v_d}{X} \left[\frac{\text{cm sec}^{-1}}{\text{V cm}^{-1}} \right] \qquad (4.148)$$

The relation between the absolute and conventional mobilities follows thus from Eqs. (4.146) and (4.148)

$$u_{\text{conv}} = \frac{v_d}{X}$$

$$= \frac{v_d}{\vec{F}} \frac{z_i e_0}{300}$$

i.e.,

$$u_{\text{conv}} = \bar{u}_{\text{abs}} z_i e_0 \frac{1}{300} \quad (4.149)$$

Thus, the conventional and absolute mobilities are simply proportional to each other, the proportionality constant being an integral multiple z_i of the electronic charge divided by 300. In the example cited above,

$$u_{\text{conv}} = \frac{2 \times 10^{-5}}{0.05}$$

$$\sim \frac{10^{-4} \text{ cm sec}^{-1}}{1 \text{ V cm}^{-1}}$$

Though the two types of mobilities are closely related, it must be stressed that the concept of absolute mobility is more general because it can be used for *any* force which determines the drift velocity of ions and not only the electric force used in the definition of conventional mobilities.

4.4.4. The Current Density Associated with the Directed Movement of Ions in Solution, in Terms of the Ionic Drift Velocities

It is the aim now to show how the concept of drift velocity can be used to obtain an expression for the ionic current density flowing through an electrolyte in response to an externally applied electric field.

Consider a transit plane of unit area normal to the direction of drift (Fig. 4.59). Both the positive and the negative ions will drift across this plane. Consider the positive ions first, and let their drift velocity be $(v_d)_+$, or simply v_+. Then, in 1 sec, all positive ions within a distance v_+ cm of the transit plane will cross it. The flux J_+ of positive ions, i.e., the number of moles of these ions arriving in 1 sec at unit area of the plane is equal to the number of moles of positive ions in a volume 1 cm² in area and v_+ cm

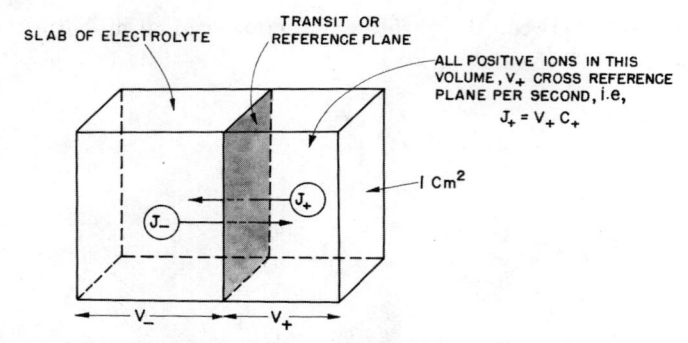

Fig. 4.59. Diagram for the derivation of a relation between the current density and the drift velocity.

in length. Hence, J_+ is equal to the volume v_+ in cubic centimeters times the concentration c_+ expressed† in moles per cubic centimeter

$$J_+ = c_+ v_+ \tag{4.150}$$

The flow of charge across the plane due to this flux of positive ions, i.e., the current density i_+ is obtained by multiplying the flux J_+ by the charge $z_+ F$ borne by 1 mole of ions

$$i_+ = z_+ F c_+ v_+ \tag{4.151}$$

This, however, is only the contribution of the positive ions. Other ionic species will make their own contributions of current density. In general, therefore, the current density due to the *j*th species will be

$$i_j = z_j F c_j v_j \tag{4.152}$$

The total current density due to the contribution of all the ionic species will therefore be

$$i = \sum_j i_j \tag{4.153}$$

$$= \sum_j z_j F c_j v_j \tag{4.154}$$

---

† Since one is concerned with the number of moles in a volume of v cm³, one must not express the concentration in moles per *liter* but in moles per cubic centimeter. The two concentrations, however, are simply related as

$$c \text{ [moles cm}^{-3}] = \frac{c}{1000} \text{ [moles liter}^{-1}]$$

If a $z:z$-valent electrolyte is taken, then $z_+ = z_- = z$ and $c = c_+ = c_-$ and one has

$$i = zFc(v_+ + v_-) \tag{4.155}$$

By recalling that the ionic drift velocities are related, through the force operating on the ions, to the ionic mobilities [Eq. (4.148)], it will be realized that Eq. (4.154) is the basic expression from which may be derived the expressions for conductance, equivalent conductivity, specific conductivity, etc.

4.4.5. The Specific and Equivalent Conductivities in Terms of the Ionic Mobilities

Let the fundamental expression [Eq. (4.148)] for the drift velocity of ions be substituted in Eq. (4.154) for current density. One obtains

$$i = \sum z_j Fc_j (u_{\text{conv}})_j X \tag{4.156}$$

or, from (4.125),

$$\sigma = \frac{i}{X} = \sum z_j Fc_j (u_{\text{conv}})_j \tag{4.157}$$

which reduces in the special case of a $z:z$-valent electrolyte to

$$\sigma = zFc[(u_{\text{conv}})_+ + (u_{\text{conv}})_-] \tag{4.158}$$

Several conclusions follow from this atomistic expression for specific conductivity. Firstly, it is obvious from this equation that the specific conductivity σ of an electrolyte cannot be a concentration-independent constant (as it is in the case of metals). It will vary because the number of moles of ions per unit volume, c, can be varied in an electrolytic solution.

Secondly, the specific conductivity can easily be related to the molar, Λ_m, and equivalent, Λ, conductivities. Take the case of a $z:z$-valent electrolyte. With Eqs. (4.158), (4.133), and (4.134), it is found that

$$\Lambda_m = \frac{\sigma}{c}$$

$$= zF[(u_{\text{conv}})_+ + (u_{\text{conv}})_-] \tag{4.159}$$

and

$$\Lambda = \frac{\Lambda_m}{z}$$

$$= F[(u_{\text{conv}})_+ + (u_{\text{conv}})_+] \tag{4.160}$$

What does Eq. (4.160) reveal? It shows that the equivalent conductivity will be a constant independent of concentration only if the electrical mobility does not vary with concentration. It will be seen however that ion–ion interactions (which have been shown in Section 3.3.8 to depend on concentration) prevent the electrical mobility from being a constant. Hence, the equivalent conductivity must be a function of concentration.

4.4.6. The Einstein Relation between the Absolute Mobility and the Diffusion Coefficient

The process of diffusion results from the random walk of ions; the process of migration (i.e., conduction) results from the drift velocity acquired by ions when they experience a force. Now, the drift of ions does not obviate their random walk; in fact, it is superimposed on their random walk. Hence, the drift and the random walk must be intimately linked. Einstein realized this and deduced a vital relation between the absolute mobility $\bar{u}_{\text{abs}}$, which is a quantitative characteristic of the drift, and the diffusion coefficient D, which is a quantitative characteristic of the random walk.

Both diffusion and conduction are nonequilibrium (irreversible) processes and are therefore not amenable to the methods of equilibrium thermodynamics or equilibrium statistical mechanics. In these latter disciplines, the concepts of time and change are absent. It is possible however to imagine a situation where the two processes oppose and balance each other and a "pseudoequilibrium" obtains. This is done as follows (Fig. 4.60).

Consider a solution of an electrolyte MX to which a certain amount

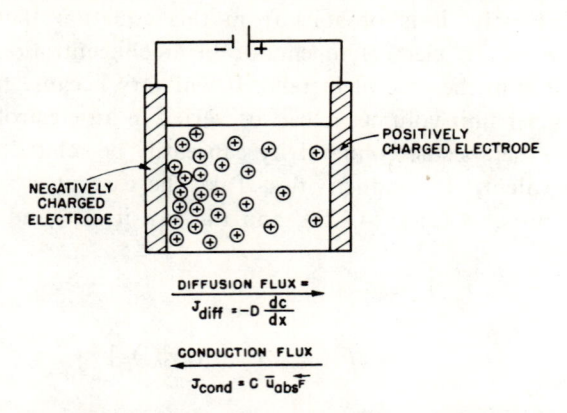

Fig. 4.60. An imaginary situation in which the applied field is adjusted so that the conduction flux of tagged ions (the only ones shown in the diagram) is exactly equal and opposite to the diffusion flux.

of radioactive M⁺ ions are added in the form of the salt MX. Further, suppose that the tracer ions are not dispersed uniformly throughout the solution; instead, let there be a concentration gradient of the tagged species such that its diffusion flux J_D is given by Fick's first law[†]

$$J_D = -D\frac{dc}{dx} \tag{4.16}$$

Now let an electric field be applied. Each tagged ion feels the field, and the drift velocity is [cf. Eq. (4.146)]

$$v_d = \bar{u}_{\text{abs}}\vec{F} \tag{4.146}$$

This drift velocity produces a current density given by[‡] [cf. Eq. (4.151)]

$$i = zFcv_d \tag{4.161}$$

i.e., a conduction flux J_C which is arrived at by dividing the conduction current density i by the charge per mole of ions

$$J_C = \frac{i}{zF} = cv_d \tag{4.162}$$

By introducing the expression (4.146) for the drift velocity into (4.162), the conduction flux becomes

$$J_C = c\bar{u}_{\text{abs}}\vec{F}$$

Now, let the applied field be imagined to be adjusted so that the conduction flux exactly compensates the diffusion flux. In other words, if the tracer ions (which are positively charged) are diffusing toward the positive electrode, then the magnitude of the applied field is such that the positively charged electrode repels the positive tracer ions to an extent that their net flux is zero. Thus,

$$J_D + J_C = 0 \tag{4.163}$$

or

$$-D\frac{dc}{dx} + c\bar{u}_{\text{abs}}\vec{F} = 0$$

i.e.,

$$\frac{dc}{dx} = \frac{c\bar{u}_{\text{abs}}\vec{F}}{D} \tag{4.164}$$

[†] All the c's in this derivation are in moles per cubic centimeter.
[‡] In Eq. (4.151), one will find v_+; the reason is that, in Section 4.4.4, the drift velocity of a positive ion, $(v_d)_+$, had been concisely written as v_+.

Under these "balanced" conditions, the situation may be regarded as tantamount to equilibrium because there is no net flux or transport of ions. Hence, the Boltzmann law can be used. The argument is that, since the potential varies along the x direction, the concentrations of ions at any distance x is given by

$$c = c^0 e^{-U/kT} \tag{4.165}$$

where U is the potential energy of an ion in the applied field and c^0 is the concentration in a region where the potential energy is zero. Differentiating this expression, one obtains

$$\frac{dc}{dx} = -c^0 e^{-U/kT} \frac{1}{kT} \frac{dU}{dx}$$

$$= -\frac{c}{kT} \frac{dU}{dx} \tag{4.166}$$

But, by the definition of force,

$$\vec{F} = -\frac{dU}{dx} \tag{4.167}$$

Hence, from (4.166) and (4.167), one obtains

$$\frac{dc}{dx} = \frac{c}{kT} \vec{F} \tag{4.168}$$

If, now, Eqs. (4.164) and (4.168) are compared, it is obvious that

$$\frac{\bar{u}_{\text{abs}}}{D} = \frac{1}{kT}$$

or

$$D = \bar{u}_{\text{abs}} kT \tag{4.169}$$

This is the Einstein relation. It is probably the most important relation in the theory of the movements and drift of ions, atoms, molecules, and other submicroscopic particles. It has been derived here in an atomistic way. But it will be recalled that, in the phenomenological treatment of the diffusion coefficient (Section 4.2.3), it was shown that

$$D = BRT \tag{4.17}$$

where B was an undetermined phenomenological coefficient. Now, if one combines (4.169) and (4.17), it is clear that

$$BRT = \bar{u}_{abs}kT$$

or

$$B = \frac{\bar{u}_{abs}kT}{RT} = \frac{\bar{u}_{abs}}{N_A} \qquad (4.170)$$

Thus, one has provided a fundamental basis for the phenomenological coefficient B; it is the absolute mobility $\bar{u}_{abs}$ divided by the Avogadro number.

The Einstein relation also permits experiments on diffusion to be linked up with other phenomena involving the mobility of ions, i.e., phenomena in which there are forces that produce drift velocities. Two such forces are the force experienced by an ion when it overcomes the viscous drag of a solution and the force arising from an applied electric field. Thus, the diffusion coefficient may be linked up to the viscosity (the Stokes–Einstein relation) and to the equivalent conductivity (the Nernst–Einstein relation).

4.4.7. What Is the Drag (or Viscous) Force Acting on an Ion in Solution?

Striking advances in science sometimes arise from seeing the common factors in two apparently dissimilar situations. One such advance was made by Einstein when he intuitively asserted the similarity between a macroscopic sphere moving in an incompressible fluid and a particle (e.g., an ion) moving in a solution (Fig. 4.61).

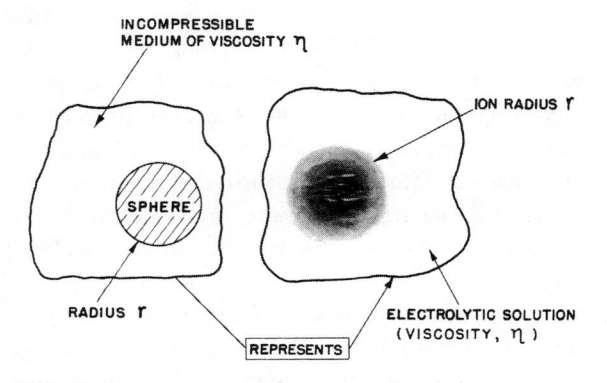

Fig. 4.61. An ion drifting in an electrolytic solution is likened to a sphere (of the same radius as the ion) moving in an incompressible medium (of the same viscosity as the electrolyte).

The macroscopic sphere experiences a viscous, or drag, force which opposes its motion. The value of the drag force depends on several factors—the velocity v and diameter d of the sphere and the viscosity η and density ϱ of the medium. These factors can all be combined together and used to define a dimensionless quantity known as the *Reynolds number* Re defined thus

$$\text{Re} = vd\frac{\varrho}{\eta} \qquad (4.171)$$

When the hydrodynamic conditions are such that this Reynolds number is much smaller than unity, Stokes showed that the drag force F opposing the sphere is given by the following relation

$$F = 6\pi r \eta v \qquad (4.172)$$

where v is the velocity of the macroscopic body. The relation is known as Stokes' law. Its derivation is lengthy and awkward, essentially because the most convenient coordinates to describe the sphere and its environment are spherical coordinates and those to describe the flow are rectangular coordinates.

The real question, of course, centers around the applicability of Stokes' law to microscopic ions moving in a structured medium in which the surrounding particles are roughly the same size as the ions. Initially, one can easily check on whether the Reynolds number *is* smaller than unity for ions drifting through an electrolyte. With the use of the values $d_{\text{ion}} \sim 10^{-8}$ cm, $v_{\text{ion}} = v_d \sim 10^{-4}$ cm sec^{-1}, $\eta \sim 10^{-2}$ poise, and $\varrho \simeq 1$, it turns out that the Reynolds number for an ion moving through an electrolyte is about 10^{-10}. Thus, the hydrodynamic condition Re $\ll 1$ required for the validity of Stokes' law is easily satisfied by an ion in solution.

But the hydrodynamic problem which Stokes solved to get $F = 6\pi r \eta v$ pertains to a sphere moving in an incompressible *continuum* fluid. This is a far cry indeed from the actuality of an ion drifting inside a discontinuous electrolyte containing particles (solvent molecules, other ions, etc.) of about the same size as the ion. Further, the ions considered may not be spherical.

From this point of view, the use of Stokes' law for the viscous force experienced by ions is a bold step. Several attempts have been made over a long time to theorize about the viscous drag on angstrom-sized particles in terms of a more realistic model than that used by Stokes. It has been shown, for example, that, if the moving particle is cylindrical and not spherical, the factor 6π should be replaced by 4π. While refraining from the none-too-easy analysis of the degree of applicability of Stokes' law to ions in electrolytes, one point must be stressed. For sufficiently small ions, Stokes' law does not have a numerical significance[†] of greater than about ±50%. Attempts to tackle the problem of the flow of ions in solution without resorting to Stokes' law do not give much better results.

4.4.8. The Stokes–Einstein Relation

During the course of diffusion, the individual particles are executing the complicated starts, accelerations, collisions, stops, and zig-zags that have come to be known as random walk. When a particle is engaged in its random walk, it is, of course, subject to the viscous-drag force exerted by its environment. But the application of Stokes' law to these detailed random motions is no easy matter because of the haphazard variation in the speed and direction of the particles. Instead, one can apply Stokes' law to the diffusional movements of ions by adopting the following artifice suggested by Einstein.

When diffusion is occurring, it can be considered that there is a driving force $-d\mu/dx$ operating on the particles. This driving force produces a steady-state diffusion flux J, corresponding to which [cf. Eq. (4.14)] one

---

[†] Stokes' law is often used in electrochemical problems, but its approximate nature is not always brought out. Apart from the validity of extrapolating from the macroscopic-sphere–continuum-fluid model of Stokes' to atomic near-spheres in a molecular liquid, another reason for the limited validity of Stokes' law arises from questions concerning the radii which should be substituted in any application of the law. These should not be the crystallographic radii, and an appraisal of the correct value implies a rather detailed knowledge of the structure of the solvation sheath (cf. Section 2.4). Further, the viscosity used in Stokes' law is the bulk average viscosity of the whole solution, whereas it is the local viscosity in the neighborhood of the ion which should be taken. The two may not be the same, because the ion's field may affect the solvent structure and hence its viscosity.

In recent years, attention has been drawn by Fuoss, Boyd, and Zwanzig to the fact that, while Stokes' law deals with uncharged spheres, the real situation involves charged ions. The existence of a charge on the moving body has the following effect on a polar solvent: The charge tends to produce an orientation of solvent dipoles in the vicinity of the ion. Since, however, the charge is moving, the dipoles, once oriented, take some finite relaxation time τ to disorient. During this relaxation time, a relaxation force operates on the ion; this relaxation force is equivalent to an additional frictional force on the ion and results in an expression for the drag force of the form

$$F = 6\pi\eta v r + 6\pi\eta v \frac{s}{\varepsilon}$$

where s is $4/9(\tau/6\pi\eta)e_0^2/r^3$, and ε is the dielectric constant of the medium.

can imagine a drift velocity v_d for the diffusing particles.[†] Since this velocity v_d is a steady-state velocity, the diffusional driving force $-d\mu/dx$ must be opposed by an equal resistive force which shall be taken to be the Stokes viscous force $6\pi r \eta v_d$. Hence,

$$-\frac{d\mu}{dx} = 6\pi r \eta v_d \qquad (4.173)$$

One can therefore define the absolute mobility $\bar{u}_{abs}$ for the diffusing particles by dividing the drift velocity by either the diffusional driving force or the equal and opposite Stokes viscous force

$$\bar{u}_{abs} = \frac{v_d}{-d\mu/dx} = \frac{v_d}{6\pi r \eta v_d} = \frac{1}{6\pi r \eta} \qquad (4.174)$$

Now, the fundamental expression (4.169) relating the diffusion coefficient and the absolute mobility can be written thus

$$\bar{u}_{abs} = \frac{D}{kT} \qquad (4.175)$$

By equating (4.174) and (4.175), the Stokes–Einstein relation is obtained

$$D = \frac{kT}{6\pi r \eta} \qquad (4.176)$$

It links the processes of diffusion and viscous flow.

The Stokes–Einstein relation proved extremely useful in the classical work of Perrin. Using an ultramicroscope, he watched the random walk of a colloidal particle, and, from the mean square distance $\langle x^2 \rangle$ traveled in a time t, he obtained the diffusion coefficient D from the relation (4.27)

$$D = \frac{\langle x^2 \rangle}{2t} \qquad (4.27)$$

The weight of the colloidal particles and their density being known, their radius r was then obtained. Then, the viscosity η of the medium can be used to obtain the Boltzmann constant

$$k = \frac{6\pi r \eta D}{T} \qquad (4.177)$$

[†] The hypothetical nature of the argument lies in the fact that, in diffusion, there is no actual force exerted on the particles. Consequently, there is not the actual force-derived component of the velocity; i.e., there is no actual drift velocity (see Section 4.2.1). Thus, the drift velocity enters the argument only as a device.

But

$$k = \frac{R}{N_A} \tag{4.178}$$

or

$$N_A = \frac{R}{k} \tag{4.179}$$

and, thus, the Avogadro number can be determined.

The use of Stokes' law also permits the derivation of a very simple relation between the viscosity of a medium and the conventional electrochemical mobility u_{conv}. Starting from the earlier derived equation

$$\bar{u}_{\text{abs}} = \frac{1}{6\pi r \eta} \tag{4.174}$$

one substitutes for the absolute mobility the expression from (4.149)

$$\bar{u}_{\text{abs}} = \frac{300 u_{\text{conv}}}{z_i e_0} \tag{4.149}$$

and gets the result

$$u_{\text{conv}} = \frac{z_i e_0}{1800 \pi r \eta} \tag{4.180}$$

This relation shows that, owing to the Stokes viscous force, the conventional mobility of an ion depends on the charge and radius[†] of the solvated ion and the viscosity of a medium. The mobility given by Eq. (4.180) is often called the *Stokes mobility*. It will be seen later that the Stokes mobility is a highly simplified expression for the mobility and ion–ion interaction effects introduce a concentration dependence which is not seen in Eq. (4.180).

4.4.9. The Nernst–Einstein Equation

Now the Einstein relation (4.169) will be used to connect the transport processes of diffusion and conduction.

The starting point is the basic equation relating the equivalent conductivity of a $z:z$-valent electrolyte to the conventional mobilities of the ions, i.e., to the drift velocities under a potential gradient of 1 V cm^{-1},

$$\Lambda = F(u_{\text{conv},+} + u_{\text{conv},-}) \tag{4.160}$$

[†] Earlier, the radius dependence of the conventional mobility was used to obtain information of the solvation number (*cf.* Section 2.4.5).

CHAPTER 4

By using the relation between the conventional and absolute mobilities, Eq. (4.160) can be written[†]

$$\Lambda = z_i e_0 F(\bar{u}_{\text{abs},+} + \bar{u}_{\text{abs},-}) \tag{4.181}$$

With the aid of the Einstein relation (4.169),

$$\bar{u}_{\text{abs},+} = \frac{D_+}{kT} \quad \text{and} \quad \bar{u}_{\text{abs},-} = \frac{D_-}{kT} \tag{4.182}$$

one can transform Eq. (4.181) into the form

$$\Lambda = \frac{ze_0 F}{kT}(D_+ + D_-) \tag{4.183}$$

This is one form of the Nernst–Einstein equation; from a knowledge of the diffusion coefficients of the individual ions, it permits a calculation of the equivalent conductivity.

A more usual form of the Nernst–Einstein equation is obtained by multiplying numerator and denominator by the Avogadro number, in which case, it is obvious that

$$\Lambda = \frac{zF^2}{RT}(D_+ + D_-) \tag{4.184}$$

4.4.10. Some Limitations of the Nernst–Einstein Relation[‡]

There were several aspects of the Stokes–Einstein relation which reduced it to being only an approximate relation between the diffusion coefficient of an ionic species and the viscosity of the medium. In addition, there were fundamental questions regarding the extrapolation of a law derived for macroscopic spheres moving in an incompressible medium to a situation involving the movement of ions in an environment of solvent molecules and other ions. In the Nernst–Einstein relation, the factors which limit its validity are more subtle.

An implicit but principal requirement for the Nernst–Einstein equation to hold is that the species involved in diffusion must also be the species responsible for conduction. Suppose now that the species M exists not only as ions M^{z+} but also as ion pairs $M^{z+}A^{z-}$ of the type described in Section 3.8.1.

[†] For simplicity, the conversion factor (1/300) of Eq. (4.149) has been omitted here. It must be introduced, however, when one makes numerical calculations.

[‡] Further discussion of this topic is contained in Section 6.4.6.

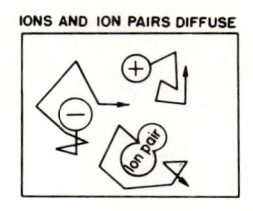

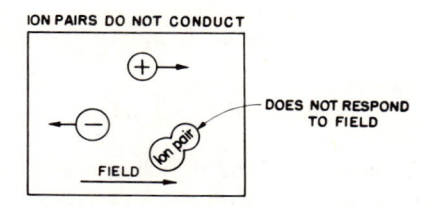

Fig. 4.62. The difference in the behavior of neutral ion pairs during diffusion and conduction.

The diffusive transport of M proceeds *both* through ions *and* ion pairs. In the conduction process, however, the situation is different (Fig. 4.62). The applied electric field only exerts a driving force on the charged particles. But an ion pair as a whole is electrically neutral; it does not feel the electric field. Thus, ion pairs are not participants in the conduction process. This point is of considerable importance in conduction in nonaqueous media (see Section 4.7.9).

In systems where ion-pair formation is possible, the mobility calculated from the diffusion coefficient, $\bar{u}_{abs} = D/kT$, is not equal to the mobility calculated from the equivalent conductivity, $\bar{u}_{abs} = u_{conv}/z_i e_0 = (\Lambda/z_i e_0)F$, and, therefore, the Nernst–Einstein equation, which is based on equating these two mobilities, may not be completely valid. In practice, one finds a degree of nonapplicability of up to about 20%.

Another important limitation on the Nernst–Einstein equation in electrolytic solutions may be approached through the following considerations. The diffusion coefficient is in general not a constant. This has been pointed out in Section 4.2.3, where the following expression was derived,

$$D = BRT\left(1 + \frac{d \ln f}{d \ln c}\right) \qquad (4.19)$$

It is clear that BRT is the value of the diffusion coefficient when the solution behaves ideally, i.e., $f = 1$; this ideal value of the diffusion coefficient shall be called D^0. Hence,

$$D = D^0\left(1 + \frac{d \ln f}{d \ln c}\right)$$
$$= D^0\left(1 + c\frac{d \log f}{dc}\right) \quad (4.185)$$

and, making use of the Debye–Hückel limiting law for the activity coefficient (*cf.* Section 3.5),

$$\log f = -Ac^{\frac{1}{2}}$$

one has

$$D = D^0(1 - \tfrac{1}{2}Ac^{\frac{1}{2}}) \quad (4.186)$$

an expression which shows how the diffusion coefficient varies with concentration. In addition, there is Kohlrausch's law

$$\Lambda = \Lambda^0 - Ac^{\frac{1}{2}} \quad (4.136)$$

where Λ^0 is the equivalent conductivity at infinite dilution, i.e., the ideal value.

From Eqs. (4.186) and (4.136), it is obvious that the diffusion coefficient D and the equivalent conductivity Λ have different dependences on concentration (Fig. 4.63). This experimentally observed fact has an important implication as far as the applicability of the Nernst–Einstein equation *in electrolytic solutions* is concerned. For, if the equation is true at one concentration, it cannot be true at another because the diffusion coefficient and the equivalent conductivity have varied to different extents in going from one concentration to the other.

The above argument brings out an important point about the limitations of the Nernst–Einstein equation. It does not matter whether the dif-

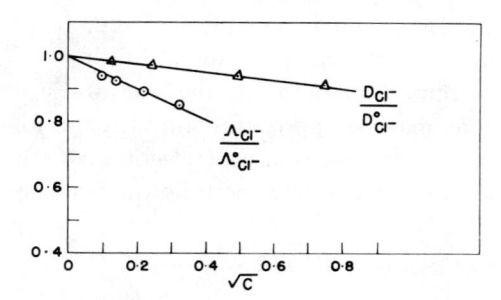

Fig. 4.63. The variation of the diffusion coefficient and the equivalent conductivity with concentration.

fusion coefficient and the equivalent conductivity vary with concentration; to introduce deviations into the Nernst–Einstein equation, D and Λ must have different concentration dependences. Now, the concentration dependence of the diffusion coefficient has been shown to be due to nonideality ($f \neq 1$), i.e., due to ion–ion interactions; and it will be shown later that the concentration dependence of the equivalent conductivity is also due to ion–ion interactions. But it is not the existence of interactions *per se* which underlies deviations from the Nernst–Einstein equation; otherwise, molten salts and ionic crystals, in which there are strong interionic forces, would show far more than the observed few per cent deviation of experimental data from values calculated by the Nernst–Einstein equation. The essential point is that the interactions must affect the diffusion coefficient and the equivalent conductivity by *different mechanisms* and, thus, to different extents. How this comes about for diffusion and conduction in solution will be seen later.

In *solutions* of electrolytes, the $c^{\frac{1}{2}}$ terms in the expressions for D and Λ tend to zero as the concentration of the electrolytic solution decreases, and the *differences* in the concentration variation of D and Λ become more and more negligible; in other words, the Nernst–Einstein equation becomes increasingly valid for electrolytic solutions with increasing dilution.

4.4.11. A Very Approximate Relation between Equivalent Conductivity and Viscosity: Walden's Rule

The Stokes–Einstein equation (4.175) connects the diffusion coefficient and the viscosity of the medium; the Nernst–Einstein equation (4.184) relates the diffusion coefficient to the equivalent conductivity. Hence, by eliminating the diffusion coefficient in these two equations, it is possible to obtain a relation between the equivalent conductivity and the viscosity of the electrolyte. The algebra is as follows:

$$D = \frac{kT}{6\pi r\eta} = \frac{RT}{zF^2}\Lambda$$

and, therefore,

$$\Lambda = \frac{zF^2 k}{6\pi R} \frac{1}{r\eta} \tag{4.187}$$

Since $F = N_A e_0$ and $k/R = 1/N_A$, one obtains

$$\Lambda = \frac{ze_0 F}{6\pi} \frac{1}{r\eta}$$

$$\Lambda\eta = \frac{\text{constant}}{r} \tag{4.188}$$

TABLE 4.14

Tests of Walden's rule. The product $\Lambda\eta$ for Potassium Iodide in Various Solvents at 25°C

Solvent	Λ	η	Λ_η
Sulfur dioxide[†]	265	0.00394	1.044
Acetonitrile	198.2	0.00345	0.684
Acetone	185.5	0.00316	0.586
Nitromethane	124.0	0.00611	0.758
Methyl alcohol	114.8	0.00546	0.627
Pyridine[‡]	71.3	0.00958	0.682
Ethyl alcohol	50.9	0.01096	0.560
Furfural	43.1	0.01490	0.642
Acetophenone	39.8	0.01620	0.644

[†] 0 °C.
[‡] 20 °C.

Hence, *if the radius of the moving (kinetic) entity in conduction, i.e., the solvated ion, can be considered the same in solvents of various viscosities,* the following relation is obtained

$$\Lambda\eta = \text{constant} = \frac{ze_0F}{6\pi r} \qquad (4.189)$$

This means that the product of the equivalent conductivity and the viscosity of the solvent for a particular electrolyte at a given temperature should be a constant (at one temperature). This is indeed what the *empirical* Walden's rule states.

Some experimental data on the product $\Lambda\eta$ are presented in Table 4.14 for solutions of potassium iodide in various solvents. Walden's rule has

TABLE 4.15

The Product Λ_η for Sodium Chloride in Various Solvents at 25°C

Solvent	Λ	η	$\Lambda\eta$
Water	126.39	0.00895	1.131
Methyl alcohol	96.9	0.00546	0.529
Ethyl alcohol	42.5	0.01096	0.466

some rough applicability in organic solvents. When, however, the $\Lambda\eta$ products for a solute dissolved, on the one hand, in water and, on the other, in organic solvents are compared, it is found that there is considerable discrepancy (Table 4.15). This should hardly come as a surprise; one should expect differences in the solvation of ions in water and in organic solvents (Section 2.8) and the resulting differences in radii of the moving ions.

4.4.12. The Rate-Process Approach to Ionic Migration

The fundamental equation for the current density (flux of charge) as a function of the drift velocity has been shown to be

$$i = zFcv_d \tag{4.161}$$

Hitherto, the drift velocity has been related to *macroscopic* forces (e.g., the Stokes viscous force $\vec{F} = 6\pi r\eta$ or the electric force $\vec{F} = ze_0X$) through the relation

$$v_d = \frac{\vec{F}}{m}\tau \tag{4.144}$$

Another approach to the drift velocity is by molecular models. The drift velocity v_d is considered the *net* velocity, i.e., the difference of the velocity $\vec{v}$ of ions in the direction of the force field and the velocity $\overleftarrow{v}$ of ions in a direction opposite to the field (Fig. 4.64). In symbols, one writes

$$v_d = \vec{v} - \overleftarrow{v} \tag{4.190}$$

Any velocity is given by the distance traveled divided by the time taken

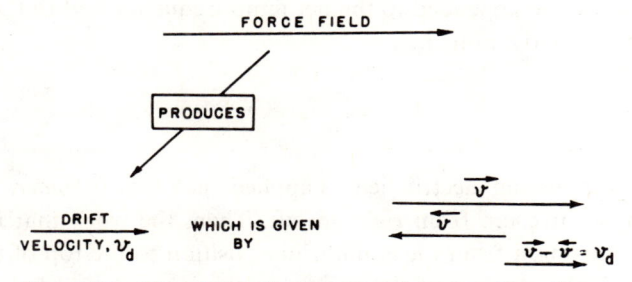

Fig. 4.64. The drift velocity v_d, can be considered made up of a velocity $\vec{v}$ in the direction of the force field and a velocity $\overleftarrow{v}$ in a direction opposite to the force field.

388 CHAPTER 4

to travel that distance. In the present case, the distance is the jump distance l, i.e., the mean distance that an ion jumps in hopping from site to site in the course of its directed random walk, and the time is the mean time τ between successive jumps. This mean time includes the time the ion may wait in a "cell" of surrounding particles as well as the actual time involved in jumping. Thus,

$$\vec{v} = \frac{\text{Mean jump distance } l}{\text{Mean time between jumps } \tau} \qquad (4.191)$$

The reciprocal of the mean time between jumps is the *net* jump frequency k, which is the number of jumps per unit of time. Hence,

$$\text{Velocity} = \text{Jump distance} \times \text{Net jump frequency}$$

or

$$\vec{v} = l \cdot \vec{k} \qquad (4.192)$$

and

$$\overleftarrow{v} = l \cdot \overleftarrow{k} \qquad (4.193)$$

For diffusion, the net jump frequency k was related to molecular quantities by viewing the ionic jumps as a *rate process* (Section 4.2.15). In this view, for an ion to jump, it must possess a certain free energy of activation to surmount the free-energy *barrier*. It was shown that the net jump frequency is given by

$$\vec{k} = \frac{kT}{h} e^{-\Delta G^{0\ddagger}/RT} \qquad (4.111)$$

To emphasize that this is the jump frequency for a pure diffusion process, in which case the ions are not subjected to an externally applied field, a subscript D will be appended to the net jump frequency and to the standard free energy of activation, i.e.,

$$\vec{k}_D = \frac{kT}{h} e^{-\Delta G_D^{0\ddagger}/RT} \qquad (4.194)$$

Now, suppose an electric field is applied such that it *hinders* the movement of a positive ion from right to left. Then, the work that is done *on* the ion in moving it from the equilibrium position to the top of the barrier (Fig. 4.65) is the product of the charge on the ion, z_+e_0, and the potential difference between the equilibrium position and the activated state, i.e., the position at the top of the barrier. Let this potential difference be a

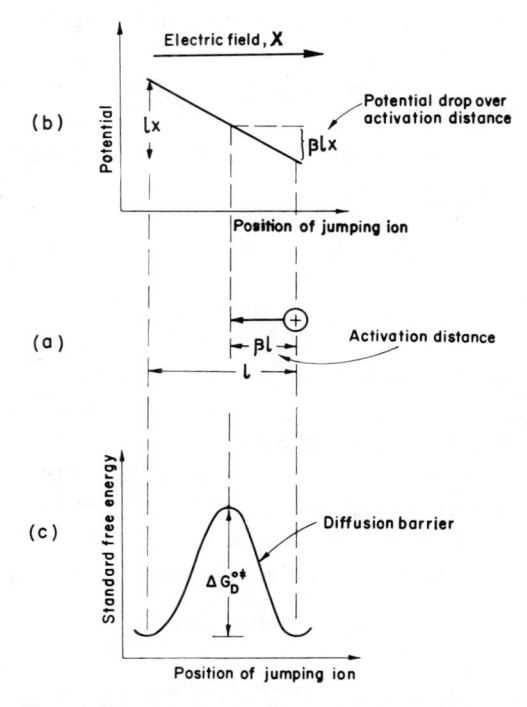

Fig. 4.65. (a) As the ion moves for the jump, it has to climb (b) the potential gradient arising from the electric field in the electrolyte, in addition to (c) the free-energy barrier for diffusion. To be activated, the ion has to climb the fraction $\beta l X$ of the total potential difference lX between the initial and final positions for the jump.

fraction β of the total potential difference (i.e., the applied electric field X times the distance l) between two equilibrium sites. Then, the electrical work done *on one* positive ion in making it climb to the top of the barrier, i.e., in *activating* it, is equal to the charge on the ion, $z_+ e_0$, times the potential difference $\beta X l$ through which it is transported. Thus, the electrical work is $z_+ e_0 \beta X l$ per ion, or $z_+ F \beta X l$ per mole of ions.

The electrical work of activation corresponds to a free-energy change. It appears therefore that there is a contribution to the total free energy of activation due to the *electrical* work done on the ion in making it climb the barrier. This electrical contribution to the free energy of activation is

$$\Delta G_e^{0\ddagger} = z_+ F \beta X l \qquad (4.195)$$

Hence, the *total* free energy of activation (for positive ions moving from right to left) is

$$\Delta G^{0\ne}_{\text{total}} = \Delta G^{0\ne}_D + \Delta G^{0\ne}_e \qquad (4.196)$$

Thus, in the presence of the field, the frequency of right → left jump is

$$\overleftarrow{k} = \frac{kT}{h} e^{-(\Delta G^{0\ne}_D + z_+ F\beta Xl)/RT} \qquad (4.197)$$

or

$$\overleftarrow{k} = k_D e^{-z_+ F\beta Xl/RT} \qquad (4.198)$$

By a similar argument, the left → right jump frequency $\overrightarrow{k}$, or the number of jumps per second from left to right, may be obtained. There are, however, two differences: When positive ions move from left to right, (1) they are moving with the field and therefore are *helped*, not *hindered*, by the field, and (2) they have to climb only through a fraction $1 - \beta$ of the barrier. Hence, the electrical work of activation is $-[z_+ F(1 - \beta)Xl]$, the minus sign indicating that the field assists the ion. Thus,

$$\overrightarrow{k} = k_D e^{z_+ F(1-\beta)Xl/RT} \qquad (4.199)$$

If the factor β is assumed to be $\frac{1}{2}$,[†] then $\beta = 1 - \beta = \frac{1}{2}$ and Eqs. (4.198) and (4.199) can be written

$$\overleftarrow{k} = k_D e^{-pX} \qquad (4.200)$$

and

$$\overrightarrow{k} = k_D e^{+pX} \qquad (4.201)$$

where, for conciseness, p is written instead of $z_+ Fl/2RT$. It follows from these equations that $\overleftarrow{k} < k_D$ and $\overrightarrow{k} > k_D$, or $\overrightarrow{k} > \overleftarrow{k}$.

In the presence of the field, therefore, the jumping frequency is anisotropic, i.e., it varies with direction. The jumping frequency of an ion in the direction of the field, $\overrightarrow{k}$, is greater than that $\overleftarrow{k}$ against the field. When, however, there is no field, the jump frequency k_D is the same in all directions and, therefore, jumps in all directions are equally likely. This is the characteristic of a random walk. The application of the field destroys the equivalence of all directions. The walk is not quite random. The field makes the ions more likely to more with it than against it. There is drift. In Eqs.

[†] This implies (*cf.* Fig. 4.65) that the energy barrier is symmetrical.

(4.200) and (4.201), the k_D is a random-walk term, the exponential factors are the perturbations due to the field, and the result is a drift. The equations are therefore a quantitative expression of the qualitative statement made in Section 4.4.1.

$$\begin{array}{c} \text{Drift due} \\ \text{to field} \end{array} = \begin{array}{c} \text{Random walk in} \\ \text{absence of field} \end{array} \times \begin{array}{c} \text{Perturbation} \\ \text{due to field} \end{array} \quad (4.202)$$

4.4.13. The Rate-Process Expression for Equivalent Conductivity

Introducing the expressions (4.200) and (4.201) for $\vec{k}$ and $\overleftarrow{k}$ into the equations for the component forward and backward velocities $\vec{v}$ and $\overleftarrow{v}$, [i.e., into Eqs. (4.192) and (4.193)] one obtains

$$\overleftarrow{v} = lk_D e^{-pX} \tag{4.203}$$

and

$$\vec{v} = lk_D e^{+pX} \tag{4.204}$$

where, as stated earlier,

$$p = \frac{z_+ Fl}{2RT}$$

The drift velocity v_d is obtained [cf. Eq. (4.190)] by subtracting (4.203) from (4.204), thus,

$$\begin{aligned} v_d = \vec{v} - \overleftarrow{v} &= lk_D e^{+pX} - lk_D e^{-pX} \\ &= lk_D(e^{+pX} - e^{-pX}) \\ &= 2lk_D \sinh pX \end{aligned} \tag{4.205}$$

The net charge transported per second across unit area, i.e., the current density i is given by Eq. (4.161), i.e.,

$$i = z_+ Fcv_d \tag{4.161}$$

Upon inserting the expression (4.205) for the drift velocity into Eq. (4.161), it is clear that

$$i = z_+ Fc(2lk_D \sinh pX) \tag{4.206}$$

A picture of the hyperbolic sine relation between the ionic current density and the electric field that would result from Eq. (4.206) is shown in Fig. 4.66.

The fundamental thinking used in the derivation of Eq. (4.206) has wide applicability. Take the case of an oxide film which grows on an elec-

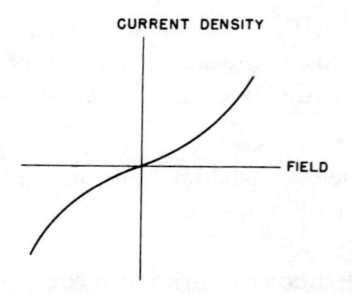

Fig. 4.66. The hyperbolic sine relation between the ionic current density and the electric field according to (4.206).

tron-sink electrode (anode). All one has to do is to consider an ionic crystal (the oxide) instead of an electrolytic solution, and all the arguments used to derive the hyperbolic sine relation (4.206) become immediately applicable to the ionic current flowing through the oxide in response to the potential gradient in the solid (Fig. 4.67). In fact, Eq. (4.206) is the basic equation describing the field-induced migration of ions in any ionic conductor.

Equation (4.206) is also formally similar to the expression for the current density due to a charge-transfer electrodic reaction occurring under the electric field present at an electrode–electrolyte interface.

In all these cases, two significant approximations can be made. One is the *high-field Tafel type* (*cf.* Section 8.2.10) of approximation, in which the absolute magnitude of the exponents $|pX|$ in Eq. (4.206), i.e., the argument pX of the hyperbolic sine in Eq. (4.206), is much greater than

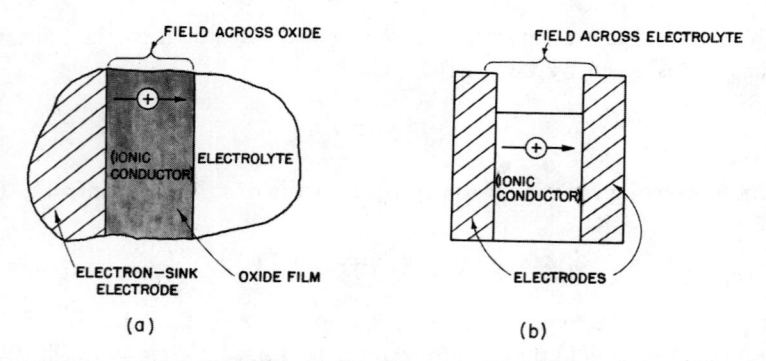

Fig. 4.67. The similarity between the ionic current flowing (a) through an oxide between an electrode and an electrolyte and (b) through an electrolyte between two electrodes.

ION TRANSPORT IN SOLUTIONS   393

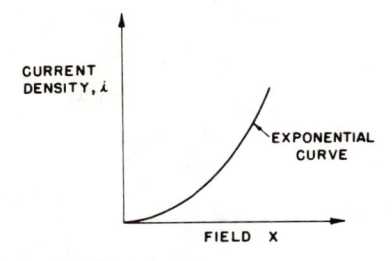

Fig. 4.68. Under high-field conditions, there is an exponential relation between ionic current density and the field across an oxide.

unity. Under this condition of $pX \gg 1$, one obtains $\sinh pX \sim e^{pX}/2$ because one can neglect e^{-pX} in comparison with e^{pX}. Thus (Fig. 4.68),

$$i = z_+ Fclk_D e^{pX} \tag{4.207}$$

i.e., the current density bears an exponential relation to the field. Such an exponential dependence of current on field is commonly observed in oxide growth, at electrode-electrolyte interfaces, but not in electrolyte solutions.

In electrolytic solutions, however, the conditions for the high-field approximation are not often observed. The applied field X is generally relatively small, in which case $pX \ll 1$ and the following approximation can be used

$$\sinh pX \sim pX = \frac{z_+ FlX}{2RT} \tag{4.208}$$

and the current density in Eq. (4.206) is approximately given (Fig. 4.69) by

$$i_+ = z_+ Fc2lk_D \frac{z_+ FlX}{2RT}$$

$$= \left(z_+^2 F^2 c \frac{l^2 k_D}{RT}\right) X \tag{4.209}$$

All the quantities within the parentheses are constants in a particular electrolyte, and, therefore,

$$i_+ = \text{constant } X \tag{4.210}$$

i.e., the ionic current density is proportional to the field. This is the *low-field* approximation. It is, in fact, the rate-process version of Ohm's law. An important point, however, has emerged; *Ohm's law is valid only for small*

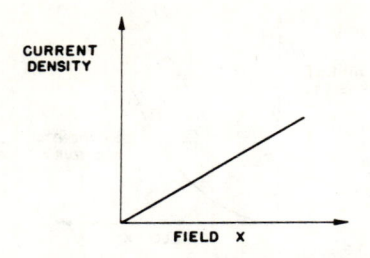

Fig. 4.69. Under low-field conditions, there is a linear relation between the ionic current density and the field in the electrolyte.

fields. Of course, this was accounted for in the phenomenological treatment of conduction where the general flux–force relation

$$J = A + BX + CX^2 + \cdots \qquad (4.118)$$

reduced to the linear relation

$$J = BX \qquad (4.119)$$

only for small fields.

4.4.14. The Total Driving Force for Ionic Transport: The Gradient of the Electrochemical Potential

In the rate-process view of conduction which has just been presented, it has been assumed that the concentration is the same throughout the electrolyte. Suppose, however, that there is a concentration gradient of a particular ionic species, say, positively charged radiotracer ions. Further, let the concentration vary continuously in the x direction (see Fig. 4.70), so that, if the concentration of positive ions at x on the left of the barrier is $(c_+)_x$, the number on the right (i.e., at $x + l$) is given by

Concentration = Concentration + Rate of change of × Distance
on the right   on the left c_+ with distance

i.e.,

$$(c_+)_{x+1} = (c_+)_x + \frac{dc_+}{dx} l \qquad (4.211)$$

In this case, there will be diffusion of the tracer ions, and, therefore, the current density i_+ is not simply given by a conduction law, i.e., by

ION TRANSPORT IN SOLUTIONS 395

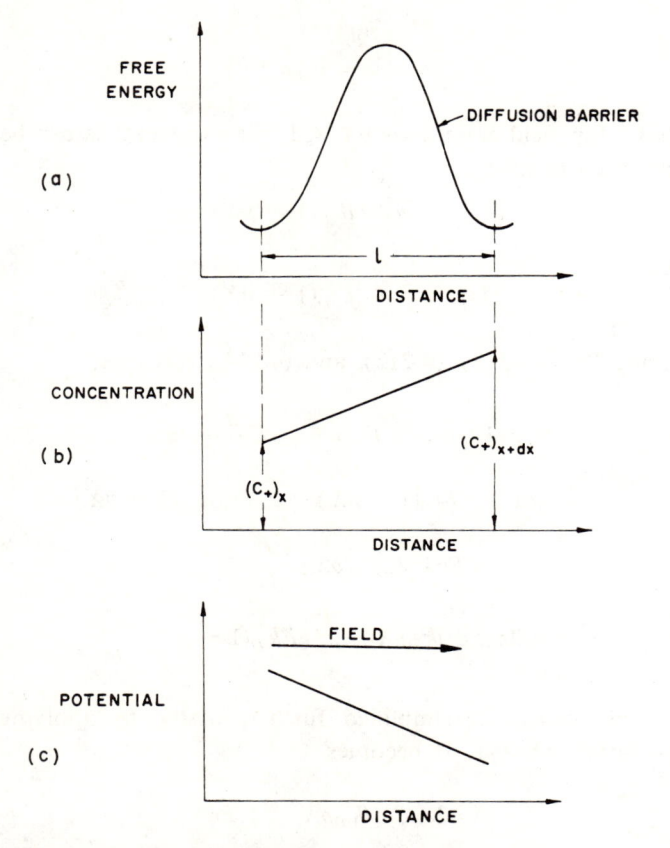

Fig. 4.70. Measurement of ions under both a concentration gradient and a potential gradient: (a) the free-energy barrier for the diffusive jump of an ion, (b) the concentration variation over the jump distance, and (c) the potential variation over the jump distance.

Eq. (4.156), which governs the situation in the absence of a concentration gradient. Instead, the expression for the current density has to be written (the subscript x in $(c_+)_x$ has been dropped for the sake of convenience)

$$i_+ = z_+ F c_+ \vec{v} - z_+ F \left(c_x + \frac{dc_+}{dx} l \right) \overleftarrow{v} \qquad (4.212)$$

But $\vec{v}$ and $\overleftarrow{v}$ have been evaluated as

$$\vec{v} = l k_D e^{pX} \qquad (4.204)$$

and
$$\overleftarrow{v} = lk_D e^{-pX} \tag{4.203}$$

Under low-field conditions $pX \ll 1$, the exponentials can be expanded and linearized to give
$$\overrightarrow{v} \sim lk_D(1 + pX) \tag{4.213}$$
and
$$\overleftarrow{v} \sim lk_D(1 - pX) \tag{4.214}$$

Combining Eqs. (4.212), (4.213), and (4.214), one gets
$$i_+ = z_+ F c_+ \overrightarrow{v} - z_+ F c_+ \overleftarrow{v} - z_+ F \frac{dc_+}{dx} l\overleftarrow{v}$$
$$= z_+ F c_+ lk_D(1 + pX) - z_+ F c_+ lk_D(1 - pX)$$
$$\quad - z_+ F l^2 k_D (1 - pX) \frac{dc_+}{dx}$$
$$= 2z_+ F c_+ lk_D pX - z_+ F l^2 k_D (1 - pX) \frac{dc_+}{dx} \tag{4.215}$$

This expression can be simplified further, firstly, by applying the low-field condition $pX \ll 1$. It becomes
$$i_+ = 2z_+ F c_+ lk_D pX - z_+ F l^2 k_D \frac{dc_+}{dx} \tag{4.216}$$

Secondly, by substituting $z_+ Fl/(2RT)$ for p, one has
$$i_+ = z^2 F^2 c_+ \frac{l^2 k_D}{RT} X - z_+ F l^2 k_D \frac{dc_+}{dx} \tag{4.217}$$

and, finally, by replacing $l^2 k_D$ with D_+ [cf. Eq. (4.113)], the result is
$$i_+ = \frac{z_+^2 F^2 c_+ D_+ X}{RT} - z_+ F D_+ \frac{dc_+}{dx} \tag{4.218}$$

To go from the current density i_+ to the flux J_+ of positive tracer ions is straightforward. Thus,
$$J_+ = \frac{i_+}{z_+ F} = \frac{D_+ c_+}{RT}(z_+ FX) - D_+ \frac{dc_+}{dx} \tag{4.219}$$

The second term on the right-hand side can be rewritten as

$$D_+ \frac{dc_+}{dx} = \frac{Dc_+}{RT} \frac{RT}{c_+} \frac{dc_+}{dx}$$

$$= \frac{Dc_+}{RT} \frac{d(RT \ln c_+)}{dx}$$

$$= \frac{Dc_+}{RT} \frac{d(\mu_+^0 + RT \ln c_+)}{dx}$$

$$= \frac{Dc_+}{RT} \frac{d\mu_+}{dx} \qquad (4.220)$$

since, according to the definition of the chemical potential for ideal solutions [cf. Eq. (3.54)],

$$\mu_+ = \mu_+^0 + RT \ln c_+$$

In addition, from Eq. (4.9), the electric field X is equal to minus the gradient of electrostatic potential, i.e.,

$$X = -\frac{d\psi}{dx} \qquad (4.9)$$

Hence,

$$J_+ = \frac{i_+}{z_+F} = -\frac{D_+c_+}{RT}\left(z_+F\frac{d\psi}{dx} + \frac{d\mu_+}{dx}\right)$$

$$= -\frac{D_+c_+}{RT}\frac{d}{dx}(zF\psi + \mu_+) \qquad (4.221)$$

This is an interesting result. The negative gradient of chemical potential, $d\mu_+/dx$, is known to be the driving force for pure diffusion; and $-z_+F(d\psi/dx) = z_+FX$, the driving force for pure conduction. But, when there is both a chemical potential (or concentration) gradient and an electric field $(-d\psi/dx)$ then the total driving force for ionic transport is the negative gradient of

$$\underset{\text{Electrostatic potential}}{zF\psi} \quad + \quad \underset{\text{Chemical potential}}{\mu_+}$$

This quantity $zF\psi + \mu_+$ could be called the *electrostatic-chemical* potential, or simply the *electrochemical potential*, of the positive ions and is denoted by the symbol $\overline{\mu_+}$. Thus,

$$\overline{\mu_+} = \mu_+ + zF\varphi \qquad (4.222)$$

398 CHAPTER 4

and the total driving force for the drift of ions is the gradient of the electrochemical potential. Thus, one can write the flux J_+ of Eq. (4.221) in the form

$$J_+ = \frac{i_+}{z_+ F} = -\frac{D_+ c_+}{RT}\frac{d\overline{\mu_+}}{dx} \tag{4.223}$$

Or, by making use of the Einstein relation,

$$D_+ = (\bar{u}_{\text{abs}})_+ kT \tag{4.169}$$

and the relation between absolute and conventional mobilities, i.e.,[†]

$$(u_{\text{conv}})_+ = (\bar{u}_{\text{abs}})_+ z_+ e_0 \tag{4.149}$$

one can rewrite Eq. (4.223) in the form

$$\begin{aligned}J_+ &= -\frac{D_+}{RT} c_+ \frac{d\overline{\mu_+}}{dx} = -\frac{(\bar{u}_{\text{abs}})_+ kT}{RT} c_+ \frac{d\overline{\mu_+}}{dx} \\ &= -\frac{(u_{\text{conv}})_+}{z_+ e_0}\frac{kT}{RT} c_+ \frac{d\overline{\mu_+}}{dx} \\ &= -\frac{(u_{\text{conv}})_+}{z_+ F} c_+ \frac{d\overline{\mu_+}}{dx} \end{aligned} \tag{4.224}$$

Expression (4.224) is known as the *Nernst–Planck flux equation*. It is an important equation for the description of the flux or flow of a species under the total driving force of an electrochemical potential. The Nernst–Planck flux expression is useful in explaining, for example, the electrodeposition of silver from silver cyanide ions. In this process, the negatively charged $[Ag(CN)_2]^-$ ions travel to the negatively charged electron source or cathode, a fact which cannot be explained by considering that the only driving force on the $[Ag(CN)_2]^-$ ions is the electric field because the electric field drives these ions *away from* the negatively charged electrode. If, however, the concentration gradient of these ions in a direction normal to the electron source is such that, in the expanded form of the Nernst–Planck equation, i.e.,

$$J = \frac{Dc}{RT} zFX - D\frac{dc}{dx} \tag{4.219}$$

[†] In Eqs. (4.149) through (4.224), the 1/300 factor has been omitted. In making numerical calculations, however, it must be taken into account. Thus, (4.224) is

$$J_+ = -\frac{300 u_{\text{conv}}}{z_+ F} c_+ \frac{d\bar{\mu}_+}{dx}$$

the second term is larger than the first, then the flux of the $[Ag(CN)_2]^-$ ions is opposite to the direction of the electric field, i.e., toward the negatively charged electrode.

Further Reading

1. A. Einstein, *Investigations on the Theory of the Brownian Movement*, Methuen, London, 1926.
2. S. Glasstone, K. J. Laidler, and H. Eyring, *Theory of the Rate Processes*, McGraw-Hill Book Company, New York, 1941.
3. R. W. Gurney, *Ionic Processes in Solution*, Dover Publications, New York, 1953.
4. R. A. Robinson and R. H. Stokes, *Electrolyte Solutions*, Butterworth's Publications, Ltd., London, 1955.
5. R. H. Boyd, "Extension of Stokes Law for Ionic Motion to Include the Effect of Dielectric Relaxation," *J. Chem. Phys.*, **35**: 19, 281 (1961).
6. R. W. Laity, *J. Chem. Educ.*, **39**: 56 (1962).
7. R. Zwanzig, "Dielectric Friction on a Moving Ion," *J. Chem. Phys.*, **38**: 1603 (1963).
8. R. P. Feynman, *Lectures on Physics*, Vol. 1, Addison-Wesley Publishing Co., Inc., Reading, Mass, 1964.

4.5. THE INTERDEPENDENCE OF IONIC DRIFTS

4.5.1. The Drift of One Ionic Species May Influence the Drift of Another

The processes of diffusion and conduction have been treated so far with the simple assumption that each ionic species drifts independently of every other one. In general, however, the assumption is not realistic for electrolytic solutions because it presupposes the absence of ionic atmospheres resulting from ion–ion interactions. One has been talking therefore of ideal laws of ionic transport and expressed them in the Nernst–Planck equation for the independent flux of a species i.

$$J_i = -\frac{(u_{\text{conv}})_i}{z_i F} c_i \frac{d\bar{\mu}_i}{dx} \qquad (4.225)$$

The time has come to free the treatment of ionic transport from the assumption of the independence of the various ionic fluxes and to consider some phenomena which depend on the fact that the drift of a species i is affected by the flows of other species present in the solution. For it is the whole society of ions that displays a transport process, and each individual ionic species takes into account what all the other species are doing. Ions interact with each other through their coulombic fields, and, thus, it will be seen that the law of electroneutrality which seeks zero excess charge in any macroscopic volume element plays a fundamental role in phenomena where ionic flows influence each other.

A stimulating approach to the problem of the interdependence of ionic drifts can be developed as follows. Since different ions have different radii, their Stokes mobilities, given by

$$(u_{\text{conv}})_i = \frac{z_i e_0}{6\pi r \eta} \qquad (4.180)$$

must be different. What are the consequences that result from the fact that different ionic species have unequal mobilities?

4.5.2. A Consequence of the Unequal Mobilities of Cations and Anions, the Transport Numbers

The current density i_i due to an ionic species i is related to the mobility $u_{\text{conv},i}$ in the following manner [*cf.* Eq. (4.156)]

$$i_i = z_i F c_i (u_{\text{conv}})_i X \qquad (4.156)$$

If, therefore, one considers a unit field $X = 1$ in an electrolyte solution containing a $z:z$-valent electrolyte (i.e., $z_+ = z_- = z$ and $c_+' = c_-' = c$), then, since $(u_{\text{conv}})_+ \neq (u_{\text{conv}})_-$, it follows that

$$i_+ \neq i_- \qquad (4.226)$$

This is a thought-provoking result. It shows that although all ions feel the externally applied electric field to the extent of their charges, some respond by migrating more than others. It also shows that, though the burden of carrying the current through the electrolytic solution falls on the whole community of ions, the burden is not shared equally among the various *species* of ions. Even if there are equal numbers of the various ions, those which have higher mobility contribute more to the communal task of transporting the current through the electrolytic solution than the ions handicapped by lower mobilities.

It is logical under these circumstances to seek a quantitative measure of the extent to which each ionic species is taxed with the job of carrying current. This quantitative measure, known as the *transport number* (Table 4.16), should obviously be defined by the *fraction* of the total current carried

ION TRANSPORT IN SOLUTIONS 401

TABLE 4.16

Transport Number of Cations in Aqueous Solutions at 25°C in 0.1N solutions

Electrolyte	HCl	LiCl	NaCl	KCl	KNO$_3$	AgNO$_3$	BaCl$_2$
Transport number of cation, t_+	0.83	0.32	0.39	0.49	0.51	0.47	0.43

by the particular ionic species, i.e.,

$$t_i = \frac{i_i}{i_T} = \frac{i_i}{\sum i_i} \qquad (4.227)$$

This definition requires that the sum of the transport numbers of all the ionic species be unity for

$$\sum t_i = \sum \frac{i_i}{\sum i_i} = 1 \qquad (4.228)$$

Thus, the conduction current carried by the species i (e.g., Na$^+$ ions in a solution containing NaCl and KCl) depends upon the current transported by all the other species. Here, then, is a clear and simple indication that the drift of the ith species depends on the drift of the other species.

For example, consider a 1:1-valent electrolyte, e.g., HCl, dissolved in water. The transport numbers will be given by[†]

$$t_{H^+} = \frac{i_{H^+}}{i_{H^+} + i_{Cl^-}} = \frac{z_{H^+} F c_{H^+} u_{H^+} X}{z_{H^+} F c_{H^+} u_{H^+} X + z_{Cl^-} F c_{Cl^-} u_{Cl^-} X}$$

But $z_{H^+} = z_{Cl^-} = 1$ and $c_{H^+} = c_{Cl^-} = c$, and, therefore,

$$t_{H^+} = \frac{u_{H^+}}{u_{H^+} + u_{Cl^-}} \qquad (4.229)$$

Similarly,

$$t_{Cl^-} = \frac{u_{Cl^-}}{u_{H^+} + u_{Cl^-}} \qquad (4.230)$$

The mobilities of the H$^+$ and Cl$^-$ ions in 0.1N HCl at 25 °C are

---

[†] To avoid cumbersome notation, the symbol u_{conv} for the conventional mobilities has been contracted to u. The absence of a bar above the u stresses that it is not the absolute mobility $\bar{u}_{\text{abs}}$.

33.71×10^{-4} and 6.84×10^{-4} cm sec^{-1}/V cm^{-1}, respectively, from which it turns out the transport numbers of the H$^+$ and Cl$^-$ ions are 0.83 and 0.17, respectively. Thus, the positive ions carry a major fraction ($\sim 83\%$) of the current.

Now suppose that, to the HCl solution, an excess of KCl is added so that the concentration of H$^+$ is about $10^{-3}M$ in comparison with a K$^+$ concentration of $1M$. The transport numbers in the mixture of electrolytes will be

$$\frac{t_{K^+}}{t_{H^+}} = \frac{c_{K^+}u_{K^+}/\sum c_i u_i}{c_{H^+}u_{H^+}/\sum c_i u_i} \tag{4.231}$$

$$= \frac{c_{K^+}}{c_{H^+}} \frac{u_{K^+}}{u_{H^+}} \tag{4.232}$$

The ratio c_{K^+}/c_{H^+} is 10^3, and the ratio of mobilities is

$$\frac{u_{K^+}}{u_{H^+}} = \frac{6 \times 10^{-4}}{30 \times 10^{-4}} = \frac{1}{5}$$

Hence,

$$\frac{t_{K^+}}{t_{H^+}} = 200$$

which means that, although the H$^+$ is about 5 times more mobile than the K$^+$ ion, it carries 200 times less current. Thus, the addition of the excess of KCl has reduced to a negligible value the fraction of the current carried by the H$^+$ ions.

In fact, the transport number of the H$^+$ ions under such circumstances is virtually zero, as shown from the following approximate calculation. (Note that the concentrations must be expressed in moles per cubic centimeter, not in moles per liter.)

$$t_{H^+} = \frac{c_{H^+}u_{H^+}}{c_{H^+}u_{H^+} + c_{K^+}u_{K^+} + c_{Cl^+}u_{Cl^+}}$$

$$= \frac{10^{-6}u_{H^+}}{10^{-6}u_{H^+} + 10^{-3}u_{K^+} + 10^{-3}u_{Cl^+}} = 10^{-3} \tag{4.233}$$

Thus, the conduction current carried by an ion depends very much on the concentration in which the other ions are present.

4.5.3. The Significance of a Transport Number of Zero

In the previous section, it has been shown that the addition of an excess of KCl makes the fraction of the migration (i.e., conduction) current

ION TRANSPORT IN SOLUTIONS 403

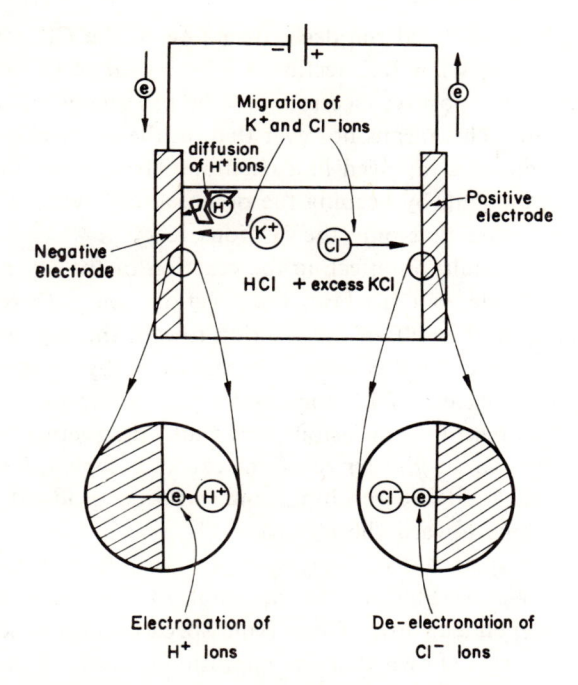

Fig. 4.71. A schematic diagram of the transport processes in an electrolyte (of HCl – KCl, with an excess of KCl) and of the reactions at the interfaces.

carried by the H$^+$ ions tend to zero. What happens if this mixture of HCl and KCl is placed between two electrodes and a potential difference applied across the cell (Fig. 4.71)?

In response to the electric field developed in the electrolyte, a migration of ions occurs and there is a conduction current in the solution. Since this conduction current is almost completely borne by K$^+$ and Cl$^-$ ions ($t_{H^+} \to 0$), there is a tendency for Cl$^-$ ions to accumulate near the positive electrode, and the K$^+$ ions, near the negative electrode. If the excess negative charge near the positive electrode and *vice versa* were to build up, then the resulting field due to lack of electroneutrality (*cf.* Section 4.3.3) would tend to bring the conduction current to a halt. It has been argued (*cf.* Section 4.3.4) however that conduction (i.e., migration) currents are sustained in an electrolyte because of charge-transfer reactions (at the electrode–electrolyte interfaces), which remove the excess charge that tends to build up near the electrodes.

In the case of the HCl + KCl electrolyte, the reaction at the positive electrode may be considered the deelectronation of the Cl⁻ ions. Further, according to Faraday's law (*cf.* Section 4.3.5), 1 g-eq of Cl⁻ ions must be de-electronated at the positive electrode for the passage of 1 F of charge in the external circuit. This means however that, at the other electrode, 1 g-eq of positive ions must be involved in a reaction. Thus, either the K⁺ or the H⁺ ions must react; but, by keeping the potential difference within certain limits, one can ensure that only the H⁺ ions react.

There is no difficulty in effecting the reaction of the layer of H⁺ near the negative electrode, but, to keep the reaction going, there must be a flux of H⁺ ions from the bulk of the solution toward the negative electrode. By what process does this flux occur? It cannot be by migration because the presence of the excess of K⁺ ions makes the transport number of H⁺ tend to zero. It is here that diffusion comes into the picture; the removal of H⁺ ions by the charge-transfer reactions causes a depletion of these ions near the electrode, and the resulting concentration gradient provokes a diffusion of H⁺ ions toward the electrode.

To provide a quantitative expression for the diffusion flux J_{H^+}, one cannot use the Nernst–Planck flux equation (4.224) because the latter describes the independent flow of one ionic species and, in the case under discussion, it has been shown that the migration current of the H⁺ ions is profoundly affected by the concentration of the K⁺ ions. A simple modification of the Nernst–Planck equation can be argued as follows.

Since conduction (i.e., migration) and diffusion are the two possible[†] modes of transport for an ionic species, the total flux J_i must be the sum of the conduction flux $(J_C)_i$ and the diffusion flux $(J_D)_i$. Thus,

$$J_i = (J_C)_i + (J_D)_i \tag{4.234}$$

The conduction flux is equal to $1/z_i F$ times the conduction current i_i borne by the particular species

$$(J_C)_i = \frac{i_i}{z_i F} \tag{4.235}$$

and the conduction current carried by the species i is related to the total conduction current $i_T = \sum i_i$ through the transport number of the species i [*cf.* Eq. (4.227)]

$$i_i = t_i \sum i_i = t_i i_T \tag{4.236}$$

[†] Throughout this chapter, another possible mode of transport, hydrodynamic flow, is not considered.

Hence,
$$(J_C)_i = \frac{t_i i_T}{z_i F} \tag{4.237}$$

Further, the diffusion flux $(J_D)_i$ is given by†

$$(J_D)_i = - Bc_i \frac{d\mu_i}{dx} \tag{4.14}$$

or, approximately, by Fick's first law

$$(J_D)_i = - D_i \frac{dc_i}{dx} \tag{4.16}$$

so that the total flux of species is

$$J_i = \frac{t_i i_T}{z_i F} - Bc_i \frac{d\mu_i}{dx} \tag{4.238}$$

or, approximately,

$$J_i = \frac{t_i i_T}{z_i F} - D_i \frac{dc_i}{dx} \tag{4.239}$$

From these modified forms of the Nernst–Planck flux equation (4.224), one can see that, even if $t_i \to 0$, it is still possible to have a flux of a species provided there is a concentration gradient, which is often brought into existence by interfacial charge-transfer reactions at the electrode-electrolyte interfaces consuming or generating the species.

From the modified Nernst–Planck flux equation (4.238), one can give a more precise definition of the transport number. If $d\mu_j/dx = 0$, in which case $dc_j/dx = 0$, then,

$$t_j = \left(\frac{z_j F J_j}{i_T}\right)_{d\mu_j/dx=0}$$

$$= \left(\frac{i_j}{i_T}\right)_{d\mu_j/dx=0} \tag{4.240}$$

It should be emphasized, therefore, that the transport number only pertains to the conduction flux, i.e., to that portion of the flux produced by an *electric field*, and any flux of an ionic species arising from a chemical potential gradient, i.e., any diffusion flux, is *not* counted in its transport number. From this definition, the transport number of a particular species

---

† The constant B has been shown in Section 4.4.6 to be equal to $\bar{u}_{\text{abs}}/N_A$.

406 CHAPTER 4

can tend to zero, $t_i \to 0$, and, at the same time, its diffusion flux can be finite.

This is an important point in electroanalytical chemistry where the general procedure is to arrange for the ions which are being analyzed to move to the electrode-electrolyte interface by diffusion only. Then, if the experimental conditions correspond to clearly defined boundary conditions (e.g., constant flux), the partial differential equation (Fick's second law) can be solved exactly to give a theoretical expression for the bulk concentration of the substance to be analyzed. In other words, the transport number of the substance being analyzed must be made to tend to zero if the solution of Fick's second law is to be applicable. This is ensured by adding some other electrolyte in such great excess that it takes on virtually the entire burden of the conduction current. The added electrolyte is known as the *indifferent* electrolyte. It is indifferent only to the electrodic reaction at the interface; it is far from indifferent to the conduction current.

4.5.4. The Diffusion Potential, Another Consequence of the Unequal Mobilities of Ions

Consider that a solution of a $z:z$-valent electrolyte (of concentration c moles liter^{-1}) is instantaneously brought into contact with water at the plane $x = 0$ (Fig. 4.72). A concentration gradient exists both for the positive ions and for the negative ions. They therefore start diffusing into the water.

Since, in general,[†] $\bar{u}_+ \neq \bar{u}_-$, let it be assumed that $\bar{u}_+ > \bar{u}_-$. With the use of the Einstein relation (4.169), it is clear that

$$D_+ = \bar{u}_+ kT \quad \text{and} \quad D_- = \bar{u}_- kT$$

or that

$$D_- < D_+$$

This means that the positive ions try to lead the negative ions in the diffusion into the water. But, when an ionic species of one charge moves faster than a species of the opposite charge, any unit volume in the water phase will receive more ions of the faster-moving variety.

Now, compare two unit volumes (Fig. 4.73), one situated at x_2 and the other at x_1, where $x_2 > x_1$, i.e., x_2 is farther than the plane of contact

---

[†] Though the subscript "abs" has been dropped, it is clear from the presence of a bar over the u's that one is referring to absolute mobilities.

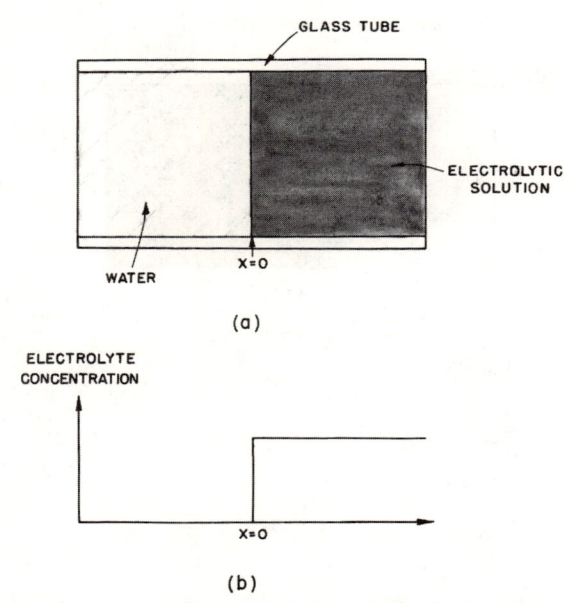

Fig. 4.72. (a) An electrolytic solution is instantaneously brought into contact with water at a plane $x = 0$; (b) the variation of the electrolyte concentration in the container at the instant of contact.

($x = 0$) of the two solutions. The positive ions are random walking faster than the negative ions, and, therefore, the greater the value of x, the greater is the ratio $c_{+,x}/c_{-,x}$.

All this is another way of saying that the center of the positive charge tends to separate from the center of the negative charge (Fig. 4.73). Hence, there is a tendency for the segregation of charge and the breakdown of the law of electroneutrality.

But, when charges of opposite sign are spatially separated, a potential difference develops. This potential difference between two unit volumes at x_2 and x_1 *opposes* the attempt at charge segregation. The faster-moving positive ions face strong opposition from the electroneutrality field, and they are slowed down. In contrast, the slower-moving negative ions are assisted by the potential difference (arising from the incipient charge separation), and they are speeded up. When a steady state is reached, the acceleration of the slow negative ions and the deceleration of the initially fast, positive ions, resulting from the electroneutrality field that develops, exactly compensate the inherent differences in mobilities. The electroneutrality

408 CHAPTER 4

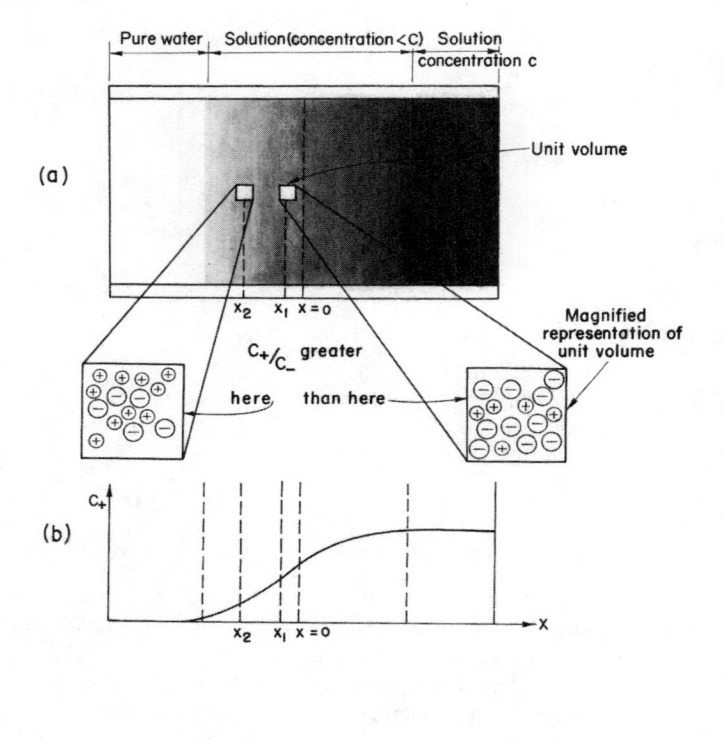

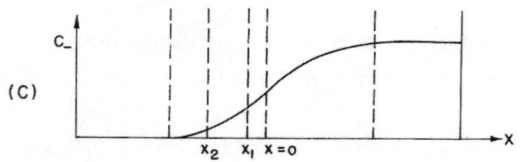

Fig. 4.73. (a) At a time $t > 0$ after the electrolyte and water are brought into contact, pure water and the electrolyte are separated by a region of mixing. In this mixing region, the c_+/c_- ratio increases from right to left because of the higher mobility of the positive ions. (b) and (c). The distance variations of the concentrations of positive and negative ions.

field is the leveler of ionic mobilities, helping and retarding ions according to their need so as to keep the situation as electroneutral as possible.

The conclusion that may be drawn from this analysis has quite profound ramifications. The basic phenomenon is that, whenever solutions of differing concentration are allowed to come into contact, diffusion occurs, there is a tendency for charge separation due to differences between ionic mobilities,

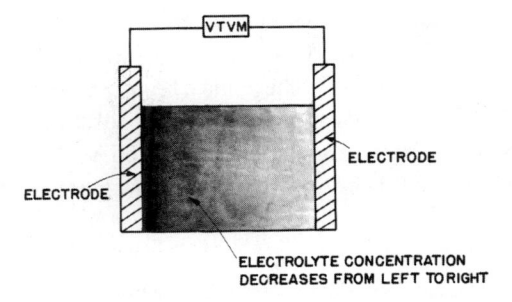

Fig. 4.74. A potential difference is registered by a vacuum-tube voltmeter (VTVM) connected to a concentration cell, i.e., to electrodes dipped in an electrolyte, the concentration of which varies from electrode to electrode.

and a potential difference develops across the interphase region in which there is a transition from the concentration of one solution to the concentration of the other.

This potential is known by the generic term *diffusion potential*. But, the precise name given to the potential varies with the situation, i.e., with the nature of the interphase region. If one ignores the interphase region and simply sticks two electrodes, one into each solution in order to "tap" the potential difference, then the whole assembly is known as a *concentration cell* (Fig. 4.74). On the other hand, if one constrains or restricts the interphase region by interposing a sintered-glass disk or any uncharged membrane between the two solutions so that the concentrations of the two

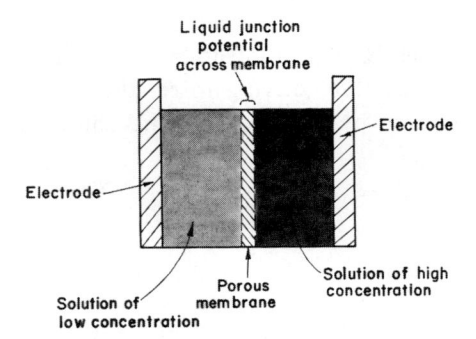

Fig. 4.75. A potential difference, the liquid-junction potential, is developed across a porous membrane introduced between two solutions of differing concentration.

solutions are uniform up to the porous material, then one has a *liquid-junction potential*[†] (Fig. 4.75). A *membrane potential* is a more complicated affair for two main reasons: (1) There may be pressure difference across the membrane producing hydrodynamic flow of the solution; (2) the membrane itself may consist of charged groups, some fixed and others exchangeable with the electrolytic solution, a situation equivalent to having sources of ions *within* the membrane.

4.5.5. Electroneutrality Coupling between the Drifts of Different Ionic Species

The picture of the development of the electroneutrality field raises a general question concerning the flow or drift of ions in an electrolytic solution. Is the flux of one ionic species dependent on the fluxes of the other species? In the diffusion experiment just discussed, is the diffusion of positive ions affected by the diffusion of negative ions? The answer to both these questions is in the affirmative.

Without doubt, the ionic flows start off as if they were completely independent, but it is this attempt to assert their freedom that leads to an incipient charge separation and the generation of an electroneutrality field. This field, which is dependent on the flows of all ionic species, curtails the independence of any one particular species. In this way, the flow of one ionic species is "coupled" to the flows of the other species.

In the absence of any interaction or coupling between flows, i.e., when the drift of any particular ionic species i is completely independent, the flow of that species i (i.e., the number of moles of i crossing per square centimeter per second) is described as follows.

The total driving force on an ionic species drifting, independent of any other ionic species, is the gradient of the electrochemical potential, $d\bar{\mu}_i/dx$. In terms of this total driving force, the expression for the total independent flux is given by the Nernst–Planck flux equation

$$J_i = \frac{i_i}{z_i F} = -\frac{u_i}{z_i F} c_i \frac{d\bar{\mu}_i}{dx} \qquad (4.224)$$

When, however, the flux of the species i is affected by the flux of the

---

[†] Now that the origin of a liquid-junction potential is understood, the method of minimizing it becomes clear. One simply chooses positive and negative ions with a negligible difference in mobilities; K^+ and Cl^- ions are the usual pair. This is the basis of the so-called " KCl salt bridge."

species j through the electroneutrality field, then another modification (*cf.* Section 4.5.3) of the Nernst–Planck flux equations has to be made. The modification that will now be described is a more thorough version than that presented earlier.

4.5.6. How Does One Represent the Interaction between Ionic Fluxes? The Onsager Phenomenological Equations

The development of an electroneutrality field introduces an interaction between flows and makes the flux of one species dependent on the fluxes of all the other species. To treat situations in which there is a coupling between the drift of one species and that of another, a general formalism will be developed. Hence, it is only when there is zero coupling or zero interaction that one can write the Nernst–Planck flux equation

$$J_i = -\frac{u_i}{z_i F} c_i \frac{d\bar{\mu}_i}{dx} \tag{4.224}$$

Once the interaction (due to the electroneutrality field) develops, a correction term is required, i.e.,

$$J_i = -\frac{u_i}{z_i F} c_i \frac{d\bar{\mu}_i}{dx} + \text{Coupling correction} \tag{4.241}$$

It is in the treatment of such interacting transport processes, or coupled flows, that the methods of near-equilibrium thermodynamics yield clear understanding of such phenomena, but only from a macroscopic or phenomenological point of view. These methods, as relevant to the present discussion, can be summarized with the following series of statements:

1. As long as the system remains close to equilibrium and the fluxes are independent, the fluxes are treated as proportional to the driving forces. Experience (Table 4.17) commends this view for diffusion [Fick's law, Eq. (4.16)], conduction [Ohm's law, Eq. (4.130)], and heat flow (Fourier's law). Thus, the *independent* flux of an ionic species 1 given by the Nernst–Planck equation (4.224) shall be written

$$J_1 = L_{11}\vec{F}_1 \tag{4.242}$$

where L_{11} is the proportionality or phenomenological constant and $\vec{F}_1$ is the driving force.

2. When there is coupling, the flux of one species (*e.g.*, J_1) is not simply proportional to its own (or, as it is called, *conjugate*) driving force

412　CHAPTER 4

TABLE 4.17
Some Linear Flux-Force Laws

Phenomenon	Flux of	Driving force	Law
Diffusion	Matter J	Concentration gradient dc/dx	Fick $J = -D\,dc/dx$
Migration (conduction)	Charge i	Potential gradient $X = -d\psi/dx$	Ohm $i = \sigma X$
Heat conduction	Heat J_{heat}	Temperature gradient	Fourier $J_{\text{heat}} = K\,dT/dx$

(i.e., $\vec{F}_1$), but it is contributed to by the driving forces on all the other particles. In symbols,

$$J_1 = L_{11}\vec{F}_1 + \text{Coupling correction}$$
$$= L_{11}\vec{F}_1 + \begin{array}{c}\text{Flux of 1 due}\\\text{to driving force}\\\text{on species 2}\end{array} + \begin{array}{c}\text{Flux of 1 due}\\\text{to driving force}\\\text{on species 3}\end{array} + \text{etc.} \quad (4.243)$$

3. The linearity or proportionality between fluxes and conjugate driving force is also valid for the contributions to the flux of one species from the forces on the *other* species. Hence, with this assumption, one can write Eq. (4.243) in the form

$$J_1 = L_{11}\vec{F}_1 + [L_{12}\vec{F}_2 + L_{13}\vec{F}_3 + \cdots] \quad (4.244)$$

4. Similar expressions are used for the fluxes of all the species in the system. If the system consists of an electrolyte dissolved in water, one has three species: positive ions, negative ions, and water. Hence, by using the symbol $+$ for the positive ions, $-$ for the negative ions, and 0 for the water, the fluxes are

$$J_+ = L_{++}\vec{F}_+ + L_{+-}\vec{F}_- + L_{+0}\vec{F}_0$$
$$J_- = L_{-+}\vec{F}_+ + L_{--}\vec{F}_- + L_{-0}\vec{F}_0 \quad (4.245)$$
$$J_0 = L_{0+}\vec{F}_+ + L_{0-}\vec{F}_- + L_{00}\vec{F}_0$$

These equations are known as the *Onsager phenomenological equations*. They represent a complete macroscopic description of the interacting flows when the system is near equilibrium.† It is clear that all the "straight" coefficients L_{ij}, where the indices are equal, $i = j$, pertain to the independent, uncoupled, fluxes. Thus $L_{++}X_+$ and $L_{--}X_-$ are the fluxes of the positive and negative ions, respectively, when there are no interactions. All other cross terms represent interactions between fluxes, e.g., $L_{+-}X_-$ represents the contribution to the flux of positive ions from the driving force on the negative ions.

5. What are the various coefficients L_{ij}? Onsager put forward the helpful *reciprocity relation*. According to this, all symmetrical coefficients are equal, i.e.,

$$L_{ij} = L_{ji} \tag{4.246}$$

This principle has the same status in nonequilibrium thermodynamics as the law of conservation of energy has in classical thermodynamics; it has not been disproved by experience.

4.5.7. An Expression for the Diffusion Potential

The expression for the diffusion potential can be obtained in a straightforward, though hardly brief, manner by using the Onsager phenomenological equations to describe the interaction flows.

Consider an electrolytic solution consisting of the ionic species M^{z+} and A^{z-} and the solvent. When a transport process involves the ions in the system, there are two ionic fluxes J_+ and J_-. Since, however, the ions are solvated, the solvent also participates in the motion of ions, and, hence, there is also a solvent flux J_0.

If, however, the solvent is considered fixed, i.e., the solvent is taken as the coordinate system or the frame of reference,‡ then one can consider ionic fluxes *relative* to the solvent. Under this condition, $J_0 = 0$, and one has only two ionic fluxes. Thus, one can describe the interacting and independent ionic drifts by the following equations

$$J_+ = L_{++}\vec{F}_+ + L_{+-}\vec{F}_- \tag{4.247}$$

$$J_- = L_{-+}\vec{F}_+ + L_{--}\vec{F}_- \tag{4.248}$$

The straight coefficients L_{++} and L_{--} represent the independent flows; and the cross coefficients L_{+-} and L_{-+}, the coupling between the flows.

---

† If the system is not near equilibrium, the flows are no longer proportional to the driving forces [*cf.* Eq. (4.243)].
‡ Coordinate systems are chosen for convenience.

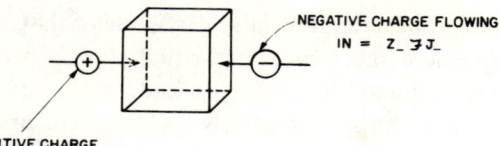

Fig. 4.76. According to the principle of electroneutrality as applied to the fluxes, the flux of positive ions into a volume element must be equal to the flux of negative ions into the volume element, so that the total negative charge is equal to the total positive charge.

The important step in the derivation of the diffusion potential is the statement that, under conditions of steady state, the electroneutrality field sees to it that the quantity of positive charge flowing into a volume element is equal in magnitude but opposite in sign to the quantity of negative charge flowing in (Fig. 4.76). That is,

$$z_+ F J_+ + z_- F J_- = 0 \tag{4.249}$$

For convenience, z_+F and z_-F are written as q_+ and q_-, respectively. Now the expressions (4.247) and (4.248) for J_+ and J_- are substituted in Eq. (4.249)

$$q_+ L_{++} \vec{F}_+ + q_+ L_{+-} \vec{F}_- + q_- L_{-+} \vec{F}_+ + q_- L_{--} \vec{F}_- = 0$$

or

$$\vec{F}_+(q_+ L_{++} + q_- L_{-+}) + \vec{F}_-(q_+ L_{+-} + q_- L_{--}) = 0 \tag{4.250}$$

Using the symbols

$$p_+ = q_+ L_{++} + q_- L_{-+} \tag{4.251}$$

and

$$p_- = q_+ L_{+-} + q_- L_{--} \tag{4.252}$$

one has

$$p_+ \vec{F}_+ + p_- \vec{F}_- = 0 \tag{4.253}$$

But what are the driving forces $\vec{F}_+$ and $\vec{F}_-$ for the independent flows of the positive and negative ions? They are the gradients of electrochemical potential (cf. Section 4.4.14)

$$\vec{F}_+ = \frac{d\mu_+}{dx} + q_+ \frac{d\psi}{dx} \tag{4.254}$$

and

$$\vec{F}_- = \frac{d\mu_-}{dx} + q_- \frac{d\psi}{dx} \tag{4.255}$$

With these expressions, Eq. (4.253) becomes

$$p_+ \frac{d\mu_+}{dx} + p_+q_+ \frac{d\psi}{dx} + p_- \frac{d\mu_-}{dx} + p_-q_- \frac{d\psi}{dx} = 0$$

or

$$-\frac{d\psi}{dx} = \frac{p_+}{p_+q_+ + p_-q_-} \frac{d\mu_+}{dx} + \frac{p_-}{p_+q_+ + p_-q_-} \frac{d\mu_-}{dx} \qquad (4.256)$$

It can be shown, however, that (*cf.* Appendix 4.4)

$$\frac{p_+}{p_+q_+ + p_-q_-} = \frac{t_+}{z_+ F} \qquad (4.257)$$

and

$$\frac{p_-}{p_+q_+ + p_-q_-} = \frac{t_-}{z_- F} \qquad (4.258)$$

where t_+ and t_- are the transport numbers of the positive and negative ions. By making use of these relations, Eq. (4.256) becomes

$$-\frac{d\psi}{dx} = \frac{t_+}{z_+ F} \frac{d\mu_+}{dx} + \frac{t_-}{z_- F} \frac{d\mu_-}{dx}$$

$$= \sum \frac{t_i}{z_i F} \frac{d\mu_i}{dx} \qquad (4.259)$$

The negative sign before the electric field shows that it is opposite in direction to the chemical potential gradients of all the diffusing ions.

If one considers (Fig. 4.77) an infinitesimal length dx parallel to the direction of the electric- and chemical-potential fields, one can obtain the electric-potential difference $d\psi$ and the chemical-potential difference $d\mu$

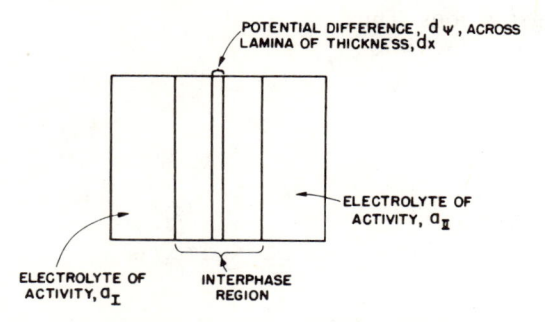

Fig. 4.77. Schematic representation of a dx-thick lamina in the interphase region between two electrolytes of activities a_I and a_II.

across the length dx

$$-\frac{d\psi}{dx}dx = \sum \frac{t_i}{z_i F}\frac{d\mu_i}{dx}dx \qquad (4.260)$$

or

$$-d\psi = \frac{1}{F}\sum \frac{t_i}{z_i}d\mu_i \qquad (4.261)$$

$$= \frac{RT}{F}\sum \frac{t_i}{z_i}d\ln a_i \qquad (4.262)$$

This is the basic equation for the diffusion potential. It has been derived here on the basis of a realistic point of view, namely, that the diffusion potential arises from the *nonequilibrium* process of diffusion.

There is however another method[†] of deriving the diffusion potential. One takes note of the fact that, when a *steady-state* electroneutrality field has developed, the system relevant to a study of the diffusion potential hangs together in a fine and delicate balance. The diffusion flux is exactly balanced by the electric flux; the concentrations and the electrostatic potential throughout the interphase region *do not vary with time*. (Remember the derivation of the Einstein relation, *cf.* Section 4.4.) In fact, one may turn a blind eye to the drift and pretend that the whole system is in equilibrium.

On this basis, one can simply equate to zero the sum of the electrical and diffusional work of transporting ions across a lamina dx of the interphase region (Fig. 4.78). If one equivalent of charge (both positive and negative ions) is taken across this lamina, the electrical work is $F\,d\psi$. But this one equivalent of charge consists of t_+/z_+ moles of positive ions and t_-/z_- moles of negative ions. Hence, the diffusional work is $d\mu_+$ per mole, or $(t_+/z_+)\,d\mu_+$ per t_+/z_+ moles of positive ions and $(t_-/z_-)\,d\mu_-$ per t_-/z_- moles of negative ions. Thus,

$$F\,d\psi + \frac{t_+}{z_+}d\mu_+ + \frac{t_-}{z_-}d\mu_- = 0 \qquad (4.263)$$

or

$$-d\psi = \frac{1}{F}\sum \frac{t_i}{z_i}d\mu_i \qquad (4.261)$$

$$= \frac{RT}{F}\sum \frac{t_i}{z_i}d\ln a_i \qquad (4.262)$$

[†] This method is based on Thomson's hypothesis according to which it is legitimate to apply equilibrium thermodynamics to the reversible parts of a steady-state, nonequilibrium process.

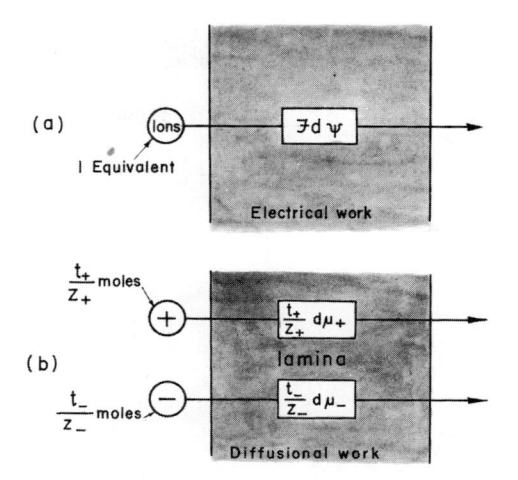

Fig. 4.78. The sum of (a) the electrical work $F\,d\psi$ and (b) the diffusional work $(t_+/z_+)d\mu_+ + (t_-/z_-)d\mu_-$ of transporting one equivalent of ions across a dx-thick lamina in the interphase region, is equal to zero.

4.5.8. The Integration of the Differential Equation for Diffusion Potentials: The Planck–Henderson Equation

An equation has been derived for the diffusion potential [*cf.* Eq. (4.261)], but it is a *differential* equation relating the infinitesimal potential difference $d\psi$ developed across an infinitesimally thick lamina dx in the interphase region. What one experimentally measures, however, is the total potential difference $\Delta\psi = \psi^0 - \psi^l$ across a transition region extending from $x = 0$ to $x = l$ (Fig. 4.79). Hence, to theorize about the measured potential differences, one has to integrate the differential equation (4.261); i.e.,

$$-\Delta\psi = \psi^0 - \psi^l = \frac{1}{F} \sum_i \int_{x=0}^{x=l} \frac{t_i}{z_i} \frac{d\mu_i}{dx} dx$$

$$= \frac{RT}{F} \sum_i \int_{x=0}^{x=l} \frac{t_i}{z_i} \frac{d\ln a_i}{dx} dx$$

$$= \frac{RT}{F} \sum_i \int_{x=0}^{x=l} \frac{t_i}{z_i} \frac{1}{f_i c_i} \frac{d(f_i c_i)}{dx} dx \quad (4.264)$$

Here lies the problem. To carry out the integration, one must know:

1. How the concentrations of all the species vary in the transition region.

418 CHAPTER 4

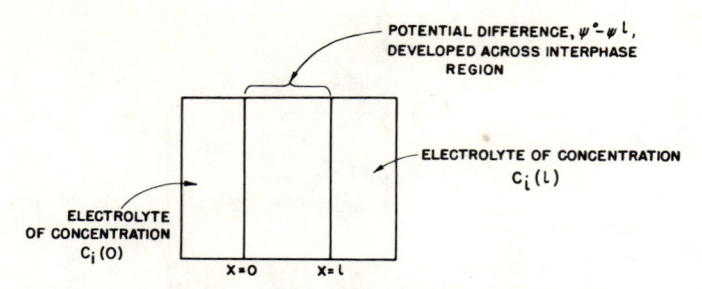

Fig. 4.79. The measured quantity is the total potential difference $\psi^0 - \psi^l$ across the whole interphase region between electrolytes of differing concentration $c_i(0)$ and $c_i(l)$.

2. How the activity coefficients f_i vary with c_i.
3. How the transport number varies with c_i.

The general case is too difficult to solve analytically, but several special cases can be solved. For example (Fig. 4.80), the activity coefficients can be taken as unity, $f_i = 1$—ideal conditions; the transport numbers t_i can be assumed constant; and a *linear* variation of concentrations with distance can be assumed. The last assumption implies that the concentration $c_i(x)$ of the *i*th species at x is related to its concentration $c_i(0)$ at $x = 0$ in the following way

$$c_i(x) = c_i(0) + k_i x \qquad (4.265)$$

and

$$\frac{dc_i}{dx} = \text{Constant} = k_i = \frac{c_i(l) - c_i(0)}{l} \qquad (4.266)$$

With the aid of these assumptions, the integration becomes simple. Thus, with $t_i \neq f(x)$, $f_i = 1$, and Eqs. (4.265) and (4.266), one has in (4.264)

$$-\Delta\psi = \frac{RT}{F} \sum_i \int_{x=0}^{x=l} \frac{t_i}{z_i} \frac{k_1\, dx}{c_i(0) + k_1 x}$$

$$= \frac{RT}{F} \sum_i \frac{t_i}{z_i} \int_{x=0}^{x=l} \frac{d[c(0) + k_1 x]}{c_i(0) + k_1 x}$$

$$= \frac{RT}{F} \sum_i \frac{t_i}{z_i} \left\{ \ln\left[c_i(0) + k_1 x\right]\right\}\Big|_{x=0}^{x=l}$$

$$= \frac{RT}{F} \sum_i \frac{t_i}{z_i} \ln \frac{c_i(l)}{c_i(0)} \qquad (4.267)$$

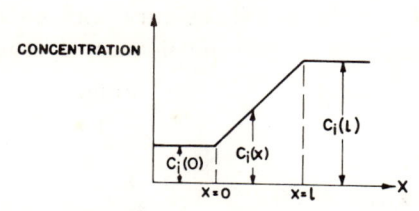

Fig. 4.80. In the derivation of the Planck–Henderson equation, a linear variation of concentration is assumed in the interphase region which commences at $x = 0$ and ends at $x = l$.

This is known as the Planck–Henderson equation for diffusion or liquid-junction potentials.

In the special case of a $z:z$-valent electrolyte, $z_+ = z_- = z$ and $c_+ = c_- = c$, Eq. (4.267) reduces to

$$-\Delta\psi = \frac{RT}{F}(t_+ - t_-) \ln \frac{c(l)}{c(0)} \qquad (4.268)$$

and, since $t_+ + t_- = 1$,

$$-\Delta\psi = (2t_+ - 1)\frac{RT}{F} \ln \frac{c(l)}{c(0)} \qquad (4.269)$$

In the highly simplified treatment of the diffusion potential which has just been presented, several drastic assumptions have been made. The one regarding the concentration variation within the transition region can be avoided. One may choose a more realistic concentration versus distance relationship either by thinking about it in more detail or by using experimental knowledge on the matter. Similarly, instead of assuming the activity coefficient to be unity, one can feed in the theoretical or experimental concentration dependence of the activity coefficients. Of course, the introduction of nonideality ($f_i \neq 1$) makes the mathematics awkward; in principle, however, the problem is understandable.

But what about the assumption of the constancy of the transport number? Is this reasonable? In the case of a $z:z$-valent electrolyte, the transport number depends on the mobilities

$$t_+ = \frac{u_+}{u_+ + u_-} \quad \text{and} \quad t_- = \frac{u_-}{u_+ + u_-}$$

Thus, the constancy of the transport numbers with concentration depends on the degree to which the mobilities vary with concentration. That is something to be dealt with in the model-oriented arguments of the next section.

Further Reading

1. M. Planck, *Ann. Physik*, **40**: 561 (1890).
2. P. Henderson, *Z. Physik. Chem. (Leipzig)*, **59**: 118 (1907); and **63**: 325 (1908).
3. L. Onsager, *Phys. Rev.*, **37**: 405 (1931); and **38**: 2265 (1931).
4. J. Meixner, *Ann. Physik*, **39**: 333 (1941).
5. K. G. Denbigh, *Thermodynamics of the Steady State*, John Wiley & Sons, Inc., New York, 1951.
6. R. A. Robinson and R. H. Stokes, *Electrolyte Solutions*, Butterworths' Publications, Ltd., London, 1955.
7. B. Baranowski, *Bull. Acad. Polon. Sci., Ser. Sci. Chim.*, **8**: 609 (1960).
8. A. Katchalsky and P. F. Curran, *Non-Equilibrium Thermodynamics in Biophysics*, Harvard University Press, Cambridge, Mass., 1965.
9. N. Lakshminarayanaiah, *Chem. Rev.*, **65**: 491 (1965).

4.6. THE INFLUENCE OF IONIC ATMOSPHERES ON IONIC MIGRATION

4.6.1. The Concentration Dependence of the Mobility of Ions

In the phenomenological treatment of conduction (Section 4.2.12), it has been stated that the equivalent conductivity Λ varies with the concentration c of the electrolyte according to the *empirical* law of Kohlrausch [Eq. (4.136)]

$$\Lambda = \Lambda^0 - Ac^{\frac{1}{2}} \qquad (4.136)$$

where A is a constant, and Λ^0 is the pristine or ungarbled value of equivalent conductivity, i.e., the value at infinite dilution.

The equivalent conductivity, however, has been related to the conventional electrochemical mobilities[†] u_+ and u_- of the current-carrying ions by the following expression

$$\Lambda = F(u_+ + u_-) \qquad (4.160)$$

from which it follows that

$$\Lambda^0 = F(u_+{}^0 + u_-{}^0) \qquad (4.270)$$

[†] To avoid cumbersome notation, the conventional mobilities are written in this section without the subscript "conv," i.e., one writes u_+ instead of $(u_{\text{conv}})_+$.

where the u^0's are the conventional mobilities (i.e., drift velocities under a field of 1 V cm^{-1}) at infinite dilution. Thus, Eq. (4.160) can be written as

$$(u_+ + u_-) = (u_+^0 + u_-^0) - \left(\frac{A}{F}\frac{1}{2}\right)(c_+^{\frac{1}{2}} + c_-^{\frac{1}{2}}) \qquad (4.271)$$

or it can be split up into two equations

$$u_+ = u_+^0 - A'c_+^{\frac{1}{2}} \qquad (4.272)$$

and

$$u_- = u_-^0 - A'c_-^{\frac{1}{2}} \qquad (4.273)$$

What is the origin of this experimentally observed dependence of ionic mobilities on concentration? Equations (4.272) and (4.273) indicate that, the more ions there are per unit volume, the more they diminish each other's mobility. In other words, at appreciable concentrations, the movement of any particular ion does not seem to be independent of the existence and motions of the other ions, and there appear to be forces of interaction between ions. This coupling between the individual drifts of ions has already been recognized, but now the discussion is intended to be on an atomistic, rather than a phenomenological, level. The interactions between ions can be succinctly expressed through the concept of the ionic cloud (Chapter 3). It is thus necessary to analyze and incorporate ion-atmosphere effects into the zero-approximation atomistic picture of conduction (Section 4.4) and, in this way, understand how the mobilities of ions depend on the concentration of the electrolyte.

Attention should be drawn to the fact that there has been a degree of inconsistency in the treatments of ionic clouds (Chapter 3) and the elementary theory of ionic drift (Section 4.4.2). When the ion atmosphere was described, the central ion was considered—from a time-average point of view—at rest. To the extent that one seeks to interpret the *equilibrium* properties of electrolytic solutions, this picture of a static central ion is quite reasonable. This is because, in the absence of a spatially directed field acting on the ions, the only ionic motion to be considered is random walk, the essential characteristic of which is that the mean distance traveled by an ion (not the mean square distance, see Section 4.2.5) is zero. The central ion can therefore be considered to remain where it is, i.e., to be at rest.

When, however, the elementary picture of ionic drift (Section 4.4.2) was sketched, the ionic cloud around the central ion was ignored. This approximation is justified only when the ion atmosphere is so tenuous that its effects on the movement of ions can be neglected. This condition of extreme tenuosity (in which there is a negligible coupling between ions) obtains increasingly as the solution tends to infinite dilution. Hence, the simple, unclouded picture of conduction (Section 4.4) is valid only at infinite dilution.

To summarize the duality of the treatment so far: When the ion atmosphere was treated in Chapter 3, the motion of the central ion was ignored and, therefore, only equilibrium properties fell within the scope of analysis; when the motion of the central ion under an applied electric field was considered, the ionic cloud (which is a convenient description of the interactions between an ion and its environment) was neglected and, therefore, only the infinite dilution conduction could be analyzed. Thus, a unified treatment of *ionic atmospheres around moving ions* is required. The central problem is: How does the interaction between an ion and its cloud affect the motion of the ion?

4.6.2. Ionic Clouds Attempt to Catch Up with Moving Ions

In the absence of a driving force (e.g., an externally applied electric field), no direction in space from the central ion is privileged. The coulombic field of the central ion has spherical symmetry, and, therefore, the probability of finding, say, a negative ion at a distance r from the reference ion is the same irrespective of the direction in which the point r lies. On this basis, it was shown that the ionic cloud was spherically symmetrical (*cf.* Section 3.8.2).

When, however, the ions are subject to a driving force (be it an electric field, a velocity field due to the flow of an electrolyte, or a chemical-potential field producing diffusion), one direction in space becomes privileged. The distribution function (which is a measure of the probability of finding an ion of a certain charge in a particular volume element) has to be lopsided, or asymmetrical. The probability depends not only on the distance of the volume element from the central ion but also on the direction in which the volume element lies in relation to the direction of ionic motion. The procedure of Chapter 3 no longer applies. One cannot simply assume a Boltzmann distribution and, for the work done to bring an ion (of charge $z_i e_0$) to the volume element under consideration, use the electrostatic work $z_i e_0 \psi$ because the electrostatic potential ψ was, in the context of Chapter 3, a function of r only and one would then imply a symmetrical distribution of function.

The rigorous but, unfortunately, mathematically difficult approach to the problem of ionic clouds around moving ions is to seek the asymmetrical-distribution functions and then work out the implications of such functions for the electric fields developed among moving ions.

A simpler approach will be followed here. This is the relaxation approach. The essence of a relaxation analysis is to consider a system in one state, then perturb it slightly with a stimulus, and analyze the *time dependence* of the response of the system to the stimulus. (It will be seen later that relaxation techniques are much used in modern studies of the mechanism of electrode reactions.)

Consider therefore the spherically symmetrical ionic cloud around a stationary central ion. Now, let the stimulus of a driving force displace the reference ion in the x direction. The erstwhile spherical symmetry of the ion atmosphere can be restored only if its contents (the ions and the solvent molecules) *immediately* readjust to the new position of the central ion. This is possible only if the movements involved in restoring spherical symmetry are instantaneous, i.e., if no frictional resistances are experienced in the course of these movements. But the readjustment of the ionic cloud involves ionic movements which are rate processes. Hence, a finite time is required to reestablish spherical symmetry.

Even if this time were available, spherical symmetry would obtain only if the central ion did not move away still farther *while* the ionic cloud was trying to readjust. But, under the influence of the externally applied field, the central ion just keeps moving on, and its ionic atmosphere never quite catches up. It is as though the part of the cloud behind the central ion is "left standing." This is because its reason for existence (the field of the central ion) has deserted it and thermal motions try to disperse this part of the ionic cloud. In front of the central ion, the cloud is being continually built up. When ions move, therefore, one has a picture of the ions' losing the part of the cloud behind them and bulding up the cloud in front of them.

4.6.3. An Egg-Shaped Ionic Cloud and the "Portable" Field on the Central Ion

The constant lead which the central ion has on its atmosphere means that the center of charge of the central ion is displaced from the center of charge of its cloud. The first implication of this argument is that the ionic cloud is no longer spherically symmetrical around the *moving* central ion. It is egg-shaped (see Fig. 4.81).

A more serious implication is that, since the center of charge on a drifting central ion does not coincide with the center of charge of its oppositely charged (egg-shaped) ionic cloud, an electrical force develops be-

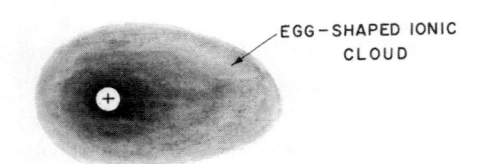

Fig. 4.81. The egg-shaped ionic cloud around a moving central ion.

tween the ion and its cloud. The development of an electric force between a moving ion and its lagging atmosphere means that the ion is then subject to an electric field. Since this field arises from the continual relaxation (or decay) of the cloud behind the ion and its buildup in front of the ion, it is known as a *relaxation field*. Notice, however, that the centers of charge of the ion and of the cloud lie on the path traced out by the moving ion (Fig. 4.82); and, consequently, the relaxation field generated by this charge separation acts in a direction precisely opposite to the direction of the driving force on the ion (e.g., the externally applied field). Hence, a moving ion, by having an egg-shaped ionic cloud, carries along its own "portable" field of force, the relaxation field, which acts to retard the central ion and decrease its mobility compared with that which it would have were it only pulled on by the externally applied field and retarded by the Stokes force [the zeroth-order theory of conductance, see Eq. (4.180)].

4.6.4. A Second Braking Effect of the Ionic Cloud on the Central Ion: The Electrophoretic Effect

The externally applied electric field acts not only on the central ion but also on its oppositely charged cloud. Consequently, the ion and its atmosphere *tend* to move in opposite directions.

This poses an interesting problem. The ionic atmosphere can be considered a charged sphere of radius $\varkappa^{-1}$ (Fig. 4.83). Under the action of an electric field, the charged sphere moves. The thickness of the ionic cloud

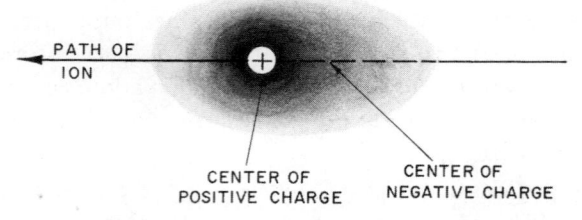

Fig. 4.82. The centers of charge of the ion and of the cloud lie on the path of the drifting ion.

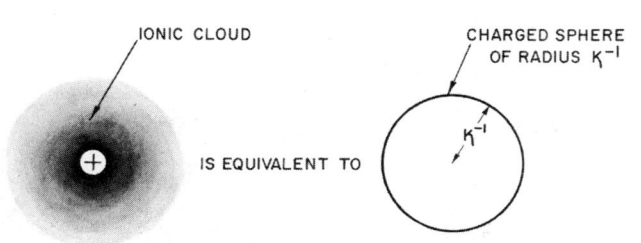

Fig. 4.83. The ionic atmosphere can be considered a charged sphere of radius $\varkappa^{-1}$.

in a millimolar solution of a 1:1-valent electrolyte is about 100 Å (see Table 3.2). One is concerned, therefore, with the migration of a fairly large "particle" under the influence of the electrical field. The term *electrophoresis* is generally used to describe the migration of particles of colloidal dimensions (10 to 10,000 Å) in an electric field. It is appropriate, therefore, to describe the migration of the ionic cloud as an *electrophoretic effect*.

The interesting point, however, is that, when the ionic cloud moves, it tries to carry along its entire baggage, the ions and the solvent molecules constituting the cloud *plus* the central ion. Thus, not only does the moving central ion attract and try to keep its cloud (the *relaxation effect*), but the moving cloud also attracts and tries to keep its central ion by means of a force which is then termed the *electrophoretic force* $\vec{F}_E$.

4.6.5. The Net Drift Velocity of an Ion Interacting with Its Atmosphere

In the elementary treatment of the migration of ions, it was assumed that the drift velocity of an ion was determined solely by the *electric force* $\vec{F}$ arising from the externally applied field.

When, however, the mutual interactions between an ion and its cloud were considered, it turned out (Sections 4.6.3 and 4.6.4) that there were two other forces operating on an ion. These extra forces consisted of: (1) the *relaxation force* $\vec{F}_R$ resulting from the distortion of the cloud around a moving ion, and (2) the electrophoretic force $\vec{F}_E$ arising from the fact that the ion shares in the electrophoretic motion of its ionic cloud. Thus, in a rigorous treatment of the migrational drift velocity of ions, one must consider a total force $\vec{F}_{\text{total}}$, which is the resultant of that due to the applied electric field together with the relaxation and electrophoretic forces (Fig. 4.84)

$$\vec{F}_{\text{total}} = \vec{F} - (\vec{F}_E + \vec{F}_R) \qquad (4.274)$$

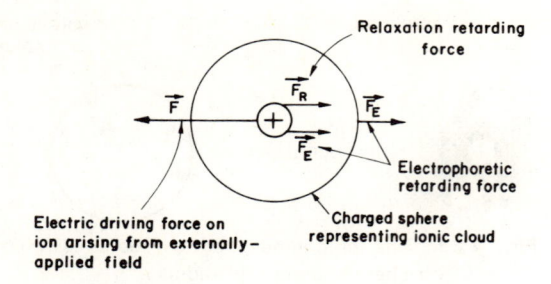

Fig. 4.84. The ion drift due to a net force which is a resultant of the electric driving force and two retarding forces, the relaxation and electrophoretic forces.

The minus sign is used because both the electrophoretic and relaxation forces act in a direction opposite to that of the externally applied field.

Since an ion is subject to a resultant, or net, force, its drift velocity, too, must be a *net* drift velocity resolvable into components. Further, since each component force should produce a component of the overall drift velocity, there must be three components of the net drift velocity. The first component, which shall be designated v^0, is the direct result of the externally applied field only and excludes the influence of interactions between the ion and the ionic cloud; the second is the electrophoretic component v_E and arises from the participation of the ion in the electrophoretic motion of its cloud; and, finally, the third component is the reaxation field component v_R originating from the relaxation force which retards the drift of the ion. Since the electrophoretic and relaxation forces act in an opposite sense to the externally applied electric field, it follows that the electrophoretic and relaxation components must diminish the overall drift velocity (Fig. 4.85), i.e.,

$$v_d = v^0 - (v_E + v_R) \tag{4.275}$$

The next step is to evaluate the electrophoretic and relaxation components of the net drift velocity of an ion.

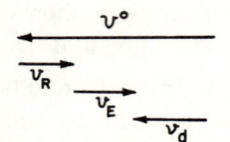

Fig. 4.85. The components of the overall drift velocity.

4.6.6. The Electrophoretic Component of the Drift Velocity

The electrophoretic component v_E of the drift velocity of an ion is equal to the electrophoretic velocity of its ionic cloud because the central ion shares in the motion of its cloud. If one ignores the asymmetry of the ionic cloud, a simple calculation of the electrophoretic velocity v_E can be made.

The ionic atmosphere is accelerated by the externally applied electric force[†] $ze_0 X/300$, but it is retarded by a Stokes viscous force. When the cloud attains a steady-state electrophoretic velocity v_E, then the viscous force is exactly equal and opposite to the electric force driving the cloud

$$ze_0 \frac{X}{300} = \text{Stokes' force on cloud} \qquad (4.276)$$

The general formula for Stokes' viscous force is $6\pi r \eta v$, where r and v are the radius and velocity of the moving sphere. In computing the viscous force on the cloud, one can substitute $\varkappa^{-1}$ for r and v_E for v in Stokes' formula. Thus,

$$ze_0 \frac{X}{300} = 6\pi \varkappa^{-1} \eta v_E \qquad (2.277)$$

from which it follows that

$$v_E = \frac{ze_0}{6\pi \varkappa^{-1} \eta} \frac{X}{300} \qquad (4.278)$$

This, therefore, is the expression for the electrophoretic contribution to the drift velocity of an ion.

4.6.7. The Procedure for Calculating the Relaxation Component of the Drift Velocity

From the familiar relation [*cf.* Eq. (4.146)],

Velocity = Absolute mobility × Force

it is clear that the relaxation component v_R of the drift velocity of an ion can be obtained by substituting for the relaxation force $\vec{F}_R$ in

$$v_R = \bar{u}_{\text{abs}}^0 \vec{F}_R \qquad (4.279)$$

The problem, therefore, is to evaluate the relaxation force.

[†] The factor 1/300 is introduced to permit the field X to be expressed in volts per centimeter rather than in electrostatic units of potential per centimeter.

428 CHAPTER 4

Since the latter arises from the distortion of the ionic cloud, one must derive a relation between the relaxation force and a quantity characterizing the distortion. It will be seen that the straightforward measure of the asymmetry of the cloud is the distance d through which the center of charge of the ion and the center of charge of the cloud are displaced.

But the distortion d of the cloud itself depends on a relaxation process in which the part of the cloud in front of the moving ion is being built up and the part at the back is decaying. Hence, the distortion d and, therefore, the relaxation force $\vec{F}_R$ must depend on the time taken by a cloud to relax, or decay.

Thus, it is necessary firstly to calculate how long an atmosphere would take to decay, then to compute the distortion parameter d, and finally to obtain an expression for the relaxation force $\vec{F}_R$. Once this force is evaluated, it can be introduced into Eq. (4.279) for the relaxation component v_R of the drift velocity.

4.6.8. How Long Does an Ion Atmosphere Take to Decay?

An idea of the time involved in the readjustment of the ionic cloud around the moving central ion can be obtained by a thought experiment suggested by Debye (Fig. 4.86). Consider a static central ion with an equilibrium, spherical ionic cloud around it. Let the central ion suddenly be discharged. This perturbation of the ion–ionic cloud system sets up a relaxation process. The ionic cloud is now at the mercy of the thermal forces which try to destroy the ordering effect previously maintained by the central ion and responsible for the creation of the cloud.

The actual mechanism by which the ions constituting the ionic atmosphere are dispersed is none other than the random-walk process described in Section 4.2. Hence, the time taken by the ionic cloud to relax or disperse may be estimated by the use of the Einstein–Smoluchowski relation (Section 4.2.6)

$$t = \frac{\langle x^2 \rangle}{2D} \qquad (4.27)$$

But what distance x is to be used? In other words, when can the ionic cloud be declared to have dispersed or relaxed? These questions may be answered by recalling the description of the ionic atmosphere where it was stated that the charge density in a dr-thick spherical shell in the cloud declines rapidly at distances greater than the Debye–Hückel length $\varkappa^{-1}$. Hence, if the ions diffuse to a distance $\varkappa^{-1}$, the central ion can be stated to have lost its cloud, and the time taken for this diffusion provides an

ION TRANSPORT IN SOLUTIONS 429

(a) CHARGED CENTRAL ION
IONIC CLOUD

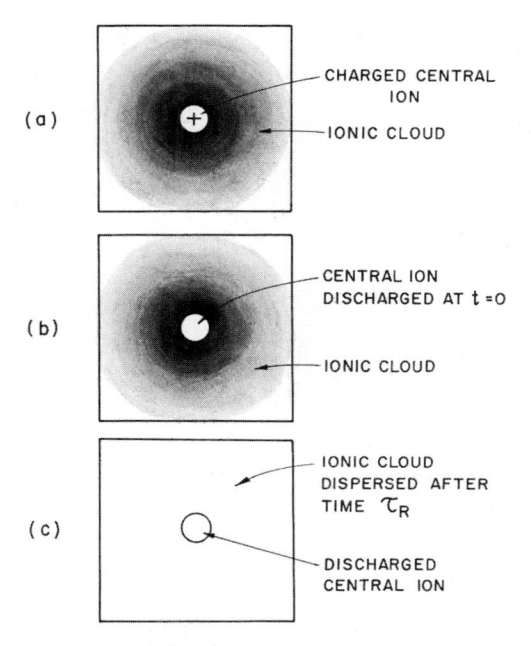

(b) CENTRAL ION DISCHARGED AT t = 0
IONIC CLOUD

(c) IONIC CLOUD DISPERSED AFTER TIME τ_R
DISCHARGED CENTRAL ION

Fig. 4.86. Debye's thought experiment to calculate the time for the ion atmosphere to relax: (a) the ionic cloud around a central ion; (b) at $t = 0$, the central ion is discharged; and (c) after time τ_R, the ion atmosphere has relaxed or dispersed.

estimate of the relaxation time τ_R. One has, by substituting $\varkappa^{-1}$ in the Einstein–Smoluchowski relation [Eq. (4.27)]

$$\tau_R = \frac{(\varkappa^{-1})^2}{2D} \tag{4.280}$$

which, with the aid of the Einstein relation $D = \bar{u}_{\text{abs}}kT$ [Eq. (4.169)], can be transformed into the expression

$$\tau_R = \frac{(\varkappa^{-1})^2}{2\bar{u}_{\text{abs}}kT} \tag{4.281}$$

4.6.9. The Quantitative Measure of the Asymmetry of the Ionic Cloud around a Moving Ion

To know how asymmetric the ionic cloud has become owing to the relaxation effect, one must calculate the distance d through which the central ion has moved in the relaxation time τ_R. This is easily done by multiplying the relaxation time τ_R by the velocity v^0 which the central ion acquires from the externally applied electric force, i.e.,

$$d = \tau_R v^0 \qquad (4.282)$$

By substituting the expression (4.281) for the relaxation time τ_R, Eq. (4.282) becomes

$$d = \frac{v^0 (\varkappa^{-1})^2}{2\bar{u}^0_{\text{abs}} kT} \qquad (4.283)$$

The center of charge of the relaxing ionic cloud coincides with the original location of the central ion; in the meantime, however, the central ion and its center of charge moves through a distance d. Hence, the centers of charge of the central ion and its ionic cloud are displaced through the distance d, which therefore is a quantitative measure of the egg-shapedness of the ion atmosphere around a moving ion.

4.6.10. The Magnitude of the Relaxation Force and the Relaxation Component of the Drift Velocity

Consider first a *static* central ion. The ion may exert an electric force on the cloud, and *vice versa*; but, at first, the net force is zero because of the spherical symmetry of the cloud around the static central ion.

When the central ion *moves*, it can be considered at a distance d from the center of its cloud. Now, the net force due to the asymmetry of the cloud is nonzero. A rough calculation of the force can be made as follows.

The relaxation force is zero when the centers of charge of the ion and its cloud coincide, and it is nonzero when they are separated. So let it be assumed in this approximate treatment that the relaxation force is proportional to d, i.e., proportional to the distance through which the ion has moved from the original center of charge of the cloud. On this basis, the relaxation force $\vec{F}_R$ will be given by the maximum total force of the atmosphere on the central ion, i.e., $z^2 e_0^2 / [\varepsilon(\varkappa^{-1})^2]$,[†] multiplied by the fraction of the radius of the cloud through which the central ion is displaced during it motion under the external field, i.e., $d/\varkappa^{-1}$. Hence, the relaxation force is

$$\vec{F}_R = \frac{z^2 e_0^2}{\varepsilon(\varkappa^{-1})^2} \frac{d}{\varkappa^{-1}} \qquad (4.284)$$

[†] This total force is obtained by considering the ionic cloud equivalent to an equal and opposite charge placed at a distance $\varkappa^{-1}$ from the central ion. Then, Coulomb's law for the force between these two charges gives the result $z^2 e_0^2 / [\varepsilon(\varkappa^{-1})^2]$.

and, using Eq. (4.283) for d, one has

$$\vec{F}_R = \frac{z^2 e_0^2}{\varepsilon(\varkappa^{-1})^2} \frac{1}{\varkappa^{-1}} \frac{v^0(\varkappa^{-1})^2}{2\bar{u}^0_{\text{abs}} kT}$$

$$= \frac{z^2 e_0^2 \varkappa}{2\varepsilon kT} \frac{v^0}{\bar{u}^0_{\text{abs}}} \qquad (4.285)$$

Since, however, the velocity v^0 arises solely from the externally applied field and excludes the influence of ion–ion interactions, the ratio $v^0/\bar{u}^0_{\text{abs}}$ is equal to the applied electric force

$$\frac{v^0}{\bar{u}^0_{\text{abs}}} = \vec{F} = \frac{z e_0}{300} X \qquad (4.286)$$

On inserting this into Eq. (4.285), it turns out that

$$\vec{F}_R = \frac{z^3 e_0^3 \varkappa}{2\varepsilon kT} \frac{X}{300} \qquad (4.287)$$

In the above treatment of the relaxation field, it has been assumed that the only motion of the central ion destroying the spherical symmetry of the ionic cloud is motion in the direction of the applied external field. This latter directed motion is in fact a drift superimposed on a random walk. But the random walk is, itself, a series of motions, and these motions are random in direction. Thus, the central ion exercises an erratic, rather than a consistent, leadership on its atmosphere.

Onsager considered the effect that this erratic character of the leadership would have on the time-average shape of the ionic cloud and therefore on the relaxation field. His final result differs from Eq. (4.287) in two respects: (1) Instead of the numerical factor $\frac{1}{2}$, there is a factor $\frac{1}{3}$; and (2) a correction factor $\omega/2z^2$ has to be introduced, the quantity ω being given by

$$\omega = z_+ z_- \frac{2q}{1 + \sqrt{q}} \qquad (4.288)$$

in which

$$q = \frac{z_+ z_-}{z_+ + z_-} \frac{\lambda_+ + \lambda_-}{z_+ \lambda_+ + z_- \lambda_-} \qquad (4.289)$$

where λ_+ and λ_- are related to mobilities of cation and anion, respectively. For symmetrical or z:z-valent electrolytes, the expression for q reduces to $\frac{1}{2}$,

TABLE 4.18

Value of ω for Different Types of Electrolytes

Type of electrolyte	ω
1:1	0.5859
2:2	2.3436

and that for ω becomes (Table 4.18)

$$\omega = z^2 \frac{1}{1 + (1/\sqrt{2})} \tag{4.290}$$

Thus, a more rigorous expression for the relaxation force is

$$\vec{F}_R = \frac{z^3 e_0^3 \varkappa}{3\varepsilon kT} \frac{\omega}{2z^2} \frac{X}{300}$$

$$= \frac{z e_0^3 \varkappa \omega}{6\varepsilon kT} \frac{X}{300} \tag{4.291}$$

Substituting the expression (4.291) for the relaxation force in the equation (4.279) for the relaxation component of the drift velocity, one gets

$$v_R = \bar{u}_{\text{abs}}^0 \frac{z e_0^3 \varkappa \omega}{6\varepsilon kT} \frac{X}{300} \tag{4.292}$$

Further, from the definition (4.149) of the conventional mobility,

$$u^0 = u_{\text{conv}}^0 = \bar{u}_{\text{abs}}^0 \frac{z e_0}{300}$$

Eq. (4.292) becomes

$$v_R = \frac{300 u^0}{z e_0} \frac{z e_0^3 \varkappa \omega}{6\varepsilon kT} \frac{X}{300}$$

$$= \frac{u^0 e_0^2 \varkappa \omega}{6\varepsilon kT} X \tag{4.293}$$

4.6.11. The Net Drift Velocity and Mobility of an Ion Subject to Ion–Ion Interactions

Now that the electrophoretic and relaxation components in the drift velocity of an ion have been evaluated, they can be introduced into Eq.

(4.275) to give

$$v_d = v^0 - (v_E + v_R) \quad (4.275)$$

$$= v^0 - \left(\frac{ze_0\varkappa}{6\pi\eta}\frac{X}{300} + \frac{u^0 e_0^2 \varkappa \omega}{6\varepsilon kT}X\right)$$

$$= v^0 - \left(\frac{ze_0}{6\pi\eta} + \frac{300 u^0 e_0^2 \omega}{6\varepsilon kT}\right)\varkappa \frac{X}{300}$$

If one divides throughout by X, then, according to the definition of the conventional mobility $u_{\text{conv}} = u = v/X$, one has

$$u = u^0 - \left(\frac{ze_0}{6\pi\eta} + \frac{300 u^0 e_0^2 \omega}{6\varepsilon kT}\right)\frac{\varkappa}{300} \quad (4.294)$$

An intelligent inspection of expression (4.294) shows that the mobility u of ions is not a constant independent of concentration. Thus, it depends on the Debye–Hückel reciprocal length $\varkappa$. But this parameter $\varkappa$ is a function of concentration (see Eq. 3.84). Hence, Eq. (4.294) shows that the mobility of ions is a function of concentration, as was suspected (Section 4.6.1) on the basis of the empirical law of Kohlrausch.

As the concentration decreases, $\varkappa^{-1}$ increases and $\varkappa$ decreases, as can be seen from Eq. (3.84). In the limit of infinite dilution ($c \to 0$), $\varkappa^{-1} \to \infty$ or $\varkappa \to 0$. Under these conditions, the second and third terms in Eq. (4.294) drop out, which leaves

$$u_{\text{limit},\,c\to 0} = u^0 \quad (4.295)$$

The quantity u^0 is therefore the mobility at infinite dilution and can be

TABLE 4.19

Transport Numbers of Cations in Aqueous Solutions at 25°C

Concentration	HCl	LiCl	NaCl	KCl
0.01N	0.8251	0.3289	0.3918	0.4902
0.02	0.8266	0.3261	0.3902	0.4901
0.05	0.8292	0.3211	0.3876	0.4899
0.1	0.8314	0.3168	0.3854	0.4898
0.2	0.8337	0.3112	0.3821	0.4894
0.5	—	0.300	—	0.4888
1.0	—	0.287	—	0.4882

considered given by the expression for the Stokes mobility (Section 4.4.8), i.e.,

$$u^0 = \frac{1}{300} \frac{ze_0}{6\pi r \eta} \quad (4.180)$$

To go back to the question which concluded the previous section (i.e., 4.5), it is now clear that, since transport numbers depend on ionic mobilities, which have now been shown to vary with concentration, the transport number must itself be a concentration-dependent quantity (Table 4.19). However, it is seen that this variation is a small one.

4.6.12. The Debye–Hückel–Onsager Equation

Now, the equivalent conductivity Λ of an electrolytic solution is simply related to the mobilities of the constituent ions [Eq. (4.160)]

$$\Lambda = F(u_+ + u_-)$$

Thus, to obtain the equivalent conductivity, one has only to write down the expression for the mobilities of the positive and negative ions, multiply both the expressions by the Faraday F, and then add up the two expressions. The result is

$$\Lambda = F(u_+ + u_-)$$

$$= F\left[u_+^0 - \frac{\varkappa}{300}\left(\frac{z_+ e_0}{6\pi\eta} + \frac{300\, e_0^2 \omega}{6\varepsilon kT} u_+^0\right)\right]$$

$$+ F\left[u_-^0 - \frac{\varkappa}{300}\left(\frac{z_- e_0}{6\pi\eta} + \frac{300\, e_0^2 \omega}{6\varepsilon kT} u_-^0\right)\right] \quad (4.296)$$

For a symmetrical electrolyte, $z_+ = z_- = z$ or $z_+ + z_- = 2z$, and, therefore, Eq. (4.296) reduces to

$$\Lambda = F(u_+^0 + u_-^0) - \left[\frac{ze_0 F\varkappa}{900\pi\eta} + \frac{e_0^2 \omega \varkappa}{6\varepsilon kT}F(u_+^0 + u_-^0)\right] \quad (4.297)$$

However, according to Eq. (4.270),

$$\Lambda^0 = F(u_+^0 + u_-^0) \quad (4.270)$$

Hence,

$$\Lambda = \Lambda^0 - \left(\frac{ze_0 F\varkappa}{900\pi\eta} + \frac{e_0^2 \omega \varkappa}{6\varepsilon kT}\Lambda^0\right) \quad (4.298)$$

ION TRANSPORT IN SOLUTIONS 435

Replacing $\varkappa$ by the familiar expression (3.84), i.e.,

$$\varkappa = \left(\frac{8\pi z^2 e_0^2 c}{\varepsilon k T}\right)^{\frac{1}{2}} \left(\frac{N_A}{1000}\right)^{\frac{1}{2}}$$

one has

$$\Lambda = \Lambda^0 - \left[\frac{ze_0 F}{900\pi\eta}\left(\frac{8\pi z^2 e_0^2 N_A}{1000\varepsilon k T}\right)^{\frac{1}{2}} + \frac{e_0^2 \omega}{6\varepsilon k T}\left(\frac{8\pi z^2 e_0^2 N_A}{1000\varepsilon k T}\right)^{\frac{1}{2}} \Lambda^0\right] c^{\frac{1}{2}} \quad (4.299)$$

This is the well-known Debye–Hückel–Onsager equation for a symmetrical electrolyte. By defining the following constants

$$A = \frac{ze_0 F}{900\pi\eta}\left(\frac{8\pi z^2 e_0^2 N_A}{1000\varepsilon k T}\right)^{\frac{1}{2}} \quad \text{and} \quad B = \frac{e_0^2 \omega}{6\varepsilon k T}\left(\frac{8\pi z e_0 N_A}{1000\varepsilon k T}\right)^{\frac{1}{2}}$$

it can also be written thus

$$\Lambda = \Lambda^0 - (A + B\Lambda^0)c^{\frac{1}{2}}$$

or

$$= \Lambda^0 - \text{constant } c^{\frac{1}{2}} \quad (4.300)$$

Thus, the theory of ionic clouds has been able to give rise to an equation which has the same form as the empirical law of Kohlrausch (Section 4.3.9).

4.6.13. The Theoretical Predictions of the Debye–Hückel–Onsager Equation versus the Observed Conductance Curves

The two constants

$$A = \frac{ze_0 F}{900\pi\eta}\left(\frac{8\pi z^2 e_0^2}{1000\varepsilon k T}\right)^{\frac{1}{2}} \quad (4.301)$$

and

$$B = \frac{e_0^2 \omega}{6\varepsilon k T}\left(\frac{8\pi z^2 e_0^2}{1000\varepsilon k T}\right)^{\frac{1}{2}} \quad (4.302)$$

in the Debye–Hückel–Onsager equation are completely determined (Table 4.20) by the valence type of the electrolyte, z, the temperature T, the dielectric constant ε, and the viscosity η of the solution and by universal constants.

The Debye–Hückel–Onsager equation has been tested against a large body of accurate experimental data. The confrontation of theory and experiment is shown in Fig. 4.87 and Table 4.21 for *aqueous* solutions of true electrolytes, i.e., substances which consisted of ions in their crystal lattices before they were dissolved in water.

TABLE 4.20
Values of the Onsager Constants for Uni-Univalent Electrolytes at 25°

Solvent	A	B
Water	60.20	0.229
Methyl alcohol	156.1	0.923
Ethyl alcohol	89.7	1.83
Acetone	32.8	1.63
Acetonitrile	22.9	0.716
Nitromethane	125.1	0.708
Nitrobenzene	44.2	0.776

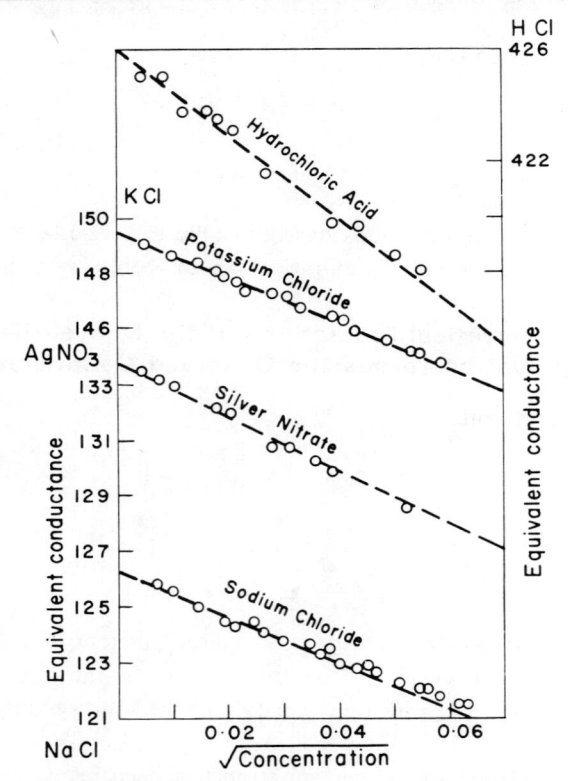

Fig. 4.87. Comparison of the equivalent conductivities of HCl and some salts predicted by the Debye–Hückel–Onsager equation (4.299) with those observed experimentally.

TABLE 4.21
Observed and Calculated Onsager Slopes in Aqueous Solutions at 25°C

Electrolyte	Observed slope	Calculated slope
LiCl	81.1	72.7
NaNO$_3$	82.4	74.3
KBr	87.9	80.2
KCNS	76.5	77.8
CsCl	76.0	80.5
MgCl$_2$	144.1	145.6
Ba(NO$_3$)$_2$	160.7	150.5
K$_2$SO$_4$	140.3	159.5

At very low concentrations ($< 0.001 N$, say), the agreement between theory and experiment is very good. There is no doubt that the theoretical equation is a satisfactory expression for the *limiting tangent* to the experimentally obtained Λ versus $c^{\frac{1}{2}}$ curves.

One cannot, however, expect the Debye–Hückel–Onsager theory of the nonequilibrium conduction properties of ionic solutions to fare better at high concentration than the corresponding Debye–Hückel theory of the

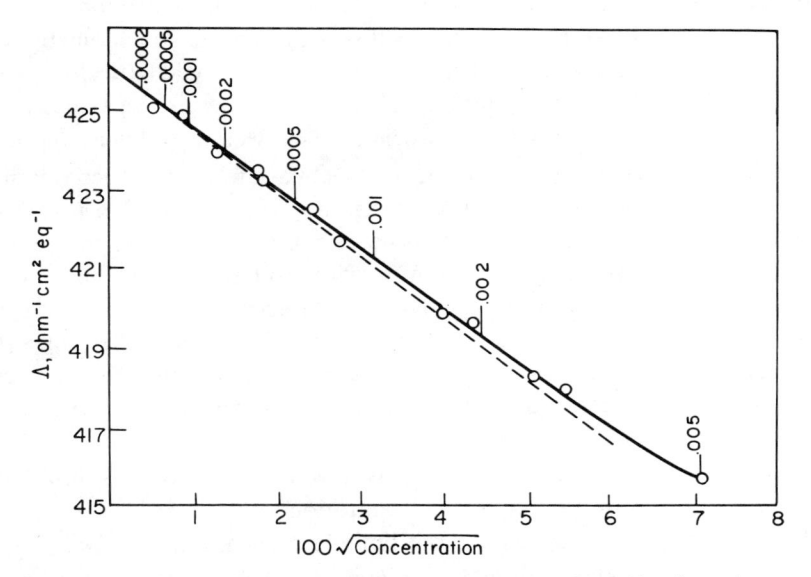

Fig. 4.88. Deviation of the predicted equivalent conductivities from those observed for HCl.

equilibrium properties, e.g., activity coefficients, of electrolytic solutions; both theories are based on the ionic-cloud concept.

In the case of the Debye–Hückel–Onsager equation, it is seen from Fig. 4.88 that, as the concentration increases (particularly above $0.001N$), the disparity between the theoretical and experimental curves widens.

4.6.14. A Theoretical Basis for Some Modifications of the Debye–Hückel–Onsager Equation

A general procedure adopted by theorists for the improvement and extension of any theory is to examine its basic assumptions critically.

The elementary theory of conduction in electrolytic solutions was based on the assumption that the charge-carrying ions do not get in each other's way, i.e., that the ionic cloud is so tenuous that it can be ignored. The removal of this assumption required the incorporation of ionic-atmosphere effects, in particular, the development of an expression for the relaxation field and the effects of the electrophoretic movement of the ionic cloud. Thus arose the Debye–Hückel–Onsager equation for equivalent conductivity. Even this theory, however, is not free of assumptions and approximations.

One obvious assumption in the Debye–Hückel–Onsager theory centers around the quantity c. The theory considers c the concentration of charge carriers in the solution. Whether c will be equal to the stoichiometric concentration c_s depends on whether every ion from the ionic lattice of the true electrolyte dissolves in the solution and remains free as a charge carrier. This condition will be met only if ion association (see Section 3.8) does not occur.

A more serious assumption in the Debye–Hückel–Onsager treatment of conductance is that of point-charge ions. (This assumption is not self-evident in the presentation adopted here, but it is explicit in the rigorous derivations of the theory.) One way of considering *finite*, nonzero ion sizes is to replace $\varkappa$ in the Debye–Hückel–Onsager equation by $\varkappa/(1 + \varkappa a)$, where a is the closet distance of approach of a cloud ion to the central ion. It may be recalled that, in the theory of activity coefficients, a similar change was required to incorporate ion-size effects in the expression for the activity coefficient (see Section 3.5).

The effect of considering that ions have a finite size is to introduce a positive term proportional to concentration and make the theoretical Λ versus $c^{\frac{1}{2}}$ curve concave from the limiting tangent up.

Finally, the relaxation and electrophoretic effects have been discussed in the simple treatment as if they were independent of each other. In fact, both effects arise from the response of the same ionic cloud, populated by the same ions and solvent molecules, to the changing field of a moving central ion (relaxation effect) and to the external field (electrophoretic effect). Since the same assembly of ions is responsible for the two effects, they cannot be independent. For instance, the electrophoretic movement of the ionic cloud must affect the relaxation field.

Fuoss and Onsager, in a series of papers from 1955 onwards, have considered the effects of finite-ion-size, ion-pairing, and electrophoretic terms in the relaxation field on the conductance of electrolytes.

They have also allowed for two other factors. Firstly, since the ionic cloud is egg shaped around a moving central ion, there will be more ions of opposite sign behind the central ion than in front of it (i.e., in the direction in which it is drifting). These oppositely charged ions are attracted to the central ion, which therefore gets more kicks from behind than from the front. The central ions thus acquire a small velocity component in the direction opposite to that of travel demanded by the external field. In other words, there is in effect a small force in the same direction as the external field. Secondly, due to the presence of an excess of oppositely charged ions around the central ion, there is a small change in the local viscosity. This change alters the mobility of the central ion, as can be seen from Stokes' laws.

The Fuoss–Onsager extended theory appears to be as far as one can go in the conductance problem while still retaining the basic concepts of the ionic cloud. It may be recalled, however, that in concentrated solutions when the mean ion–ion distance becomes of the order of ionic dimensions, the ionic-atmosphere concept becomes rather untenable (see Section 3.9). One must think of ion clusters and quasi-lattices. The problems of conductance in concentrated solutions and in the most concentrated solutions of all—pure electrolytes devoid of water—demand an altogether different approach, to which some attention will be given in Chapter 6.

Further Reading

1. P. Debye and E. Hückel, *Z. Physik*, **24**: 185 (1923).
2. L. Onsager, *Z. Physik*, **28**: 277 (1927).
3. H. Falkenhagen, *Z. Physik*, **32**: 365 and 745 (1931).
4. R. M. Fuoss and F. Accasina, *Electrolytic Conductance*, Interscience Publishers, Inc., New York, 1959.
5. H. L. Friedman, *Ionic Solution Theory*, Interscience Publishers, Inc., New York, 1962.
6. R. M. Fuoss and L. Onsager, *J. Phys. Chem.*, **66**: 1722 (1962); **67**: 621 (1963); and **68**: 1 (1964).

440 CHAPTER 4

7. R. M. Fuoss, L. Onsager, and J. F. Skinner, *J. Phys. Chem.*, **69**: 2581 (1965).
8. H. L. Friedman, "A New Theory of Conductance," in: B. E. Conway and R. G. Barradas, eds., *Chemical Physics of Ionic Solutions*, John Wiley & Sons, Inc., New York, 1966.

4.7. NONAQUEOUS SOLUTIONS: A NEW FRONTIER IN IONICS?

4.7.1. Water Is the Most Plentiful Solvent

Not only is water the most plentiful solvent, it is also a most successful and useful solvent, and there are several good reasons which support this description. Firstly, the dissolution of true electrolytes occurs by solvation (Chapter 2) and, therefore, depends on the free energy of solvation. A sizable fraction of this free energy depends on the Born contribution $(N_A z_i^2 e_0^2/2r)(1 - (1/\varepsilon))$. It follows that, the greater the dielectric constant of the solvent is, the greater is its ability to dissolve true electrolytes. Since water has a particularly high dielectric constant (Table 4.22), it is a successful solvent for true electrolytes.

A second advantage of water is that, in addition to being able to dissolve electrolytes by the physical forces involved in solvation, it is also able to undergo *chemical* proton-transfer reactions with potential electrolytes and thus produce ionic solutions (Chapter 5). Now, water is able both to donate

TABLE 4.22

Dielectric Constants of Some Solvents
(Temperature 25°C unless Otherwise Noted)

Solvent	Dielectric constant
Water	78.30
Acetone	20.70
Acetonitrile	36.70
Ammonia (-34 °C)	22.00
Benzene	2.27
Dimethyl acetamide	37.78
Dimethyl sulfoxide	46.70
Dioxan	2.21
Ethanol	24.30
Ethylene diamine	12.90
Hydrogen cyanide (16 °C)	118.30
Pyridine	12.00
Sulfuric acid	101.00

TABLE 4.23
Boiling Points of Some Solvents

Solvent	Boiling point, °C
Water	100.00
Acetone	56.20
Benzene	80.10
1:1 Dichloroethane	57.00
Methanol	64.96

protons to, and to receive protons from, molecules of potential electrolytes. Thus, water can function as both a source and a sink for protons and can consequently enter into ion-forming reactions with a particularly large range of substances. This is why potential electrolytes often react best with water as a partner in the proton-transfer reactions.

Finally, water is stable both chemically and physically at ambient temperature, unlike many organic solvents, which tend to evaporate (Table 4.23) or decompose slowly with time.

On the whole, therefore, ionics is best practiced in water. Nevertheless, there are also good reasons why nonaqueous solutions of electrolytes are often of interest.

4.7.2. Water Is Often Not an Ideal Solvent

If water were the ideal solvent, there would be no need in technology to consider other solvents. But, in many situations, water is hardly the ideal solvent. Take the electrolytic production of sodium metal, for example. If an *aqueous* solution of a sodium salt is taken in an electrolytic cell and a current passed between two electrodes, then all that will happen at the cathode is the liberation of hydrogen gas; there will be no electrodeposition of sodium (*cf.* Chapter 8). Hence, sodium cannot be electrowon from aqueous solutions. This is why the electrolytic extraction of sodium has taken place from molten sodium hydroxide, i.e., from a medium free of hydrogen. But this process requires the system to be kept molten ($\sim 600°C$), and it, therefore, compels the use of high-temperature technology with its associated materials problems. It would be a boon to industry if one could use a low temperature conducting solution having the capacity to maintain sodium ions in a nonaqueous solvent free of ionizing hydrogen. This argument is valid for many other metals which are extracted today by electrodic reactions in fused salts at high temperatures with the attendant difficulties of corrosion and heat losses.

Another vast field awaiting the development of nonaqueous electrochemistry is that of energy storage for automobiles. Many reasons, e.g., the growing danger of pollution from automobile exhausts, the increasing concentration of CO_2 (with the consequences of a warmup of the atmosphere), the accelerating consumption of oil reserves, etc., make the search for an alternative to the internal combustion engine a necessity. Nuclear reactors with their attendant shielding problems will always be too heavy for the relatively small power needed in road vehicles. Thus, there would be attractive advantages to be obtained by the development of a fumeless, vibration-free electric power source. However, the presently available cheap electrochemical device—the lead–acid battery—is too heavy for the electric energy which it stores to offer reasonable performance with convenient distance range between recharging. Electrochemical-energy storers available today which do have a sufficiently high energy capacity per unit weight offer difficulties because of their expense. The highest energy density theoretically conceivable is in a storage device which utilizes the dissolution of lithium or beryllium. But aqueous electrolytes are debarred because, in them, these metals corrode wastefully rather than dissolve with useful power production. So one answer to the need for an electrochemically powered transport system is the development of a nonaqueous electrochemical-energy storage system incorporating alkali or alkaline-earth metal electrodes.

Many other examples could be cited in which the presence of water as a solvent is a nuisance. In all these cases, there may exist an important future for applications using nonaqueous solutions.

However, a nonaqueous solution must be able to conduct electricity if it is going to be useful. What, therefore, determines the conductivity of a nonaqueous solution? The case of nonaqueous solutions in which the charge carriers are produced by chemical reactions between molecules of the solvent and the potential electrolyte will be dealt with in Chapter 5. Here, the theoretical principles involved in the conductance behavior of *true* electrolytes in nonaqueous solvents will be sketched.

4.7.3. The Debye–Hückel–Onsager Theory for Nonaqueous Solutions

An examination of the Debye–Hückel–Onsager theory in Section 4.6 shows that it is in no way wedded to the particular solvent water. Does experiment support the predicted Λ versus $c^{\frac{1}{2}}$ curve in nonaqueous solutions, too?

ION TRANSPORT IN SOLUTIONS 443

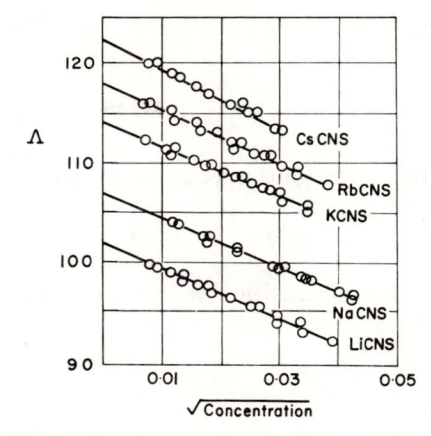

Fig. 4.89. Change in the equivalent conductivity of some alkali cyanates with concentration in methyl alcohol.

Figure 4.89 shows the variation of the equivalent conductivity *versus* concentration for a number of alkali sulphocyanates[†] in a methanol solvent. The agreement with the theoretical predictions demonstrates the applicability of the Debye–Hückel–Onsager equation.

When one switches from water to some nonaqueous solvent, the magnitudes of several quantities in the Debye–Hückel–Onsager equation alter, sometimes drastically, even if one considers the same true electrolyte in all these solvents. These quantities are the viscosity and the dielectric constant of the medium and the distance of closest approach of the solvated ions, i.e., the sum of the radii of the solvated ions. As a result, the mobilities of the ions at infinite dilution, the slope of the Λ versus $c^{\frac{1}{2}}$ curve, and, lastly, the concentration c of *free* ions cause the conductance behavior of an electrolyte to vary when one goes over from water to nonaqueous solvent. These effects will now be considered in detail.

4.7.4. The Solvent Effect on the Mobility at Infinite Dilution

At infinite dilution, neither the relaxation nor the electrophoretic effects are operative on the drift of ions; both these effects depend for their existence on a finite-sized ionic cloud. Under these special conditions, therefore,

---

[†] Abundant work exists on such electrolytes, rather than on, say, NaCl, because electrolytes with ions larger than Cl^- ions are more soluble in organic solvents than are the chlorides.

the infinite-dilution mobility can be considered given by the Stokes mobility

$$u^0_{\text{conv}} = \frac{ze_0}{6\pi r\eta} \frac{1}{300} \qquad (4.180)$$

Hence, considering the same ionic species in several solvents, one has

$$u^0_{\text{conv}} r\eta = \text{constant} \qquad (4.303)$$

If the radius r of the solvated ion is independent of the solvent, then one can approximate Eq. (4.303) to†

$$u^0\eta = \text{constant} \qquad (4.304)$$

Hence, an increase in the viscosity of the medium leads to a decrease in the infinite-dilution mobility, and *vice versa*.

Table 4.14 contains data on the equivalent conductivity, the viscosity, and the Walden constant $\Lambda^0\eta$ for several electrolytes and several solvents. It is seen that: (1) Eq. (4.304) is a fair approximation in many solvents and (2) its validity is better for solvents other than water.

In fact, however, the radius of the kinetic entity may change in going from one solvent to the other, essentially because of changes in the structure of the solvation sheath. Sometimes, these solvation effects on the radius may be as much as 100%. Hence, it is only to this rough degree that one can use the approximate equation (4.304).

In some cases, the changes in the radii of the solvated ions are mainly due to the changes in the sizes of the solvent molecules in the solvation sheath. Thus, in the case of the solvents water, methanol, and ethanol, the size of the three solvent molecules increases in the order

$$\text{Water} \to \text{Methanol} \to \text{Ethanol}$$

Since the radius of the solvated ions should also increase in the same order, it follows from Eq. (4.303) that the mobility or equivalent conductivity at infinite dilution should increase from ethanol to methanol to water. This is indeed what is observed (*cf.* Table 4.15).

One should be careful in using a simple Walden's rule $\Lambda\eta = \text{constant}$, which assumes that the radii of the moving ions are independent of the

---

† Actually, Eq. (4.304) containing the mobility is a form of Walden's rule [*cf.* Eq. (4.189)] which contains the equivalent conductivity. Since $\Lambda^0 = F[(u_{\text{conv}})_+^0 + (u_{\text{conv}})_-^0]$—*cf.* Eq. (4.270)—, it is easy to transform (4.304) into the usual form of Walden's rule.

solvent. Rather, one should use a generalized Walden's rule, namely,

$$u^0 r\eta = \text{constant} \tag{4.303}$$

where r is the radius of the ionic entity concerned in a given solvent.

4.7.5. The Slope of the Λ versus $c^{\frac{1}{2}}$ Curve as a Function of the Solvent

If one takes the generalized Walden's rule (4.303) and calculates (from $\Lambda^0 = Fu^0$) the equivalent conductivity at *infinite dilution* for a number of nonaqueous solutions, it turns out that the values of Λ^0 in such solutions are relatively high. They are near those of water and are in some cases greater than those of water.

One might naïvely conclude from this fact that, in using nonaqueous solutions instead of aqueous solutions in an electrochemical system, the conductivity presents no problem. Unfortunately, this is not the case. The crucial quantity which often determines the feasibility of using nonaqueous solutions in practical electrochemical systems is the *specific conductivity σ at a finite concentration*, not the equivalent conductivity Λ^0 at infinite dilution. The point is that it is the specific conductivity which, in conjunction with the electrode geometry, determines the electrolyte resistance in an electrochemical system. This electrolyte resistance is an important factor in the operation of an electrochemical system because the extent to which useful power is diverted into the wasteful heating of the solution depends on I^2R, where I is the current passing through the electrolyte; hence, R must be reduced or the σ increased.

Now, σ is related to the equivalent conductivity Λ at the same concentration

$$\sigma = \Lambda zc \tag{4.305}$$

But Λ varies with concentration; this is what the Debye–Hückel–Onsager equation was all about. Hence, to understand the specific conductivity σ at a concentration c, it is not adequate to know the equivalent conductivity under a hypothetical condition of infinite dilution. One must be able to calculate,[†] for the nonaqueous solution, the equivalent conductivity at finite concentrations, utilizing the Λ^0 value and the theoretical slope of the Λ

[†] Of course, the validity of the calculation depends upon whether the theoretical expression for the equivalent conductivity (*e.g.*, the Debye–Huckel–Onsager equation) is valid in the given concentration range.

versus $c^{\frac{1}{2}}$ curve. This will be possible if one knows the values of the constants A and B in the Debye–Hückel–Onsager equation

$$\Lambda = \Lambda^0 - (A + B\Lambda^0)c^{\frac{1}{2}}$$

where

$$A = \frac{ze_0 F}{900\pi\eta}\left(\frac{8\pi z^2 e_0^2}{\varepsilon kT}\right)^{\frac{1}{2}}\left(\frac{N_A}{1000}\right)^{\frac{1}{2}}$$

and

$$B = \frac{e_0^2 \omega}{6\varepsilon kT}\left(\frac{8\pi z^2 e_0^2}{\varepsilon kT}\right)^{\frac{1}{2}}\left(\frac{N_A}{1000}\right)^{\frac{1}{2}}$$

When one looks at the above expressions for A and B, it becomes obvious that, as ε decreases, A and B increase; the result is that Λ and therefore σ decrease. Physically, this corresponds to the stronger interionic interaction's arising as the ε is reduced.

So the question of the specific conductivity of nonaqueous solutions vis-à-vis aqueous solutions hinges on whether the dielectric constant of nonaqueous solvents is lower or higher than that of water. Table 4.22 shows that many nonaqueous solvents have ε's considerably lower than that of water. There are some notable exceptions, namely, the hydrogen-bonded liquids.

Thus, on the whole, the effect of increase of electrolyte concentration on lowering the equivalent conductance is much greater in nonaqueous than in aqueous solutions. The result is that the specific conductivity of nonaqueous solutions containing practical electrolyte concentrations is far less than the specific conductivity of aqueous solutions at the same electrolyte concentration (Table 4.24 and Fig. 4.90).

TABLE 4.24

Specific Conductivity of Electrolytes in Aqueous and Nonaqueous Solvents at the Same Concentration

Electrolyte	Concentration, moles liter^{-1}	Specific conductivity, ohm^{-1} cm^{-1} at 25 °C in	
		Water	Nonaqueous solvent
HCl	0.1	391.32 × 10^{-4}	(methanol) 122.5 × 10^{-4}
			(ethanol) 35.43 × 10^{-4}
			(n-propanol) 8.80 × 10^{-4}
NaCl	0.01	11.85 × 10^{-4}	(methanol) 7.671 × 10^{-4}
KCl	0.01	14.127 × 10^{-4}	(methanol) 8.232 × 10^{-4}

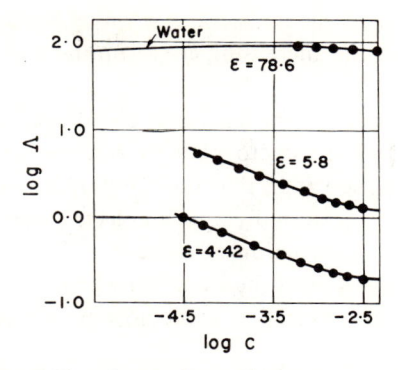

Fig. 4.90. Comparison of the concentration dependence of the equivalent conductivity of tetraisoamylammonium nitrate dissolved in water and in water–dioxane mixtures.

In summary, it is the lower dielectric constants of the typical nonaqueous solvent which causes a far greater decrease in equivalent conductivity with increase of concentration than that which takes place in typical aqueous solutions over a similar concentration range. Even if the infinite-dilution value Λ^0 makes a nonaqueous electrochemical system look hopeful, the practically important value of the specific conductivity is nearly always much less than in the corresponding aqueous solution. That is the sad aspect of nonaqueous solutions.

4.7.6. The Effect of the Solvent on the Concentration of Free Ions: Ion Association

The concentration c which appears in the Debye–Hückel–Onsager equation pertains only to the *free* ions. This concentration becomes equal to the analytical concentration, which shall be designated here as c_a, only if every ion from the ionic lattice from which the electrolyte was produced is stabilized in solution as an independent mobile charge carrier; i.e., $c \neq c_a$ if there is ion-pair formation. Whether ion-pair formation occurs or not depends on the relative values of a, the closest distance of approach of oppositely charged ions and of the Bjerrum parameter $q = (z_+ z_- e_0^2/2kT)1/\varepsilon$. When $a < q$, the condition for ion-pair formation is satisfied, and, when $a > q$, the ions remain free.

From the expression for q, it is clear that the lower the dielectric constant of the solvent is, the larger is the magnitude of q. Hence, when one replaces water by a nonaqueous solvent, the likelihood of ion-pair formation increases because of the increasing q (assuming that a does not increase in proportion to q).

It has already been emphasized that an ion pair, taken as a whole, is electrically neutral and ceases to play its role in the ionic cloud (Section 3.8). For the same reason (i.e., that the ion pair is uncharged), the ion pair does not respond to an externally applied electric field. Hence, ion pairs do not participate in the conduction of current.

A quantitative analysis of the extent to which ion-pair formation affects the conductivity of an electrolyte must now be considered.

4.7.7. The Effect of Ion Association upon Conductivity

In treating the thermodynamic consequences of ion-pair formation (Section 3.8.4), it has been shown that the association constant K_A for the equilibrium between free ions and ion pairs is given by

$$K_A = \frac{\theta}{(1-\theta)^2} \frac{1}{c_a} \frac{f_{IP}}{f_\pm^2} \qquad (3.172)$$

where θ is the fraction of ions which are associated, c_a is the analytical concentration of the electrolyte, $f_\pm$ is the mean activity coefficient, and f_{IP} is the activity coefficient for the ion pairs. Since neutral ion pairs are not involved in the ion–ion interactions responsible for activity coefficients deviating from unity, it is reasonable to assume that $f_{IP} \sim 1$, in which case,

$$K_A(1-\theta)^2 c_a f_\pm^2 = \theta \qquad (4.306)$$

A relation between θ and the conductivity of the electrolyte will now be developed. The specific conductivity has been shown [cf. Eq. (4.307)] to be related to the concentration of mobile charge carriers (i.e., of free ions) in the following way

$$\sigma = zF(u_+ + u_-)c_{\text{free ions}} \qquad (4.307)$$

One can rewrite this equation in the form

$$\sigma = zF(u_+ + u_-)\frac{c_{\text{free ions}}}{c_a} c_a \qquad (4.308)$$

Since $c_{\text{free ions}}/c_a$ is the fraction of ions which are *not* associated (i.e., are free), it is equal to unity minus the fraction of ions which *are* associated.

Hence,
$$\frac{c_{\text{free ions}}}{c_a} = 1 - \theta \tag{4.309}$$

and, using this result in equation

$$\sigma = zF(u_+ + u_-)(1 - \theta)c_a \tag{4.310}$$

or, from the definition of equivalent conductivity, i.e.,

$$\Lambda = \frac{\sigma}{zc_a} \tag{4.305}$$

one can write

$$\Lambda = F(u_+ + u_-)(1 - \theta) \tag{4.311}$$

If there is no ion association, i.e., $\theta = 0$, then one can define a quantity $\Lambda_{\theta=0}$, which is given by [cf. Eq. (4.311)]

$$\Lambda_{\theta=0} = F(u_+ + u_-) \tag{4.312}$$

By dividing Eq. (4.311) by (4.312), the result is

$$1 - \theta = \frac{\Lambda}{\Lambda_{\theta=0}} \tag{4.313}$$

and

$$\theta = 1 - \frac{\Lambda}{\Lambda_{\theta=0}} \tag{4.314}$$

Introducing these expressions for θ and $1 - \theta$ into Eq. (4.306), one finds that

$$K_A \frac{\Lambda^2}{\Lambda_{\theta=0}^2} c_a f_\pm^2 = 1 - \frac{\Lambda}{\Lambda_{\theta=0}}$$

or

$$\frac{1}{\Lambda} = \frac{1}{\Lambda_{\theta=0}} + \frac{K_A f_\pm^2}{\Lambda_{\theta=0}^2} \Lambda c_a \tag{4.315}$$

Though Eq. (4.315) relates the equivalent conductivity to the electrolyte concentration, it contains the unevaluated quantity $\Lambda_{\theta=0}$. By combining Eqs. (4.310) and (4.312), one gets

$$\Lambda_{\theta=0} = F(u_+ + u_-) = \frac{\sigma}{z[(1 - \theta)c_a]} \tag{4.316}$$

from which it is clear that $\Lambda_{\theta=0}$ is the equivalent conductivity of a solution in which there is no ion association but in which the concentration is $(1 - \theta)c_a$. Thus, for small concentrations (see Section 4.6.12), one can express $\Lambda_{\theta=0}$ by the Debye–Hückel–Onsager equation (4.299), taking care to use the concentration $(1 - \theta)c_a$. Thus,

$$\Lambda_{\theta=0} = \Lambda^0 - (A + B\Lambda^0)(1 - \theta)^{\frac{1}{2}}c_a^{\frac{1}{2}} \quad (4.317)$$

which can be written in the form (*cf.* Appendix 4.5)

$$\Lambda_{\theta=0} = \Lambda^0 - (A + B\Lambda^0)(1 - \theta)^{\frac{1}{2}}c_a^{\frac{1}{2}} = \Lambda^0 Z \quad (4.318)$$

where Z is the continued fraction

$$Z = 1 - z\{1 - z[1 - z(\cdots)^{-\frac{1}{2}}]^{-\frac{1}{2}}\}^{-\frac{1}{2}} \quad (4.319)$$

with

$$z = \frac{(A + B\Lambda^0)c_a^{\frac{1}{2}}\Lambda^{\frac{1}{2}}}{\Lambda^{0\frac{3}{2}}} \quad (4.320)$$

Introducing expression (4.318) for $\Lambda_{\theta=0}$ into Eq. (4.315), one has

$$\frac{1}{\Lambda} = \frac{1}{\Lambda^0 Z} + \frac{K_A f_\pm^2}{(\Lambda^0)^2 Z^2} \Lambda c_a$$

or

$$\frac{Z}{\Lambda} = \frac{1}{\Lambda^0} + \frac{K_A}{(\Lambda^0)^2} \frac{\Lambda c_a f_\pm^2}{Z} \quad (4.321)$$

This is an interesting result. It can be seen from (4.321) that the association of ions into ion pairs has entirely changed the form of the equivalent conductivity *versus* concentration curve. In the absence of significant association, Λ was linearly dependent on $c^{\frac{1}{2}}$, as empirically shown by Kohlrausch. When, however, there is considerable ion-pair formation (as would be the case in nonaqueous solvents of low dielectric constant), instead of the Kohlrausch law, one finds that, when Z/Λ is plotted against $\Lambda f_\pm^2 c_a / Z$, a straight line is obtained with slope $K_A/(\Lambda^0)^2$ and intercept $1/\Lambda^0$. Figure 4.91 shows the experimental demonstration of this conductance behavior.

4.7.8. Even Triple Ions Can Be Formed in Nonaqueous Solutions

When the dielectric constant of the nonaqueous solvent goes below about 15, ions can associate not only in ion pairs but also in ion triplets. This comes about by one of the ions (e.g., M$^+$) of an ion pair M$^+\cdots$A$^-$

ION TRANSPORT IN SOLUTIONS 451

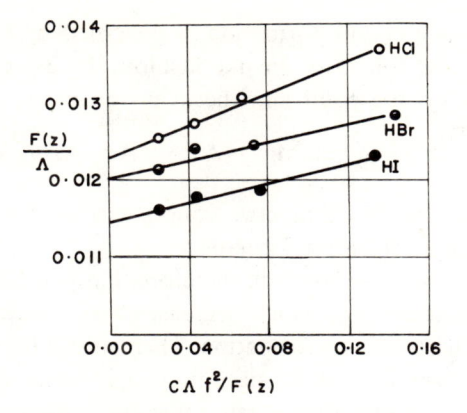

Fig. 4.91. Plots of Eq. (4.321) for the hydrogen halides in ethyl alcohol.

coulombically attracting a free ion A^- strongly enough to overcome the thermal forces of dissociation

$$A^- + M^+ \cdots A^- \rightleftharpoons A^- \cdots M^+ \cdots A^-$$

From the conductance point of view, ion pairs and triple ions behave quite differently. The former, being uncharged, do not respond to an external field; the latter are charged and respond to the external field by drifting and contributing to the conductance.

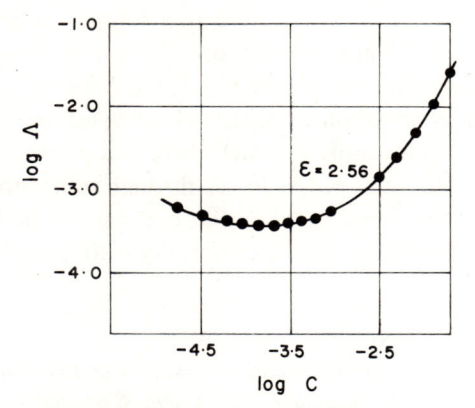

Fig. 4.92. Minimum in the curve for equivalent conductivity versus concentration in the case of tetraisoamylammonium nitrate in a water-dioxane mixture of dielectric constant of $\varepsilon = 2.56$.

The extent of ion-pair formation is governed by the equilibrium between free ions and ion pairs. In like fashion, the extent of triple-ion formation depends on the eqilibrium between ion pairs and triple ions.

$$M^+ + A^- \rightleftharpoons M^+\cdots A^- \overset{A^-}{\rightleftharpoons} A^-\cdots M^+\cdots A^-$$

Thus, the greater the stoichiometric concentration, the greater is the ion-pair formation and triple-ion formation.

With increasing concentration, therefore, ion-pair formation dominates the equivalent conductivity, which decreases with increasing concentration faster than had there been no formation. Then, at still higher concentrations, when triple-ion formation starts becoming significant, the equivalent conductivity starts increasing after passing through a minimum. This behavior has been experimentally demonstrated (Fig. 4.92).

4.7.9. Some Conclusions about the Conductance of Nonaqueous Solutions of True Electrolytes

The change from aqueous to nonaqueous solutions of *true* electrolytes results in characteristic effects on the conductance. The *order of magnitude* of the equivalent conductivity at infinite dilution is approximately the same in both types of solutions. But the slope of the equivalent-conductivity *versus* concentration curve is drastically more negative in nonaqueous solutions than in the corresponding aqueous solutions. This means that the actual specific conductivity σ, which is the only significant quantity as far as the conducting power of an actual solution is concerned, is much lower for nonaqueous solutions. Ion-pair formation worsens the conductance situation; triple-ion formation is a slight help.

Thus, nonaqueous solutions of true electrolytes are not to be regarded optimistically for applications in which there is a premium on high specific conductivity and minimum power losses through resistance heating. If nonaqueous solutions are going to be useful, one has to think of solutions of *potential* electrolytes which interact chemically with the solvent. (Chapter 5).

Further Reading

1. N. Bjerrum, *Kgl. Danske Videnskab. Selskab Mat-Fys. Medd.*, **7** (9) (1933).
2. R. M. Fuoss and C. A. Kraus, *J. Am. Chem. Soc.*, **55**: 476 (1933).
3. R. A. Robinson and R. H. Stokes, *Electrolyte Solutions*, Butterworths' Publications, Ltd., London, 1955.
4. H. S. Harned and B. B. Owen, *The Physical Chemistry of Electrolytic Solutions*, Reinhold Publishing Corp., New York, 1957.

5. R. M. Fuoss and F. Accasina, *Electrolytic Conductance*, Interscience Publishers, Inc., New York, 1959.
6. C. B. Monk, *Electrolytic Dissociation*, Academic Press, London, 1961.
7. C. W. Davies, *Ion Association*, Butterworths' Publications, Ltd., London, 1962.
8. J. E. Prue, "Ion Association and Solvation," in: B. E. Conway and R. G. Barradas, eds., *Chemical Physics of Ionic Solutions*, John Wiley & Sons, Inc., New York, 1966.

Appendix 4.1. The Mean Square Distance Traveled by a Random-Walking Particle

In the one-dimensional random-walk problem, the expression for $\langle x^2 \rangle$ is found by mathematical induction as follows. Consider that, after $N-1$ steps, the sailor has progressed a distance x_{N-1}. If he takes one *more* step, the distance x_N from the origin will be either

$$x_N = x_{N-1} + l \tag{A4.1.1}$$

or

$$x_N = x_{N-1} - l \tag{A4.1.2}$$

Squaring both sides of Eqs. (A4.1.1) and (A4.1.2), one obtains

$$x_N^2 = x_{N-1}^2 + l^2 + 2x_{N-1}l \tag{A4.1.3}$$

and

$$x_N^2 = x_{N-1}^2 + l^2 - 2x_{N-1}l \tag{A4.1.4}$$

The average of these two possibilities must be

$$x_N^2 = x_{N-1}^2 + l^2 \tag{A4.1.5}$$

This is the result for x_N^2 when the distance traveled after $N-1$ steps is exactly x_{N-1}. In general, however, one can only expect, for the value of the square of the distance at the $(N-1)$th step, an averaged value $\langle x_{N-1}^2 \rangle$, in which case one must write

$$\langle x_N^2 \rangle = \langle x_{N-1}^2 \rangle + l^2 \tag{A4.1.6}$$

Now, at the start of the random walk, i.e., after zero steps, the progress is given by

$$\langle x_0^2 \rangle = 0 \tag{A4.1.7}$$

After one step, it is

$$\langle x_1^2 \rangle = 1l^2 = l^2 \tag{A4.1.8}$$

After two steps, from Aq. (A4.1.6), one has

$$\langle x_2^2 \rangle = \langle x_1^2 \rangle + l^2 \tag{A4.1.9}$$

and, using Eq. (A4.1.8),
$$\langle x_2^2 \rangle = l^2 + l^2 = 2l^2 \qquad (A4.1.10)$$

Similarly,
$$\langle x_3^2 \rangle = \langle x_2^2 \rangle + l^2$$
$$= 2l^2 + l^2$$
$$= 3l^2 \qquad (A4.1.11)$$

Hence, in general,
$$\langle x_N^2 \rangle = Nl^2 \qquad (A4.1.12)$$

This equation has been derived for one-dimensional random walk, but it can be shown to be valid for three-dimensional random flights, too.

The mean square distance $\langle x^2 \rangle$ which a particle travels depends upon the time of travel in the following manner. The number of steps, N, obviously increases with time and is proportional to it, i.e.,

$$N = kt \qquad (A4.1.13)$$

where k is the constant of proportionality. Hence, by combining (A4.1.12) and (A4.1.13),
$$\langle x^2 \rangle = ktl^2 \qquad (A4.1.14)$$

which may be written
$$\langle x^2 \rangle = \alpha t \qquad (A4.1.15)$$

where α is a proportionality constant to be evaluated in the Einstein–Smoluchowski equation.

Appendix 4.2. The Laplace Transform of a Constant

The Laplace transform $\bar{\alpha}$ of a constant α is, by definition (4.33), given by

$$\bar{\alpha} = \int_0^\infty e^{-pz} \alpha \, dz$$

$$\bar{\alpha} = \alpha \left[\frac{e^{-pz}}{-p} \right]_0^\infty$$

$$= \alpha \left[\frac{1}{e^{pz}(-p)} \right]_{z=\infty} - \alpha \left[\frac{1}{e^{pz}(-p)} \right]_{z=0}$$

$$= 0 - \left(\frac{\alpha}{-p} \right)$$

$$= \frac{\alpha}{p}$$

Hence, the Laplace transform of a constant is equal to that same constant divided by p.

ION TRANSPORT IN SOLUTIONS   455

Appendix 4.3. A Few Elementary Ideas on the Theory of Rate Processes

1. In any rate process, the initial state (condition) of the system is characterized by a certain configuration (arrangement) of the entities (ions, atoms, sites, etc.) involved. Similarly, the final state is characterized by a certain (but different from the initial) configuration of the entities. Thus, in the case of an ionic jump, the initial state of the system is

$$\boxed{\text{Ion}}_1 \;\leftarrow l \rightarrow\; \boxed{\phantom{\text{Ion}}}_2$$

where the subscripts 1 and 2 serve to differentiate the sites, and the final state is

$$\boxed{\phantom{\text{Ion}}}_1 \;\leftarrow l \rightarrow\; \boxed{\text{Ion}}_2$$

2. When the initial or final states are considered, the position of the ion is described in "either-or" language: The ion is in either site 1 or site 2. *During the jump*, however, the ion is in intermediate positions, and the jumping process can be described by giving the two distances, d_{1-I} and d_{I-2}, which separate the ion I from the two sites.

3. As the ion jumps from site 1 to site 2, the distance d_{1-I} increases from zero in the initial state to the mean jump distance l in the final state. While d_{1-I} is fairly small, one can reasonably [see Fig. A4.3.1(a) and (b)] say that the system is in the *initial state* (the jump has not yet occurred). But there is a limit to this view. The quantitative change of the distances d_{1-I} and d_{I-2} leads to a qualitative change in the configuration of the system from a configuration characterizing the initial state to one characterizing the final state. There is a *critical configuration* beyond which the configuration of the ion and the sites [Fig. A4.3.1(c)] more closely resembles the system in its final state (after the jump), i.e., d_{I-2} small, than in its initial state (before the jump), i.e., d_{1-I} small. This critical configuration is known as the *activated complex* (for reasons which will be explained). When the initial and final states are mirror images, then the activated complex,

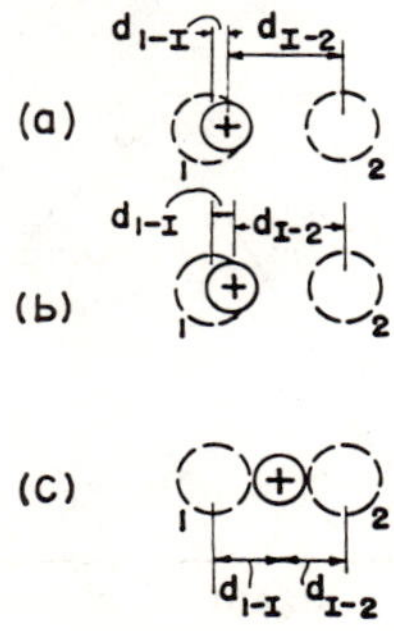

(a)

(b)

(c)

Fig. A4.3.1.

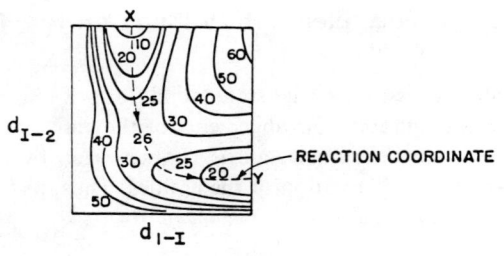

Fig. A4.3.2.

or critical configuration, must correspond to the halfway point; when these states are not mirror symmetrical, one has to get the configuration of the activated complex by considering the dependence of the energy of the system upon the distances between the reacting entities.

4. Each configuration (during the process of jumping) corresponds to a particular energy for the system. Thus, for every value of d_{1-I} and d_{I-2}, there is a particular value of the energy of the system. If these energies are plotted as a function of the distance variables, then one obtains a three-dimensional potential-energy surface (Fig. A4.3.2). Across this surface, there are an infinite number of paths representing the passage of the system from the initial state or the final state. The easiest of these paths is termed the *reaction coordinate*. If the surface is cut along the reaction coordinate, then a potential-energy versus distance diagram (Fig. A4.3.3) is obtained. It is necessary to be clear that any point on this diagram *represents* the potential energy that the system has for the corresponding configuration of the entities involved. Thus, in the case of an ionic jump from one site to another, the jump can be described by a continuous change of the coordinates of the ion in relation to the two sites. Hence, this jump is *represented* by the motion of a point along the potential-energy curve from the minimum on the left-hand side to the minimum on the right-hand side.

5. It can be seen from the potential-energy diagram that, as the ion starts moving away from one site (toward the other), the potential energy of the system starts increasing and the point representing the energy of the system (on the

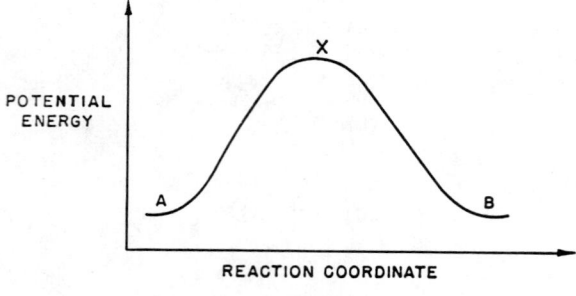

Fig. A4.3.3.

potential-energy–distance diagram) starts "climbing" a potential-energy hill Near the halfway point, the potential energy attains a maximum and then starts decreasing, and the representative point goes over the *peak* of the curve. The peak corresponds to the configuration of the system known as the *activated complex*, i.e., to the critical configuration which the system must attain before the process (the diffusion jump, in this case) can be said to occur. This is why it is said that the system has to climb a potential-energy *barrier* and has to be activated with a critical *activation energy* $\Delta H^{\ddagger}$ for the process to occur, be it diffusion or a chemical reaction. If the entropy change $\Delta S^{\ddagger}$ in going from the initial to this activated state is also taken into account, then the rate process can be considered to need a critical free energy of activation $\Delta G^{\ddagger} = \Delta H^{\ddagger} - T\Delta S^{\ddagger}$.

6. The system in the activated state, i.e., the activated complex, has one exceptional property: If one distance between the constituents increases (d_{1-I} in the diffusion-jump case), there is a breakdown of the activated complex into the system corresponding to the final state, i.e., the system tumbles down the hillside toward the final state. Why does d_{1-I} change? One may consider the entities □₁ and I in a sort of "bond" which vibrates. It is this vibration which leads to decomposition of the activated complex.

7. The vibration leading to breakdown of the activated complex takes place with a certain vibration frequency ν which has associated with it, according to Planck's law, an energy E given by

$$E = h\nu \qquad (A4.3.1)$$

where h is the Planck constant. But a vibration in the activated complex is a relative displacement of the entities constituting the system. The displacement can also be viewed as a *translation* of one of the entities in relation to the rest of the system (displacement of the ion I in relation to site 1). The translation has associated with it a translational energy which, according to the kinetic theory,[†] is $\tfrac{1}{2}kT$ per entity or kT for a pair of entities (*cf.* the calculation of the translational energy in Bjerrum's ion-pair theory). Hence, when the vibration of the 1–I bond is viewed as a translation, the associated energy is E given by

$$E = kT \qquad (A4.3.2)$$

Since Eqs. (A4.3.1) and (A4.3.2) arise from two ways of looking at the same process, the two expressions for the energy must be equal, i.e.,

$$h\nu = kT \qquad (A4.3.3)$$

The particular vibration frequency ν of the activated complex leading to decomposition (the production of the final state) is therefore given by the expression

$$\nu = \frac{kT}{h} \qquad (A4.3.4)$$

[†] *Note*: The k here is the Boltzmann constant.

458   CHAPTER 4

8. At this stage, a central feature of the rate-process theory enters the picture. The activated complexes are considered in equilibrium with the reactants

$$\text{Reactants} \rightleftharpoons \text{Activated complexes}$$

Hence, the law of mass action can be used to give

$$\frac{c_{AC}}{c_R} = K = e^{-\Delta G^{0\neq}/RT} \tag{A4.3.5}$$

where c_{AC} and c_R are the concentration of activated complexes and reactants, K is the equilibrium constant, and $\Delta G^{0\neq}$ is the standard free energy of activation. Considering unit concentration of reactants, $c_R = 1$, one has

$$c_{AC} = e^{-\Delta G^{0\neq}/RT} \tag{A4.3.6}$$

9. The number of times per second, k, that the rate process occurs is equal to the concentration of activated complexes, c_{AC}, times their vibration frequency v. Hence,

$$\vec{k} = c_{AC} v \tag{A4.3.7}$$

and, inserting the expressions for $\dot{c}_{AC}$ and v from Eqs. (A4.3.6) and (A4.3.4), one finds that

$$\vec{k} = \frac{kT}{h} e^{-\Delta G^{0\neq}/RT} \tag{A4.3.8}$$

This is the expression for the rate constant of a rate process, e.g., of ionic jumping.

Appendix 4.4. The Derivation of Equations (4.257) and (4.258)

According to notation [cf. Eqs. (4.251) and (4.252)],

$$p_+ = q_+ L_{++} + q_- L_{-+} \tag{A4.4.1}$$

and

$$p_- = q_+ L_{+-} + q_- L_{--} \tag{A4.4.2}$$

Hence, one can carry out the following expansions

$$\frac{p_+}{p_+ q_+ + p_- q_-} = \frac{q_+ L_{++} + q_- L_{-+}}{q_+(q_+ L_{++} + q_- L_{-+}) + q_-(q_+ L_{+-} + q_- L_{--})}$$

$$= \frac{q_+ L_{++} + q_- L_{-+}}{q_+(q_+ L_{++} + q_- L_{+-}) + q_-(q_+ L_{-+} + q_- L_{--})}$$

$$= \frac{q_+ L_{++} \dfrac{d\varphi}{dx} + q_- L_{-+} \dfrac{d\varphi}{dx}}{q_+\left(q_+ L_{++} \dfrac{d\varphi}{dx} + q_- L_{+-} \dfrac{d\varphi}{dx}\right) + q_-\left(q_+ L_{-+} \dfrac{d\varphi}{dx} + q_- L_{--} \dfrac{d\varphi}{dx}\right)} \tag{A4.4.3}$$

ION TRANSPORT IN SOLUTIONS 459

Now, it has been stated [*cf.* Eq. (4.247)] that

$$J_+ = L_{++}\vec{F}_+ + L_{+-}\vec{F}_- \tag{A4.4.4}$$

and that [*cf.* Eqs. (4.254) and (4.255)]

$$\vec{F}_+ = \frac{d\mu_+}{dx} + q_+ \frac{d\varphi}{dx} \tag{A4.4.5}$$

and

$$\vec{F}_- = \frac{d\mu_-}{dx} + q_- \frac{d\varphi}{dx} \tag{A4.4.6}$$

Hence, substituting for $\vec{F}_+$ and $\vec{F}_-$ in (A4.4.4) and setting $d\mu/dx = 0$, one has

$$(J_+)_{d\mu/dx=0} = q_+ L_{++} \frac{d\varphi}{dx} + q_- L_{+-} \frac{d\varphi}{dx} \tag{A4.4.7}$$

Similarly,

$$(J_-)_{d\mu/dx=0} = q_+ L_{-+} \frac{d\varphi}{dx} + q_- L_{--} \frac{d\varphi}{dx} \tag{A4.4.8}$$

In terms of these expressions, Eq. (A4.4.3) becomes

$$\frac{p_+}{p_+ q_+ + p_- q_-} = \left(\frac{J_+}{q_+ J_+ + q_- J_-} \right)_{d\mu/dx=0}$$

$$= \frac{1}{q_+} \left(\frac{q_+ J_+}{q_+ J_+ + q_- J_-} \right)_{d\mu/dx=0} \tag{A4.4.9}$$

But, by notation,

$$q_+ = z_+ F \quad \text{and} \quad q_- = z_- F \tag{A4.4.10}$$

and, therefore, Eq. (A4.4.8) can be rewritten as

$$\frac{p_+}{p_+ q_+ + p_- q_-} = \frac{1}{z_+ F} \left(\frac{z_+ F J_+}{z_+ F J_+ + z_- F J_-} \right)_{d\mu/dx=0} \tag{A4.4.11}$$

Further, according to the relation between current density and flux,

$$i_+ = z_+ F J_+$$

and
$$\tag{A4.4.12}$$
$$i_- = z_- F J_-$$

Using these relations in Eq. (A4.4.10), one has

$$\frac{p_+}{p_+ q_+ + p_- q_-} = \frac{1}{z_+ F} \left(\frac{i_+}{i_+ + i_-} \right)_{d\mu/dx=0} \tag{A4.4.13}$$

By definition, however [cf. Eq. (4.240)],

$$t_+ = \left(\frac{i_+}{i_+ + i_-}\right)_{d\mu/dx=0} \quad \text{(A4.4.14)}$$

Hence, by combining Eqs. (A4.4.12) and (A4.4.13), the result is

$$\frac{p_+}{p_+q_+ + p_-q_-} = \frac{t_+}{z_+F} \quad \text{(A4.4.15)}$$

Similarly, it can be shown that

$$\frac{p_-}{p_+q_- + p_-q_-} = \frac{t_-}{z_-F} \quad \text{(A4.4.16)}$$

Appendix 4.5. The Derivation of Equation (4.318)

One can rewrite Eq. (4.317), namely,

$$\Lambda_{\theta=0} = \Lambda^0 - (A + B\Lambda^0)(1-\theta)^{\frac{1}{2}} c_a^{\frac{1}{2}} \quad \text{(A4.5.1)}$$

in the form

$$\Lambda_{\theta=0} = \Lambda^0 \left[1 - \frac{1}{\Lambda^0} m c_a^{\frac{1}{2}} (1-\theta)^{\frac{1}{2}}\right] \quad \text{(A4.5.2)}$$

where, for conciseness, the symbol m is used instead of $(A + B\Lambda^0)$.

It has been shown, however, that [cf. Eq. (4.313)]

$$1 - \theta = \frac{\Lambda}{\Lambda_{\theta=0}} \quad \text{(A4.5.3)}$$

If this relation is used in Eq. (A4.5.2), one gets

$$\Lambda_{\theta=0} = \Lambda^0 \left[1 - \frac{m c_a^{\frac{1}{2}}}{\Lambda^0} \left(\frac{\Lambda}{\Lambda_{\theta=0}}\right)^{\frac{1}{2}}\right] \quad \text{(A4.5.4)}$$

and, substituting for $\Lambda_{\theta=0}$ in the right-hand side from Eq. (A4.5.2), the result is

$$\Lambda_{\theta=0} = \Lambda^0 \left\{1 - \frac{m(\Lambda c_a)^{\frac{1}{2}}}{\Lambda^{0\frac{3}{2}}} \frac{1}{[1 - (m c_a^{\frac{1}{2}}/\Lambda^0)(1-\theta)^{\frac{1}{2}}]^{\frac{1}{2}}}\right\} \quad \text{(A4.5.5)}$$

One has again been left with $(1-\theta)^{\frac{1}{2}}$ on the right-hand side, and, thus, one again substitutes from Eq. (A4.5.2). This process of substitution can be repeated *ad infinitum* to give the result

$$\Lambda_{\theta=0} = \Lambda^0 (1 - z\{1 - z[1 - z(\ldots)^{-\frac{1}{2}}]^{-\frac{1}{2}}\})^{-\frac{1}{2}} \quad \text{(A4.5.6)}$$

$$= \Lambda^0 Z \quad \text{(A4.5.7)}$$

where

$$Z = \frac{m(\Lambda c_a)^{\frac{1}{2}}}{\Lambda^{0\frac{3}{2}}} = \frac{(A + B\Lambda^0) c_a^{\frac{1}{2}} \Lambda^{\frac{1}{2}}}{\Lambda^{0\frac{3}{2}}} \quad \text{(A4.5.8)}$$

CHAPTER 5
PROTONS IN SOLUTION

5.1. THE CASE OF THE NONCONFORMING ION: THE PROTON

The proton is formed when a hydrogen atom is stripped of its electron. Its electronic structure is singular: It has no electrons and, therefore, no structure. It is a pure nucleus. Devoid of an electron shell, the center of charge of a proton can approach far closer to a neighboring ion or atom than can any other ion or atom. The proton, therefore, has few steric restrictions in chemical reactions. Other ions bear the burden of electron shells and are of angstrom dimensions ($\sim 10^{-8}$ cm); the proton is of fermi dimensions ($\sim 10^{-13}$ cm). It is the smallest ion, and also the lightest. Indeed, it is an elementary particle—a structural building block of matter. The proton is a nonconformist ion.

Such an abnormal ion must have abnormal properties. It has been this knowledge which has dictated the exclusion of the proton from the treatment of ion–solvent (Chapter 2) and ion–ion interactions (Chapter 3) and the treatment of the movement of ions (Chapter 4). It will be noticed that, while the solvation of metal ions and various anions has been discussed in detail, little attention has been paid to the question of the *solvation of a proton*. Similarly, in treating ion–ion interactions, the only ions considered were those pulled out from an ionic crystal by ion–solvent forces. No attention was paid to potential electrolytes, such as acetic acid, which yield solutions containing hydrogen ions by a mechanism in which a molecule of the acid reacts with water.

It is intended in this chapter to give protons the special treatment they deserve. The first step will be to consider the interaction between protons and solvent molecules and to go into the details of proton solvation. Then, the mechanism by which hydrated protons move in solution will be explored. Here, there is a special feature, for protons migrate by a mechanism fundamentally different from that used by other ions. Thus, the latter retain their identity and separateness as kinetic entities in their passage through the solvent, whereas it will be seen that protons are handed on from water molecule to water molecule and, when they are waiting to be passed on, they are indistinguishable from the protons of the host solvent. Finally, the interaction between hydrated protons and other ions will be treated, with less emphasis on the long-range electrostatic effects characteristic of the Debye–Hückel approach than on the short-range chemical interactions which are the basis of the behavior of acids and bases.

5.2. PROTON SOLVATION

5.2.1. What Is the Condition of the Proton in Solution?

The ionization energy required to remove an electron from a hydrogen atom is much larger than the corresponding energy for the formation of a univalent ion of any other element (Table 5.1). This fact reflects the extreme attraction which the proton has for electrons. Thus, protons tend to form covalent bonds by sharing electron pairs. They also tend to form hydrogen bonds, which, as a first approximation, may be considered an electrostatic attraction between a proton and an *un*shared pair of electrons.

TABLE 5.1

Ionization Energies of the First-Group Elements

Atomic number	Element	Ionization energy, kcal mole^{-1}
1	H	313
3	Li	124
11	Na	118
19	K	100
37	Rb	96
55	Cs	90

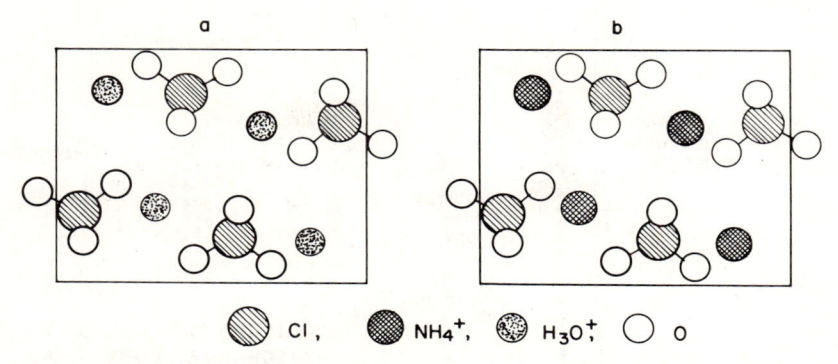

Fig. 5.1. The isomorphous nature of (a) perchloric acid monohydrate and (b) ammonium perchlorate.

All this suggests that it is highly unlikely that *free* protons are stable in solution. Then, in what form do they exist?

Suppose one considers substances like nitric acid or perchloric acid. Normally, these are liquids, but it is possible to prepare solid forms of these substances—the so-called "acid hydrates." The solids can then be subjected to X-ray examination designed to reveal their crystalline structure.

It turns out that perchloric acid monohydrate $HClO_4 \cdot H_2O$ has the same structure as ammonium perchlorate NH_4ClO_4, i.e., the two substances are isomorphous (Fig. 5.1). It is known, however, that ammonium perchlorate is an ionic crystal consisting of an assembly of NH_4^+ and ClO_4^- ions. Hence, the implication of the fact that $HClO_4 \cdot H_2O$ is isomorphous with NH_4ClO_4 is that ClO_4^- ions must exist in the acid monohydrate and the remaining proton and water molecules are associated in the way that a proton is linked with an ammonia molecule to give NH_4^+. It is difficult, therefore, to avoid the conclusion that the isomorphous character of the $HClO_4 \cdot H_2O$ and NH_4ClO_4 is based on there being H_3O^+ ions in the monohydrate of perchloric acid corresponding to NH_4^+ ions in its perchlorate. Further, since an NH_4^+ ion is known as an ammonium ion, it is appropriate to call the H_3O^+ ion a *hydronium* ion.

What does this X-ray evidence add up to? That protons are associated with water molecules (or water molecules are protonated) to form H_3O^+ ions. But the X-ray studies were performed on *solids*, and there is no guarantee that H_3O^+ ions exist in a solution of perchloric acid. This is true; yet the solution froze into the solid, and it seems likely that the ions existing in the liquid phase survived the phase transition without structural changes. So, by an inferential argument, one is led to believe that H_3O^+ ions exist in solution.

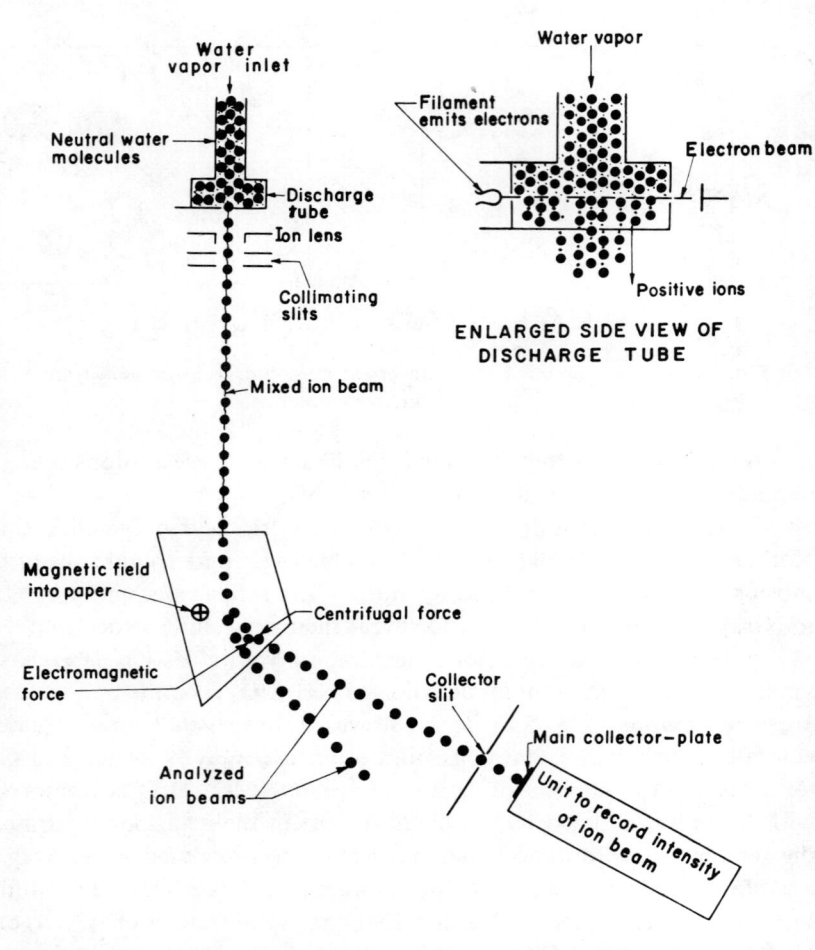

Fig. 5.2. Schematic representation of mass spectrometer to determine ionic species in water vapor.

There are other indirect indications that the H_3O^+ ion is the mode of existence of protons in solution. For example, consider the studies in the gas phase. Water vapor is taken in an evacuated tube and a glow discharge struck between two electrodes in the tube (Fig. 5.2). The electrons from the negative electrode strike the water molecules and ionize them, and these ions interact. To know what are the products of all these events, the vapor can be routed for analysis to a mass spectrometer. The report on the analysis shows clearly that the most abundant ionic species is H_3O^+. Of course, there are other species, such as $H_5O_2^+$, $H_7O_3^+$, $H_9O_4^+$, etc. (i.e.,

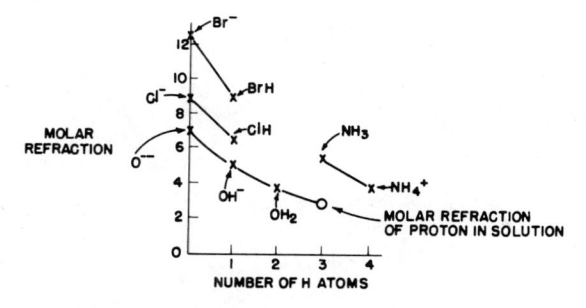

Fig. 5.3. Molar refraction of isoelectronic series of ions and molecules as a function of the number of hydrogen atoms in the ion or molecule. For the hydrogen ion (i.e., the proton in solution) to fall on the lowest curve, the number of hydrogen atoms in the hydrogen ion must be three.

H_3O^+ associated with 1, 2, 3, etc., water molecules), but to smaller extents.

But, all this evidence for H_3O^+ ions in the solid and gas phases does not *prove* that they exist in solution. Why not, therefore, use direct analytical methods? The obvious tool to use is infrared spectroscopy, and the use of this technique does permit the identification of H_3O^+ species in solution (Falk and Giguère). A further clear piece of evidence comes from some investigations of molar refraction. The refractive index of a solution is a property which indicates the size and shape of the ions present. It is found that, if a series of molar refractions are measured for an isoelectronic series of ions and molecules (i.e., ions and molecules with the same number of electrons), then the "proton in solution" only fits into the theory in the right position if it is counted, not as a proton, but as an entity with three hydrogen atoms, i.e., as an H_3O^+ ion (fig. 5.3).

To summarize: In solution, protons are unstable as isolated entities because their field is too powerful and their affinity for an electron pair too strong. The result is that protons interact with water. Protons most probably exist in solution as H_3O^+ ions, but the evidence is not all that certain. The fuller meaning of the statement, "exist as H_3O^+ ions," will become clear when one considers the mechanism of proton drift in solution.

The shape and size of the H_3O^+ ion is fairly well established from NMR data on solid acid hydrates. The H_3O^+ ion is a rather flat trigonal pyramidal structure with the hydrogens at the corners of the pyramid and the oxygen in the middle (Fig. 5.4). The O—H bond is found to be about 1.02 Å in length, the proton–proton distance is about 1.72 Å, and the H—O—H angle is 115°. The whole structure is similar to that of the ammonia molecule.

466   CHAPTER 5

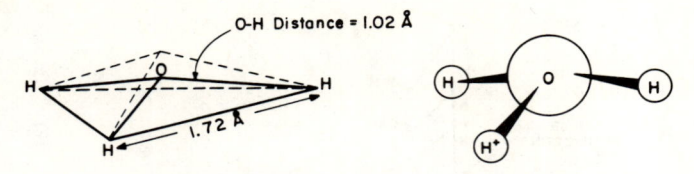

Fig. 5.4. The trigonal-pyramid structure of the hydrogen ion H_3O^+.

5.2.2. Proton Affinity

The formation of an H_3O^+ ion is an expression, on the one hand, of the instability of a free proton in solution and, on the other, of the affinity of a water molecule for a proton. What is the energy change involved in the protonation of water, i.e., in the reaction

$$H_2O + H^+ \rightarrow H_3O^+$$

To get at this energy, one has to resort to thermodynamic cycles. The cycle suggested by Sherman is shown in Figure 5.5. Since all the quantities

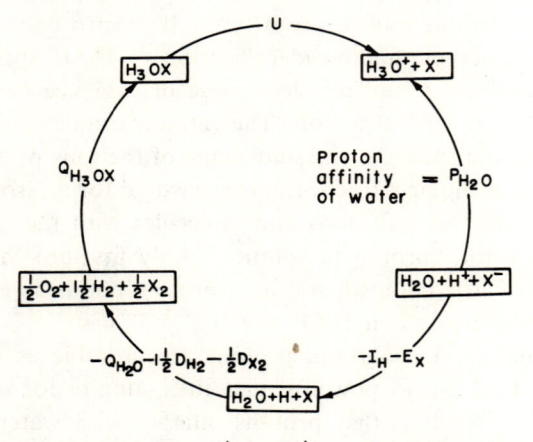

$$Q_{H_3OX} + U - Q_{H_2O} - 1\tfrac{1}{2}D_{H_2} - \tfrac{1}{2}D_{X_2} - I_H - E_X = P_{H_2O}$$

Fig. 5.5. Thermodynamic cycle for the calculation of the proton affinity p_{H_2O} of water. Here, Q_{H_3OX} is the heat of formation of H_3OX (X is a halide ion); U is the lattice energy of the halide, Q_{H_2O} is the heat of formation of water; D_{H_2} and D_{X_2} are the dissociation energies of the hydrogen and halogen molecules, respectively; I_H is the ionization energy of hydrogen atoms; and E_X is the electronegativity of the halogen.

PROTONS IN SOLUTION 467

in the cycle, except the proton affinity, can be obtained directly from experiment or indirectly from other cycles, the one unknown, i.e., the proton affinity of water, can easily be determined; it turns out to be about 170 kcal mole^{-1}.

It is this large energy which makes the existence of free protons in solution unlikely. Their hypothetical concentration in aqueous solutions at room temperature turns out to be about 10^{-150} g of ions per liter, about as zero as one can get.

5.2.3. The Overall Heat of Hydration of a Proton

The energy change associated with the interaction between a proton and a single water molecule has been calculated as 170 kcal mole^{-1}. Is this the hydration energy of a proton? No. If one examines (Fig. 5.5) the thermodynamic cycle used for the calculation, it will be seen that the H_3O^+ ion formed by the interaction of H^+ and H_2O has been left in the gas phase. But, the heat of hydration of a proton (*cf.* Chapter 2) is the heat evolved when the proton is transferred from the gas phase into the solvent. Hence, it is not enough to know the heat of interaction between a proton and a water molecule; one must know the heat of interaction between the H_3O^+ ion and the solvent.

It must now be recalled (*cf.* Section 2.3.11) that there is experimental support for Eq. (2.60)

$$\Delta H_{M_i^+}[\text{rel}] - \Delta H_{X_i^+}[\text{rel}] = -2\Delta H_{H^+}[\text{abs}] - 10 + \frac{4N_A z_i e_0 p_w}{(r_i + 1.38)^3} \quad (5.1)$$

where $\Delta H_{M_i^+}[\text{rel}]$ and $\Delta H_{X_i^-}[\text{rel}]$ are the experimentally known relative heats of hydration of a pair of oppositely charged ions M_i^+ and X_i^- which have the same radius r_i; and $\Delta H_{H^+}[\text{abs}]$ is the absolute heat of hydration of the proton (i.e., the heat of interaction of the H^+ not only with one molecule but with all of the surrounding water). Thus, by following Halliwell and Nyburg and plotting the experimental values of $\frac{1}{2}(\Delta H_{M_i^+}[\text{rel}] - \Delta H_{X_i^-}[\text{rel}])$ versus $(r_i + 1.38)^{-3}$ and then extrapolating the resulting straight line to infinite radius (*cf.* Fig. 5.6), the intercept is equal to $-\Delta H_{H^+}[\text{abs}] - 5$. By following this procedure, it was shown[†] that the absolute heat of hydration of the proton ion, $\Delta H_{H^+}[\text{abs}]$, is -266 kcal mole^{-1}.

[†] It will be recalled (*cf.* Section 2.3.11) that the value of -266 kcal mole^{-1} for the absolute heat of hydration for the proton depends upon the approximation that the cavity-formation and structure-breaking contributions to the absolute heat of hydration of ions are independent of the radii of the ions.

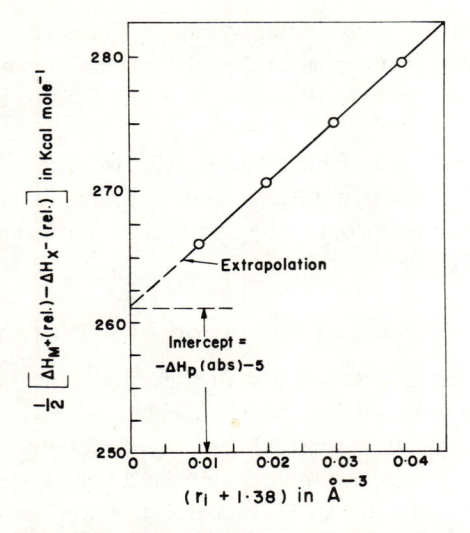

Fig. 5.6. Procedure for getting the absolute heat of hydration of the hydrogen ion. A plot of half the difference of the relative heats of hydration of oppositely charged but equiradii ions *versus* $(r_i + 1.38)^{-3}$ is extrapolated to give an intercept which is equal to $-H_{H^+}(\text{abs}) - 5$.

Now, if the proton affinity (Section 5.2.2) is subtracted from the absolute (or total) hydration heat of a proton, one gets approximately 90 kcal mole^{-1}. This, then, would be the heat of hydration of an H_3O^+ ion. It seems to be roughly the same as the heat of hydration of a K^+ ion (see Chapter 2, Table 2.15). Is this reasonable? Yes, because the radii of the H_3O^+ and K^+ ions are roughly the same.

5.2.4. The Coordination Number of a Proton

The fact that the overall hydration energy of a proton (266 kcal mole^{-1}) is larger than its energy of interaction with one water molecule (170 kcal mole^{-1}) indicates that the proton is engaged in interactions with more than one water molecule, or, in other words, that an H_3O^+ ion, itself, is hydrated. But how many water molecules are engaged in this interaction with an H_3O^+ ion?

The number n of water molecules which interact with an H_3O^+ ion can be obtained by the following considerations. The H_3O^+ ions have two main effects on the water structure. Firstly, they cause some structure breaking,

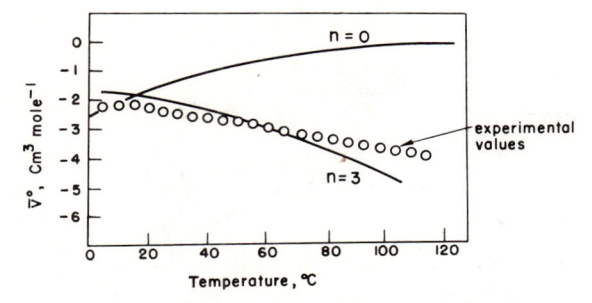

Fig. 5.7. Comparison between the experimental variation of the molal volume of water with temperature and the theoretical variation calculated on the basis that the H_3O^+ ion is associated with $n = 0$ and $n = 3$ molecules.

due to which there is a decrease in molar volume of the solvent (broken-up water occupies less space than does the same water in a structured system). Secondly, the hydration of H_3O^+ ions influences the way in which an increase in temperature affects the total volume. In the coordination shell of an H_3O^+ ion, the n water molecules are compressed owing to the electric field of the ion (electrostriction, *cf.* Section 2.4.5) and, thus, the molal volume is decreased.

Thus, by measuring molal volume (or density) as a function of temperature and comparing it with curves calculated on the basis that $n = 0$ or $n = 3$ molecules of water coordinating an H_3O^+ ion, the coordination number can be determined. It turns out that the experimental curves (Fig. 5.7) are best fitted by having $n = 3$, which means that the H_3O^+ ion is associated with three water molecules in an $(H_9O_4)^+$ group. This group has been detected in the gas phase. It involves a tetrahedral group of one H_3O^+ ion and three water molecules (see Fig. 5.8).

One can now summarize one's knowledge of the condition of a proton in solution. (1) Free protons are highly unlikely in water and aqueous solutions; (2) the association of a proton with a water molecule to form an H_3O^+ ion is accompanied by a large energy release—thus, the H_3O^+ ion appears to be the stable mode of existence of a proton in solution; and (3) the H_3O^+ ion itself interacts with three water molecules to form an $(H_9O_4)^+$ cluster.†

---

† When the methods of determining primary hydration numbers (Section 2.4.1) are used in solutions containing hydrogen ions, one obtains correspondingly a value of about four.

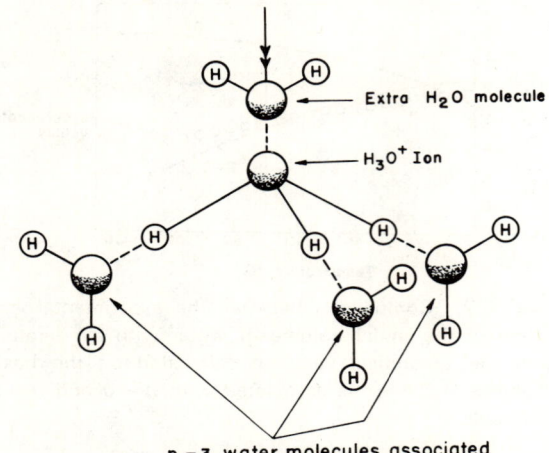

Fig. 5.8. Schematic configuration of $H_9O_4^+$ group shown with an extra H_2O molecule electrostatically bound.

Further Reading

1. J. Sherman, *Chem. Rev.*, **11**: 98 (1932).
2. V. N. Kondratyev and N. D. Sokolov, *Zh. Fiz. Khim.*, **29**: 1265 (1955).
3. M. Falk and P. A. Giguère, *Can. J. Chem.*, **35**: 1195 (1957).
4. R. P. Bell, *The Proton in Chemistry*, Cornell University Press, Ithaca, N.Y., 1959.
5. H. F. Halliwell and S. C. Nyburg, *Trans. Faraday Soc.*, **56**: 1126 (1963).
6. B. E. Conway, "Proton Solvation and Proton Transfer Processes in Solution," in: J. O'M. Bockris and B. E. Conway, eds., Vol. 3, Chap. 2, *Modern Aspects of Electrochemistry*, Butterworth's Publications, Ltd., London, 1964.
7. L. L. Schaleger, P. Salomaa, and F. A. Long, in: *Chemical Physics of Ionic Solutions*, John Wiley and Sons, Inc., New York, 1966.
8. P. Kebarle, *J. Am. Chem. Soc.* **89**: 6393 (1967).

5.3. PROTON TRANSPORT

5.3.1. The Abnormal Mobility of a Proton

What one has been talking about so far are the equilibrium interactions between a proton and an aqueous solvent. It is time now to think of a nonequilibrium process such as proton conductance under an applied electric field. One can begin the discussion with the question: When there is proton

TABLE 5.2
Mobilities of Charge Carriers in Liquids and Solids

Particle	Mobility, $cm^2 \, sec^{-1} \, V^{-1}$
Ions (e.g., K⁺) in water	$\sim 5 \times 10^{-4}$
Proton in water	3×10^{-3}
Ion (e.g., Li⁺) in ice	$\ll 10^{-8}$
Proton in ice	10^{-1}–1

transport, what is the entity which actually does the moving? Is it the proton p, the hydrogen (or hydronium) ion H_3O^+, or the $H_9O_4^+$ cluster?

One way of resolving this question of the kinetic entity in proton drift is to proceed by assuming that, whichever of the above species is the kinetic entity, it drifts in the same way as other ions (e.g., K⁺) do (Chapter 4). The resulting predictions can then be compared with the facts. The starting point therefore is the picture of ions moving through the solvent at a velocity which assumes a steady-state value when the electric driving force ($ze_0 X$) is exactly balanced by the Stokes viscous force ($6\pi r \eta v$). Thus, the ionic mobility at concentrations so small that interionic forces are negligible is given by the familiar expression based on Stokes' law

$$u_0 = \frac{ze_0}{1800\pi r \eta} \qquad (4.180)$$

But, what radius should one use? Free protons do not exist to any significant extent in water and aqueous solutions, so it is logical to ignore them in the first instance. Perhaps it is best to see what happens if the radius

TABLE 5.3
Transport Numbers of Cations in 0.01N Chloride Solutions at 25°C

Ion	Transport number
Hydrogen ion	0.8251
Li⁺	0.3289
Na⁺	0.3918
K⁺	0.4902
Ba⁺⁺	0.440
La⁺⁺⁺	0.4625

TABLE 5.4
Equivalent Conductivity at Infinite Dilution of HCl and LiCl at 20°C

	Solvent			
	Water	Methanol	Ethanol	n-Propanol
Λ_{HCl}	426.2	192	84.3	22
Λ_{LiCl}	115.0	90.9	38.0	18
$\Lambda_{HCl} - \Lambda_{LiCl}$	311.2	101.2	46.3	4

of the hydronium ion H_3O^+ is used in the above expression for the limiting mobility. Now, hydronium ions have roughly the same radii as potassium ions, so one would expect that their mobilities would be approximately the same, i.e., about 7.6×10^{-4} cm² sec⁻¹ V⁻¹.

The facts of proton transport, however, constitute a curious and interesting story. Most important of all (see Table 5.2), *the mobility of protons is abnormally high.* For most other ions in solution, the mobility is of the order of 5×10^{-4} cm² sec⁻¹ V⁻¹; protons migrate with a limiting mobility of about 36×10^{-4} cm² sec⁻¹ V⁻¹. This abnormal mobility is reflected in the transport number, too. Other positive ions in a 1:1-valent electrolyte carry about 50% of the total current; hydrogen ions, however, transport as much as 80% of the total current (Table 5.3).

A number of other facts concerning proton mobility are as surprising. Firstly, the abnormally high mobility is diminished to normal values if water is replaced as solvent by some other substance. For example, in n-propanol the conductance is quite normal (Table 5.4); and, in methanol–water or ethanol–water mixtures, the abnormality in the conductance decreases with increasing alcohol content until, at about 80% alcohol, the abnormal mobility is reduced almost to zero (Fig. 5.9). Secondly, the ratio of the excess (or abnormal) mobilities between the hydrogen and deuterium ions in water is 1.4 at 25 °C (Table 5.5)—much more than might be expected

TABLE 5.5
The Mobilities at Infinite Dilution of Hydrogen and Deuterium Ions

Ion	Infinite dilution mobility
Hydrogen ion in H_2O	36.2×10^{-4}
Deuterium ion in D_2O	25.1×10^{-4}

PROTONS IN SOLUTION 473

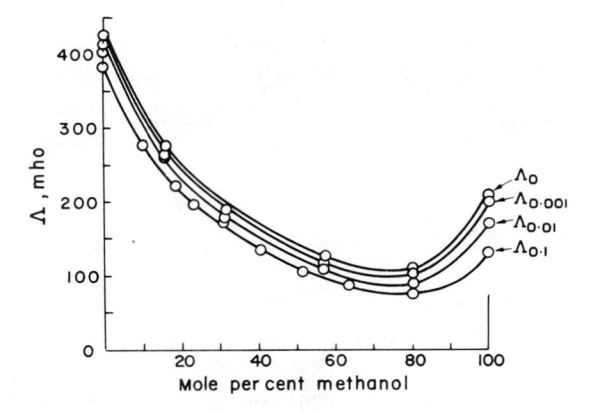

Fig. 5.9. The equivalent conductivity of HCl in methanol-water mixtures decreases with increasing alochol content and reaches a minimum at about 80 mole % of alcohol. At this concentration, the abnormal mobility, i.e., the mobility of the H_3O^+ ion compared with that of the K^+ ion (which is of similar size), is almost reduced to zero.

TABLE 5.6

Variation of Heat of Activation for Proton Transport in Aqueous Solution

Temperature, °K	Mean heat of activation between temperatures shown, kcal mole^{-1}
273	
	2.822
291	
	2.482
298	
	1.907
323	
	1.430
348	
	0.886
373	
	0.636
401	
	0.211
429	

474 CHAPTER 5

on a Stokes' law basis since the two ions are virtually the same size. Thirdly, the proton mobility shows a temperature coefficient which indicates that the heat of activation decreases with temperature (Table 5.6); in the case of most other ions, a single activation energy is good enough over an appreciable range of temperature.

5.3.2. Protons Conduct by a Chain Mechanism

One fundamental conclusion can be drawn from this anomalous behavior of migrating protons. Protons must conduct by a mechanism which is radically different from that used by other ions. In particular, one cannot use for protons the conventional Stokes' law approach with the limiting mobility of ions' being decided by their radius and the viscosity of the medium. This is because the predicted mobility is far too low if one assumes migration of the stable H_3O^+ ion; and, of course, if one assumes bare protons drifting by a Stokes mechanism, the calculated mobility comes out to be much too high because of the extremely small size of the proton. Obviously, therefore, the atomistic picture of proton conduction has to be sketched anew but not with the Stokes' law-oriented model used to explain the drift of other ions.

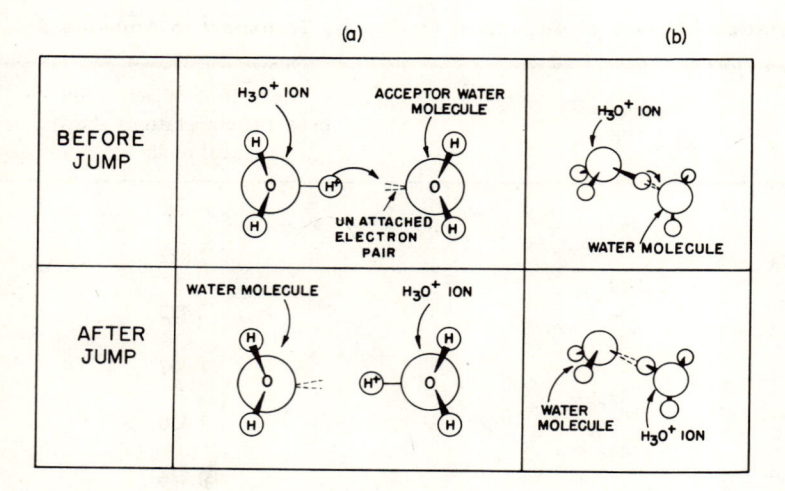

Fig. 5.10. Two schematic views (a) and (b) of a water molecule adjacent to an H_3O^+ ion. The free electron pair (orbital) of the O of the water molecule is oriented along the O (of H_3O^+)—H^+—O (of H_2O) line. The jump of the proton H^+ from the H_3O^+ ion to the water molecule converts the water molecule into an H_3O^+ ion and the H_3O^+ ion into a water molecule.

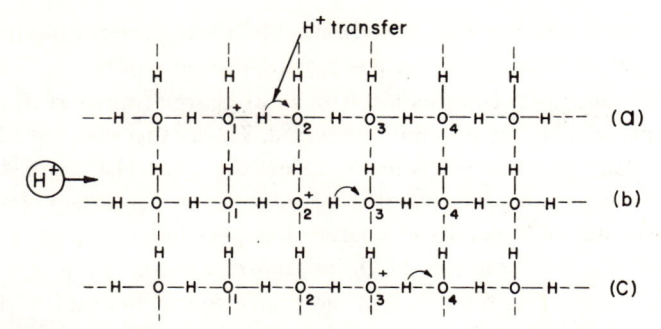

Fig. 5.11. If there are a series of proton jumps down a line of water molecules, the net result is equivalent to the migration of an H_3O^+ ion (indicated by a charge on the oxygen) along the line.

To approach an alternative view, one can start by considering the environment of a hydrogen ion in solution. The chances are that there is a water molecule next to the hydrogen ion. Assume also that the mutual orientation of the H_3O^+ ion and the H_2O molecule are as shown in Fig. 5.10. Notice that there is an attractive unattached orbital a short distance away from one of the protons of the H_3O^+ ion. Suppose, therefore, that the proton jumps from the charged H_3O^+ ion at the left to the neutral H_2O at the right. What is the net result of this jump? The erstwhile-neutral particle at the right has become a charged H_3O^+ ion, and the erstwhile-charged particle at the left has become a neutral water molecule. So the proton jump has produced an effect which is equivalent to the translational movement of an H_3O^+ ion and, therefore, to the movement of charge.

Now consider a whole row of water molecules, and initially let the hydrogen ion be at the extreme left (Fig. 5.11). If the proton transfer is from O_1 to O_2, O_2 to O_3,..., right down the line, then the net effect is *equivalent* to a H_3O^+ ion's moving from left to right. Further, it is also equivalent to the migration of charge from left to right.

If the proton transfers are in response to an electric field, a proton conduction mechanism has been evolved. But this is an entirely different kind of mechanism for charge transport in solution[†] from that discussed previously. There is no movement of an ion as a whole. One water molecule

---

[†] It is interesting to note that Grotthus suggested, in 1806, the idea of the transfer of charge down a chain of particles as a general mechanism of conduction in ionic solutions. His view was later surpassed, but it becomes cogent again just for this special case of proton migration.

passes on a proton to the next water molecule, which passes on a proton to the next molecule, and so on—a passing-the-proton game.

The difference between this mechanism of charge transport in solution and the former one discussed shall be stressed. Thus, other ions have to move bodily from one point in the solution to another; the H_3O^+ ion does not have to move as a whole to transport charge. But the repetition of elementary proton jumps not only results in charge transport but also gives the same effect (if not the same rate) as H_3O^+ migration.

Another interesting point is that the proton jump mechanism does not violate the idea that protons exist in solution as H_3O^+ ions (coordinated by other water molecules) and not as free protons. For it will be noticed that the hopping protons have no independent existence except *during* the actual jumps, and these will be shown (Section 5.3.11) to take up little time in comparison with the time the protons remain attached to one or another water molecule.

5.3.3. Classical Proton Jumps and Proton Mobility

An alternative mechanism of charge transport in solution, specially made for protons, has been suggested. Now one must consider some quantitative aspects of the picture to see whether they fit the facts well enough to justify acceptance of the model. The crucial task, of course, is to obtain a recognition of the factors which decide the mobility of the protons because it is the value of this quantity which provides the major anomaly in comparison with values expected from the previous model.

If the new model outlined in this chapter is correct, the mobility obviously depends on the rate at which protons jump from water molecule to water molecule. The basic theory for this rate of particle jump should be the theory of rate processes (Appendix 4.3). It will be recalled that, for a successful jump to occur, the proton has to be energized with a critical activation energy before it can climb an energy barrier which represents the change of energy of the system (two water molecules plus the jumping proton) as the proton moves from the first water molecule to the second. Once the activation energy is known, the mobility can be estimated (see Appendix 4.3).

The rigorous way of obtaining the activation energy is to calculate the energy of the system as the proton jumps from one water molecule to the next and then plot a potential energy surface. This is a difficult, laborious task. There is, however, an easier, if more approximate, approach. In the initial state, the proton p is bonded to one particular water molecule (H_2O),

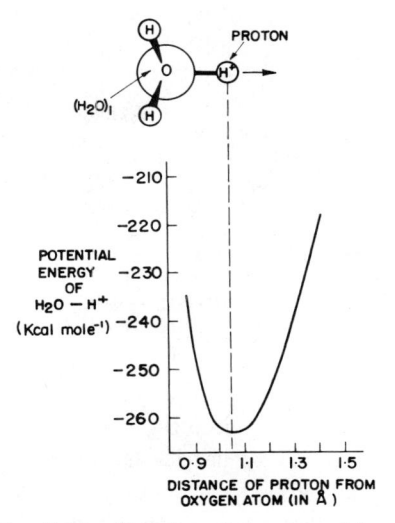

Fig. 5.12. Variation of the potential energy of the $(H_2O)_1$-p system with the stretching of the $(H_2O)_1$-p bond.

and, as an approximation, the $(H_2O)_1$-p system may be treated as a pseudo-diatomic system. As the $(H_2O)_1$-p bond is stretched, the energy of the system increases. Since the stretching of diatomic molecules has been studied in spectroscopy, one may as well borrow from these studies the expression for the energy U_r of the diatomic system as a function of the distance apart, r, of the two "atoms." This expression is known as the *Morse equation*

$$U_r = D_0\{1 - \exp[-a(r - r_e)]\}^2$$

where D_0 is the energy to dissociate the H_2O-p bond, a is a constant, and r_e is the equilibrium distance between the p and H_2O in H_3O^+. For D_0, the overall hydration energy of the proton is used, i.e., about 266 kcal mole^{-1}; and, for r_e, the distance is 0.98 Å.

Thus, one obtains (Fig. 5.12) the potential energy of the $(H_2O)_1$-p system as the proton moves away from the water molecule. The energy *versus* distance plot is known as a *Morse curve*. Because of the symmetry of the system, one has an identical curve for the approach of the proton to the second water molecule.

The two Morse curves, one for the $(H_2O)_1$-p and the other for p-$(H_2O)_2$, are then put together. The problem of the relative positioning of the minima of the two Morse curves is met by assuming a reasonable model based on

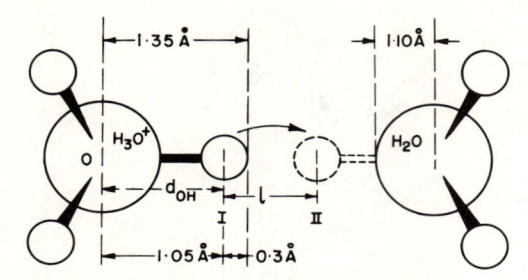

Fig. 5.13. Model for proton transfer between an H_3O^+ ion and a favorably oriented H_2O molecule.

a knowledge of bond lengths, etc., such as those shown in Fig. 5.13. When the two minima are thus fixed, it is seen that the two Morse curves intersect (Fig. 5.14) and a potential-energy barrier has been synthesized.[†]

This then is the energy barrier for the jump of a proton from one water molecule to another over a potential-energy barrier. From the barrier, one can calculate the energy of activation and thus the rate of proton transfers. It turns out that the mobility calculated according to this model has the right order of magnitude in comparison with experiment.

5.3.4. Do Proton Jumps Obey Classical Laws?

Numerical agreement between the calculated and observed values of a principal experimental quantity is an important step in giving credence to a model. But it is not a sufficient one. In addition, the model has to be in accord with all the other experimental observations. Thus, in the context of the question of proton mobilities, the model which has been described above must explain the observed facts on the ratio of the mobilities of the H_3O^+ and D_3O^+ ions and on the temperature dependence of the mobility of the proton.

When the comparison is made, it turns out that the model of simple classical jumps of protons outlined above predicts that H_3O^+ ions in H_2O should travel about 14 times faster than D_3O^+ in D_2O in comparison with the experimental ratio $u_{H_3O^+}/u_{D_3O^+}$ of about 1.4. Further, as the temperature is raised from T_2 to T_1, the distance between the minima in Fig. 5.14 is

---

[†] The procedure used involves a number of unstated approximations. Thus, systems interact near the intersection (Fig. 5.14) so that there is not a sharp linear type of intersection, as shown, but a rounding-off, and resultant reduction of the heat of activation. Further, the zero-point energy levels are not shown in Fig. 5.14, and no discussion has been given of the effect of the environment on the motions calculated.

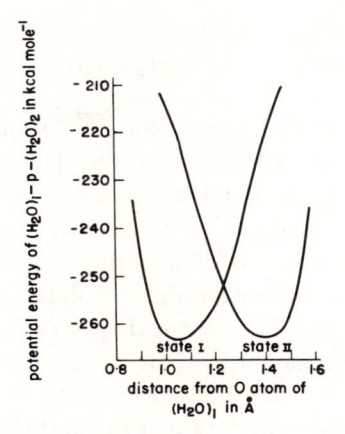

Fig. 5.14. Potential-energy barrier for proton transfer from H_3O^+ to H_2O.

increased, which leads to an *increase* in the energy of activation (minimum to crossing point) for proton jumps (Fig. 5.15). Experimental observations do indicate that the heat of activation changes with temperature; but it is found that the heat of activation *decreases* with temperature, i.e., it changes in a direction opposite to that predicted.

Thus, after the initial encouragement at finding a model which gives the correct order of magnitude from the anomalous proton mobility, one begins to have doubts and hence second thoughts.

The rate process theory presented in Chapter 4 and Appendix 4.3 was developed for processes involving ions which are massive compared with protons. Is it right to use this classical theory to treat the transfer of a relatively light, elementary particle such as the proton? Thus, in the mobility calculation of the previous section one insisted upon the proton's climbing

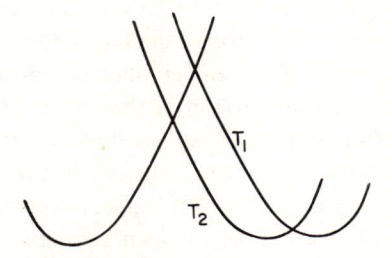

Fig. 5.15. Schematic diagram to show that the activation energy for proton transfer depends upon the distance between the minima of the Morse curves.

over the barrier for, in the usual treatment of rate processes, the only way the moving particles can cross the energy barrier is by having the necessary energy to *surmount* it.

Classical mechanics is an approximation valid for objects which are large compared with their de Broglie wavelengths. Particles which are sufficiently small and light do not behave according to classical laws. Their motions are best described by the laws of quantum mechanics. One of these laws is that the position of a particle cannot be defined with certainty;[†] one can only ascribe a certain probability to the particle's being in a certain region of space. This means that, even if a particle has less than the critical energy to surmount the barrier, there is a finite probability that it crosses the barrier. It is customary to refer to this phenomenon as *quantum-mechanical tunneling*; one says that the particle has leaked through the barrier rather than climbed over it.

It can be shown that the velocity with which tunneling occurs increases exponentially as the mass of the particle decreases. This means that tunneling is particularly important for light particles, in particular, therefore, for electrons. (It will be seen in Chapter 8 that tunneling of electrons from an electrode to particles in solutions constitutes the basis of charge transfer at electrified interfaces.) Can tunneling be significant also for the proton jumps which are suggested to be the basis of proton conduction in solutions?

5.3.5. Quantum-Mechanical Proton Jumps and Proton Mobility

The calculation of the rate of quantum-mechanical proton transfer from hydrogen ions to water molecules was carried out in detail in a classical paper by Bernal and Fowler and published in the same paper containing their structural theory of ion–solvent interactions. Although the calculation is of historic importance to the theory of proton conduction in aqueous solution, the result proved to be rather disappointing—protons tunnel too fast. In fact, the mobility of protons calculated on the basis of tunnel transfers is about one hundred times faster than the observed value.

One can summarize the position at this stage. Migration of H_3O^+ ions by a conventional Stokes mechanism yields too low a mobility, and thus stresses the need for an alternative model for the mechanism of proton migration. Thus, a chain mechanism was conceived with an elementary step of proton transfer between hydrogen ions and water molecules. It is of importance, however, to understand what individual step in this migration

---

[†] The criterion of when to use quantum mechanics has been given in Chapter 1.

mechanism controls the rate of proton transfer. Classical proton transfer gives the right order of magnitude for the mobility of protons but disagrees with other experimental facts. Fundamental considerations suggest the probable presence of quantum-mechanical tunnel transfers. Tunnel transfer through the barrier is possible, and calculations indicate that it is easier than classical over-the-barrier transfer. Hence, tunneling should be considered *the* predominant mode of proton transfer from hydrogen ions to water molecules. But calculation shows it is too fast to explain the observed mobility. The analysis is apparently incomplete.

5.3.6. Water Reorientation, a Prerequisite for Proton Jumps

It has been assumed so far that the only factor in proton transfer is the actual jump itself and that no preconditions must be satisfied. However, from Fig. 5.16, it is clear that a proton from an H_3O^+ ion cannot jump indiscriminately to a water molecule. It can only jump when the oxygen atom of the H_2O which receives the jumping proton is in a favorable orientation. The point is that the oxygen of the water molecule can form bonds with protons only along certain directions in space. These directions correspond to the directions of the vacant orbitals. Thus, the proton jump can occur only if the acceptor water molecules orient so that the oxygens, the proton, and the unshared electron pair of the acceptor oxygen line up. Then, the proton jumps along the pathway leading to the oxygen orbital. It is as though the journey of the proton from the H_3O^+ ion to the water molecule starts with the tunnel through the barrier but is complete only when, on emerging from the tunnel, the proton crosses a swing bridge. But the latter obstacle in the journey can be overcome only if the swing bridge is in line.

Now the analysis of proton mobility can be completed. The tunneling

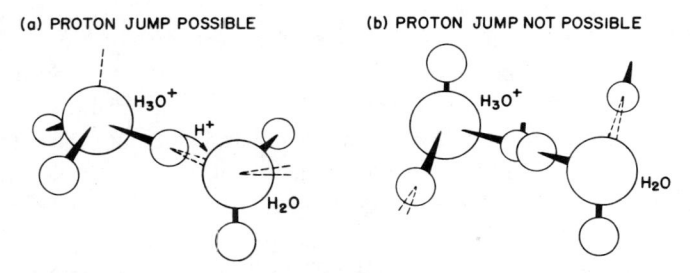

Fig. 5.16. Schematic diagram to show that the transfer of a proton from an H_3O^+ ion to a water molecule is possible in configuration (a), but not in configuration (b).

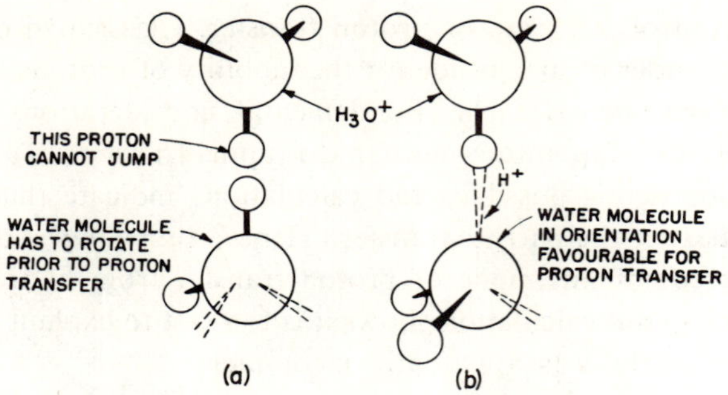

Fig. 5.17. In (a), the proton from the H_3O^+ ion cannot jump to the water molecule because the latter is in an unfavorable orientation. When the water molecule reorients and is in a configuration corresponding to (b), the proton jump can occur.

of protons will give too high a mobility only if the water molecules are always oriented correctly. If, however, the orientation is "incorrect," then the protons will have to wait till the water molecule turns around and, only after this orientation process is complete, can the jump occur (Fig. 5.17). So there are two processes which must cooperate to give the successful proton transfers which are the basis of proton mobility. First, there is water reorientation, and, then, the proton tunneling. Hence, the rate of proton transfers will be limited by whichever of the two processes is slower. One must therefore investigate whether water reorientation is the *rate-determining step* in the process of proton transfer (for the tunneling through the barrier has already been shown to be too fast to be consistent with the observed mobility).

5.3.7. The Rate of Water Reorientation and Proton Mobility

The calculation of the specific rate of water reorientation is quite a complex task. This is because one cannot consider the reorienting water molecule as an isolated entity. If that were so, then one could work on the basis of the rate of rotation of a gas molecule and calculate the rate. However, the water molecule is hydrogen bonded to other water molecules, and, therefore, the reorientation involves the stretching and breaking of the hydrogen bonds which bind the molecules together.

The calculation was nevertheless first performed by Conway *et al.* in 1956. At first, this calculation proved a disappointment because, if the tunneling protons had to wait for the water molecules to turn around and

PROTONS IN SOLUTION   483

get into the right position every time a proton arrived on the scene (i.e., turn around to present a free orbital to the next oncoming proton), then the predicted mobility came out low compared with experiment. The analogy has been made to a swing bridge which has to be in a position to receive the proton tunneling through the barrier. There are two ways in which the bridge could swing into position. It could do so spontaneously by random motions; this would correspond to the acceptor water molecule's reorienting by the thermal motions of the molecular world. It was this thermal

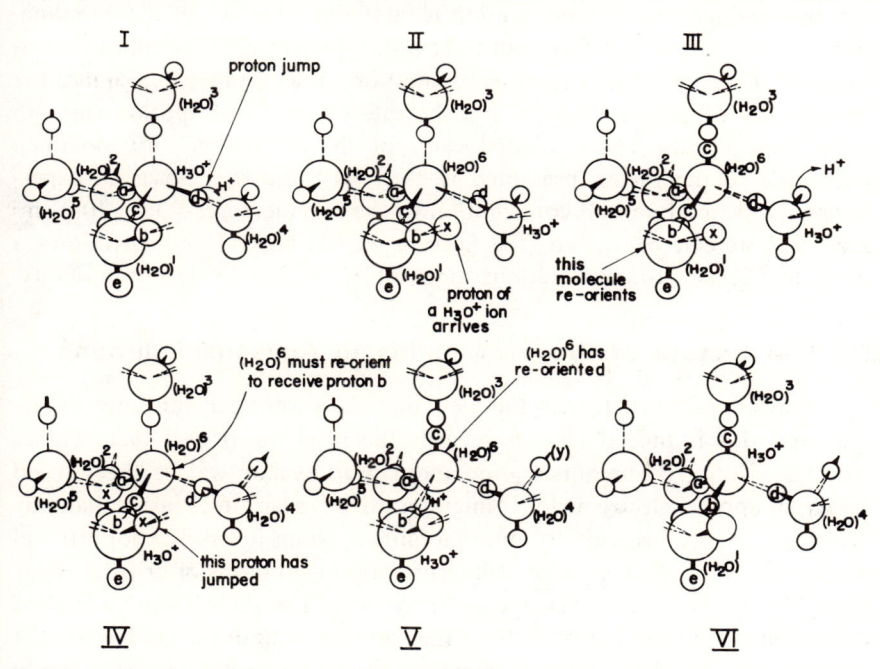

Fig. 5.18. A detailed schematic representation of the proton mobility model in which proton tunneling has to be preceded by an induced reorientation of the acceptor water molecule. I. The proton d of the central H_3O^+ ion is transferred to $(H_2O)^4$; thus the H_3O^+ ion becomes $(H_2O)^6$ and $(H_2O)^4$ becomes an H_3O^+ ion. II. The proton X of an H_3O^+ ion (not shown in the diagram) arrives next to $(H_2O)^1$ but cannot be transferred to $(H_2O)^1$ because the latter is in an unfavorable orientation. III. The water molecule $(H_2O)^1$ reorients so that the H atom b which was blocking the jump of X is moved away to face H atom c of $(H_2O)^6$. Another event is the jump of proton Y of the H_3O^+ ion to an adjacent water molecule (not shown in diagram). IV. Proton X jumps to $(H_2O)^1$ which now becomes an H_3O^+ ion. Note that this newly created H_3O^+ ion cannot transfer proton b to the central water molecule $(H_2O)^6$ because the last is unfavorably oriented. V. The central water molecule $(H_2O)^6$ reorients, and proton b can now be transferred to it. VI. After the jump of proton b, the central water molecule has again become an H_3O^+ ion, and the H_3O^+ion, a water molecule $(H_2O)^1$; cf. step I.

reorientation mechanism that Conway *et al.* found too slow to account for the mobility. Alternatively, it could be automated to turn at the approach of a vehicle: the approach of the proton can trigger a mechanism which turns the water molecule around and makes it ready to meet the tunneling proton. It is precisely this *field-induced reorientation* of the water molecules that is necessary to explain the proton mobility. What happens is that, as the proton approaches the dipole, its coulombic field attracts the water dipole and drags it around into position (Fig. 5.18). Though the rate of *this* field-induced water reorientation is faster than the rate of the spontaneous thermal rotation, it turns out to be much slower than the proton tunneling rate. Thus, it is the field-induced rotation of water that determines the overall rate of proton transfer and the rate of proton migration through aqueous solutions. The estimated value of the proton mobility obtained according to the theory presented is 28×10^{-4} and that observed experimentally as an abnormal contribution to the conductance is 36×10^{-4} cm^2 sec^{-1} V^{-1}. It can be shown (see Section 5.3.9) that this model allows a consistent series of interpretations for most of the facts of proton mobility.

5.3.8. A Picture of Proton Mobility in Aqueous Solutions

This chapter began with the view that the proton was an unorthodox ion, and this is indeed true as regards the mechanism by which it goes through solutions. The outstanding anomaly to explain was its sheer speed under an applied electric field. It migrates about seven times faster than any other ion. As one is used to ionic mobilities' being inversely proportional to the radius, the first attempt at interpretation is in terms of drifting H_3O^+ ions, but the mobility turns out to be too low. The temptation to explain the mobility by thinking of drifting free protons cannot be yielded to, not only because anything but a momentary nonattachment to water is surely very unlikely for the proton, but also because, were it to be regarded in this way, the model would predict too high a mobility compared with what is observed.

So, the ordinary Stokes mechanism of ions drifting through the solution does not tie up with experiment. One has to turn to other possibilities, and an attractive one is a chain mechanism in which protons leap from the oxygen of an H_3O^+ ion to the oxygen of an adjacent water molecule. The model in which proton mobility is controlled by classical over-the-barrier proton jumps between water molecules does not give results consistent with all the experimental facts. Further, though the proton is much heavier than the electron, it is still small and light enough (compared with other

TABLE 5.7
Comparison of Different Models of Proton Mobility

Model and rate-determining step	Absolute mobility, cm^2 sec^{-1} V^{-1}	u_{H^+}/u_{D^+}
Classical proton transfer	1.24×10^{-2}	6.1
Proton tunneling	75.8×10^{-2}	13.9
Field-assisted reorientation of water molecules followed by fast tunneling	0.28×10^{-2}	1.42

ions) to leak through the barrier in quantum-mechanical fashion. This tunneling, with its noncommon sense laws, turns out to be a much faster way of jumping than the classical path of going over the barrier. If, however, it is assumed that quantum-mechanical proton jumps determine proton mobility, then the predicted overall mobility proves to be too high and one must think of some precondition for proton transfer which limits the tunneling rate. This other step is the rotation of a water molecule so that its free orbital is in line with the proton which wishes to tunnel to it (particles which tunnel must have a bond, or acceptor, state to tunnel *to*). Calculation shows that *this* model does turn out to be in reasonable accord with the facts—not too slow as is the Stokes transport and not too fast as is simple tunnel transfer unfettered by consideration of the availability of acceptor states. So the model which provides the best interpretation of the high mobility of protons is that in which they tunnel from an H_3O^+ ion to a water molecule but the tunneling rate is limited by the rate at which the acceptor water molecules reorient so that their free orbitals face the tunneling protons (Table 5.7).

One further remark must be added to this picture. For 96% of its life, a proton exists attached to a water molecule, i.e., the H_3O^+ ion has an independent existence. Hence, in addition to the water-reorientation-followed-by-proton-tunneling mechanism just described, there is also the usual Stokes type of drift of H_3O^+ ions. In fact, the Stokes transport of H_3O^+ ions accounts for about 20% of the total mobility of protons.

5.3.9. The Rate-Determining Water-Rotation Model of Proton Mobility and the Other Anomalous Facts

The numerical *value* of the high mobility of protons has been explained reasonably well. What of the other anomalies listed in Section 5.3.1?

The anomalous decrease of the heat of activation with increase of temperature follows from the model because the mechanism depends on the fact that there is available a hydrogen-bonded liquid with a considerable degree of ordering in its structure. Increase of temperature causes increase of disorder in the water structure, and, consequently, there are, on the average, fewer H bonds to break when water molecules reorient. Since the water reorientation becomes easier as the temperature increases, the heat of activation becomes smaller. An explanation can also be advanced for the decrease of the anomalous mobility of the proton in the presence of added alcohol solvents: The new bonding possibilities which arise when alcohol is added to water make the alcohol reorientation more difficult than that for water and thus cause a fall in the proton mobility by the tunneling and solvent-orientation method.

It must be noted, however, that any proton-conduction mechanism in which the rate-determining step involves the breaking of hydrogen bonds would succeed equally well in explaining the above two anomalies; they are therefore not stringent tests on the model. In contrast, the 1.4 ratio of the mobilities of hydrogen and deuterium is a diagnostic criterion for the various models of proton mobility, and, therefore, the correct calculation of this ratio (as the model indeed makes possible) is strong evidence in favor of a mechanism involving water reorientation as a precondition for proton tunneling. There are, in addition, several other phenomena which can be quantitatively interpreted in terms of the water-reorientation-tunnel-transfer model. Mention may be made of the anomalous mobility of hydroxyl OH^- ions compared with that of protons; and interpretation of the mobility of NH_4^+ and NH_2^- in NH_3. These various interpretive successes in related areas increase the plausibility of the water-rotation-controlled, proton-tunneling mechanism in explaining the abnormal mobility of protons.

5.3.10. Proton Mobility in Ice

It is the matter of the proton mobility in ice which has been argued by Conway *et al.* as providing the most striking evidence in favor of the water-rotation, proton-tunneling model. The fact is: Ice exhibits a higher proton mobility than water does. Does this mean that water molecules turn faster in ice than in water? This appears most unlikely. Is the model wrong, then? The following fact is the bridge to enlightenment on the matter. The *number* of protons per cubic centimeter in ice is much less than that in the usual systems in which measurements are made, namely, water or aqueous solutions. This means that, when one measures the proton mobility in ice, the *flux* of protons across a given piece of the system is less than that in water or in aqueous solutions. Now, recall what was said about swing bridges which swing spontaneously but gradually back into place and the modern ones which are automated to be ready, after a pause, when a vehicle approaches them.

Suppose the number of vehicles approaching the bridge grew very much smaller than had been the case when it was arranged for the bridge to be back in place when a vehicle arrived. Then it would have time to swing back by itself without help from some extra mechanism. There would be no pause for the swingback accelerated by an approaching vehicle. Thus, for protons in ice, there are so few protons that the thermal rotation rate of the water molecules in ice is sufficient to bring the molecules back into the position to receive the tunneling protons. They are always ready in time in ice. If one may mix metaphors a little, for protons which so infrequently come through the tunnels in the ice, the lights are always green; the protons do not have to wait for the orbitals to turn around and be ready to receive them, i.e., their mobility is no longer determined by the rate of water rotation. Hence, the tunneling itself becomes the rate-determining step; but this tunneling is a much faster process than that of water rotation, so that it leads to higher mobilities of protons in ice than in the water or aqueous solutions in which the water rotation is rate determining.

5.3.11. The Existence of the Hydronium Ion from the Point of View of Proton Mobility

The rapid proton transfers from H_3O^+ ions to water molecules clearly imply that H_3O^+ ions cannot be *permanent* entities in aqueous solution. The moment an H_3O^+ ion loses its proton to a neighboring water molecule, it is transformed into a water molecule. However, the time taken in the transfer process is about 10^{-14} sec, while the time an H_3O^+ ion waits for its adjacent water molecule to reorient is about 2.4×10^{-13} sec. Since, therefore, the proton spends some 96% of its time linked to a definite molecule in an H_3O^+ species rather than as a tunneling proton, it can be concluded that H^+ ions in aqueous solution have an *effective existence* as H_3O^+.

5.3.12. Why Is the Mechanism of Proton Mobility So Important?

Protons are central to the electrochemistry of solutions; they are the basis of the concepts of pH and of acids and bases. But that is not the only reason why their state in solution is so important. Another reason stems from the central question: How are electric charges transported in solid biological systems? A great many biological systems require concepts connected with electrochemical charge transfer for an explanation of their behavior. A difficulty is in explaining how electric current gets carried within the macromolecules. Now, in the systems concerned, there is much H bonding; recall the protein helix held together by hydrogen bonds. It is a speculation, but perhaps one of interest, that the charge transport in such systems is by protons, perhaps tunneling protons, not electrons. Hence, any mechanism by which protons get carried along particularly quickly, especially an H-bonded solid (e.g., ice), has important implications for a large area of molecular biology.

Further Reading

1. J. D. Bernal and R. H. Fowler, *J. Chem. Phys.*, **1**: 515 (1933).
2. B. E. Conway, J. O'M. Bockris, and H. Linton, *J. Chem. Phys.*, **24**: 834 (1956).
3. M. Eigen and L. de Maeyer, *Proc. Roy. Soc. (London)*, **A247**: 505 (1958).
4. B. E. Conway and J. O'M. Bockris, *J. Chem. Phys.*, **31**: 1133 (1959).
5. B. E. Conway, "Proton Solvation and Proton Transfer Processes in Solution," in: J. O'M. Bockris and B. E. Conway, eds., *Modern Aspects of Electrochemistry*, Vol. 3, Butterworth's Publications, Ltd., London, 1964.
6. B. E. Conway and M. Salomon, in: B. E. Conway and R. G. Barradas, eds., *Chemical Physics of Ionic Solutions*, John Wiley and Sons, Inc., New York, 1966.
7. R. G. Wawro and T. J. Swift, *J. Am. Chem. Soc.*, **90**: 2792 (1968).
8. T. A. Stephenson, T. J. Swift, and J. B. Spencer, *J. Am. Chem. Soc.*, **90**: 4291 (1968).

5.4. HOMOGENEOUS PROTON-TRANSFER REACTIONS AND POTENTIAL ELECTROLYTES

5.4.1. Acids Produce Hydrogen Ions and Bases Produce Hydroxyl Ions: The Initial View

It is common knowledge that aqueous solutions of a certain class of substance, called *acid*, corrode so-called "active" metals and turn blue litmus paper red. In 1887, Arrhenius suggested that this characteristic behavior of the class of substances called acids arises from the common characteristic that they produce hydrogen ions in solution. Correspondingly, Arrhenius suggested that hydroxyl ions are responsible for the well-known properties of aqueous solutions of alkalis or bases, namely the ability to turn red litmus paper blue and neutralize acids.

According to this classical view, therefore, acids are substances which dissolve in water to give hydrogen ions, and bases are substances which dissolve to give hydroxyl ions. It follows that, in order to understand acids and bases, one must understand the mechanism of the production of hydrogen ions and hydroxyl ions.

5.4.2. Acids Are Proton Donors, and Bases Are Proton Acceptors: The Brönsted View

A convenient starting point for understanding the mechanism of the production of hydrogen and hydroxyl ions is the mechanism of proton conduction in aqueous solutions. The elementary step in this mechanism turned out to consist of proton transfers between H_3O^+ ions and water molecules. Notice, however, that the water molecule which accepts the jumping proton is converted into a hydronium ion. One can now introduce an obvious terminology. In the transfer of a proton from an H_3O^+ ion to a water molecule, the H_3O^+ ion can be termed the *proton donor*, and the H_2O, the *proton acceptor* (Fig. 5.19).

The H_3O^+ ion is not the only possible donor of a proton to a water molecule; there are many other substances which can act as proton donors. Consider, for example, acetic acid and sulphuric acid. Reactions such as those shown in Fig. 5.20, in which there are proton transfers to water molecules, result in the production of H_3O^+ ions. But substances which dissolve in water to produce H_3O^+ ions are historically known as *acids*. At this stage, therefore, one would define an acid as a donor of protons to water.

Just as one wondered whether the H_3O^+ ion was the only possible proton donor, it is necessary to consider whether water molecules are the only proton acceptors. Then, it becomes clear that, since free protons are unstable in solutions, one substance cannot donate protons unless there is an-

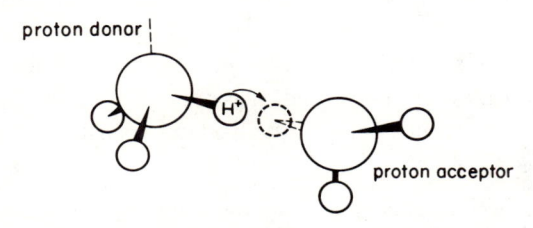

Fig. 5.19. The proton donor and proton acceptor in the proton transfer from an H_3O^+ ion to a water molecule.

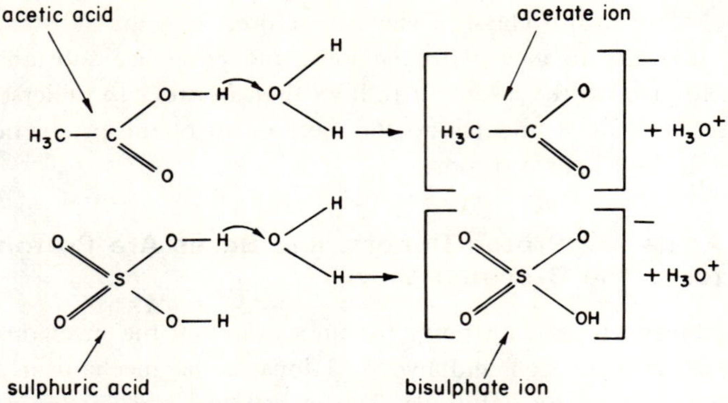

Fig. 5.20. The production of H_3O^+ ions by the transfer of protons from proton donors to water molecules.

other substance to accept them. Hence, the reactants in any proton-transfer reaction must consist of one proton donor and one proton acceptor.

Consider the following reaction between ammonia and water

$$H_2O + NH_3 \rightarrow OH^- + NH_4^+ \qquad (5.3)$$

It will be noticed that, in this proton-transfer reaction, water molecules are the proton donors and the ammonia molecules act as proton acceptors. Further, the result of the reaction is the production of hydroxyl ions. This is interesting because Arrhenius ascribed the characteristic behavior of bases to hydroxyl ions. So one can at this point define a base as a substance which accepts protons from water molecules in a proton-transfer reaction.

The classical (or Arrhenius) picture of acids and bases has been presented above as proton-transfer reactions *involving water*: Acids donate protons to water molecules to produce H_3O^+ ions, and bases accept protons from water molecules to form OH^- ions.[†] The trouble with this picture is that it is linked too firmly with the solvent water. For instance, what is to be done with the dissolution of acetic acid in liquid ammonia (Fig. 5.21)? It is obvious that this reaction involves proton transfer between a proton donor and a proton acceptor. In fact, it is the analogue in liquid ammonia of what happens when acetic acid is dissolved in the particular solvent, water.

[†] Substances which can function either as proton donors (i.e., acids) or as proton acceptors (i.e., bases) are termed *amphoteric* (or *amphiprotic*). One of the reasons why water is such a widely used solvent is that it is amphoteric and therefore enters into proton-transfer reactions with either proton donors or proton acceptors.

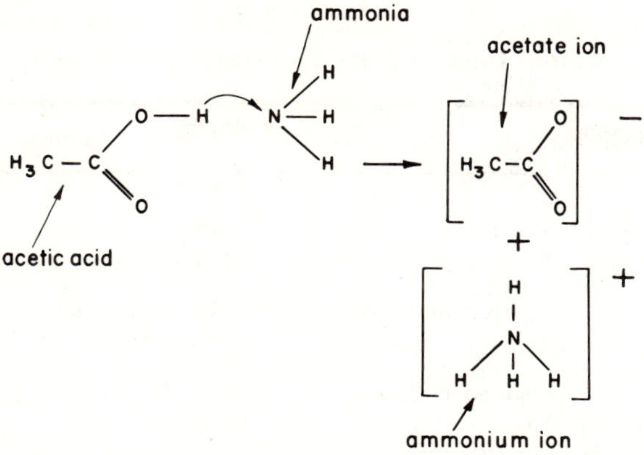

Fig. 5.21. The proton transfer reaction between acetic acid and liquid ammonia.

The fundamental similarities between proton-transfer reactions in various solvents led Brönsted to suggest that the definition of acids and bases be freed from the solvent water. He proposed that, irrespective of the solvent in which the proton-transfer reaction takes place, the proton donor be termed an acid and the proton acceptor, a base.

5.4.3. The Dissolution of Potential Electrolytes and Other Types of Proton-Transfer Reactions

In the introduction to the treatment of ion–solvent interactions, it was stressed that electrolytes dissolve in a solvent by two fundamentally different mechanisms: True electrolytes undergo physical interactions with the solvent molecules and form solvated ions; potential electrolytes engage in chemical reactions with the solvent molecules and give rise to ionic solutions. It is now seen that the chemical reactions which potential electrolytes engage in are proton-transfer reactions of a special type in which an uncharged potential-electrolyte molecule reacts with an uncharged solvent molecule to produce a pair of oppositely charged ions.

This is only one of the types of proton-transfer reactions. There are others. For example, an ion and a neutral molecule can react to produce another ion and another neutral molecule

$$NH_4^+ + H_2O \rightarrow NH_3 + H_3O^+ \tag{5.4}$$

Thus, the destruction of the electrical field of the NH_4^+ ion is compensated for by the generation of the electrical field of the H_3O^+ ion. This is in con-

TABLE 5.8
Some Types of Proton-Transfer Reactions

Reactants	Products
1. Two uncharged molecules, e.g., HCl and H_2O	A pair of oppositely charged ions, e.g., $H_3O^+ + Cl^-$
2. An ion and an uncharged molecule, e.g., $NH_4^+ + H_2O$	An ion and an uncharged particle, e.g., $NH_3 + H_3O^+$
3. An ion and an uncharged molecule, e.g., $HCO_3^- + H_2O$	A pair of oppositely charged ions, e.g., $CO_3^{--} + H_3O^+$

trast to the proton-transfer reaction involving a potential electrolyte where the net result of the reaction is the creation of two species giving rise to ionic fields.

Or an ion can react with a neutral molecule to produce two ions, e.g.,

$$HSO_4^- + H_2O \rightarrow SO_4^{--} + H_3O^+ \tag{5.5}$$

In this type of proton-transfer reaction, two ionic field-giving species are generated, while one disappears.

It is seen that the fundamental difference between the above mentioned types of proton-transfer reactions arises from the electrostatic work involved in the occurrence of the reaction, i.e., the work associated with generating or destroying the electrostatic fields of the charged species (Table 5.8). More will be said later on this electrostatic contribution to the free-energy change in proton-transfer reactions.

Another interesting type of proton-transfer reaction is that involving two molecules of an amphoteric solvent, i.e., a solvent which can function either as an acid or as a base. The reaction consists in one solvent molecule's being a proton donor and another solvent molecule's being a proton acceptor. For example, one may have

$$H_2O + H_2O \rightarrow H_3O^+ + OH^- \tag{5.6}$$

or

$$C_2H_5OH + C_2H_5OH \rightarrow C_2H_5OH_2^+ + C_2H_5O^- \tag{5.7}$$

It is because of such *autodissociation* reactions that even highly purified amphoteric solvents display some electrical conductivity.

5.4.4. An Important Consequence of the Brönsted View: Conjugate Acid–Base Pairs

An important consequence of the Brönsted definition of acids and bases comes out by examining a particular proton-transfer reaction in detail. Consider, for example, the dissolution of acetic acid in water

$$CH_3COOH + H_2O \rightarrow CH_3COO^- + H_3O^+ \quad (5.8)$$

What is the result of the proton-transfer reaction? The acetic acid molecule (i.e., the proton donor, or acid) has been deprotonated, and the water molecule (i.e., the proton acceptor, or base) has been protonated. In fact, one can *conceptually* break up the overall proton-transfer reaction into one deprotonation reaction

$$CH_3COOH \rightarrow CH_3COO^- + p \quad (5.9)$$

and one protonation reaction

$$H_2O + p \rightarrow H_3O^+ \quad (5.10)$$

where p stands for proton. It is clear that, according to these partial reactions, the acetic acid has behaved as an acid (it is a proton donor) and the water molecule, as a base (it is a proton acceptor).

If, however, a reaction can take place in one direction, it can also run the opposite way (*cf.* the principle of microscopic reversibility). Thus, when the partial reactions proceed in the reverse direction, one has

$$H_3O^+ \rightarrow H_2O + p \quad (5.11)$$

$$CH_3COO^- + p \rightarrow CH_3COOH \quad (5.12)$$

which add up to give the overall reaction

$$CH_3COO^- + H_3O^+ \rightarrow CH_3COOH + H_2O \quad (5.13)$$

This is a proton-transfer reaction in which the acetate ion behaves as a proton acceptor (base) and the H_3O^+ ion as a proton donor (acid).

Thus, the reaction between the acid CH_3COOH and the base H_2O results in the production of another acid H_3O^+ and another base CH_3COO^-. In general therefore, an acid A_1 and a base B_2 react to give another acid A_2 and another base B_1

$$A_1 + B_2 \rightarrow A_2 + B_1 \quad (5.14)$$

This reaction can be broken up into two partial reactions

$$A_1 \rightarrow B_1 + p \tag{5.15}$$

and

$$B_2 + p \rightarrow A_2 \tag{5.16}$$

The species A_1 and B_1 are the protonated and deprotonated forms of the same substance and are called a *conjugate acid–base pair*. It is seen that, in a proton-transfer reaction, two acid–base pairs are involved, i.e., A_1 and B_1 and A_2 and B_2.

The homogeneous *proton*-transfer (or acid–base) reactions described above are similar to homogeneous *electron*-transfer reactions in that the overall electron-transfer reaction, e.g.,

$$Fe^{+++} + Ce^{+++} \rightarrow Fe^{++} + Ce^{++++} \tag{5.17}$$

can be decomposed into one electronation reaction

$$Fe^{+++} + e \rightarrow Fe^{++} \tag{5.18}$$

and one deelectronation reaction

$$Ce^{+++} \rightarrow Ce^{++++} + e \tag{5.19}$$

Further, each of these partial reactions involves the electronated (reduced) and de-electronated (oxidized) forms of a substance, or what is called a *redox couple*.

5.4.5. The Absolute Strength of an Acid or a Base

The attainment of equilibrium in proton-transfer reactions is very rapid, and, therefore, it is valid to represent such reactions as at equilibrium:

$$A_1 + B_2 \rightleftharpoons B_1 + A_2 \tag{5.20}$$

By considering the conjugate acid–base pairs involved in the above reaction, one can think of the partial equilibria

$$A_1 \rightleftharpoons B_1 + p \tag{5.21}$$

and

$$B_2 + p \rightleftharpoons A_2 \tag{5.22}$$

and the corresponding equilibrium constants

$$K_{a,A_1} = \frac{a_{B_1} a_p}{a_{A_1}} \tag{5.23}$$

and
$$K_{a,B_2} = \frac{a_{A_2}}{a_{B_2}a_p} \tag{5.24}$$

It is obvious that K_{a,A_1} is an absolute measure of the tendency of A_1 to donate (or lose) a proton and K_{a,B_2} is an absolute measure of the tendency of B_2 to accept a proton. Since the acidity of a substance is equivalent to its proton-donating tendency, the quantity K_{a,A_1} may be termed the *absolute strength of the acid* A_1, and, since the basicity of a substance is equivalent to its proton-accepting tendency, K_{a,B_2} may be termed the *absolute strength of the base* B_2.

Further, by considering the product $K_{a,A_1}K_{a,B_2}$, one has

$$K_r = K_{a,A_1}K_{a,B_2} = \frac{a_{A_2}a_{B_1}}{a_{A_1}a_{B_2}} \tag{5.25}$$

which shows that the product $K_{a,A_1}K_{a,B_2} = K_r$ is, in fact, the equilibrium constant for the reaction $A_1 + B_2 \rightleftharpoons B_1 + A_2$.

Thus, if one could determine the absolute acid strength K_{a,A_1} of A_1 and the absolute base strength K_{a,B_2} of B_2, one could calculate K_r and thus decide how far the equilibrium $A_1 + B_2 \rightleftharpoons B_1 + A_2$ is pushed to the right. It is clear, for instance, that the stronger the acid A_1 (i.e., the larger the value of K_{a,A_1}) and the stronger the base B_2 (i.e., the larger the value of K_{a,B_2}), the more will the position of the equilibrium be shifted to the right.

The attempt to determine K_{a,A_1} and K_{a,B_2} would succeed if the partial equilibriums (5.21) and (5.22) involving protons as reactants or products are realizable in practice. It has already been stated, however, that the concentration of *free* protons in solution is far too small to permit them to be recognized as reaction products or reactants. Hence, the partial equilibriums (5.21) and (5.22) for conjugate acid–base pairs are experimentally unrealizable as separate entities and are, therefore, purely hypothetical.

One must perforce conclude that the absolute strengths of acids and bases cannot be measured and a scale of absolute acid or base strengths has no operational significance.

5.4.6. The Relative Strengths of Acids and Bases

Since *absolute* acid (or base) strengths cannot be measured and since it is only equilibriums of the type $A_1 + B_2 \rightleftharpoons B_1 + A_2$ that are experimentally realizable, an obvious step is to accept a particular acid–base pair as a standard and then develop a scale of relative strengths.

To define the *relative strength of acids*, it is customary to accept the solvent as the standard proton acceptor. Thus, for aqueous solutions, the water molecule is taken as the standard proton acceptor, in which case the proton-transfer equilibrium must be written as

$$A + H_2O \rightleftharpoons B + H_3O^+ \tag{5.26}$$

and the relative strength of the acid A_1 is by definition

$$K_{r,A} = K_{a,A} K_{a,H_2O \text{ as base}} = \frac{a_B a_{H_3O^+}}{a_A a_{H_2O}} \tag{5.27}$$

In very dilute solutions, activities can be replaced by concentrations, and one can set $a_{H_2O} = 1$. Then one obtains the *acid dissociation constant* $K_{c,A}$ defined thus

$$K_{c,A} = \frac{c_B c_{H_3O^+}}{c_A} \tag{5.28}$$

The acid dissociation constant of A is a quantitative measure of the power of the acid A to donate a proton *to a water molecule*.

Now, it is experimentally found that the dissociation constants of acids vary over several powers of 10 (Table 5.9). Hence, instead of remembering dissociation constants in the form $K_{c,A} = \alpha \times 10^{-\beta} = 10^{-x}$, where x is the exponent, it is convenient to represent dissociation constants on a logarithmic scale thus

$$x = -\log_{10} K_{c,A} \tag{5.29}$$

The x's of different acids are the *powers* of $K_{c,A}$ and are, therefore, fairly logically referred to as pK_A values, i.e.,

$$pK_A = -\log_{10} K_{c,A} = x \tag{5.30}$$

In terms of pK_A, Eq. (5.28) becomes

$$pK_A = -\log_{10} c_{H_3O^+} - \log \frac{c_B}{c_A} \tag{5.31}$$

Since hydrogen ion concentrations, too, vary over several powers of 10, one can refer to the *powers of hydrogen*-ion concentration by the symbol pH defined thus[†] (Table 5.10)

$$pH = -\log_{10} c_{H_3O^+} \tag{5.32}$$

[†] Strictly speaking, one must define pH in terms of the activity of H_3O^+ ions.

TABLE 5.9
Dissociation Constants of Some Acids

Acid	Base produced	K_c
1. Very weak acids, $K_{c,A} < 10^{-7}$		
HCO_3^-	$CO_3{=}$	4.69×10^{-11}
HIO	IO^-	1×10^{-11}
C_6H_5OH	$C_6H_5O^-$	1.20×10^{-10}
HBrO	BrO^-	2×10^{-9}
H_2PO_4	$HPO_4{=}$	6.22×10^{-8}
2. Weak acids, $K_{c,A} < 10^{-2}$		
HClO	ClO^-	9.56×10^{-7}
H_2CO_3	HCO_3^-	4.5×10^{-7}
$C(CH_3)_3 \cdot COOH$	$C(CH_3)_3 \cdot COO^-$	8.91×10^{-6}
CH_3COOH	CH_3COO^-	1.75×10^{-5}
$C_6H_5 \cdot COOH$	$C_6H_5 \cdot COO^-$	6.12×10^{-5}
HCOOH	$HCOO^-$	1.77×10^{-4}
$CH_2Cl \cdot COOH$	$CH_2Cl \cdot COO^-$	1.34×10^{-3}
3. Strong acids, $K_{c,A} \sim 10^3$		
H_2SO_4	HSO_4	10^3
$(HSO_4^-$	$SO{=}$	$1.01 \times 10^{-2})$
HCl	Cl	10^3
$HClO_3$	ClO_3^-	10^3
4. Very strong acids, $K_{c,A} \sim 10^8$		
$HClO_4$	ClO_4^-	10^8

TABLE 5.10
Hydrogen Ion Concentrations and pH

Hydrogen ion concentration	pH	Hydrogen ion concentration	pH
10^{-1} (0.1N HCl)	1	10^{-8}	8
10^{-2}	2	10^{-9}	9
10^{-3}	3	10^{-10}	10
10^{-4}	4	10^{-11}	11
10^{-5}	5	10^{-12}	12
10^{-6}	6	10^{-13}	13
10^{-7} (pure water)	7	10^{-14} (1N NaOH)	14

TABLE 5.11

The pK Values of Some Acids

Acid	pK
HCO_3^-	10.33
HIO	11
C_6H_5OH	9.92
HBrO	9.92
$H_2PO_4^-$	8.7
HClO	6.02
H_2CO_3	6.35
$C(CH_3)_3 \cdot COOH$	5.03
$CH_3 \cdot COOH$	4.8
C_6H_5COOH	4.2
HCOOH	3.75
$CH_2Cl \cdot COOH$	2.87
H_2SO_4	−3
HSO_4^-	2.0
HCl	−3
$HClO_3$	−3
$HClO_4$	−8

and write Eq. (5.31) in the form

$$pK_A = pH - \log \frac{c_B}{c_A} \qquad (5.33)$$

Since low pK_A values indicate large values of the acid dissociation constant $K_{c,A} = c_B c_{H_3O^+}/c_A$, Eq. (5.33) shows that the lower the value of pK_A, the more is the conversion of the acid A to its base form B with the concomitant production of H_3O^+ ions. Strong acids therefore have small pK_A values (i.e., large K_A values, Table 5.11).

The consideration of the strength of bases proceeds on lines similar to that for acids. That is, the relative base strength is defined by taking the solvent as the standard proton *donor*. In aqueous solution, this means that the water molecule is the proton donor and the proton-transfer reaction is

$$H_2O + B \rightleftharpoons OH^- + A \qquad (5.34)$$

PROTONS IN SOLUTION 499

TABLE 5.12
Autodissociation Constants of Some Solvents

Reaction	Temperature	K
$2NH_3 \rightleftharpoons NH_4^+ + NH_2^-$	$-33\ °C$	10^{-22}
$2CH_3OH \rightleftharpoons CH_3OH_2^+ + CH_3O^-$	$25\ °C$	10^{-17}
$2H_2O \rightleftharpoons H_3O^+ + OH^-$	$25\ °C$	10^{-14}
$2CH_3COOH \rightleftharpoons CH_3COOH_2^+ + CH_3COO^-$	$25\ °C$	10^{-13}
$2H_2SO_4 \rightleftharpoons H_3O^+ + HS_2O_7^-$	$10\ °C$	7×10^{-5}
$2HNO_3 \rightleftharpoons NO_2^+ + NO_3^- + H_2O$	$25\ °C$	2×10^{-2}

The relative strength $K_{r,B}$ of the base B is then defined as

$$K_{r,B} = \frac{a_{OH^-} a_A}{a_B a_{H_2O}} \qquad (5.35)$$

or, by replacing activities by concentrations and setting $a_{H_2O} = 1$, one has the analogue for bases of the acid dissociation constant, i.e.,

$$K_{c,B} = \frac{c_{OH^-} c_A}{c_B}$$

$$= \frac{c_A}{c_B c_{H_3O^+}} c_{OH^-} c_{H_3O^+}$$

$$= \frac{1}{K_{c,A}} c_{OH^-} c_{H_3O^+} \qquad (5.36)$$

Now, the equilibrium constant for the autodissociation of water (Table 5.12), i.e.,

$$H_2O + H_2O \rightleftharpoons OH^- + H_3O^+$$

is termed the *ionic product* of water K_W and is given by

$$K_W = c_{OH^-} c_{H_3O^+} \qquad (5.37)$$

On combining Eqs. (5.36) and (5.37), the result is

$$K_{c,B} = \frac{K_W}{K_{c,A}} \qquad (5.38)$$

or

$$pK_B = pK_W - pK_A \qquad (5.39)$$

TABLE 5.13

pK$_B$ Values of Some Bases and pK$_A$ Values of the Conjugate Acids

Base	pK$_B$	Conjugate acid	pK$_A$
Ammonia	4.76	NH_4^+	9.24
Methylamine	3.30	$CH_3NH_3^+$	10.70
Dimethylamine	3.13	$(CH_3)_2NH_2^+$	10.87
Trimethylamine	4.13	$(CH_3)_3NH^+$	9.87
Ethylamine	3.25	$C_2H_5NH_3^+$	10.75
Diethylamine	2.90	$(C_2H_5)_2NH_2^+$	11.10
Triethylamine	3.20	$(C_2H_5)_3NH^+$	10.80
Aniline	9.39	$C_6H_5NH_3^+$	4.61
Benzylamine	4.63	$(C_6H_5NH_2)_3NH^+$	9.37
Diphenylamine	13.16	$(C_6H_5)_2NH_2^+$	0.84
Pyridine	8.80	$C_5H_5NH^+$	5.2

where

$$pK_B = -\log_{10} K_{c,B} \quad (5.40)$$

and

$$pK_W = -\log_{10} K_W \quad (5.41)$$

Equation (5.39) shows that there is a simple relation between the pK$_B$ value of the base B (Table 5.13) and the pK$_A$ value of the acid A in the proton-transfer reaction

$$H_2O + B \rightleftharpoons OH^- + A \quad (5.34)$$

For example, consider the reaction

$$H_2O + NH_3 \rightleftharpoons OH^- + NH_4^+ \quad (5.42)$$

where NH_3 and NH_4^+ are the conjugate base–acid pair. For this equilibrium Eq. (5.39) becomes

$$pK_{NH_3} = pK_W - pK_{NH_4^+} \quad (5.43)$$

which indicates that one can determine the pK of the acid NH_4^+ if one knows the pK of NH_3, or *vice versa*.

5.4.7. Proton Free-Energy Levels

Since proton-transfer reactions are the basis of the behavior of acids and bases, the relative strengths of acids and bases must arise from the

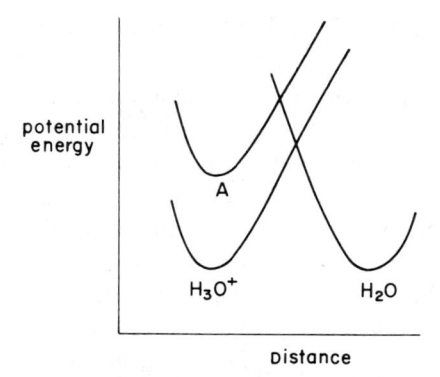

Fig. 5.22. Schematic potential-energy barriers for proton transfers from an H_3O^+ ion to a water molecule and from an acid A to a water molecule.

differing energies of the reactants and products. This aspect will now be discussed.

Consider the transfer of a proton from an H_3O^+ ion to a water molecule. This transfer process can be described by a potential-energy barrier synthesized from two Morse curves (Fig. 5.14). One curve represents the energy change as the proton leaves the proton donor H_3O^+; and the other, the energy change as the proton approaches the proton acceptor H_2O.

Such curves can be drawn for the transfer of a proton from *any* proton donor to a water molecule.† Suppose one compares the potential barrier for two proton-transfer reactions, e.g., that from an H_3O^+ ion to a water molecule and that from some acid A to a water molecule (Fig. 5.22)

$$H_3O^+ \xrightarrow{\text{proton}} H_2O$$

$$A \xrightarrow{\text{proton}} H_2O$$

The Morse curve for the energy changes as a proton approaches the water molecule (the proton acceptor) will be the same in both proton-transfer reactions. But the curve for the removal of the proton from the H_3O^+ ion will not be the same as that for its removal from the acid A because it is

---

† Since one is considering the overall energy change between the initial and final states, the type of mechanics to be used in considering the transfer—classical or quantal—does not affect the issue. It is when proton-transfer *rates* are considered that the mechanism of transfer becomes important.

unlikely that the proton affinities of H_2O and B (the conjugate base of A) will be equal.

This argument has an important consequence: The minima of the deprotonation curves for H_3O^+ and A are not at the same level in relation to the minimum of the curve for the protonation of a water molecule, but the difference between the minima of the deprotonation (of H_3O^+ or A) and the protonation (of H_2O) curves defines the difference in energy between the initial state and the final state of the proton-transfer reaction. If one also takes into account the entropy difference (usually small), one will obtain the chemical or nonelectrostatic contribution ΔG_{ch} to the standard free-energy change ΔG^0 of the proton-transfer reaction. Since the reactant and product species may be charged, one must also consider the electrostatic free energy ΔG^0_{el} of charging or discharging the ions. (The method of calculating the electrostatic contribution to the total free-energy change is shown in Section 5.4.9.)

The total free-energy change ΔG^0, which is therefore made up of two contributions

$$\Delta G^0 = \Delta G^0_{ch} + \Delta G^0_{el} \tag{5.44}$$

represents the net work done to pull out a proton from the proton donor, transfer it to a water molecule, and then separate the resulting species. One could use the terminology suggested by Gurney and call the ΔG^0's "proton free-energy levels" in the proton donors H_3O^+, A_1, A_2, etc., defined in relation to the water molecule as the standard proton acceptor. These proton free-energy levels can also be represented on a diagram which would reveal, on examination, whether the jump of a proton to a water molecule would lead to a decrease of free energy and would thus tend to occur spontaneously (Fig. 5.23).

At present, the statistical mechanical calculation of the ΔG^0's for proton-transfer reactions involves so many approximations that it is preferable to obtain them in the following way.

The standard free-energy change for the transfer of a proton from an acid to a water molecule is simply related to the equilibrium constant for the reaction

$$A + H_2O \rightleftharpoons H_3O^+ + B \tag{5.45}$$

through the Van't Hoff reaction isotherm

$$K_{r.A} = e^{-\Delta G^0/RT} = \frac{a_B a_{H_3O^+}}{a_A a_{H_2O}} \tag{5.46}$$

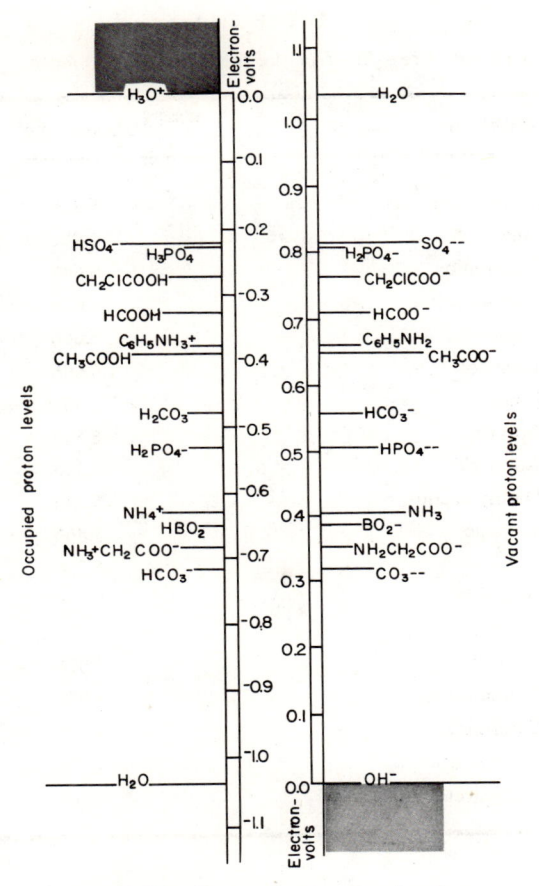

Fig. 5.23. Proton free-energy levels.

and, if conditions of infinite dilution are considered, then activities are equal to concentrations and $a_{H_2O} \sim 1$, i.e.,

$$K_{c,A} = \frac{c_B c_{H_3O^+}}{c_A} = e^{-\Delta G^0/RT} \tag{5.47}$$

Hence,

$$\Delta G^0 = -RT \ln K_{c,A}$$
$$= (-\log_{10} K_{c,A}) 2.303 RT \tag{5.48}$$

and, from the definition of pK_A [cf. Eq. (5.30)],

$$\Delta G^0 = (2.303 RT) \, pK \tag{5.49}$$

TABLE 5.14
Proton Free-Energy Levels of Some Acids

Acid	ΔG^0, kcal mole^{-1}
Trimethylacetic	6.856
Acetic	6.486
Propionic	6.646
Hexoic	6.626
Isobutyric	6.606
Isohexoic	6.606
Valeric	6.606
Butyric	6.566
Isoullesic	6.516
Diethylacetic	6.456
Succinic	5.866
Lactic	5.256
Glycollic	5.216
Formic	5.106
Iodoacetic	4.326
Bromoacetic	3.956
Chloroacetic	3.906
Fluoroacetic	3.526
Cyanoacetic	3.366

i.e., the standard free energy for the transfer of a proton from the acid to a water molecule is proportional to the pK of the acid.

Thus, the concept of proton free-energy levels, expressed in terms of ΔG^0 (Table 5.14), is simply related to the pK value and, therefore, to the strength of an acid relative to a particular solvent functioning as a proton acceptor.

5.4.8. The Primary Effect of the Solvent upon the Relative Strength of an Acid

An obvious conclusion arises from a consideration of the potential-energy curves for proton transfer from an acid to the solvent: the lower the minimum of the solvent curve in relation to that for the acid, the greater is the free-energy change for the proton-transfer reaction and, hence, the greater is the relative strength of the acid. This conclusion can also be reached from Eq. (5.27), which can be written as

$$K_{r,A} = K_{a,A} K_{a,S} \tag{5.50}$$

for the reaction

$$A + S \rightleftharpoons B + SH^+ \tag{5.51}$$

It is clear that, the greater the absolute base strength $K_{a,S}$ of the solvent (i.e., the greater its proton-accepting tendency), the greater is the relative acid strength $K_{r,A}$ of the acid A. Hence, a particular acid may appear strong in one solvent and weak in another.

For example, HCl is a strong acid in aqueous solutions because the proton-transfer reaction

$$HCl + H_2O \rightleftharpoons H_3O^+ + Cl^- \tag{5.51}$$

proceeds almost completely to the right (H_2O acting as a strong base); but it is a weak acid in glacial acetic acid because the interaction

$$HCl + CH_3COOH \rightleftharpoons CH_3C(OH)_2^+ + Cl^- \tag{5.52}$$

occurs only to a small extent. The reason, then, is that water is a stronger base than acetic acid.

Suppose now one considers several acids $A_1, A_2, A_3, \ldots$, and suppose

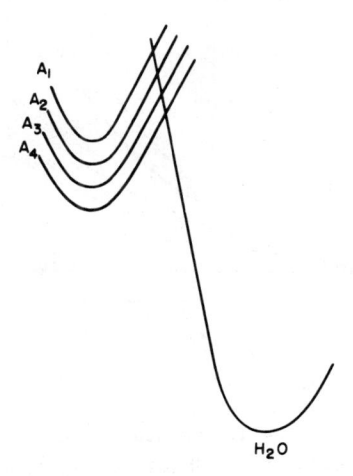

Fig. 5.24. The potential-energy barriers for proton transfer from strong acids to a weakly acidic solvent.

506 CHAPTER 5

the minima of their proton-donating curves lie far above that of the proton-accepting curve of the solvent (Fig. 5.24). Then these various acids will react almost completely with the solvent, i.e., the position of the equilibrium

$$A_i + S \rightleftharpoons B_i + SH^+ \tag{5.53}$$

will occur so much to the right that, effectively, all the acid will be in its base form B_i. Under these circumstances, all the acids A_1, A_2,... will appear equally strong (Fig. 5.25). This is what happens with aqueous solutions of the strong acids $HClO_4$, HBr, H_2SO_4, HCl, and NHO_3. The reactions such as

$$HNO_3 + H_2O \rightleftharpoons H_3O^+ + NO_3^- \tag{5.54}$$

proceed almost completely to the right, and, in solutions less concentrated than $2M$, all these acids appear equally strong, i.e., the differences in their relative acid strengths cannot be distinguished. This is because the behavior of the aqueous solutions is determined by the H_3O^+ ion, which is formed virtually to the same extent in all these solutions. One says that, because

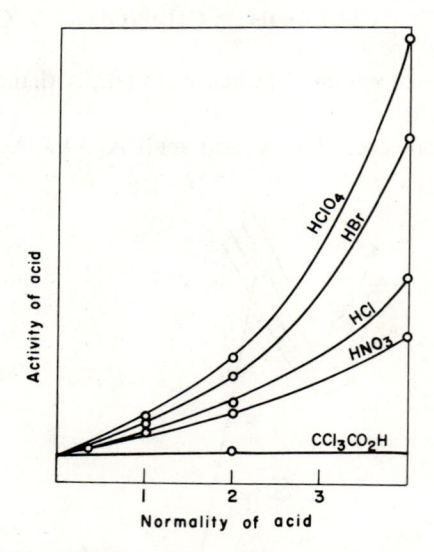

Fig. 5.25. A strongly basic solvent levels out the differences in the relative strengths of strong acids. The relative strengths of the acids are indicated in the diagram by their activity in catalyzing a chemical reaction. Below $2M$ concentration, the activities of the various acids are almost the same.

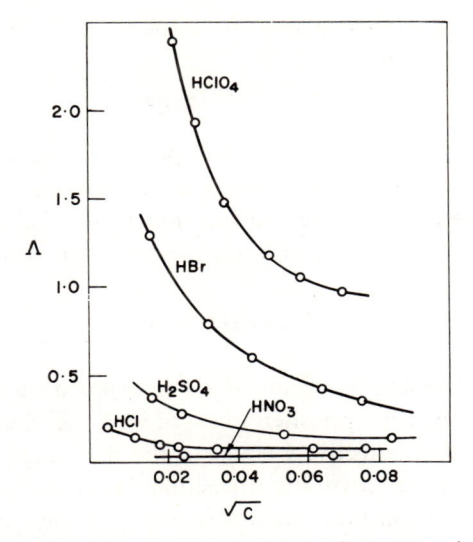

Fig. 5.26. A weakly basic (i.e., strongly acidic) solvent brings out the differences in the relative strengths of strong acids. In the diagram, the relative strengths of the various acids are indicated by their equivalent conductivities in glacial acetic acid.

the aqueous solvent has a low absolute acid strength (a strong base strength), it has leveled out—the so-called *leveling* effect—the differences in the relative strengths of the various acids.

To bring out these differences, it is necessary to use a solvent which has a greater absolute *acid* strength (or a lesser base strength). This is equivalent to choosing a solvent which has a protonation curve, the minimum of which lies close enough to the minima of the deprotonation curves to make the difference in their minima significant. Thus, $HClO_4$, HBr, H_2SO_4, HCl, and HNO_3 display quite different relative strengths (i.e., ionic concentrations, as indicated by the equivalent conductivities) when dissolved in glacial acetic acid (Fig. 5.26), a more acid, less basic solvent than water.

5.4.9. A Secondary (Electrostatic) Effect of the Solvent on the Relative Strength of Acids

It has been demonstrated that the solvent plays an essential role in determining the relative strength of an acid. The key characteristic of the solvent is its absolute base strength $K_{a,S}$, which determines the relative strength $K_{r,A}$ of an acid through the Eq. (5.50)

$$K_{r,A} = K_{a,A}K_{a,S} \qquad (5.50)$$

But, in addition to functioning as a proton acceptor in a proton-transfer reaction, the solvent also serves as a dielectric medium, which affects the electrostatic aspects of the reaction in the following way.

Suppose, for instance, the proton-transfer reaction is of the potential-electrolyte type

$$HA + S \rightleftharpoons SH^+ + A^- \qquad (5.55)$$

in which an uncharged acid molecule HA interacts with a neutral solvent molecule S to form a pair of oppositely charged ions. The dielectric constant of the medium, which is effectively equal to the dielectric constant of the solvent, determines the electrostatic attraction between the pair of ions; the larger the dielectric constant, the less strongly do the ions SH^+ and A^- attract each other and the less is the tendency of the backward reaction. If, therefore, one considers two solvents which do not differ considerably in their absolute acid strengths, one would expect the relative strength of the acid to be greater in the solvent with a greater dielectric constant. This is indeed the case with the uncharged carboxylic acids when one uses alcohol instead of water as the solvent.

When, however, an ion and an uncharged molecule react to form another ion and uncharged molecule, e.g.,

$$NH_4^+ + S \rightleftharpoons SH^+ + NH_3 \qquad (5.56)$$

the proton-transfer reaction does not require the separation of oppositely charged ions, and, therefore, the dielectric constant of the medium does not have a significant effect on the relative strength of the acid.

To make these considerations quantitative, it is necessary to isolate the dielectric-constant effect of a solvent from its proton-accepting properties. This can be done as follows.

If two acids A_1 and A_2 are considered in the same solvent, then the *ratio* of their relative strength is given by [*cf.* (5.50)]

$$\frac{K_{r,A_1}}{K_{r,A_2}} = \frac{K_{a,A_1}K_{a,S}}{K_{a,A_2}K_{a,S}}$$

$$= \frac{K_{a,A_1}}{K_{a,A_2}} \qquad (5.57)$$

PROTONS IN SOLUTION   509

Thus, if the absolute base strength $K_{a,S}$ of a solvent is the only characteristic of the solvent which affects the relative strengths of acids, then the ratio of the relative strengths of a pair of acids should be independent of the solvent. If, on the other hand, it is found that the ratio $K_{r,A_1}/K_{r,A_2}$ varies with the solvent, then the variation must be ascribed to the influence of the dielectric constant of the medium.

It is easy to derive a relationship between the relative acid-strength ratio $K_{r,A_1}/K_{r,A_2}$ and the dielectric constant of the medium. The relative acid-strength ratio can be elaborated upon thus

$$\frac{K_{r,A_1}}{K_{r,A_2}} = \frac{K_{a,A_1}}{K_{a,A_2}} = \frac{a_{B_1}a_p/a_{A_1}}{a_{B_2}a_p/a_{A_2}}$$

$$= \frac{a_{B_1}a_{A_2}}{a_{A_1}a_{B_2}} \tag{5.58}$$

from which it is clear that $K_{r,A_1}/K_{r,A_2}$ is the equilibrium constant for the reaction

$$A_1 + B_2 \rightleftharpoons B_1 + A_2 \tag{5.59}$$

This reaction may be written in the form

$$HA_1 + A_2^- \rightleftharpoons A_1^- + HA_2 \tag{5.60}$$

in order that the electrostatic effects may be seen more easily. The equilibrium constant ratio $K_{r,A_1}/K_{r,A_2}$ is given by

$$- RT \ln \frac{K_{r,A_1}}{K_{r,A_2}} = \Delta G^0 \tag{5.61}$$

where ΔG^0 is the standard free energy change for reaction (5.60). This ΔG^0 includes the electrostatic contribution ΔG^0_{el}, arising from the disappearance and appearance of charges in the reaction, as well as a nonelectrostatic, or chemical, contribution ΔG^0_{ch}, arising from the relative levels of the minima in the proton-transfer barrier (*cf.* Section 5.4.7). Hence,

$$- RT \ln \frac{K_{r,A_1}}{K_{r,A_2}} = \Delta G^0_{ch} + \Delta G^0_{el} \tag{5.62}$$

To evaluate the electrostatic contribution to the free-energy change in the proton-transfer reaction, one can assume that the ions A_1^- and A_2^- are spherical, and set ΔG^0_{el} equal to the Born charging work for the singly

510   CHAPTER 5

charged A_1^- ion minus the Born discharging work for the singly charged A_2^- ion, i.e.,

$$\Delta G_{el}^0 = \frac{N_A e_0^2}{2\varepsilon r_{A_1^-}} - \frac{N_A e_0^2}{2\varepsilon r_{A_2^-}}$$

$$= \frac{N_A e_0^2}{2\varepsilon}\left(\frac{1}{r_{A_1^-}} - \frac{1}{r_{A_2^-}}\right) \quad (5.63)$$

where $r_{A_1^-}$ and $r_{A_2^-}$ are the radii of the A_1^- and A_2^- ions and the other terms have their usual connotation (*cf.* Section 2.2.5).

Introducing this expression into Eq. (5.62), one has

$$\ln \frac{K_{r,A_1}}{K_{r,A_2}} = -\frac{\Delta G_{ch}^0}{RT} - \frac{N_A e_0^2}{2\varepsilon RT}\left(\frac{1}{r_{A_1^-}} - \frac{1}{r_{A_2^-}}\right) \quad (5.64)$$

Thus, the plot of the logarithm of the ratio of the relative strengths of two acids *versus* the reciprocal of the dielectric constant should be a straight line of slope

$$-\frac{N_A e_0^2}{2RT}\left(\frac{1}{r_{A_1^-}} - \frac{1}{r_{A_2^-}}\right)$$

the sign of the slope being positive when $r_{A_2^-} < r_{A_1^-}$ and negative when $r_{A_1^-} < r_{A_2^-}$. Further, by extrapolating the $\ln(K_{r,A_1}/K_{r,A_2})$ versus $1/\varepsilon$ straight line to infinite dielectric constant [i.e., $(1/\varepsilon) \to 0$], one should obtain the

TABLE 5.15

Ratios of the Relative Strengths of Substituted Benzoic Acids to the Relative Strengths of Benzoic Acid in Various Solvents at 25°C

Derivative of benzoic acid	Value of log K_r relative to log K_r for benzoic acid			
	H_2O, $\varepsilon = 78.5$	CH_3OH, $\varepsilon = 31.5$	C_2H_5OH, $\varepsilon = 24.2$	$n\text{-}C_4HOH$, $\varepsilon = 17.4$
p-Hydroxy	−0.36	−0.53	−0.55	−0.57
p-Methyl	−0.17	−0.18	−0.18	−0.19
p-Chloro	0.22	0.34	0.42	0.39
m-Chloro	0.38	0.59	0.63	0.59
o-Chloro	1.28	1.21	1.12	1.08
o-Nitro	2.03	1.83	1.77	1.78

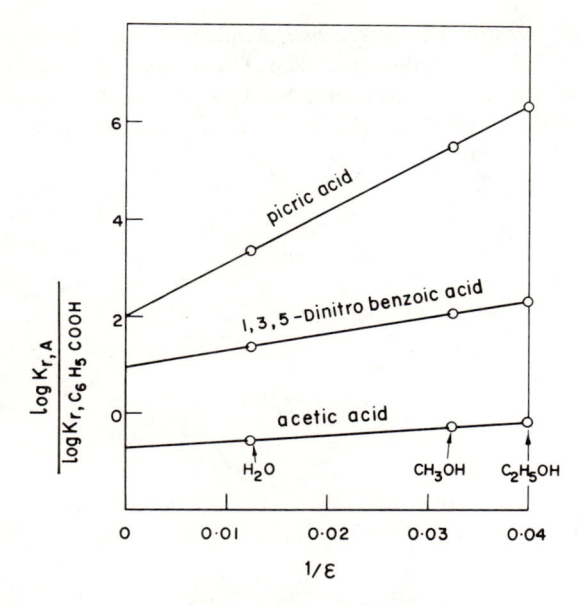

Fig. 5.27. The dielectric constant; variation of the ratio of log K_r for an acid to log K_r for benzoic acid.

intrinsic acid-strength ratio $(K_{r,A_1}/K_{r,A_2})_{\varepsilon\to\infty}$ which is free of electrostatic effects. Hence,

$$\ln \frac{K_{r,A_1}}{K_{r,A_2}} = -\ln \left(\frac{K_{r,A_1}}{K_{r,A_2}}\right)_{\varepsilon\to\infty} - \frac{N_A e_0^2}{\varepsilon RT}\left(\frac{1}{r_{A_1^-}} - \frac{1}{r_{A_2^-}}\right) \quad (5.65)$$

These predictions are borne out in practice, as may be seen from Fig. 5.27 and Table 5.15. It is seen, therefore, that while the base strength of a solvent is the primary factor in determining the relative strengths of a given acid in various solvents, the solvents' dielectric constants do have a significant secondary effect due to the changing solvation energies of the ions in the various solvents.

Further Reading

1. W. F. K. Wynne-Jones, *Proc. Roy. Soc. (London),* **A140**: 440 (1933).
2. R. W. Gurney, *Ionic Processes in Solution,* McGraw-Hill Book Company, New York, 1953.
3. R. P. Bell, *The Proton in Chemistry,* Cornell University Press, Ithaca, N.Y., 1959.

4. C. B. Monk, *Electrolytic Dissociation*, Academic Press, Inc., New York, 1961.
5. E. Grunwald and S. Meiboom, *J. Am. Chem. Soc.*, **85**: 204 (1963).
6. L. L. Schaleger, P. Salomaa, and F. A. Long, B. E. Conway and R. G. Barradas, eds., in: *Chemical Physics of Ionic Solutions*, John Wiley & Sons, Inc., New York, 1966.

CHAPTER 6
IONIC LIQUIDS

6.1. INTRODUCTION

6.1.1. The Limiting Case of Zero Solvent: Pure Liquid Electrolytes

Modern electrochemistry is concerned not only with systems based on aqueous solutions but also with water-free systems (see Section 4.7). Indeed, it is in such systems that many important electrochemical processes are carried out, e.g., the production of aluminum, sodium, and magnesium.

The rationale behind the use of (and the search for) other media shall be restated (see also Section 4.7.2). In aqueous media, electrode reactions involving water and its ionic constituents (hydrogen ions and hydroxyl ions) sometimes altogether supplant the desired electrochemical process (e.g., the electrowinning of magnesium). Further, in technologies based on the conversion of electrical energy into chemical change and *vice versa* (e.g., the dissolution of a metal in an electrochemical energy-conversion device), the desired rate of conversion may require media more conducting than aqueous solutions.

Some of the difficulties associated with carrying out processes in aqueous solutions can be sometimes overcome by using nonaqueous solvents consisting usually of organic substances, e.g., acetonitrile, to which is added some solute which dissociates in that solvent. But this often is not a good approach because of the low specific conductances of such solutions (see Section 4.7.9).

514   CHAPTER 6

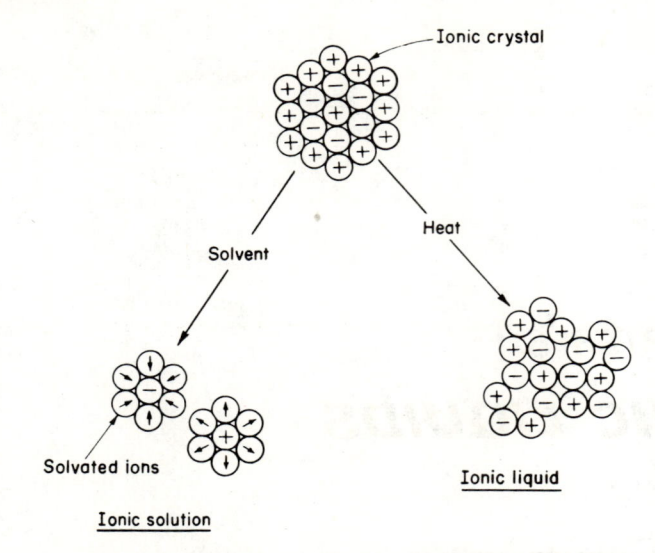

Fig. 6.1. An ionic crystal can be dismantled either by the action of a solvent or by the action of heat.

So the question arises: Why have a solvent at all? This limiting case of an aqueous or a nonaqueous ionic solution from which all the solvent is removed is a *pure liquid electrolyte*. Conceptually, this definition is accurate. Operationally, however, if one removes solvent molecules from a solution, for example, by evaporation, one is left with ionic crystals, pure *solid* electrolyte. A further step in conversion from the solid to the pure liquid form is necessary.

6.1.2. The Thermal Dismantling of an Ionic Lattice

The process of dissolution of a true electrolyte was described in Chapter 2. The basic picture is that the ions in an erstwhile-rigid ionic lattice succumb to the strong attraction[†] of the solvent molecules and follow them into solution, executing a random walk there as free, stable solvated ions. The result is an ionic solution which has the ability to conduct electricity by the drift of ions. The disassembly of the ionic lattice was achieved by the solvent overcoming the coulombic cohesive forces holding together the ions in the regular arrangement called a lattice (Fig. 6.1).

[†] In the case of aqueous solutions, the forces are essentially ion–dipole and ion–quadrupole in character.

TABLE 6.1
Specific Conductivities of Solid and Molten NaCl

	Specific conductivity, $\text{ohm}^{-1}\,\text{cm}^{-1}$
Solid NaCl	1×10^{-3} at 800 °C
Molten NaCl[†]	3.9 at 900 °C

[†] Melting point of NaCl is 801 °C.

A solvent, however, is not the only agency which can dismantle an ionic lattice. Heat energy, too, can overcome the cohesive forces and disrupt the ordered arrangement of ions in a crystal (Fig. 6.1). This process of melting results in the pure liquid electrolyte, a system having a conductance several orders of magnitude larger than that of the corresponding solid (Table 6.1).

6.1.3. Some Features of Ionic Liquids (Pure Liquid Electrolytes)

A common type of ionic lattice is that of a crystalline salt. One such ionic lattice encountered in everyday life is sodium chloride. Molten sodium chloride is a typical liquid electrolyte. It displays the characteristics of many liquid electrolytes.[†]

An appreciation of the properties of liquid electrolytes can be gained by a comparison between molten ice (water) and molten sodium chloride (Table 6.2). Both liquids are clear and colorless. Their viscosities, thermal conductivities, and surface tensions are not very different.

In fact, one can go further and make the following statement: Most molten salts look like water and, near their melting points, have viscosities, thermal conductivities, and surface tensions of the same orders of magnitude as those of water. In general, however, fused salts are stable as liquids only at relatively high temperatures (300 to 1250 °C) (Table 6.3).

One can quote exceptions to these generalizations. The tetraalkylammonium salts are liquid at much lower temperatures (Table 6.4).

[†] The terms *pure liquid electrolyte*, *ionic liquid*, *fused salt*, and *molten salt* are used synonymously.

TABLE 6.2
Comparison of Some Properties of Water and Molten NaCl

	Water 25 °C	Molten NaCl 850 °C
Viscosity, millipoise	8.95	12.5
Refractive index	1.332	1.408
Diffusion coefficient, $cm^2\,sec^{-1}$	3×10^{-5}	Na^+ 1.53×10^{-4} Cl^- 0.83×10^{-4}
Surface tension dynes cm^{-1}	72	111.8
Density	1.00	1.539

TABLE 6.3
Melting Points of Some Inorganic Salts

Salt	Melting point, °C	Salt	Melting point, °C
$AgNO_3$	210	$PbCl_2$	501
$HgBr_2$	238	$CdCl_2$	568
$LiNO_3$	254	$LiCl$	610
$ZnCl_2$	275	$CaCl$	646
$HgCl_2$	277	NaI	651
$NaNO_3$	310	$MgCl_2$	714
KNO_3	337	KCl	776
$PbBr_2$	373	$NaCl$	808
$AgBr$	434	Na_2CO_3	858

TABLE 6.4
Melting Points of Some Tetraalkylammonium Salts

Salt	Melting point, °C
Tetramethylammonium bromide	230
Tetrabutylammonium iodide	144
Tetrapropylammonium iodide	280

TABLE 6.5
Conductivities of Molten Salts and Water

Substance	Temperature, °C	Specific conductivity, ohm^{-1} cm^{-1}
H$_2$O	18	4×10^{-8}
LiCl melt	710	6.221
NaCl melt	908	3.903
KCl melt	872	2.407

6.1.4. Liquid Electrolytes Are Ionic Liquids

The crucial difference between the molten salts and molten ice lies in the values of the specific conductivity (Table 6.5). Fused salts have about 10^8 times better specific conductivity than water.

The temptation to ascribe the high conductance of fused salts to conduction by electrons must be rejected. Thus the conductivity of a molten salt is high compared to that of water; but it is some ten thousand times lower than that of a liquid metal, such as mercury (Table 6.6).

Fused salts conduct by the drift of ions. They are, in fact, ionic liquids. Pure liquid electrolytes therefore are liquids containing only ions, the ions being free or associated (see Section 6.5).

Another class of ionic liquids is the *molten oxides*. These are highly conducting liquids formed by the addition of a *metal* oxide (e.g., Li$_2$O) to the oxide of a *nonmetal* (e.g., SiO$_2$). Some properties of the molten oxides are shown in Table 6.7.

To develop a perspective on the properties of liquid electrolytes, a tabulation has been made of some properties of water, liquid sodium,

TABLE 6.6
Conductivities of Molten NaCl and of Mercury

Substance	Temperature, °C	Specific conductivity, ohm^{-1} cm^{-1}
Hg	20	1.1×10^4
NaCl melt	908	3.903

TABLE 6.7

Some Properties of Molten Oxides near the Melting Point

Molten oxide	Temp., °C	Density, g cm^{-3}	Surface tension, dyne cm^{-1}	Viscosity, centipoise	Specific conductivity ohm^{-1} cm^{-1}
Li$_2$O·SiO$_2$	1250	2.07	354	2.88×10^2	5.5 (1750 °C)
Li$_2$O·1½SiO$_2$	1100	2.13	331	5.02×10^3	
Li$_2$O·2SiO$_2$	1100	2.16	319	1.78×10^4	2.5 (1750 °C)
Na$_2$O·SiO$_2$	1100	2.23	300	1.19×10^3	4.8 (1750 °C)
Na$_2$O·2SiO$_2$	900	2.28	289	3.33×10^5	2.1 (1750 °C)
Na$_2$O·3SiO$_2$	900	2.23	282	1.99×10^6	
K$_2$O·2SiO$_2$	1100	2.20	220	1.08×10^5	1.5 (1750 °C)
CaO·SiO$_2$	1550	—	400	2.73×10^2	0.8 (1750 °C)

an aqueous solution of NaCl, crystalline NaCl, fused NaCl, and a mixture of fused Na$_2$O and SiO$_2$ (Table 6.8).

6.1.5. The Fundamental Problems in Pure Liquid Electrolytes

In dealing with aqueous and nonaqueous solutions of electrolytes, the procedure was, firstly, to seek a picture of the time-average structure of the electrolytic solution and, secondly, to understand the basic laws of ionic movements. The picture that emerged was of ions and solvent molecules interacting together to form solvated ions; of ions interacting with each other to form ionic clouds and associated ion pairs, or complexes; and of all these entities executing an aimless random walk at equilibrium, which becomes a directed drift under an external field.

The problems in pure liquid electrolytes are analogous, though perhaps more difficult to treat mathematically.

The first problem can be defined as follows: What idealized *model* could best approximate a solvent-free system of charged particles forming a highly conducting ionic liquid?

In the case of the aqueous solution, it was easy to understand the drift of ions at the behest of the applied electric field. Positive and negative ions, separated by stretches of water, drift in opposite directions (Fig. 6.2). In a pure ionic liquid, however, there is no water for the ions to float in; the ions drift among themselves (Fig. 6.3). When an external field is switched on, how is it that the ions are able to move past each other? Will not the very large interionic forces make them stick together, somewhat in the

TABLE 6.8
Comparison of the Properties of Various Liquids

Properties	Water, 25 °C	Liquid sodium at mp	1M sodium chloride soln at 25 °C	Liquid sodium chloride at mp	Na$_2$O·SiO$_2$
Melting point, °C	0.0	97.83	−3.37	801	1088
Vapor pressure, mm Hg	23.756	9.842×10^{-8}	16.8, 20 °C 30.8, 30 °C	3.45×10^{-1}	
Molar volume, cm^3	18.07	24.76	17.80	30.47	55.36
Density, g cm^{-3}	0.997	0.927	1.0369	1.5555	2.250, 1200 °C
Compressibility, 10^6 cm^2 atm^{-1}	46.3055 Isoth.	18.88 Isoth.	40.08 Isoth.	29.08 Isoth.	59.579 adiabatic, 1200°C
Diffusion coefficient, cm^2 sec^{-1}	3.0×10^{-5}	2.344×10^{-7} 1.584×10^{-7}	$D_{Na+} = 1.25 \times 10^{-5}$ $D_{Cl-} = 1.77 \times 10^{-5}$	$D_{Na+} = 1.53 \times 10^{-4}$ $D_{Cl} = 0.83 \times 10^{-4}$ 850 °C	
Surface tension dyne cm^{-1}	71.97	192.2	74.3	113.3	294, 1200 °C
Viscosity, centipoise	0.895	0.690	1.0582	1.67	980, 1100 °C
Specific electric conductance, mho cm^{-1}	4.0×10^{-8}	1.04×10^{5}	0.8576 mole cm^{-3}	3.58	4.8, 1750 °C
Refractive index	1.333	0.04	1.3426	1.408, 850 °C	1.52, solid, room temp.

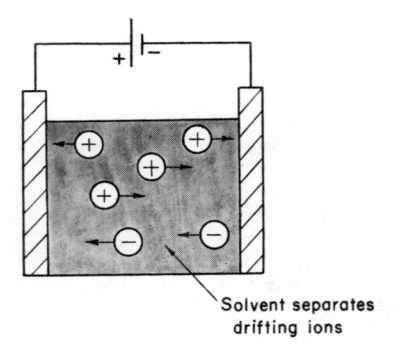

Fig. 6.2. In an aqueous solution, the solvent separates the drifting ions.

manner of the poorly conducting ionic lattices (of the solid state)? The situation appears puzzling.

In fact, what is the essential difference between the solid form and the liquid form of an ensemble of particles? This is a question which is relevant to all processes of fusion, e.g., the process of solid argon's[†] melting to form a liquid. In the case of ionic liquids, the problem is more acute. One must explain the great fluidity and corresponding high conductivity in a liquid containing *charged particles* in contact.

The second problem concerns an understanding of the *sharing* of transport duties (e.g., the carrying of current) in pure liquid electrolytes. In aqueous solutions (*cf.* Section 4.5.2), it was possible to comprehend the *relative* movements of ions in the sense that one ionic species could move with greater agility and therefore transport more electricity than another species until a concentration gradient is set up and the resulting diffusion evens out the movements. In fused salts, this comprehension is less easy to acquire. At first, it is even difficult to see how one can retain the concept of transport numbers when there is no reference medium (such as the water in aqueous solutions) in which ions can drift. The point is: Any movement of one ion relative to the other affects the situation of the whole liquid. Thus, a theory of transport numbers in liquid electrolytes must differ from that for aqueous and nonaqueous solutions.

Thirdly, there exists the problem of complex ions. In aqueous and nonaqueous solutions, it is possible to regard the entity: ion–ion atmosphere as a type of incipient complex in which the mean distance between oppositely

---

[†] This is a relatively simple solid from the point of view of the forces between the uncharged particles.

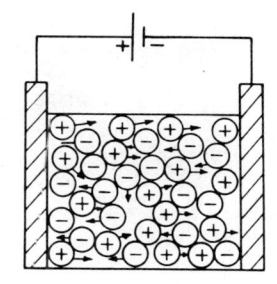

Fig. 6.3. In an ionic liquid, there is no solvent separating the drifting ions.

charged ions becomes smaller with increasing electrolyte concentration. Eventually, the ions come sufficiently close to withstand the thermal forces which tend to separate them. They remain coulombically stuck together as ion pairs for appreciable times and sometimes bond together chemically as complex ions (see Section 3.8).

For the ionic liquids, however, the situation is different for all the ions are always in contact. By definition, i.e., zero solvent, nothing separates the ions. This absence of solvent causes conceptual problems regarding the existence of complex ions in ionic liquids.

Consider a particular ion associated with another to form a vibrating complex. The ion is also in contact with, and jostled continually by, neighboring ions which are exactly like its partner in the complex (Fig. 6.4). Which is the partner and which the neighbor; which is the vibration and which the collision? A distinction between these two types of contacts constitutes one of the problems in this field.

In aqueous solutions, the situation is clarified by the solvent. This solvent keeps the complex ions apart at mean distances, defines them as

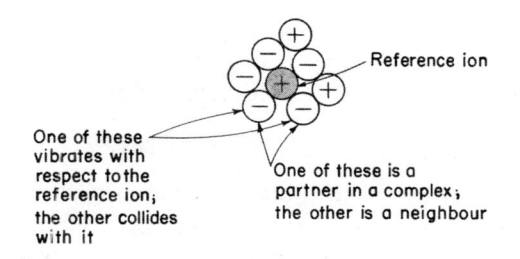

Fig. 6.4. The problem of distinguishin a neighboring ion colliding with the reference ion from a ligand (i.e., a partner in complex formation) vibrating in relation to the reference ion.

522 CHAPTER 6

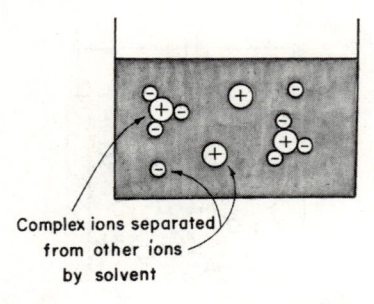

Fig. 6.5. In an aqueous solution, the complex ion is spatially separated from the other ions.

(Complex ions separated from other ions by solvent)

independent stable entities, and permits probing radiation (e.g., visible light) to pick them out from the surroundings (Fig. 6.5).

The concept of complex ions is therefore more subtle in ionic liquids than in aqueous solutions. It has even been asked: Is the concept valid at all?

Further Reading

1. H. Bloom, in: J. O'M. Bockris, ed., *Modern Aspects of Electrochemistry*, Chap. 3, Vol. 2, Butterworths' Scientific Publications, Ltd., London, 1959.
2. Iu. K. Delimarskii and B. F. Markov, *Electrochemistry of Fused Salts*, The Sigma Press, Publishers, Inc., Washington, D.C., 1961.
3. H. Bloom, *Pure Appl. Chem.*, **7**: 389 (1963).
4. B. R. Sundheim, ed. *Fused Salts*, McGraw-Hill Book Company, New York, 1964.
5. M. Blander, ed. *Molten Salt Chemistry*, Interscience Publishers, Inc., New York, 1964.
6. E. A. Ukshe, *Russ. Chem. Rev.*, **34**: 141 (1965).
7. H. Bloom, *The Chemistry of Molten Salts*, W. A. Benjamin, Inc., New York, 1967.

6.2. MODELS OF SIMPLE IONIC LIQUIDS

6.2.1. The Origin of Liquid Electrolyte Models

The structure of a system can often be understood from the way that system arose. What are the origins of a pure liquid electrolyte? It can be conceived as being formed either by the fusion of a solid ionic lattice or by condensation from a vapor of ions. Two types of models for pure liquid electrolytes can therefore be developed: gas-oriented models or lattice-oriented models (Fig. 6.6).

IONIC LIQUIDS   523

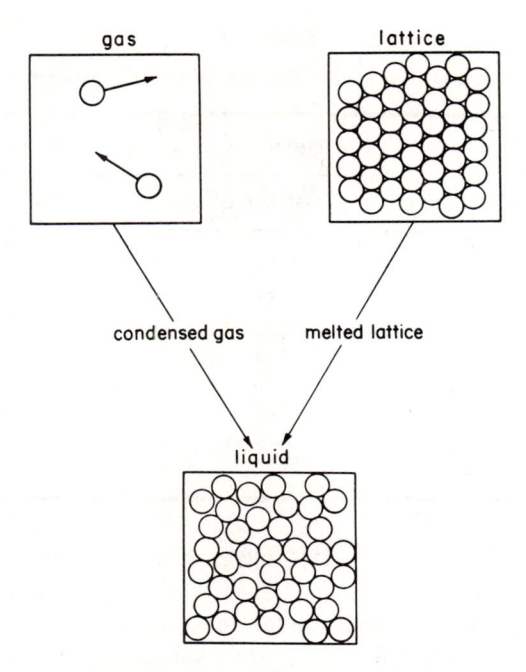

Fig. 6.6. The two types of models for pure liquid electrolytes: gas-oriented and lattice-oriented models.

6.2.2. Lattice-Oriented Models

6.2.2a. The Experimental Basis for Model Building. One's first impression of a liquid (its fluidity, conformity to the shape of the containing vessel, etc.) would suggest that its structure has nothing to do with that of the crystal from which it was obtained by melting.

If, however, a beam of monochromatic X rays is made incident on the liquid electrolyte, the scattered beam has an interesting story to tell. The ions are almost at the same internuclear distances in a fused salt as in the ionic crystal (actually, at a slightly lesser distance) (Table 6.9). This is the memory that a fused salt retains of the ionic lattice which gave birth to it in the melting process. The X-ray patterns also indicate that, in the liquid state, the local order extends over a very short distance (a few angstroms). It is as if the fused salt forgets how to continue the ordered arrangement of ions of the parent lattice (Fig. 6.7).

6.2.2b. The Need to Pour Empty Space into a Fused Salt. There is another important fact about the melting process. When many ionic lattices

TABLE 6.9

Internuclear Distances in an Ionic Crystal and the Corresponding Fused Salt

Salt	Distance between oppositely charged ions, Å	
	Crystal, mp	Molten salt
LiCl	2.66	2.47
LiBr	2.85	2.68
LiI	3.12	2.85
NaI	3.35	3.15
KCl	3.26	3.10
CsCl	3.57	3.53
CsBr	3.72	3.55
CsI	3.94	3.85

are melted, there is a 10 to 25% increase in the volume of the system (Table 6.10).

This volume increase is of fundamental importance to one who wishes to conceptualize models for ionic liquids because one is faced with an

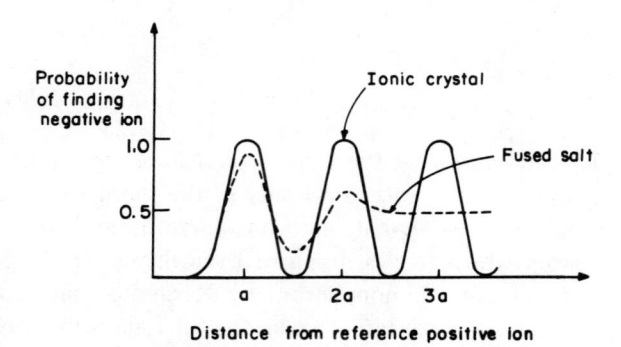

Fig. 6.7. Schematic diagram to show short-range and long-range order in an ionic crystal as opposed to only short-range order in a fused salt. In an ideal ionic crystal, if one takes a reference positive ion, there is a certainty of finding a negative ion at the lattice distance or a multiple of this distance; in a fused salt, there is a high probability of finding a negative ion one distance away; but within two or three lattice distances away, the probability becomes half, i.e., a negative ion is only half as likely as a positive ion. Thus, in a fused salt, there is no long-range order.

TABLE 6.10

Volume Change on Fusion

Substance	% Increase of volume on fusion
NaCl	25
NaF	24
NaI	19
KCl	17
KBr	17
KI	16
RbCl	14
$CdCl_2$	20
$CdBr_2$	28
$NaNO_3$	11

apparent contradiction. From the increase in volume, one would think that the mean distance apart of the ions in a liquid electrolyte should be greater than in its parent crystal. On the other hand, from the fact that the ions in a fused salt are slightly closer together than in the solid lattice, one would think that there should be a small volume decrease upon fusion.

The clue to the resolution of this apparent contradiction lies in the mobility of ions in simple molten salts, which is several orders of magnitude ($\sim 10^3$ times) greater than that in the crystalline state (Table 6.1). It has been seen (see Section 4.2.16) that ionic movements can be thought of as occurring in elementary steps, each of which requires the jump of an ion into an ionless region or site. Hence, the high mobility of ions in liquid electrolytes implies that the number of ionless regions (or vacant sites) is far greater in an ionic liquid than in an ionic crystal.

It can be concluded, therefore, that a liquid electrolyte can occupy more volume than the corresponding ionic crystal and, at the same time, preserve approximately the same short-range order, but only if empty space is introduced into it.[†]

[†] In the case of some salts, the volume changes on fusion are much smaller than are indicated in Table 6.10. Thus, calcium, strontium, and barium halides have volume changes which are about a fifth of the volume changes for the corresponding alkali halides. This is because such salts crystallize in a form which already contains plenty of open space in the solid lattice. When these open-lattice salts are fused, or melted, there is need for a smaller volume increase than is the case for the space-filled lattices of the alkali halides.

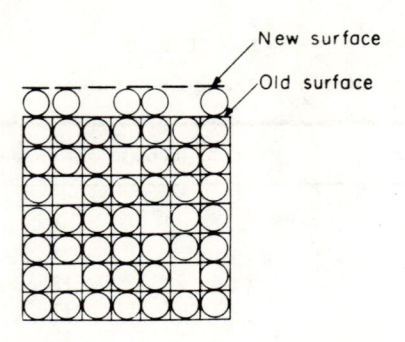

Fig. 6.8. In the vacancy model, empty space is created in the system when ions move from lattice sites to surface sites, i.e., when Schottky defects are produced.

How is this emptiness (injected into the liquid electrolyte) to be conceptualized? The different models of fused salts involve different ways of considering the empty space. Thus, the emptiness is described as vacancies, holes, free volume, etc. The differences are not purely semantic. Some of these models will now be described.

A model can only reproduce what, in the opinion of the designer of the model, are considered to be the *essential* features of the system. There should be no surprise, therefore, if a given model fails to reproduce all aspects of the behavior of fused salts. But one has to determine whether one of the various models is markedly more successful in its quantitative prediction of phenomena than are the other competing models.

6.2.2c. The Vacancy Model: A Fused Salt Is an Ionic Lattice with Numerous Vacancies. The simplest model of an ionic liquid is the quasi-lattice or vacancy model originated by Frenkel and developed by Stillinger. The picture is that of an ionic lattice, into which are injected vacancies of a particular type known as *Schottky defects*. A Schottky defect is produced in the following process. Ions are removed from lattice sites in the interior to the surface of the crystal (Fig. 6.8). Thus, vacancies are produced inside the system and, simultaneously, there is volume expansion through the advance of the frontier (surface).[†]

As the temperature of the solid lattice is increased, the number of Schottky vacancies increases until, at the melting point, there is a qualitative

---

[†] It will be seen later that the ion does not, in fact, jump from within the liquid to the surface in order to make room for the hole. The actual process involves a train of ions each of which moves outward by only a small amount.

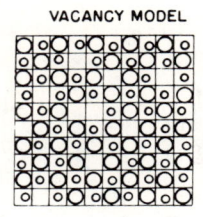

Fig. 6.9. The vacancies occur at lattice sites in the quasi-lattice model.

change in degree of order; the disturbance of the lattice is so great that long-range order disappears. In this model, a sufficient number of Schottky vacancies is introduced to account for the volume increase on fusion.

The location and size of the vacancies in a fused salt follow as a natural consequence from the assumption that the empty space consists of Schottky defects. The vacancies are about the same size as the positive and negative ions of the ionic lattice. Further, the vacancies occur at lattice sites. This is the characteristic feature of the quasi-lattice model (Fig. 6.9).

6.2.2d. The Hole Model: A Fused Salt Is Full of Holes like Swiss Cheese. In the Schottky-vacancy model, the parent ionic lattice exerted a dominating influence on the sizes and positions of the vacancies in the liquid electrolyte. The hole model proposed by Fürth represents an emancipation from that influence. The description of the empty space is not in lattice language, i.e., in the language of a three-dimensional array of points.

In the hole model, it is considered that the sizes and spatial location of the empty regions in the fused salt are random. These randomly located and variable-sized vacancies are called *holes* (Fig. 6.10). Thus, the liberation from lattice concepts leads to a fundamentally different model.

What is the process by which holes are produced? It is by a process somewhat analogous to the formation of a vacancy in a crystal. The displacement of an ion from a lattice site produces a vacancy at its former

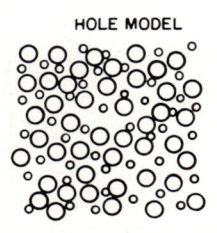

Fig. 6.10. The hole model with randomly located and variable-sized holes in the liquid.

(a) Before hole is formed   **(b) After hole is formed**

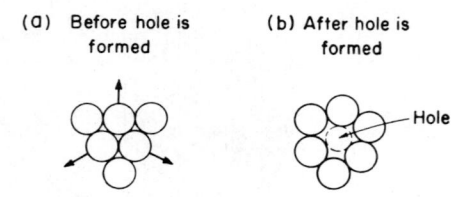

— Hole

Fig. 6.11. The formation of a hole in a liquid by the relative displacement of ions in contact.

site. In the case of the vacancy, however, the ion is removed so far from the original site that the displaced ion can be forgotten altogether. Suppose instead that, in the course of thermal motion, some of the ions constituting a cluster are displaced relative to each other but only by small amounts. Then, a hole is produced between them (see Fig. 6.11). Its size depends on the extent of displacement, which must be random because thermal motions are random. The hole size must therefore be a random quantity. Further, since thermal motions occur everywhere in the liquid electrolyte, holes can appear and disappear anywhere in this liquid.

An equivalent description (Fig. 6.12) is that holes occur by fluctuations in local density, i.e., in n/V, the number of ions per unit volume in a given locality of the liquid. The volume V is constant; hence, the density can change only through a change in n, i.e., by the ions' moving farther apart (the hole size increases) or closer together (the hole size decreases).

The holes in a liquid electrolyte resemble the holes in Swiss cheese. This is why, in the matter of their randomness of size and location, the hole theory has been referred to in homely terms as the "Swiss cheese model." What the cheese represents, however, is the time-average picture of the holes. In the model, holes are continuously forming and disappearing, moving, coalescing to form larger holes, and diminishing into smaller ones.

(a) Before hole is formed   **(b) After hole is formed**

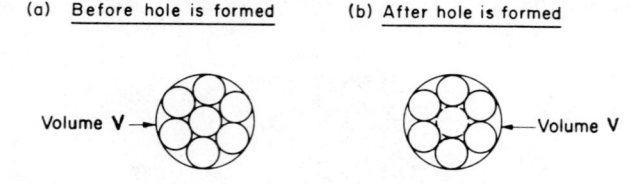

Volume V →   ← Volume V

Fig. 6.12. The formation of a hole can also be looked at in terms of the number of ions occupying a volume V. In (a) seven particles occupy the volume V before the hole is formed, and, in (b), six particles occupy the same volume after hole formation.

6.2.3. Gas-Oriented Models for Liquid Electrolytes

6.2.3a. The Cell-Theory Approach. A particle in a dilute gas has available for its motion the whole volume occupied by the gas, i.e., the entire volume of the container. As the pressure is increased, the spatial domain accessible to each particle is decreased but, as long as the system is a gas, every particle continues to possess the freedom to move in any part of the available volume.

When, however, there is a phase transition from the gaseous to the liquid state, the freedom (of motion) of a particle is largely curtailed. The neighboring particles (of any particle) conspire to confine the central particle in a "cell." The conspiracy is never completely successful; the central particle does manage occasionally to break out. Nevertheless, it spends much of its time imprisoned within its cell.

The cell theory of liquids (derived from a compressed-gas point of view) proceeds on the basis that, if a liquid consists of N particles, then the volume occupied by the liquid can be conceptually divided into N identical cells. Now, cells are larger than their occupants; hence, the confined particle is free to move *within* the cell. It has a *free volume* available for its motion. This free volume v_f is equal to the average volume available to each particle, $\bar{v}$, minus the volume v_0 of the particle considered an incompressible hard sphere (Fig. 6.13). (The average volume per particle is simply equal to the total volume V of the liquid divided by the number N of particles.) In symbols,

$$v_f = \bar{v} - v_0 \tag{6.1}$$

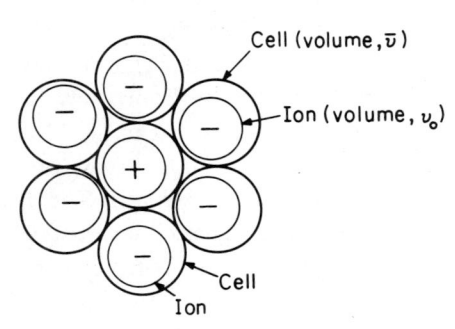

$$= \frac{V}{N} - v_0 \tag{6.2}$$

Fig. 6.13. The free volume v_f available for the motion of an ion is equal to the average volume $\bar{v}$ per ion minus the hard-sphere volume v_0, of the ion.

The cell theory accomplishes the task of picturing how the freedom of a particle in the gas phase diminishes upon its confinement in the liquid state. The decrease of freedom occurs through a restriction of the free volume accessible for its motion. However, the cell model runs into three basic difficulties.

Firstly, the cell model restricts the motion of a particle to the confines of its cell. How then can the model explain transport properties? Transport requires that particles be able to migrate from cell to cell. Secondly, by insisting that a particular particle be sentenced to confinement in a particular cell, the disorder (randomness) accompanying the exchange of particles between cells is ruled out. The result is that the calculated entropy of fusion—which is a measure of the increase in randomness predicted by the model as its version of what really happens on melting—becomes in fact much smaller than that observed.

More decisive in counting against the model than either of these difficulties, however, is the third one, which arises from the apparent contradiction of volume expansion on melting, along with a small decrease in mean internuclear distance (*cf.* Section 6.2.2b). On the basis of a cell model for a liquid, the only explanation for the positive volume increase on fusion is to have all the ions move away from each other, i.e., an increase in average cell volume. But this, of course, would mean an *increase* in the average internuclear distance on fusion (as long as multiple occupancy of cells is prohibited)—an increase which contradicts experiment (*cf.* Section 6.2.2a).

6.2.3b. The Free Volume Belongs to the Liquid and Not to the Particles: The Liquid Free-Volume Model. In the cell model, the liquid was divided into *identical* cells, each cell marking the boundaries of the motion of a particle and thus possessing a *cell* free volume. The difficulties inherent in this model (see Section 6.2.3a) have been overcome in the liquid free-volume model developed by Cohen and Turnbull.

This progress is achieved as follows. The liquid first appropriates the available free volume; hence the term *liquid* free volume. The liquid then distributes its free volume among the N particles. The distribution, however, is not done equally to each molecule. There is a statistical distribution of free volumes. In other words, N_1 particles are assigned a free volume v_1; N_2 particles, a free volume v_2; etc. In general, N_i particles possess a free volume v_i.

In this model, the free volume of a particle is not its inalienable property. Thermal forces are responsible for the statistical distribution of free volumes. Hence, these same forces cause the free volume of a particle to fluctuate in time.

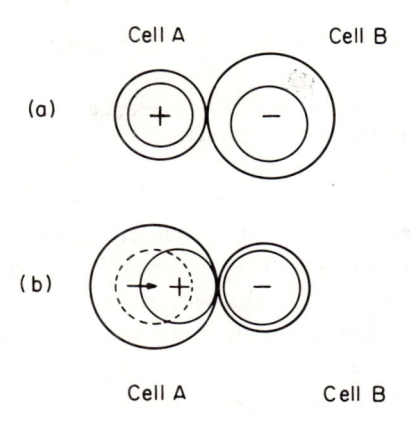

Fig. 6.14. In the liquid free-volume model, the motion of an ion toward another is accompanied by an expansion of the cell of the former and a contraction of the cell of the latter. (a) The positions of a positive and a negative ion and the sizes of their cells before motion of the positive toward the negative ion; (b) the situation after motion.

The movement of a particle from one position to another implies the expansion of the cell of the moving particle and the contraction of the neighboring cell (see Fig. 6.14). The expansion of the first cell (and the associated energy increase) is just compensated for by the contraction of the neighboring cell (and the associated energy decrease). Hence, there is a zero net energy change involved in one cell's expanding and its neighboring one's contracting as a result of the movement of a particle. In this way, the liquid free-volume model can accommodate the movement of particles, i.e., transport properties such as diffusion.

Once a movement of a particle is permitted and cell free volumes fluctuate, there is sufficient randomness introduced into the model to explain the experimental fusion entropies.

The volume increase that occurs on fusion implies an increase in the total liquid free volume. This total free-volume increase does not however imply any increase in mean internuclear distance. In all parts of the liquid, except those occupied by free space, the particles have the same internuclear distances as in the crystalline state.

Thus, by setting the cell free volumes into thermal fluctuations, the liquid free-volume model can overcome some of the difficulties (see Section 6.2.3a) which troubled the simple cell theory.

6.2.4. A Summary of the Models for Liquid Electrolytes

A simple ionic liquid can be considered to arise either from the fusion of an ionic lattice or from the condensation of a vapor of ions. Thus, there are lattice-based and gas-based models for a fused salt. The central fact which a model must attempt to explain is the volume increase upon fusion, which is usually observed along with a retention or even diminution of the mean interionic distances. This volume increase without a corresponding increase of mean interionic distance suggests that, in the process of fusion, empty space is introduced into the ionic liquid. The mode of description of this empty space is what differentiates one model from another.

The vacancy model treats a fused salt as an ionic crystal with Schottky defects which arise when ions from lattice positions move to the crystal surface. The movement to the surface accounts for the volume expansion, and the creation of vacancies inside the system accounts for the empty space. The vacancy model thus requires vacancies to be of the same size as ions and to occur at lattice sites.

In the hole model, empty space is considered to arise from thermally generated fluctuations in local density. The holes which constitute the empty space are random in size and location. At any one instant of time, the holes would make a section through the liquid electrolyte appear like one through Swiss cheese. It is important to realize that holes are constantly undergoing changes in size with new ones' being formed and old ones' destroyed. Further, the holes can also move by the mechanism of having an ion jump into one hole and thus create a new hole in the place it has just vacated.

The cell theory approach is based on considering the territory available for the motion of a particle. The transition from the gaseous state to the liquid state is accompanied by a drastic reduction in this territory. In the gas phase, the entire volume of the container is accessible to a particle; in the liquid state, the particle is confined to a cell. Within the cell, however, the particle has a free volume available for its motion. The cell theory prohibits multiple occupancy of cells and, therefore, cannot explain the volume increase upon fusion.

The liquid free-volume theory argues against identical cells and equal free volumes for the particles. It is the liquid as a whole which has a certain free volume, and there is a statistical distribution of the free volumes in the cells. Thus, the cells expand and contract, and the mean distance between particles need not increase to provide for the increase in the liquid free volume which accounts for the volume increase upon fusion.

Now that the basic models for a simple ionic liquid have been qualitatively described, the next task is to consider which model makes the most successful predictions of experimental results. At first, however, one should describe the more experiment-consistent models in quantitative terms. Since, as will be seen later, the hole model appears to yield more qualitative and quantitative agreement with experiment than do the other models, the treatment from here on will be characterized by emphasis on the hole model.

Further Reading

1. H. Eyring, *J. Chem. Phys.*, **4**: 283 (1936); **5**: 896 (1937); *J. Phys. Chem.*, **41**, 249 (1937); *J. Chem. Phys.*, **9**: 393 (1941); *Proc. Nat. Acad. Sci. U.S.*, **44**: 683 (1958); **46**: 333 (1960).
2. W. Altar, *J. Chem. Phys.*, **5**: 577 (1937).
3. R. Fürth, *Proc. Cambridge Phil. Soc.*, 252, 276, 281 (1941).
4. M. Born and H. S. Green, *Proc. Roy. Soc. (London)*, **A188**: 10 (1946).
5. J. Frenkel, *Kinetic Theory of Liquids*, Oxford University Press, New York, 1946.
6. M. H. Cohen and D. Turnbull, *J. Phys. Chem.*, **29**: 1049 (1958); **31**: 1164 (1959); **34**: 120 (1961).
7. H. Reiss, H. L. Frisch, E. Helfand, and J. L. Lebowitz, *J. Chem. Phys.*, **32**: 119 (1960).
8. F. H. Stillinger, Chapter 1 in: M. Blander, ed., *Molten Salt Chemistry*, Interscience Publishers, Inc., New York, 1964.
9. H. Bloom, Chapter 1 in: B. R. Sundheim, ed., *Fused Salts*, McGraw-Hill Book Company, New York, 1964.
10. K. D. Luks and H. T. Davis, *Eng. Chem. Ind. Fundamentals*, **6**: 194 (1967).

6.3. QUANTIFICATION OF THE HOLE MODEL FOR LIQUID ELECTROLYTES

6.3.1. An Expression for the Probability That a Hole Has a Radius between r and $r + dr$

To quantify the hole model, it is necessary to calculate a distribution function for hole sizes. As a first step toward this calculation, one can consider a particular hole in a liquid electrolyte and ask: What are the quantities (or variables) which are required to describe this hole? This problem was resolved by a formulation published by Fürth in 1941.

Since a hole in a liquid can move about like an ion or other particle, the dynamical state of a hole is specified in the same way that one describes the dynamical state of a material particle. Thus, one must specify three position and three momentum coordinates: x, y, z, and p_x, p_y, p_z. There

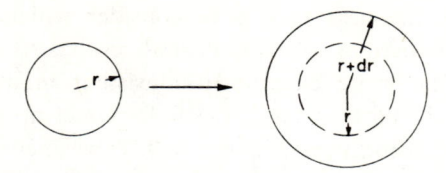

Fig. 6.15. The breathing motion of a hole involves its radial expansion.

is, however, an extra feature of the motion of a hole, which is not possessed by material particles. This feature concerns what is called the *breathing motion* of a hole (Fig. 6.15), i.e., the contraction and expansion of the hole. To characterize this breathing motion, it is sufficient to specify the hole radius r and the radial momentum p_r corresponding to the breathing motion. Hence, to characterize a hole completely, it is necessary to specify eight quantities: x, y, z, p_x, p_y, p_z, r, and p_r, whereas the first six only are adequate to describe the state of motion of a particle.

According to the usual equations of classical statistical mechanics which are used to express velocities and momenta distributed in three dimensions, the probability P that the location of a hole is between x and $x + dx$, y and $y + dy$, z and $z + dz$, that its translational momenta lie between p_x and $p_x + dp_x$, p_y and $p_y + dp_y$, p_z and $p_z + dp_z$, that its breathing momentum is between p_r and $p_r + dp_r$, and, finally, that its radius is between r and $r + dr$, is proportional to the Boltzmann probability factor, i.e.,

$$P\, dx\, dy\, dz\, dp_x\, dp_y\, dp_z\, dr\, dp_r \propto e^{-E/kT} \cdot dx\, dy\, dz\, dp_x\, dp_y\, dp_z\, dr\, dp_r \quad (6.3)$$

where E is the total energy of the hole.

Since the desired distribution function only concerns the radii (or sizes) of holes, it is sufficient to have the probability that the hole radius is between r and $r + dr$ *irrespective of the location and the translational and breathing momentum of the hole*. This probability $P\, dr$ of the hole radius' being between r and $r + dr$ is obtained from Eq. (6.3) by integrating over all possible values of the location, and of the translational and breathing momentum of the hole, i.e.,

$$P\, dr \propto \left(\int \int \int \int \int \int \int e^{-E/kT} \cdot dx\, dy\, dz\, dp_x\, dp_y\, dp_z\, dp_r \right) dr \quad (6.4)$$

But the total energy of the hole does not depend upon its position, i.e., E is independent of x, y, and z. Hence,

$$P\, dr \propto \left(\int \int \int \int e^{-E/kT} \cdot dp_x\, dp_y\, dp_z\, dp_r \right)\left(\int \int \int dx\, dy\, dz \right) dr \quad (6.5)$$

Further,
$$\int\int\int dx\,dy\,dz = V \tag{6.6}$$

the volume of the liquid. Thus, by incorporating this V into the proportionality constant implicitly associated with Eq. (6.5), one has

$$P\,dr \propto \left(\int\int\int\int e^{-E/kT} \cdot dp_x\,dp_y\,dp_z\,dp_r\right)dr \tag{6.7}$$

Now the total energy E consists of the potential energy W of the hole (i.e., the work required to form the hole) plus its kinetic energy. This kinetic energy is given by

$$\frac{p_x^2}{2m_1} + \frac{p_y^2}{2m_1} + \frac{p_z^2}{2m_1} + \frac{p_r^2}{2m_2} \tag{6.8}$$

where m_1 is the apparent mass† of the hole in its translational motions, and m_2 is the apparent mass† in its breathing motion. Hence,

$$E = W + \frac{p_x^2}{2m_1} + \frac{p_y^2}{2m_1} + \frac{p_z^2}{2m_1} + \frac{p_r^2}{2m_2} \tag{6.9}$$

Inserting this value of E into the expression (6.7) for the probability of the hole's having a radius between r and $r + dr$, one has

$$P\,dr \propto dr\,e^{-W/kT}\left(\int_{-\infty}^{\infty} e^{-p_x^2/2m_1kT}dp_x\right)\left(\int_{-\infty}^{\infty} e^{-p_y^2/2m_1kT}dp_y\right)$$
$$\times \left(\int_{-\infty}^{\infty} e^{-p_z^2/2m_1kT}dp_z\right)\left(\int_{-\infty}^{\infty} e^{-p_r^2/2m_2kT}dp_r\right) \tag{6.10}$$

From the standard integral

$$\int_{-\infty}^{\infty} e^{-ax^2}dx = \sqrt{\frac{\pi}{a}} \tag{6.11}$$

it is clear that

$$\int_{-\infty}^{\infty} e^{-p_x^2/2m_1kT}dp_x = \int_{-\infty}^{\infty} e^{-p_y^2/2m_1kT}dp_y = \int_{-\infty}^{\infty} e^{-p_z^2/2m_1kT}dp_z = (2\pi m_1 kT)^{\frac{1}{2}} \tag{6.12}$$

---

† Any entity that moves displays the property of inertia, i.e., resistance to a change of its state of rest or uniform motion. That is, the entity has a mass. If the entity is not material (a hole is a region where, in fact, there is no material), one refers to an apparent mass. Holes in semiconductors have apparent masses like holes in liquids. The inertia of the hole arises as a result of the displacement of the liquid around the hole as it moves, which gives rise to a dissipation of energy (Appendix 6.1.)

536 CHAPTER 6

and

$$\int_{-\infty}^{\infty} e^{-p_r^2/2m_2kT} dp_r = (2\pi m_2 kT)^{\frac{1}{2}} \tag{6.13}$$

By using these values of the integrals in Eq. (6.10), the result is

$$P\,dr \propto (2\pi kT)^2 m_1^{\frac{3}{2}} \, m_2^{\frac{1}{2}} e^{-W/kT} \, dr \tag{6.14}$$

It can be shown, however, that (*cf.* Appendix 6.1)

$$m_1 = \tfrac{2}{3}\pi r^3 \varrho \tag{6.15}$$

and

$$m_2 = 4\pi r^3 \varrho \tag{6.16}$$

Hence, after taking all quantities which are radius independent into the proportionality constant A, one has, by combining Eqs. (6.14), (6.15), and (6.16),

$$P\,dr = Ar^6 e^{-W/kT}\,dr \tag{6.17}$$

The evaluation of the constant is achieved through the following argument. The probability that a hole has some radius must be unity (a probability of unity for an event corresponds to the certainty of its occurrence). Equation (6.17) expresses the probability of the radius of the hole lying between r and $r + dr$. Similarly, one can write down the probabilities of the radius' being between r_1 and $r_1 + dr$, between r_2 and $r_2 + dr$, etc. If all these probabilities for r from zero to infinity are summed up (or integrated), then the sum must be unity, i.e.,

$$\int_0^{\infty} P\,dr = 1 = \int_0^{\infty} Ar^6 e^{-W/kT}\,dr \tag{6.18}$$

However, to carry out this integration, one must know whether W is a function of r, i.e., one must understand what determines the work of formation (or the potential energy) of a hole of radius r.

6.3.2. The Fürth Approach to the Work of Hole Formation

An ingenious calculation of the work of hole formation was made by Fürth, who treated holes in liquids, the sizes of which are thermally distributed, in an article published in an erudite but little-read university journal, an act which delayed recognition of the model. A hole in a fused salt is considered to behave similarly to a bubble in a liquid (Fig. 6.16). There is a net pressure acting on the surface of a bubble. The surrounding liquid exerts a hydrostatic pressure P_I on the bubble surface. Inside the bubble,

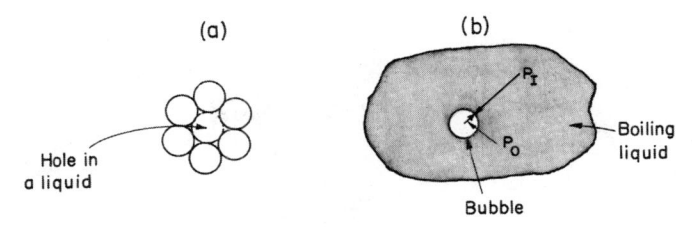

Fig. 6.16. The basis of the hole model of Fürth is the analogy between (a) a hole in a liquid and (b) a bubble in a liquid. An inward pressure F_I and an outward pressure P_O act on the bubble surface.

however, there exists included vapor, which exerts an outward pressure P_O on the surface. The net pressure is therefore $P_I - P_O$. Further, surface tension operates in the direction of reducing the surface area and, therefore, the surface energy of the bubble.

The total work required to increase the bubble size consists of two parts, the volume work $(P_I - P_O)V$ and the surface work γA, where $P_I - P_O$, V, γ, and A are the net pressure on the bubble surface, the increase of volume, the surface tension, and the increase of surface area of the bubble, respectively. Thus, the work done in making a bubble grow to a size having radius r is

$$W = (P_I - P_O)\tfrac{4}{3}\pi r^3 + 4\pi r^2 \gamma \tag{6.19}$$

Simple numerical calculations show that the first term, i.e., the volume term, is negligible compared to the second, or surface, term for bubbles of less than about 10^{-5} cm in diameter. Hence, on neglecting the volume expansion work, the work of bubble formation (i.e., the work of increasing its radius from 0 to r) reduces to

$$W = 4\pi r^2 \gamma \tag{6.20}$$

This expression can also be obtained from the general equation (6.19) by setting $P_I = P_O$. This equality between P_I and P_O represents the condition that the liquid is boiling. The analogy between a hole and a bubble consists, therefore, in assuming that the work of hole formation is given by the expression for the work of bubble formation in a liquid.

6.3.3. The Distribution Function for the Sizes of the Holes in a Liquid Electrolyte

Now that an expression for the work of hole formation has been obtained, it can be inserted into Eq. (6.18), which must be integrated to evaluate the constant A. One has

$$\int_0^\infty P\, dr = 1 = \int_0^\infty Ar^6 e^{-ar^2}\, dr \tag{6.21}$$

where

$$a = \frac{4\pi\gamma}{kT}$$

To carry out the integration, the following standard formula is used

$$\int_0^\infty r^{2n} e^{-ar^2}\, dr = \frac{1 \times 3 \times 5 \cdots (2n-1)}{2^{n+1}a^n} \times \sqrt{\frac{\pi}{a}}$$

where n is a positive integer.

The integral in Eq. (6.21) corresponds to the standard formula with $n = 3$, and, therefore,

$$\int_0^\infty P\, dr = 1 = A \int_0^\infty r^{2 \times 3} e^{-ar^2}\, dr$$

$$= A \frac{1 \times 3 \times 5}{2^4 a^3} \frac{\pi^{\frac{1}{2}}}{a^{\frac{1}{2}}}$$

$$= A \frac{15\pi^{\frac{1}{2}}}{16} \frac{1}{a^{7/2}} \tag{6.22}$$

Hence, the constant in Eq. (6.21) is given by

$$A = \frac{16}{15\pi^{\frac{1}{2}}} a^{7/2} = \frac{16}{15\pi^{\frac{1}{2}}} \left(\frac{4\pi\gamma}{kT}\right)^{7/2} \tag{6.23}$$

and the expression (6.17) for the probability of the existence of a hole of a radius between r and $r + dr$ becomes

$$P\, dr = \frac{16}{15\pi^{\frac{1}{2}}} a^{7/2} r^6 e^{-ar^2}\, dr \tag{6.24}$$

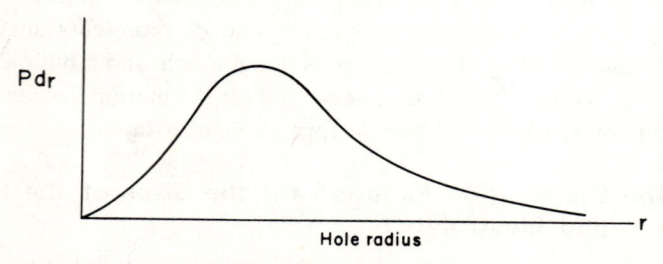

Fig. 6.17. How the probability $P\, dr$ that a hole has a radius between r and $r + dr$ varies with r.

This is the basic distribution function (Fig. 6.17) from which the average hole volume and radius can be obtained.

6.3.4. What Is the Average Size of a Hole?

The average radius $\langle r \rangle$ of a hole is obtained from Eq. (6.24) by multiplying the probability (of the hole radius being between r and $r + dr$) by the radius of the hole and integrating this product over all possible values of r. This is, in fact, the general method of getting average values of a quantity when the probability is given. Thus, the average hole radius is

$$\langle r \rangle = \int_0^\infty r P \, dr$$

$$= \int_0^\infty r \, \frac{16}{15\pi^{\frac{1}{2}}} a^{7/2} r^6 e^{-ar^2} \, dr$$

$$= \frac{16}{15\pi^{\frac{1}{2}}} a^{7/2} \int_0^\infty r^7 e^{-ar^2} \, dr \tag{6.25}$$

The integral in Eq. (6.25) can be evaluated by using the substitution $t = ar^2$

$$\int_0^\infty r^7 e^{-ar^2} \, dr \xrightarrow[t^3/a^3 = r^6]{t = ar^2,\, dt/2a = r\, dr} \frac{1}{2a^4} \int_0^\infty t^3 e^{-t} \, dt \tag{6.26}$$

which leads to

$$\langle r \rangle = \frac{8}{15\pi^{\frac{1}{2}}} \frac{1}{a^{\frac{1}{2}}} \int_0^\infty t^3 e^{-t} \, dt \tag{6.27}$$

The integral $\int_0^\infty t^3 e^{-t} \, dt$ is the gamma function $\Gamma(3 + 1)$ (Appendix 6.2) and is equal to 3! from

$$\Gamma(x + 1) = \int_0^\infty t^x e^{-t} \, dt = x! \tag{6.28}$$

Hence, Eq. (6.27) becomes

$$\langle r \rangle = \frac{8}{15\pi^{\frac{1}{2}}} \left(\frac{kT}{\gamma}\right)^{\frac{1}{2}} \frac{1}{(4\pi)^{\frac{1}{2}}} \times 3 \times 2 \times 1$$

$$= \frac{8}{5\pi} \left(\frac{kT}{\gamma}\right)^{\frac{1}{2}}$$

$$= 0.51 \left(\frac{kT}{\gamma}\right)^{\frac{1}{2}} \tag{6.29}$$

540   CHAPTER 6

Since the average surface area of a hole calculated by this procedure gives

$$4\pi \langle r^2 \rangle = 3.5 \frac{kT}{\gamma} \tag{6.30}$$

then, from Eqs. (6.29) and (6.30) one obtains

$$\frac{4\pi \langle r^2 \rangle}{\langle r \rangle^2} = \frac{3.5 kT/\gamma}{[0.51(kT/\gamma)^{\frac{1}{2}}]^2} = 13.5 \tag{6.31}$$

If all the holes were of the same size, this ratio would be $4\pi \simeq 12.6$. The value given by Eq. (6.31) differs from 4π by about 8%, which shows that one obtains a fairly good approximation by taking the holes to be of the same size.

What value of surface tension is to be used? The hole theory boldly uses the macroscopic value with suitable correction for the curvature of holes of atomic dimensions. Careful theoretical analysis appears to support the use of the macroscopic value of surface tension in a molecular model.

What typical values of mean hole radius does Eq. (6.29) yield (Table 6.11)? By using the macroscopic surface-tension value, it is found from Eq. (6.29) that the average radius of a hole in molten KCl at 900 °C is 2.1 Å. The mean ionic radius, however, is 1.6 Å.

A typical hole therefore is roughly the same size as an ion and can accommodate an ion. This result is all the more remarkable because of the process by which it has been attained. One began by considering that

TABLE 6.11

Mean Hole Radius for Various Molten Salts at 900°C

Molten salt	Surface tension, dyne cm^{-1}	Mean hole volume, Å^3	Mean hole radius, Å
NaCl	107.1	32	1.7
NaBr	90.5	41.7	1.9
KCl	89.5	42.3	1.9
KBr	77.3	52.7	2.1
CsCl	72.7	51.8	2.2
NaI	66.4	66.2	2.3
KI	60.3	76.6	2.3

a liquid electrolyte was a liquid *continuum* interspersed by holes of random size and location. Holes in a fused salt could be treated somewhat like bubbles in a boiling liquid. Thus, the work of hole formation was taken to be equal to the work of expanding the surface area of a bubble in a boiling liquid. With the use of this expression and simple probability arguments, the average hole radius was calculated.

At a fixed temperature, the only parameter determining the mean hole size is the surface tension. Though one is aiming at a microscopic (structural) explanation of the behavior of ionic liquids, one goes ahead and uses the macroscopic value of surface tension. Thereafter, the mean hole radius turns out to have the same order of magnitude as the mean ionic radius. The connection with an experimental fact is through the surface tension. The provision from the theory of an indication of *molecular*-sized holes supports the applicability of the approach. How can bubbles in liquids be used as the basis of a calculation of liquid properties? The answer shall be given by the degree of ability of such an approach to predict facts—for example, the compressibility and coefficient of expansion. The fact that it indicates just the size of holes needed for ions to jump into and diffuse is an encouraging indication.

Further Reading

1. W. Altar, *J. Chem. Phys.*, **5**: 577 (1937).
2. R. Fürth, *Proc. Cambridge Phil. Soc.*, 252, 276, 281 (1941).
3. N. E. Richards, *Proc. Roy. Soc. (London)*, **A241**: 44 (1957).
4. G. W. Hooper, Discussions Faraday Soc., **32**: 318 (1962).
5. F. H. Stillinger, Chapter 1 in: M. Blander, ed., *Molten Salt Chemistry*, Interscience Publishers, Inc., New York, 1964.
6. K. D. Luks and H. T. Davis, *Ind. Eng. Chem. Fundamentals*, **6**: 194 (1967).

6.4. TRANSPORT PHENOMENA IN LIQUID ELECTROLYTES

6.4.1. Some Simplifying Features of Transport in Fused Salts

An important characteristic of liquid ionic systems is that they lack an inert solvent; they are *pure* electrolytes. Owing to this characteristic, some aspects of transport phenomena in pure molten salts are simpler than similar phenomena in aqueous solutions.

Thus, there is no concentration variable taken into account in the consideration of transport phenomena in a pure liquid electrolyte. Hence, there cannot be a concentration gradient in a *pure* fused salt, and, without a concentration gradient, there cannot be net diffusion.† In an aqueous solution, on the other hand, it is possible to have a concentration gradient for the solute and thus have diffusion.

Another consequence of the absence of a solvent is that the mean ion–ion interaction field is constant (at constant temperature) in a pure liquid electrolyte. In ionic solutions, however, the extent of ion–ion interaction is a variable quantity. It depends on the amount of solvent dissolving a given quantity of solute, i.e., on the solute concentration.

6.4.2. Diffusion in Fused Salts

6.4.2a. Self-Diffusion in Pure Liquid Electrolytes: It May Be Revealed by Introducing Isotopes. In the absence of a solvent, it is meaningless to consider a *pure* liquid electrolyte, e.g., NaCl, as having different amounts of NaCl in different regions.

The possibility might be considered that the system could be made to have more ions of one species, e.g., Na$^+$, in one region than in another. However, this is impossible because any attempt of a single ionic species to accumulate in one region and decrease in another is promptly stopped by the electric field which develops as a consequence of the separation of charges. Overall electroneutrality must prevail, i.e., there can be no congregation of an ionic species in one part of the liquid.

An electric field that results from incipient charge separation and reduces the applied field is also set up in aqueous solutions, but, here, owing to the much smaller number of ions per unit volume, the potential difference arising from charge separation is spread over macroscopic distances, say, microns, as, e.g., in the case of a liquid-junction potential (Section 4.5.4). The restricting field, therefore, is much smaller in aqueous solutions than in pure liquid electrolytes.

Fortunately, electroneutrality only requires that the total positive *charge* in a certain region is equal to the total negative *charge*. Suppose therefore that, in liquid sodium chloride electrolyte, a certain percentage of the Na$^+$ ions is replaced by one of a radioactive isotope of sodium. There is no difference between the Na^{22} and Na^{23} as far as the principle of electroneutrality is concerned; it is only required that the number of Na$^+$ ions plus the number of tagged Na*$^+$ ions are equal to the total number of Cl$^-$

---

† The addition of tracer ions is not considered here because one can look upon a liquid electrolyte containing tracer ions as a *mixture* of pure electrolytes—one pure electrolyte, e.g., NaCl, without tracer ions, and the other, e.g., Na*Cl, with tracer ions. In mixtures of pure electrolytes there can be a concentration variable, e.g., of the tracer ions.

⊕ Tagged positive ion
⊕ Non-radioactive positive ion
⊖ Non-radioactive negative ion

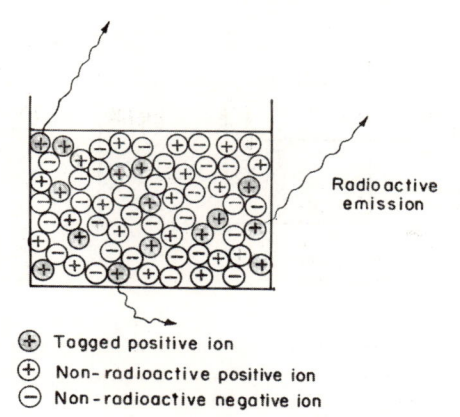

Fig. 6.18. The principle of electroneutrality is satisfied if the number of tagged positive ions plus the number of nonradiactive positive ions is equal to the total number of negative ions.

ions (Fig. 6.18). But the labeled Na*+ ions and the nonradioactive Na+ are completely different entities from the point of view of a counter; only the former produce the scintillations.

Herein lies a method of manifesting the diffusion of ions in pure ionic liquids. One takes a pure liquid electrolyte, say, NaCl, and brings it into contact with a melt containing the same salt but with a certain proportion of radioactive ions, say, NaCl with radioactive Na*+ ions. There is a negligible concentration gradient for Na+ ions, but a concentration gradient for the tracer Na*+ ions has been created. Diffusion of the tracer commences (Fig. 6.19).

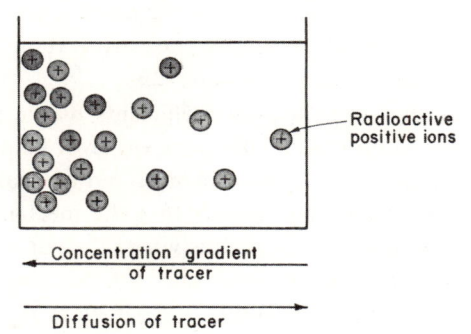

Fig. 6.19. The existence of a concentration gradient for tracer ions produces diffusion of the tracer, i.e., tracer diffusion.

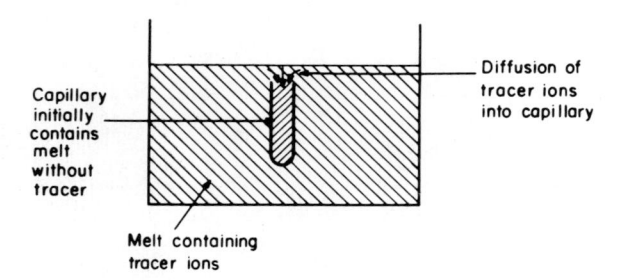

Fig. 6.20. A schematic of an experiment to study tracer diffusion. A capillary containing inactive melt is dipped into a reservoir of melt containing tracer ions. Tracer ions diffuse into the capillary.

If, therefore, a capillary containing inactive melt is suddenly introduced into a large reservoir of tracer-containing melt at $t = 0$, then diffusion of the tracer into the capillary starts (Fig. 6.20). At time t, the experiment can be terminated by withdrawing the capillary from the reservoir. The total amount of tracer in the capillary can be measured by a detector of the radioactivity. From the study of the diffusion problem and the experimentally determined average tracer concentration in the capillary, the diffusion coefficient of the Na^{*+} ions is then calculated.

Since the tracer ions (e.g., Na^{*+}) diffuse among particles (e.g., Na^{+}) which are *chemically* just like themselves, one often refers to the phenomenon as *self* diffusion (*tracer* diffusion is a more explanatory term) and to the diffusion coefficient thus determined a the *self*-diffusion coefficient.

6.4.2b. Results of Self-Diffusion Experiments. Self-diffusion coefficient studies with fused salts really began to gather momentum after radioisotopes became widely available, i.e., after about 1950. Some of the available data are presented in Table 6.12.

It can be seen that the diffusion coefficients of these liquid electrolytes (near their melting points) are of the same order of magnitude ($\sim 10^{-5}$ cm² sec^{-1}) as for liquid inert gases, liquid metals, and normal room-temperature liquids (Table 6.13). This fact suggests that the mechanism of diffusion is the same in all *simple liquids*, i.e., liquids where the particles do not associate into pairs, triplets, or network structures (*cf.* Sections 6.6 and 6.7), etc. The order of magnitude of the diffusion coefficient has evidently more to do with the liquid state than with the chemical nature of the liquid for, in the case of crystalline substances, the diffusion coefficient ranges (Table 6.14) over about four orders of magnitude (10^{-7} to 10^{-11} cm² sec^{-1}).

TABLE 6.12

Tracer-Diffusion Coefficients

Molten salt	Tracer ion	Temperature, °C	Tracer diffusion coefficient, cm² sec⁻¹
NaCl	^{22}Na+	840	9.6×10^{-5}
	^{36}Cl−	840	6.7×10^{-5}
RbCl	^{86}Rb+	740	4.7×10^{-5}
	^{36}Cl−	740	4.2×10^{-5}
CsCl	^{134}Cs+	670	3.5×10^{-5}
	^{36}Cl−	670	3.8×10^{-5}
ZnCl$_2$	^{65}Zn++	600	0.6×10^{-5}
	^{36}Cl−	600	0.4×10^{-5}
BaCl$_2$	^{140}Ba++	1000	1.84×10^{-5}
	^{36}Cl−	1000	2.99×10^{-5}

An expected feature of the results on tracer diffusion is that the diffusion coefficient varies with temperature. The temperature dependence observed experimentally can be expressed in the usual exponential way

$$D = D_0 \exp\left(-\frac{E_D}{RT}\right) \qquad (6.32)$$

where D_0 is found to depend little on substance and temperature, and E_D is the activation energy for self-diffusion (Fig. 6.21). A tabulation of some preexponential factors and the corresponding energies of activation for diffusion is given in Table 6.15.

TABLE 6.13

Self-Diffusion Coefficients of Various Types of Substance

Type of substance	Example	Temperature °C	Diffusion coefficient cm² sec⁻¹
Liquid inert gas	Ar	−173	3.70×10^{-5}
Room temperature liquid	CCl$_2$	25	1.41×10^{-5}
Liquid metal	Zn	420	2.07×10^{-5}
Molten salt	NaCl	840	D_{Na+} 9.6×10^{-5}
			D_{Cl-} 6.7×10^{-5}

TABLE 6.14

Tracer Diffusion Coefficients of Crystalline Substances near the Melting Point

Substance	Tracer	D, cm^2 sec^{-1}	Temperature, °C
Na	Na^{22}	1.7×10^{-7}	97
Ag	Ag^{110}	2.8×10^{-8}	900
NaCl	Na^{22}	4.0×10^{-9}	727
PbS	ThB	1.4×10^{-9}	1043
Pb	ThB	5.5×10^{-10}	324
PbI$_2$	ThB	7.7×10^{-11}	315

In some cases, a deviation (Fig. 6.22) occurs from the straight-line log D versus $1/T$ plots expected on the basis of the empirical exponential law for the diffusion coefficient [Eq. (6.32)]. An example of such a deviating liquid electrolyte is molten ZnCl$_2$, but, in the case of this substance,

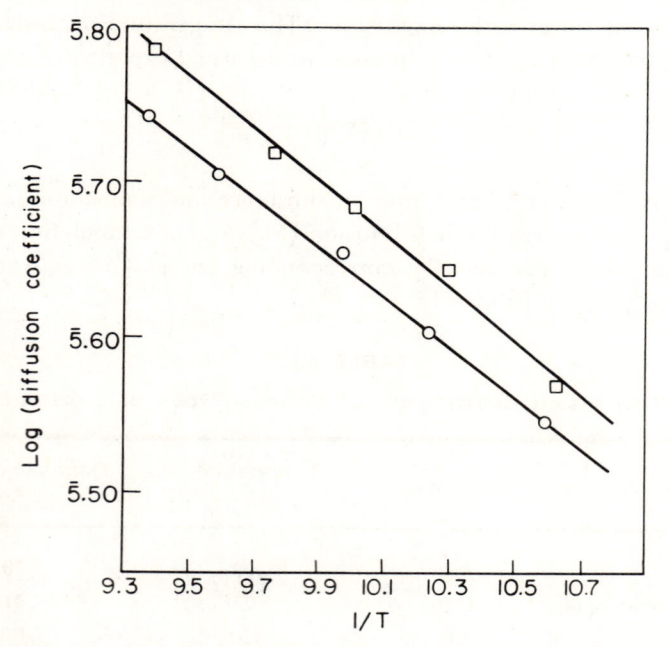

Fig. 6.21. The straight-line plot of log D versus $1/T$ observed in the case of the diffusion of Cs^{134} (○) and Cl^{36} (□) in molten CsCl.

TABLE 6.15

Energies of Activation and Preexponential Factors for Self-Diffusion in Molten Group I and II Chlorides

Molten salt/tracer	$10^3 \times D_0^*$, cm² sec⁻¹	E_a^*, kcal mole⁻¹	Temp. range, °C
NaCl/Na²²	2.1	7.14 ± 0.25	820–1020
NaCl/Cl³⁶	1.9	7.43 ± 0.84	826–1035
KCl/K⁴²	1.8	6.88 ± 0.51	798–983
KCl/Cl³⁶	1.8	7.13 ± 0.49	794–987
CaCl₂/Ca⁴⁵	0.38	6.13 ± 0.66	783–1004
CaCl₂/Cl³⁶	1.9	8.86 ± 0.96	787–1019
SrCl₂/Sr⁸⁹	0.21	5.38 ± 0.97	921–1120
SrCl₂/Cl³⁶	0.77	6.88 ± 0.72	912–1157
BaCl₂/Ba¹⁴⁰	0.64	8.96 ± 1.23	994–1207
BaCl₂/Cl³⁶	2.0	9.48 ± 1.02	993–1203
CdCl₂/Cd¹¹⁵ᵐ	1.1	6.84 ± 0.62	607–806
CdCl₂/Cl³⁶	1.1	6.80 ± 1.01	607–802

marked structural changes have been noted with increasing temperature, a possible explanation for the deviation from the straight-line log D versus $1/T$ plot.

The activation energy for self-diffusion is usually a constant, independent of temperature. It is, however, characteristic of the particular liquid electrolyte. The dependence of the activation energy for self-diffusion on the nature of the fused salt has been experimentally found to be expressible in a simple form (Fig. 6.23). The following relation is approximately applicable

$$E_D = 3.7 RT_m \tag{6.33}$$

where T_m is the melting point.

It is important to emphasize here that this same relation is valid for many liquids other than the pure nonassociated liquid electrolytes, including the liquid inert gases and the liquid metals.

6.4.3. The Viscosity of Molten Salts

Only in recent years has attention been paid to the flow of pure liquid electrolytes (Table 6.16). The still-incomplete examination shows that viscosity varies with temperature in a way strictly analogous to that of self-

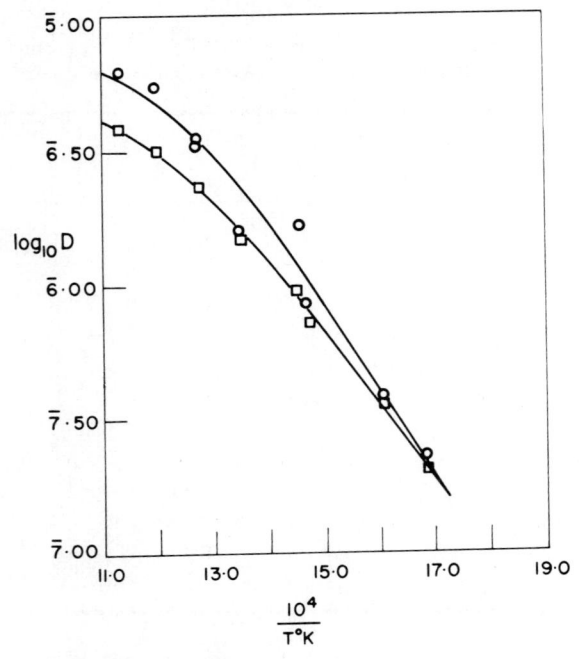

Fig. 6.22. An example of a log D versus $1/T$ plot which is not a straight line. The curve is for the diffusion of Zn^{65} (○) and Cl^{36} (□) in molten $ZnCl_2$.

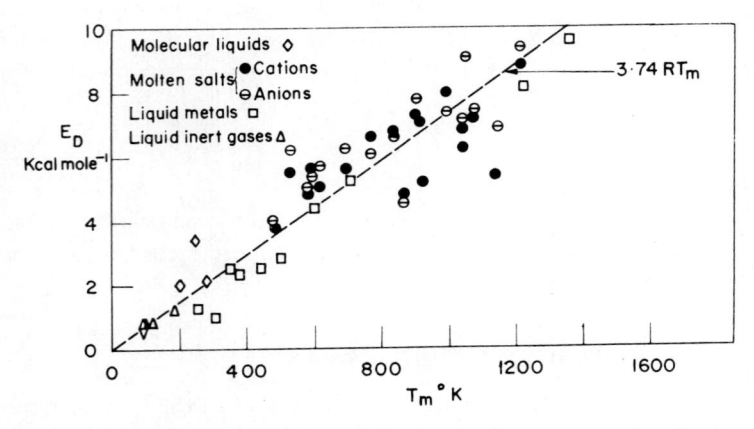

Fig. 6.23. The dependence of the experimental energy of activation for self-diffusion on the melting point.

IONIC LIQUIDS   549

TABLE 6.16
Viscosities of Fused Salts

Fused salt	Temperature, °C	Viscosity, centipoise
$CdCl_2$	600	2.31
$CdBr_2$	600	2.61
$PbBr_2$	550	2.98
$PbCl_2$	600	2.75
AgCl	600	1.66
AgBr	600	2.27
NaCl	900	1.05
NaBr	755	1.43
KCl	773	1.51
KBr	730	1.57

diffusion. For simple, unassociated liquid electrolytes, the temperature dependence is given by an empirical equation

$$\eta = \eta_0 e^{+E_\eta/RT}, \tag{6.34}$$

where η_0 is a constant analogous to D_0, and E_η is the energy of activation for viscous flow (Fig. 6.24 and Table 6.17).

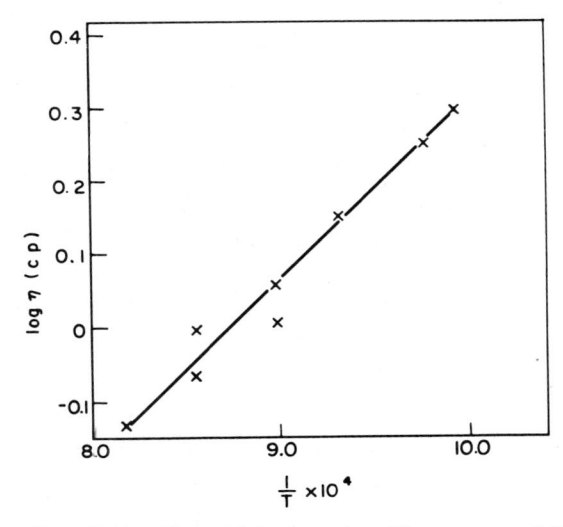

Fig. 6.24. The straight-line plot of log η versus $1/T$ for viscosity of a molten salt.

TABLE 6.17

Temperature Dependence of Viscosity for Simple Molten Electrolytes

Molten salt	E_η, kcal mole^{-1}	Molten salt	E_η, kcal mole^{-1}
LiCl	8.8	KOH	6.1
LiBr	6.0	AgCl	2.9
LiI	4.4	AgBr	3.1
LiNO$_3$	4.2	AgI	5.8
NaCl	9.1	AgNO$_3$	3.1
NaBr	8.0	TlNO$_3$	3.6
NaI	7.4	CuCl	4.2
NaNO$_3$	4.0	MgCl$_2$	6.5
NaNO$_2$	4.0	CaCl$_2$	9.5
NaOH	5.5	BaCl$_2$	9.0
NaCNS	5.8	PbCl$_2$	6.7
KCl	7.8	PbBr$_2$	6.2
KBr	7.5	CdCl$_2$	4.0
KI	9.2	CdBr$_2$	4.5
KNO$_3$	3.7	NH$_4$NO$_3$	4.6

This expression is formally analogous to Eq. (6.32) for the dependence of the self-diffusion coefficient upon temperature. In fact, for simple liquid electrolytes, the experimental activation energy for viscous flow is given by an expression (Fig. 6.25) identical to that for self-diffusion, i.e.,

$$E_\eta = E_D = 3.7RT_m \quad (6.35)$$

This implies that the basic factors determining viscous flow and self-diffusion are the same.

Simple ionic liquids have viscosities in the range of 1 to 5 centipoises. When, however, there is association of ions into aggregates, as, for example, in ZnCl$_2$ near the melting point, the viscous force resisting flow of the melt increases. Such *complex* ionic liquids are discussed later (Section 6.6).

6.4.4. What Is the Validity of the Stokes–Einstein Relation in Ionic Liquids?

All transport processes (viscous flow, diffusion, conduction of electricity) involve ionic movements and ionic drift in preferred direction; they must, therefore, be interrelated. A relationship between the phenomena of

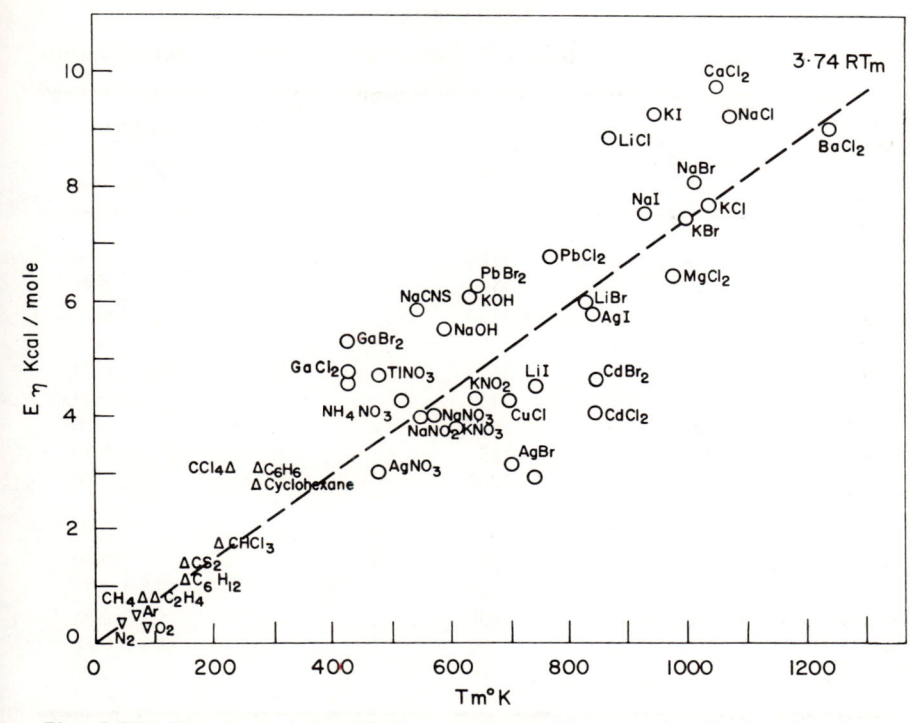

Fig. 6.25. The dependence of the experimental energy of activation for viscous flow on the melting point.

diffusion and viscosity is contained in the Stokes–Einstein equation (4.176).

$$D = \frac{kT}{6\pi r \eta} \tag{6.36}$$

The validity of this equation in ionic *solutions* has been discussed in Section 4.4.8. It is therefore of interest to inquire about the applicability of the relation in ionic liquids, i.e., fused salts. To make a test, the experimental values of the self-diffusion coefficient D^* and the viscosity η are used in conjunction with the known crystal radii of the ions. The product $D^*\eta/T$ has been tabulated in Table 6.18, and the plot of $D^*\eta/T$ versus $1/r$ is presented in Fig. 6.26, where the line of slope $k/6\pi$ corresponds to exact agreement with the Stokes–Einstein relation.[†]

[†] The essential applicability of this phenomenological equation is well shown by using the numerical comparison of $D\eta/T = k/6\pi r$. The right-hand side is 0.7×10^9 for $r = 3$ Å, and the mean of the experimental values is 0.6×10^9.

TABLE 6.18
The $D\eta/T$ of Some Molten Salts for Testing the Stokes–Einstein Relation

Molten salt	Tracer	Temperature, °K	$D_i \times 10^5$, cm² sec⁻¹	$\dfrac{D_i\eta}{T} \times 10^9$, dyne deg⁻¹
NaCl	Li[6]	1180	13.2	1.09
NaCl	Na[22]	1180	10.2	0.85
NaCl	K[42]	1180	9.7	0.81
NaCl	Rb[86]	1180	9.2	0.76
NaCl	Cs[134]	1180	8.9	0.74
KCl	K[42]	1150	8.9	0.63
NaI	Na[22]	1026	9.5	1.20
$CaCl_2$	Ca[45]	1154	2.6	0.79
$SrCl_2$	Sr[89]	1260	2.4	0.70
$BaCl_2$	Ba[140]	1356	2.4	0.69
$CdCl_2$	Cd[115m]	925	2.7	0.58
$PbCl_2$	Pb[210]	851	1.5	0.59
$LiNO_3$	Li[6]	581	2.1	1.77
$NaNO_3$	Na[22]	638	2.6	0.88
KNO_3	K[42]	667	2.1	0.68
$AgNO_3$	Ag[110]	534	1.5	0.80

It can be seen that there is a significant fit. The anions, particularly those of the Group II halides, are not very consistent with the Stokes–Einstein relation. But the poor fit is offset by the better Stokes–Einstein behavior of the cations. The relatively good fit of the cations tempts one to conclude that there is a particular reason, in the case of anions only, which leads them into deviations. Some attempts have been made to elucidate this reason. For instance, it has been suggested that, since the anions are larger than the cations, they require greater local rearrangements at a site before they can jump into it, i.e., greater entropies of activation (Appendix 4.3).

The Stokes–Einstein relation is based on Stokes' law in hydrodynamics according to which the viscous force experienced by a *large* sphere moving in an incompressible *continuum* is $6\pi r\eta v$ (cf. Section 4.4.8). Hence, the Stokes–Einstein relation depends on the view that an *ion* moving in an electrolyte experiences a Stokes force $6\pi r\eta v$, even though the ions do not move in a continuum but among particles which are of approximately the same dimensions as the ions themselves. In view of the "far-flungness"

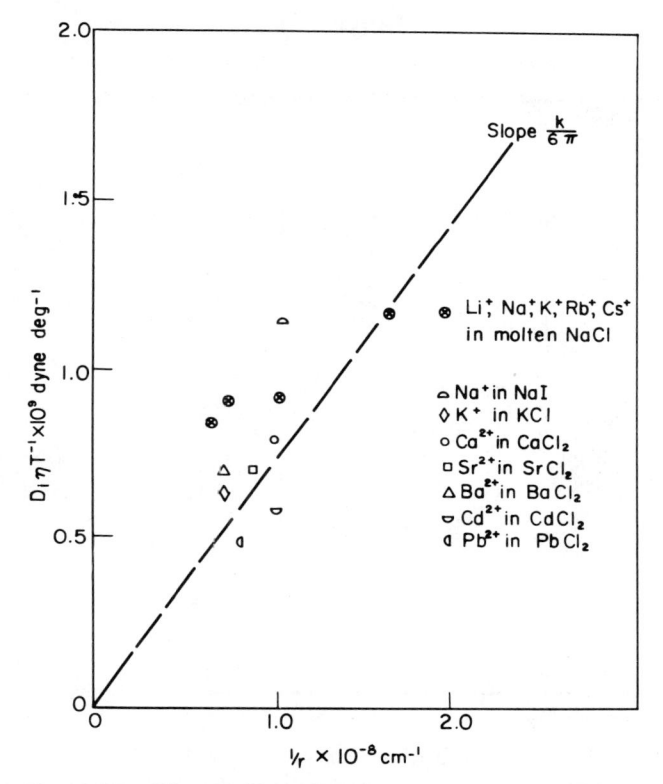

Fig. 6.26. When D/T is plotted against $1/r$, a straight line of slope $k/6\pi$ should be obtained if the Stokes–Einstein relation is applicable to molten salts. The experimental points are indicated in the figure to show the degree of applicability of the Stokes–Einstein relation to molten salts.

of the similarity between an ion in a structured medium and a sphere in an incompressible continuum, the rough applicability (in fused salts) of the Stokes–Einstein equation is somewhat unexpected.

6.4.5. The Conductivity of Pure Liquid Electrolytes

The electrical conductance of molten salts is the easiest transport property to measure. In addition, knowledge of the order of magnitude of the equivalent conductivity of a pure substance was used as an important criterion of the nature of the bonding present. For these reasons, the electrical conductance of ionic liquids has been the subject of numerous studies.

The equivalent conductivities of some of the fused chlorides are given

TABLE 6.19

Equivalent Conductivities of Molten Chlorides

HCl $\sim 10^{-6}$					
LiCl 166	BeCl$_2$ 0.086	BCl$_3$ 0	CCl$_4$ 0		
NaCl 133.5	MgCl$_2$ 28.8	AlCl$_3$ 15×10^{-6}	SiCl$_4$ 0	PCl$_5$ 0	
KCl 103.5	CaCl$_2$ 51.9	ScCl$_3$ 15	TiCl$_4$ 0	VCl$_4$ 0	
RbCl 78.2	SrCl$_2$ 55.7	YCl$_3$ 9.5	ZrCl$_4$	NbCl$_5$ $x = 2 \times 10^{-7}$	MoCl$_5$ $x = 1.8 \times 10^{-6}$
CsCl 66.7	BaCl$_2$ 64.6	LaCl$_3$ 29.0	HfCl$_4$	TaCl$_5$ $x = 3 \times 10^{-7}$	WCl$_6$ $x = 2 \times 10^{-6}$
			ThCl$_4$ 16		UCl$_4$ $x = 0.34$

in Table 6.19, where the substances have been arranged according to the periodic table. The heavy line zigzagging across the table separates the ionic from the covalent chlorides. This structural difference is shown up sharply in the orders of magnitude of the equivalent conductivities.

Two further correlations emerge from Table 6.19. Firstly, the equivalent conductivity decreases with increasing size of the cation (Table 6.20);

TABLE 6.20

Dependence of Equivalent Conductivity upon Cationic Radius

Molten salt	Radius of cation, Å	Equivalent conductivity
LiCl	0.68	183
NaCl	0.94	150
KCl	1.33	120
RbCl	1.47	94
CsCl	1.67	86

TABLE 6.21
Dependence of Equivalent Conductivity upon Valency of Cation

Molten salt	Valency of cation	Equivalent conductivity
NaCl	1	150
MgCl$_2$	2	35
AlCl$_3$	3	15.1
SiCl$_4$	4	<0.1

secondly, there is a decrease in equivalent conductivity in going from the monovalent to the divalent and then to the trivalent chlorides (Table 6.21), probably because of an increase in covalent character in this order.

As with the other transport properties, the specific (or equivalent) conductivity of fused salts varies with temperature.† For most pure liquid electrolytes, the experimental log Λ versus $1/T$ plots are essentially linear (Fig. 6.27). This implies the usual exponential dependence of a transport property upon temperature

$$\Lambda = \Lambda_0 e^{-E_\Lambda/RT} \qquad (6.37)$$

For some substances, the plots are slightly curved. In these cases, structural changes (e.g., the breaking-up of polymer networks) occur with change of temperature.

When the activation energies for conduction are computed from the log Λ versus $1/T$ plots, it is seen (Table 6.22) that they are lower than the activation energies for viscous flow and self-diffusion, i.e.,

$$E_\Lambda < E_D \sim E_\eta \qquad (6.38)$$

6.4.6. The Nernst–Einstein Relation in Ionic Liquids

6.4.6a. The Nernst–Einstein Relation: Its Degree of Applicability.

Just as the Stokes–Einstein equation gives the relation between the transport of momentum (viscous flow) and the transport of matter (diffusion), the connection between the transport processes of diffusion and conduction

---

† A convenient means of comparing different salts is to use "corresponding temperatures"; usually 1.05 or 1.10 times the value of the melting point in degrees Kelvin is used for this purpose.

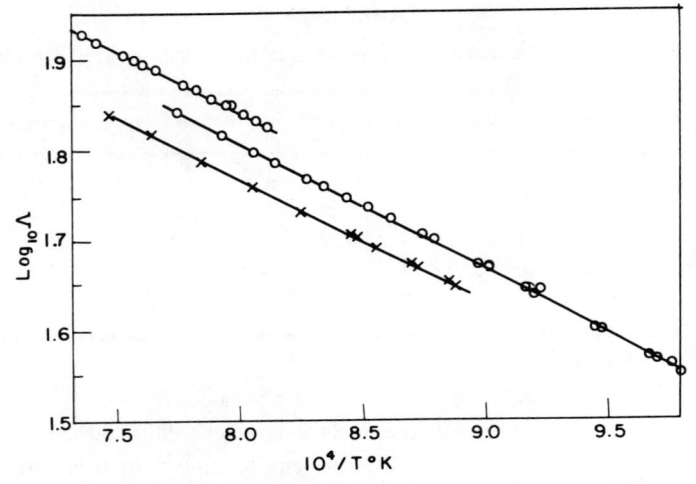

Fig. 6.27. The straight-line plot of $\log \Lambda$ versus $1/T$.

leads to the Nernst–Einstein equation (*cf.* Section 4.4.9), i.e.,

$$\Lambda = \frac{F^2}{RT}(D_+ + D_-) \tag{6.39}$$

for 1:1 electrolytes. The more general expression is

$$\Lambda = \frac{F^2}{RT}\sum_i z_i D_i \tag{6.40}$$

for asymmetrical electrolytes.

TABLE 6.22

Experimental Energies of Activation for Various Transport Processes

Molten salt	E_Λ	E_D, kcal mole^{-1}	E_η
LiCl	2.06	—	8.7
NaCl	2.92	7.3	9.1
NaNO$_3$	3.12	5.0	4
KCl	3.36	7.0	7.8
KNO$_3$	3.15	5.6	3.6
CdCl$_2$	2.20	6.8	4

TABLE 6.23
Test of the Nernst–Einstein Relation for Equivalent Conductivity of Molten NaCl

	Equivalent conductivity			
	1093 °K	1143 °K	1193 °K	1293 °K
Observed	138	147	155	171
Calculated from Eq. (6.39)	159	177	198	240

The testing of the Nernst–Einstein relation can be carried out by using the experimentally determined tracer-diffusion coefficients D_i to calculate the equivalent conductivity Λ and then comparing this theoretical value with the experimentally observed Λ. It is found that the values of Λ calculated by Eq. (6.39) are distinctly greater (by $\sim$ 10 to 50%) than the measured values (see Table 6.23 and Fig. 6.28).

6.4.6b. The Gross View of Deviations from the Nernst–Einstein Equation. It is important to understand the conditions under which there are deviations from the Nernst–Einstein relations because such deviations throw light on what is happening inside ionic liquids. The ultimate aim, of course, is to acquire a molecular view of the deviations; but, in the first instance, it is appropriate to define, in phenomenological terms, the general basis for deviations from the Nernst–Einstein relations. This phenomenological "explanation" can be derived from the near-equilibrium thermodynamic approach which has been described in Sections 4.5.6 and 4.5.7, and has been applied to fused salts by Sundheim.

The fluxes (moles flowing per square centimeter per second) of the positive and negative ions in a fused salt can be represented by [cf. Eqs. (4.247) and (4.248)]

$$J_+ = L_{++}\vec{F}_+ + L_{+-}\vec{F}_- \tag{6.41}$$

$$J_- = L_{-+}\vec{F}_+ + L_{--}\vec{F}_- \tag{6.42}$$

where $\vec{F}_+$ and $\vec{F}_-$ are the driving forces on the positive and negative ions, respectively; L_{++} and L_{--} are the straight coefficients governing the independent flows of these ions; and L_{+-} and L_{-+} are the cross coefficients

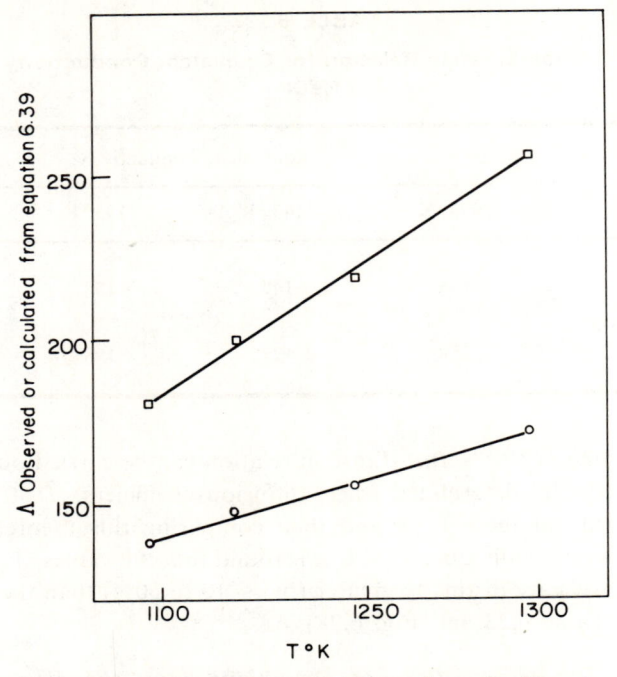

Fig. 6.28. Plot to show deviations from the Nernst–Einstein equation; (○) observed equivalent conductivity of molten NaCl and (□) calculated from Eq. (6.39).

governing the influence of the driving force $\vec{F}_-$ on the flow J_+ of the positive ions and *vice versa*.

What is the consequence of setting the cross coefficients equal to zero, i.e.,

$$L_{+-} = L_{-+} = 0 \tag{6.43}$$

With the use of condition (6.43) in Eqs. (6.41) and (6.42), the result is

$$J_+ = L_{++}\vec{F}_+ \tag{6.44}$$

and

$$J_- = L_{--}\vec{F}_- \tag{6.45}$$

When $L_{+-} = L_{-+} = 0$, the positive and negative ions drift independently of each other, i.e., the driving force $\vec{F}_-$, acting on the negative ions, does *not* affect the flux J_+ of positive ions, and *vice versa*. But it has been shown that the total driving force on an ionic species is the gradient of electrochemical potential [cf. Eq. (4.222)]

$$\vec{F}_+ = -\frac{d\bar{\mu}_+}{dx}$$

$$= -\frac{d\mu_+}{dx} - z_+ F \frac{d\varphi}{dx}$$

$$= -\frac{RT}{c_+}\frac{dc_+}{dx} - z_+ FX \qquad (6.46)$$

Hence, by combining Eqs. (6.46) and (6.44), one has

$$J_+ = -\left(L_{++}\frac{RT}{c_+}\right)\frac{dc_+}{dx} - (L_{++}z_+F)X \qquad (6.47)$$

Now, the independent flow of an ionic species has been shown to be given by the Nernst–Planck flux equation (4.224), i.e.,

$$J_+ = -D_+\frac{dc_+}{dx} - \frac{D_+c_+}{RT}z_+FX \qquad (6.48)$$

or, since [cf. Section 4.4.3 and Eqs. (4.149) and (4.169)]

$$D_+ = (u_{\text{abs}})_+ kT = \frac{(u_{\text{conv}})_+ kT}{z_+ e_0} \qquad (6.49)$$

hence, from (6.49) and (6.48),

$$J_+ = -D_+\frac{dc_+}{dx} - (u_{\text{conv}})c_+X \qquad (6.50)$$

By equating the coefficients of dc_+/dx and X from Eqs. (6.47) and (6.50) the net result is

$$L_{++} = \frac{D_+c_+}{RT} = \frac{(u_{\text{conv}})_+ c_+}{z_+ F} \qquad (6.51)$$

or

$$(u_{\text{conv}})_+ = \frac{z_+ F D_+}{RT} \qquad (6.52)$$

Similarly, it can be shown that

$$(u_{\text{conv}})_- = \frac{z_- F D_-}{RT} \qquad (6.53)$$

By combining Eqs. (6.52) and (6.53) and utilizing the relation

$$(u_{\text{conv}})_+ + (u_{\text{conv}})_- = \frac{\Lambda}{F} \qquad (6.54)$$

the final result is

$$\Lambda = \frac{F^2}{RT}(z_+ D_+ + z_- D_-) \tag{6.55}$$

$$= \frac{F^2}{RT} \sum_i z_i D_i \tag{6.56}$$

which is the Nernst–Einstein relation.

It is clear, therefore, that the Nernst–Einstein relation is obeyed when $L_{+-} = L_{-+} = 0$, i.e., when the flow of each species is unaffected by the driving forces on the other species.

Various terms are in use to describe the cross coefficients. For instance, they have been called "coupling" coefficients or "interaction" coefficients, though these terms mislead one into thinking that the presence of interactions between particles inevitably leads to deviations from the Nernst–Einstein relation. It must be emphasized, however, that such coefficients do not have (and cannot have) literal mechanistic implications for there may be abundant interactions between particles and yet the flows may be independent, $L_{+-} = L_{-+} = 0$, i.e., the flows may conform to the Nernst–Einstein relation. The only significance of the cross coefficients is that, provided by Eqs. (6.41) and (6.42): L_{+-}, for example, is a measure of the extent to which the flux of positive ions is affected by the gradient of electrochemical potential of the negative ions.

Since the cross coefficients do not contribute to a mechanistic (hence, structure-indicating) interpretation of the deviations from the Nernst–Einstein equation, it may be asked: Why use them? In the subject under discussion, i.e., deviations from the Nernst–Einstein equation, it will be shown in the next section that rationalization at a molecular level exists; hence, phenomenological coefficients are, strictly, unnecessary. Nevertheless, the nonequilibrium thermodynamic approach has been referred to here because it represents a convenient shorthand way of expressing the conditions under which the Nernst–Einstein relation is valid.

6.4.6c. Possible Molecular Mechanisms for Nernst–Einstein Deviations. Since phenomenological treatments cannot throw light on the molecular mechanisms responsible for the deviation from the Nernst–Einstein equation, one must consider the atomistic origin of the cross coefficient. In other words, how does the drift of one ionic species induce the drift of another species?

A possible answer emerges from the fact that the observed conductivity is always *less* (Table 6.23) than that calculated from the sum of the diffusion coefficients, i.e., from the Nernst–Einstein relation [Eq. (6.39)]. Now, conductive transport depends only on the *charged* species because it is only charged particles that respond to an external field. If, therefore, two species of opposite charge unite, either permanently or temporarily, to give an uncharged entity, then they will not contribute to the conduction flux (Fig. 6.29). They will, however, contribute to the diffusion flux. There will be currentless diffusion, and the conductivity calculated from the sum of the diffusion coefficients will always exceed the observed value. Currentless diffusion, therefore, will lead to a deviation from the Nernst–Einstein relation. Notice that, in currentless diffusion, the diffusion flow of i particles induces a flow of j particles with which they (or a fraction of them) are united. Thus, the mechanism of currentless diffusion is in accord with the thermodynamic version of the deviation from the Nernst–Einstein relation.

What sort of situations lead to currentless diffusion? One situation is that in which a stable, uncharged entity is formed. Thus, if the formation of an MX* species occurs in a system consisting of M^+, X^-, and X^{*-} ions, then the diffusion flux of X^{*-} consists of two components, the flux of MX* and that of X^{*-}. But the conduction flux of X* is due to the charged X^{*-} only. Now, the Nernst–Einstein equation is valid only when the *same* charged particles are involved in diffusion and conduction. Hence, there will be a deviation in the case of the formation of complex ions and stable ion pairs.

It was suggested by Borucka et al. in 1957 that a *permanent* association of positive and negative ions is not a necessary basis for a breakdown of

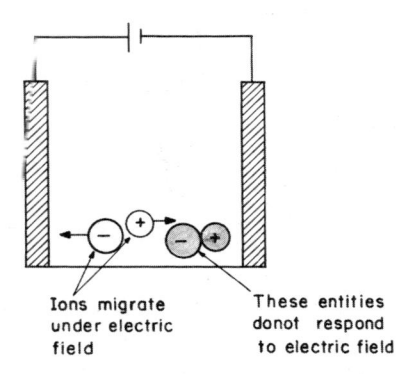

Fig. 6.29. An entity formed by the temporary or permanent association of a pair of oppositely charged ions is electrically neutral and therefore does not migrate under an electric field.

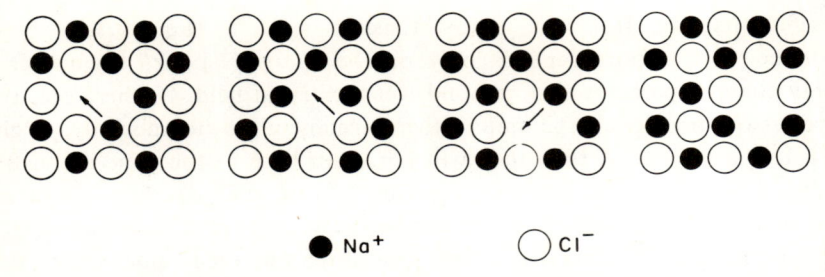

Fig. 6.30. Schematic diagrams to indicate how diffusive displacement can occur through a coordinated movement of a pair of ions into a paired vacancy.

the Nernst–Einstein equation. The only requirement is that diffusion should occur partly through the displacement of entities which have (*during jumps*) a zero net charge and thus do not contribute to conduction. The entity may be, for instance, a pair of oppositely charged ions, in which case the diffusive displacement occurs by a coordinated movement of such a pair of ions into a paired vacancy (Fig. 6.30), i.e., a vacancy which is large enough to accept a positive and a negative ion. The pair of oppositely charged ions which jumps into a coupled vacancy is neutral as a whole, and, therefore, such coordinated jumps do not play a part in the conduction process, which is determined only by the separate and uncoordinated movements of single ions.

Thus, the experimentally observed diffusive flux of either of the ionic species is made up of two contributions, the diffusive flux occurring through the independent jumps of ions and that occurring through paired jumps. Thus, taking the example of the diffusion of Na$^+$ in an NaCl melt, one has[†]

$$J'_{Na^+} = J_{Na^+,\text{ind}} + J_{NaCl} \qquad (6.57)$$

where J'_{Na^+} is the experimentally observed flux of Na$^+$ (primes refer here to experimental quantities), $J_{Na^+,\text{ind}}$ is the diffusion of Na$^+$ by independent jumps, and J_{NaCl} is the flux due to coordinated jumps of Na$^+$ and Cl$^-$ ions

---

[†] The subscript NaCl must not be taken to mean that there are entities in the melt which might be considered "molecules" of sodium chloride. The NaCl does not refer to Na$^+$ and Cl$^-$ ions which are bound together like an ion pair in aqueous solution; rather, it refers to a pair of Na$^+$ and Cl$^-$ *ions* which undergo a coordinated jump into a paired vacancy during the short time for which they momentarily exist in contact. They do not contribute to the conductance because their jumps are directed in not by the externally applied field but by the distribution of the paired vacancies which exist before the ions jump as a pair.

IONIC LIQUIDS

into paired vacancies

$$J'_{Na^+} = D'_{Na^+} \frac{dc_{Na^+}}{dx} \tag{6.58}$$

$$J_{Na^+,ind} = D_{Na^+,ind} \frac{dc_{Na^+}}{dx} \tag{6.59}$$

and

$$J_{NaCl} = D_{NaCl} \frac{dc_{Na^+}}{dx} \tag{6.60}$$

Adding Eqs. (6.59) and (6.60), one has

$$J_{Na^+,ind} + J_{NaCl} = (D_{Na^+,ind} + D_{NaCl}) \frac{dc_{Na^+}}{dx} \tag{6.61}$$

But, from Eqs. (6.57) and (6.58),

$$J_{Na^+,ind} + J_{NaCl} = J'_{Na^+} = D'_{Na^+} \frac{dc_{Na^+}}{dx} \tag{6.62}$$

Hence,

$$D'_{Na^+} = D_{Na^+,ind} + D_{NaCl} \tag{6.63}$$

Similarly,

$$D'_{Cl^-} = D_{Cl^-,ind} + D_{NaCl} \tag{6.64}$$

On adding, it is clear that

$$(D_{Na^+,ind} + D_{Cl^-,ind}) = (D'_{Na^+} + D'_{Cl^-}) - 2D_{NaCl} \tag{6.65}$$

The Na^+ and Cl^- ions which make coordinated jumps into paired vacancies, i.e., the NaCl species, contribute to diffusion but not to conduction since such a coordinated pair is effectively neutral. Hence, the Nernst–Einstein equation is only applicable to the ions which jump independently, i.e.,

$$D_{Na^+,ind} + D_{Cl^-,ind} = \frac{RT}{zF^2} \Lambda' \tag{6.66}$$

where Λ' is the experimentally observed equivalent conductivity of molten NaCl. Making use of Eqs. (6.66) and (6.65), one has

$$\frac{RT}{zF^2} \Lambda' = (D'_{Na^+} + D'_{Cl^-}) - 2D_{NaCl} \tag{6.67}$$

or

$$\Lambda' = \frac{zF^2}{RT} (D'_{Na^+} + D'_{Cl^-}) - \frac{2zF^2}{RT} D_{NaCl} \tag{6.68}$$

The first term on the right-hand side corresponds to the value of the equivalent conductivity which would be calculated on the basis of the experimentally observed diffusion coefficients. Using the symbol Λ_{calc} for this calculated value, i.e.,

$$\Lambda_{\text{calc}} = \frac{zF^2}{RT}(D'_{\text{Na}^+} + D'_{\text{Cl}^-}) \tag{6.69}$$

one has

$$\Lambda' = \Lambda_{\text{calc}} - \frac{2zF^2}{RT} D_{\text{NaCl}} \tag{6.70}$$

which shows that the experimental value of the equivalent conductivity is always less than that calculated from a Nernst–Einstein equation based on experimental diffusion coefficients. This is what seems to be observed (Table 6.23).

6.4.7. Transport Numbers in Pure Liquid Electrolytes

6.4.7a. Some Ideas about Transport Numbers in Fused Salts. The concept and determination of transport numbers in pure liquid electrolytes is one of the most interesting—and most confusing—aspects of the electrochemistry of fused salts.

The concept has been referred to in Section 4.5.2. The transport number t_i of an ionic species i is the quantitative answer to the question: What fraction of the total current $I [= \Sigma\, i_i]$ passing through electrolyte is transported by the particular ionic species i? In symbols (*cf.* Section 4.5.2):

$$t_i = \frac{i_i}{I} = \frac{i_i}{\Sigma\, i_i} \tag{6.71}$$

In the case of $z:z$-valent salts, the transport number is simply given by

$$t_i = \frac{u_i}{\Sigma\, u_i}\,^\dagger \tag{6.72}$$

or, for a *pure* liquid electrolyte consisting of one cationic and one anionic species,

$$t_+ = \frac{u_+}{u_+ + u_-} \quad \text{and} \quad t_- = \frac{u_-}{u_+ + u_-} \tag{6.73}$$

It was seen (Section 4.5.7) that, in aqueous solutions, the solvent could not be relegated to the status of an unobtrusive background. The solvent molecules, by entering into the solvation sheaths of ions, participated in

---

† The coordinate system with which these mobilities are measured is considered later on.

IONIC LIQUIDS 565

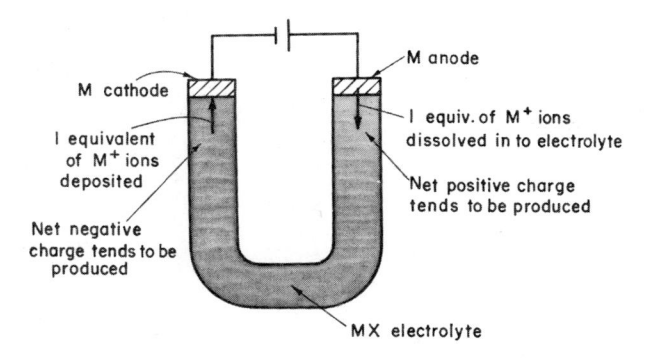

Fig. 6.31. Schematic U-tube setup with M electrodes and MX electrolyte. When 1 F of electricity is passed through the system, one equivalent of M⁺ ions is deposited at M cathode and one equivalent of ions is produced at M anode. Hence, negative charge tends to be produced near cathode and positive charge near the anode.

their drift. Thus, in addition to the flows of the positive and negative ions, there was a flux of the solvent.

This complication of solvent flux is absent in pure ionic liquids. There is, however, a very interesting effect when a current is passed through a fused salt.

Consider that a fused salt MX is taken as the ionic conductor in a U-tube and two M electrodes are introduced into the system as shown in Fig. 6.31. Let the consequences of the passage of one faraday of electricity be analyzed. Near the cathode, one equivalent of M⁺ ions will be removed from the system by deposition on the cathode; and, near the anode, one equivalent of M⁺ ions will be "pumped" into the system. Since one equivalent of M⁺ has been added and another equivalent has been removed, the total quantity of M⁺ ions in the system is unchanged.

Is the system perturbed by the passage of a faraday of charge? Yes, because, near the cathode, one equivalent of M⁺ ions has been removed, which has created a local *excess* of negative charge. This local unbalance of electroneutrality creates a local electric field.[†] A similar argument can be used for the anode region.

[†] This unbalance of electroneutrality and creation of field should not be confused with that arising from the *presence* of the electrode, which causes an anisotropy in the forces on the particles in the electrode–electrolyte interphase region. That anisotropy *also* produces an unbalance of electroneutrality and an electric double layer (Chapter 7) with a field across the interface.

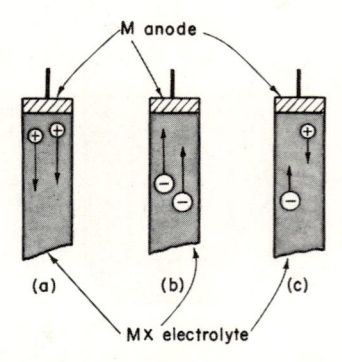

Fig. 6.32. The tendency for electroneutrality to be upset near electrodes is avoided in one of three ways. For example, near the anode, where positive charge tends to be produced, (a) positive ions can migrate away from the anode, (b) negative ions can migrate toward the anode, and (c) both (a) and (b) processes can occur to various extents.

How do the ions of the liquid electrolyte respond to this perturbation, i.e., this creation of local fields? The ions start drifting under the influence of the fields so that the initial state of electroneutrality and zero field is restored. How do the positive and negative ions share this responsibility of moving to annul the unbalance of charges—more anions than cations near the cathode and *vice versa*. It is to be noted (Fig. 6.32) that the original electroneutral situation can be restored (1) by only cations moving in the anode-to-cathode direction; (2) by only anions moving in the cathode-to-anode direction; and (3) by both cations and anions moving in opposite directions to different extents. But these possibilities represent different values of the transport numbers which are the fractions of the total field-induced ionic drift arising from the various species.

6.4.7b. The Measurement of Transport Numbers in Liquid Electrolytes. Let t_+ and t_- be the transport numbers of the M^+ and X^- ions of the fused salt. The *changes* of the numbers of equivalents of M^+ and X^- near the two electrodes are shown in Table 6.24 based on Fig. 6.31. The analysis of the changes leads to a most interesting result. The passage of 1 F of charge is equivalent to *transferring* t_- equivalents of the fused salt MX from the cathode region to the anode region (Fig. 6.33).

In the case of aqueous solutions, the ever-plentiful solvent could absorb this t_- equivalent of MX and register the transfer as a concentration change

TABLE 6.24

Changes in MX at Electrodes in Transport Experiment

Electrode material	Anode compartment, Metal M	Cathode compartment, Metal M
Electrode process per faraday passed	1 g eq M$^+$ dissolved	1 g eq M$^+$ deposited
Move out of compartment	t_+ g eq M$^+$	t_- g eq X$^-$
Move into compartment	t_- g eq X$^-$	t_+ g eq M$^+$
Net change of M$^+$	$1 - t_+$ g eq gained $= t_-$	$1 - t_+$ g eq. lost $= t_-$
Net change of MX	t_- g eq gained	t_- g eq lost
Mass change of MX in case of molten salt	$t_- M_{MX}$ g gained, where M_{MX} is eq. wt. of MX	$t_- M_{MX}$ g lost, where M_{MX} is eq. wt. of MX
Concentration change in case of aqueous solution	t_-/V_A g eq liter^{-1} increase, where V_A is vol. of anode compartment (1.)	t_-/V_C g eq liter^{-1} decrease where V_C is vol. of cathode compart. (1.)

of magnitude, t_-/V equivalents per liter, where V is the volume of the compartment. But, a pure molten salt has no concentration variable. Hence, the transfer leads to a *mass* increase near the anode, as was first suggested by Schwartz (1912) and proved mathematically by Sundheim (1956).

In molten salts, therefore, it is the change in *mass* in a compartment which reveals transport numbers; in aqueous solutions, it was the change

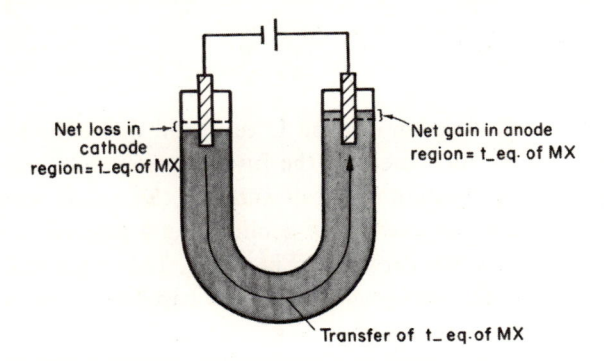

Fig. 6.33. As a result of the passage of 1 F of electricity, t_- equivalents of MX electrolyte are transferred from the cathode region (where the electrolyte level falls) to the anode region (where the level rises).

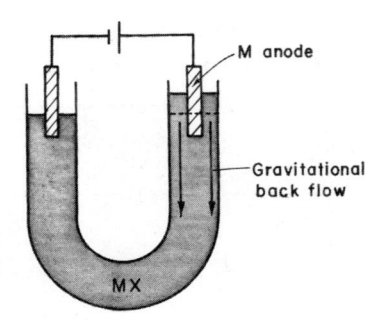

Fig. 6.34. The difference in electrolyte levels produced by the passage of 1 F of electricity leads to a gravitational flow of the electrolyte.

in concentration. However, the experiment, unless performed properly, provides information only on the change in mass, not on the transport property.

What future has this mass increase? Left alone, the mass increase is short lived and the transport experiment fails. This is because gravitational flow of the molten salt from the anode to cathode tends to equalize the amounts of MX in the two tubes (Fig. 6.34). The liquid levels must equalize.

The first step, therefore, in determining transport numbers in pure ionic liquids is to prevent the gravity flow from masking the transfer of electrolyte. If the hydrodynamic backflow cannot be prevented, it must at least be taken into account. The general procedure is to minimize the gravitational flow by interposing a membrane between anode and cathode (Fig. 6.35). However, serious objections have been raised to the use of a membrane, owing to hydrodynamic interferences between this and the moving liquid.

It is also possible to open out the U tube and make the whole liquid "lie down" so that the movement of the fused salt occurs at one level and not against gravity. The amount of salt entering the anode region is then indicated, for example, by a sliver of molten metal pushed along by the movement of the salt in the capillary (Fig. 6.36). This method is also subject to difficulties, for the movement of salts in capillary tubes is often not smooth but jerky.

This experiment demonstrates directly that, when electricity is passed through a fused salt, there is a movement of the salt as a whole. In other words, the mass center of the liquid electrolyte moves. Now, the ions also are drifting with certain mobilities, i.e., velocities under unit field. But

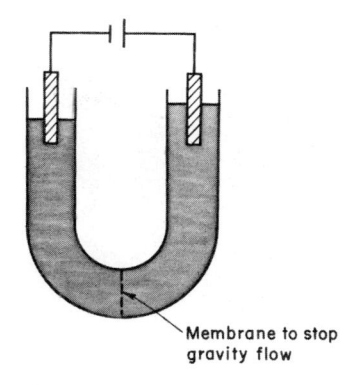

Fig. 6.35. The gravitational flow can be minimized by interposing a membrane between the anode and the cathode.

velocities with respect to what? One must define a coordinate system, or frame of reference, in relation to which the velocities (distances traversed in unit time) are reckoned. Though the laws of physics are independent of the choice of the coordinate system—the principle of relativity—all coordinate systems are not equally convenient. In fused salt it has been found convenient to use the mass center of the moving liquid electrolyte as the frame of reference, even though this choice, while providing a simple basis for computations, suffers from difficulties.

Even the elementary presentation just given makes it clear that transport-number measurements in fused salts are based on the transfer of the fused salt from the anode to the cathode compartments. The quantities measured are weight changes, the motion of indicator bubbles, the volume changes, etc. Some basic experimental setups shown in Fig. 6.37 include the

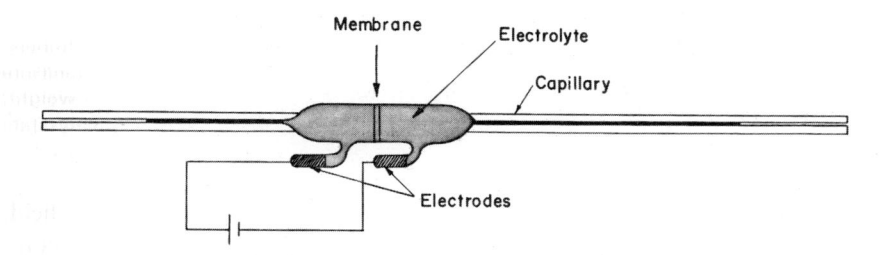

Fig. 6.36. A simple arrangement by which gravitational flow is avoided by the displacement of the electrolyte from the cathode to the anode region occurs at one level. The change in position of the melt in the capillary indicates the amount of electrolyte displaced.

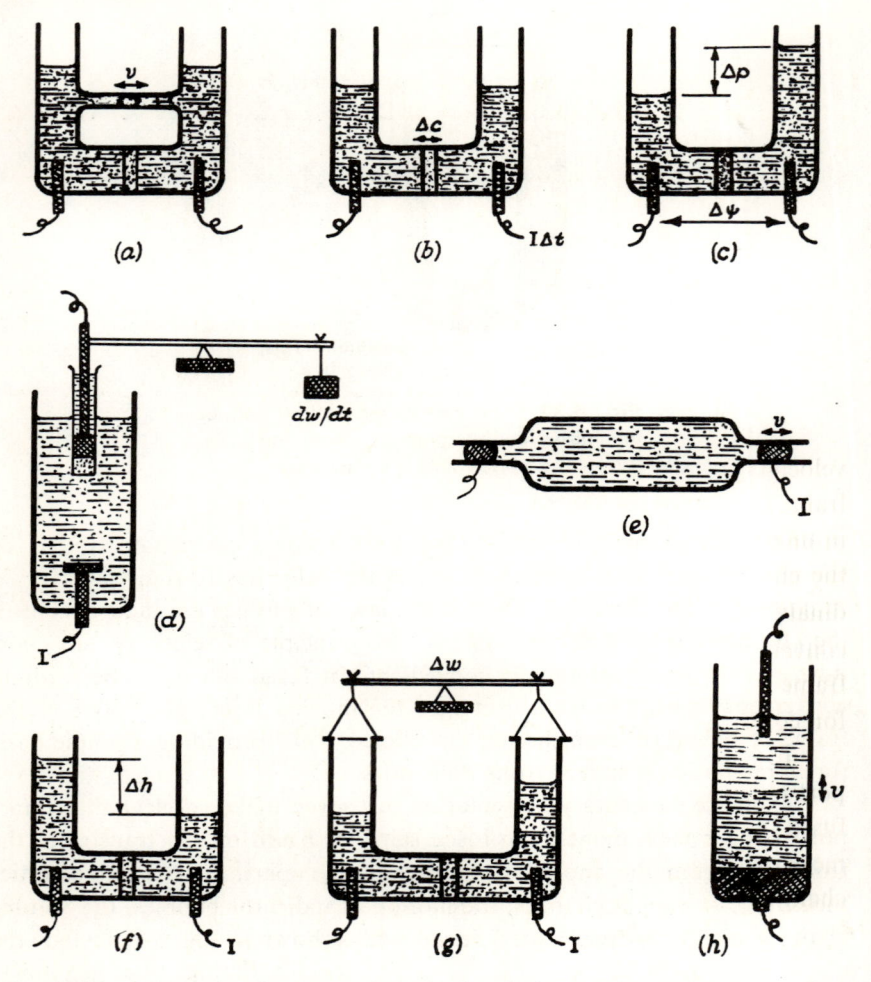

Fig. 6.37. Schematic diagrams of methods of determining transport numbers: (a) Measure velocity of the bubble; (b) measure transfer of the tracer; (c) measure the potential difference due to pressure difference; (d) measure the change in weight; (e) measure the transport of liquid metal electrodes; (f) measure the steady-state level; (g) measure the change in weight; (h) measure the moving boundary.

apparatus of Duke and Laity, Bloom, and other pioneers in this field.

The migration of the electrolyte from the anode to the cathode compartments can also be followed by using radioactive tracers and tracking their drift. Since isotopic analysis methods are sensitive to *trace* concentrations, there is no need to wait for the electrolyte migration to be large enough for visual detection.

TABLE 6.25
Some Transport Number Results for Fused Salts

Molten salt	Transport number of cation
$LiNO_3$	0.84
$NaNO_3$	0.71
$AgNO_3$	0.72
NaCl	0.87
KCl	0.77
AgCl	0.54

The results of some transport-number measurements are given in Table 6.25.

6.4.7c. A Radiotracer Method of Calculating Transport Numbers in Molten Salts. In the discussion of the applicability of the Nernst–Einstein equation to fused salts, it was pointed out that the deviations could be ascribed to the paired jump of ions resulting in a currentless diffusion.

With fused NaCl as an example, it has been shown that there is a simple relation between the experimentally determined equivalent conductivity Λ' and the experimental diffusion coefficients of the Na^+ and Cl^- as indicated by radiotracer Na^+ and Cl^-. The relation is

$$\Lambda' = \frac{zF^2}{RT}(D'_{Na^+} + D'_{Cl^-}) - \frac{2zF^2}{RT}D_{NaCl} \tag{6.68}$$

From this expression, it is clear that one can determine D_{NaCl}. Knowing D_{NaCl}, one can obtain the diffusion coefficients $D_{Na^+,ind}$ and $D_{Cl^-,ind}$, of the independently jumping Na^+ and Cl^- ions from the relations (6.63) and (6.64), i.e.,

$$D_{Na^+,ind} = D'_{Na^+} - D_{NaCl} \tag{6.74}$$

and

$$D_{Cl^-,ind} = D'_{Cl^-} - D_{NaCl} \tag{6.75}$$

Further, by using the Einstein relation [Eq. (4.169)] and the relation between absolute and conventional mobilities, one has

$$(u_{conv})_{Na^+} = z_{Na^+}e_0(\bar{u}_{abs})_{Na^+} = z_{Na^+}e_0\frac{D_{Na^+,ind}}{kT}$$

$$= \frac{z_+F}{RT}D_{Na^+,ind} \tag{6.76}$$

TABLE 6.26
Comparison of the Transport Number of Na⁺ in Molten NaCl Calculated from Equation (6.78) with Measured Values

	Transport number of Na⁺ in NaCl
Calculated from Eq. (6.78)	0.71
Measured values	$\begin{cases} 0.62 \\ 0.87 \end{cases}$

and similarly,

$$(u_{\text{conv}})_{\text{Cl}^-} = \frac{z_+ F}{RT} D_{\text{Cl}^-,\text{ind}} \tag{6.77}$$

With these values of mobilities, the transport numbers can easily be calculated from the standard formulas

$$t_+ = \frac{u_+}{u_+ + u_-} \quad \text{and} \quad t_- = \frac{u_-}{u_+ + u_-} \tag{6.73}$$

which, in the case of NaCl ($z_+ = z_- = 1$), reduces to

$$t_{\text{Na}^+} = \frac{D_{\text{Na}^+,\text{ind}}}{D_{\text{Na}^+,\text{ind}} + D_{\text{Cl}^-,\text{ind}}} \quad \text{and} \quad t_{\text{Cl}^-} = \frac{D_{\text{Cl}^-,\text{ind}}}{D_{\text{Na}^+,\text{ind}} + D_{\text{Cl}^-,\text{ind}}} \tag{6.78}$$

The comparison between transport numbers calculated in this way and those obtained by some of the experimental methods used is shown in Table 6.26.

6.4.7d. A Stokes'-Law Approach to a Rough Estimate of Transport Numbers. When an ion (of charge $z_i e_0$ and radius r_i) drifts in an electrolyte of viscosity η, the Stokes mobility is given by [cf. Eq. (4.180)]

$$u_{\text{conv}} = \frac{z_i e_0}{1800\pi\eta r_i}$$

$$= \frac{z_i k}{r_i} \tag{6.79}$$

where

$$k = \frac{e_0}{1800\pi\eta} \tag{6.80}$$

Upon introducing Stokes' mobilities into the expression (6.73) for transport

TABLE 6.27

Testing of Stokes' Law on Transport Numbers

	Transport number of Cl$^-$ in PbCl$_2$
Calculated from Eq. (6.81)	0.25
Measured	0.37

numbers, the result is

$$t_+ = \frac{u_+}{u_+ + u_-} = \frac{z_+k/r_+}{(z_+k/r_+) + (z_-k/r_-)}$$

$$= \frac{z_+r_-}{z_+r_- + z_-r_+} \tag{6.81}$$

Or, for 1:1 electrolytes,

$$t_+ = \frac{r_-}{r_+ + r_-} \tag{6.82}$$

For the radii of the ions, the crystallographic values can be used and, thus, the transport numbers calculated approximately. Table 6.27 shows the comparison between observed and estimated values. The reasonable agreement indicates that this approach is useful to provide *a first rough approximation* of the transport numbers (for it has been seen in Section 6.4 that Stokes' law has only partial applicability to molten electrolytes).

Further Reading

1. W. Klemm and W. Biltz, *Z. Anorg. Allgem. Chem.*, **152**: 255 and 267 (1926).
2. F. R. Duke and R. Laity, *J. Am. Chem. Soc.*, **76**: 4046 (1954); *J. Phys. Chem.*, **59**: 549 (1955).
3. H. Bloom and N. J. Doull, *J. Phys. Chem.*, **60**: 620 (1956).
4. B. R. Sundheim, *J. Phys. Chem.*, **60**: 1381 (1956); **61**: 485 (1957).
5. A. Z. Borucka and J. A. Kitchener, *Proc. Roy. Soc. (London)*, **A241**: 554 (1957).
6. G. J. Janz, C. Solomons, and H. J. Gardner, *Chem. Rev.*, **58**, 461 (1958).
7. R. W. Laity, *Ann. N.Y. Acad. Sci.*, **79**: 997 (1960).
8. T. B. Reddy, *Electrochem. Technol.*, **1**: 325 (1963).
9. A. Klemm, in M. Blander, ed., *Molten Salt Chemistry*, Interscience Publishers, Inc., New York, 1964.
10. B. R. Sundheim, Chapter 3 in: B. R. Sundheim, ed., *Fused Salts*, McGraw-Hill Book Company, New York, 1964.
11. W. K. Behl and J. J. Egan, *J. Phys. Chem.*, **71**: 1764 (1967).

6.5. THE ATOMISTIC VIEW OF TRANSPORT PROCESSES IN SIMPLE IONIC LIQUIDS

6.5.1. Holes and Transport Processes

Some facts about transport processes in molten salts have been mentioned (Section 6.4). The degree to which the hole model (Section 6.2) serves to give an approximate rational basis for these must now be examined. However, at first, it is necessary to cast the hole model into a form suitable for the prediction of transport properties.

The starting point for the derivation of expressions for transport properties on the basis of the hole model is the molecular-kinetic expression (Appendix 6.3) for the viscosity η of a fluid, i.e.,

$$\eta = 2nm\langle\omega\rangle\lambda \tag{6.83}$$

where n and m are the number per unit volume and the mass of the particles of the fluid, $\langle\omega\rangle$ is the mean velocity of the particles, and λ is their mean free path.

The quantity λ is linked to the picture of the phenomenon of viscosity in the kinetic theory of gases. According to this picture (Fig. 6.38), a fluid in motion is considered to consist of layers (of fluid) lying parallel to the direction of flow. (The slipping and sliding of these layers against each other provides the macroscopic explanation of viscosity). When particles jump between neighboring layers, there is momentum transfer between these layers, the cause of viscous drag (Fig. 6.39). In this picture, $\langle\omega\rangle$ is the component of the average velocity of the particles in a direction normal to the layers.

Irrespective of whether the gas as a whole is in motion or not, the particles constituting the gas are executing random motion. The particles of a *flowing* gas have a drift superimposed upon this random walk. It is by the random walk of the particles from one layer to another that the momentum transfer between layers is carried on. This momentum transfer manifests itself as the viscosity of the fluid.

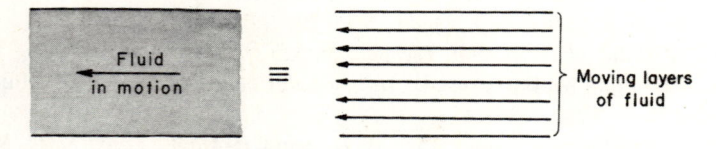

Fig. 6.38. A fluid in motion is considered equivalent to moving layers of fluid, the layers lying parallel to the flow direction.

IONIC LIQUIDS   575

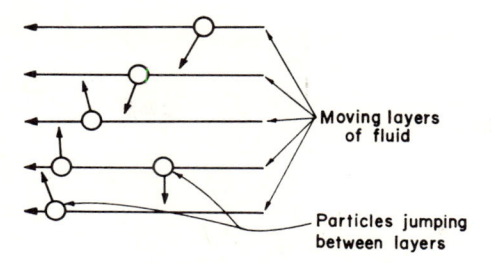

Fig. 6.39. Viscous forces are considered to arise from the momentum transferred between moving fluid layers when particles jump from one layer to another.

Holes also move. As argued earlier (*cf.* Section 6.3.1), anything which moves at finite velocities must have an inertial resistance to motion, i.e., a mass (see Appendix 6.1). Holes, therefore, have masses and momenta.

According to the hole theory, holes play the role in pure ionic liquids which molecules play in gases. Thus, it is considered that the random walk of holes between adjacent layers results in momentum transfer and, therefore, viscous drag in a moving fused salt (Fig. 6.40). The expression for the viscosity of an ionic liquid, on the basis of this model, is

$$\eta = 2n_h m_h \langle \omega \rangle \lambda_h \tag{6.84}$$

where n_h and m_h are the number per unit volume and the apparent mass for translational motion of the holes.

Now, the velocity component $\langle \omega_h \rangle$ is given by the ratio of λ_h, the mean distance between collisions (i.e., the mean free path), to τ, the mean time between collisions,

$$\langle \omega_h \rangle = \frac{\lambda_h}{\tau} \quad \text{or} \quad \langle \omega_h \rangle \lambda_h = \langle \omega_h \rangle^2 \tau \tag{6.85}$$

Fig. 6.40. According to the hole model, viscous drag arises from the momentum transferred between moving fluid layers when holes jump from one layer to another.

576 CHAPTER 6

Hence, the viscosity can be written as follows

$$\eta = 2n_h\tau(m_h\langle\omega_h\rangle^2) \tag{6.86}$$

The theorem of the equipartition of energy can now be applied to the one-dimensional motion referred to by $\langle\omega_h\rangle$

$$\tfrac{1}{2}m_h\langle\omega_h\rangle^2 \sim \tfrac{1}{2}kT \tag{6.87}$$

and, using this equation, one has

$$\eta = 2n_h kT\tau \tag{6.88}$$

6.5.2. What Is the Mean Lifetime of Holes in Fused Salts?

The parameter τ now invites consideration. In the gas phase, τ is simply the mean time between collisions. What is the significance of τ in an ionic liquid?

In a liquid, on the present model, τ would be the mean lifetime of a hole, i.e., the average time between creation and destruction of a hole through thermal fluctuation. To calculate this, one may use the formula for the number of particles escaping from the surface of a body per unit time per unit area into empty space, i.e.,

$$a = c\left(\frac{kT}{2\pi m}\right)^{\frac{1}{2}} e^{-A/RT} \tag{6.89}$$

where c is the number of particles per unit volume, m is the mass of a particle, and A is the work necessary to remove a mole of particles from the surface to an infinite distance. Here, A is also the work necessary for a mole of particles lying on the surface of the hole to be released into its interior.

In a time t, $4\pi\langle r_h\rangle at$ particles escape from the exterior into a spherical hole of radius $\langle r_h\rangle$. The hole will be filled by these particles if this number is equal to the number of particles in a sphere of radius $\langle r_h\rangle^3 c$, $\tfrac{4}{3}\pi\langle r_h\rangle$. Then, the time for destruction of the hole is

$$\tau = \frac{1}{3}\frac{\langle r_h\rangle c}{a} \tag{6.90}$$

Obviously, this is also the time for hole formation, and the lifetime is consequently

$$\tau = \frac{\langle r_h\rangle}{3}\left(\frac{2\pi m}{kT}\right)^{\frac{1}{2}} e^{A/RT} \tag{6.91}$$

This is the mean lifetime of a hole. It is seen, therefore, that the hole theory represents a Swiss-cheese sort of model of a liquid, with *flickering* holes.

But now, before one leaves expression (6.91), it is well to note the simplicity with which A has been treated. It is the heat term associated with getting a hole "unmade," with collapsing the hole, the negative of this work of forming the hole. But it has not yet been said how this will be calculated, and what terms go into this. Such a calculation will be one test which will be made of the hole theory here.

6.5.3. Expression for Viscosity in Terms of Holes

By inserting the expression [Eq. (6.91)] for the mean lifetime of a hole into Eq. (6.88), one obtains the hole-theory expression for viscosity. Thus,

$$\eta = 2n_h kT \left[\frac{1}{3} \langle r_h \rangle \left(\frac{2\pi m}{kT} \right)^{\frac{1}{2}} e^{A/RT} \right]$$

or

$$= \frac{2}{3} n_h \langle r_h \rangle (2\pi mkT)^{\frac{1}{2}} e^{A/RT} \qquad (6.92)$$

There are two quantities on the right-hand side of Eq. (6.92) which need discussion. They are n_h, the number of holes per unit volume of the liquid, and A, which occurs in the Boltzmann factor $\exp(-A/RT)$, for the probability of a successful filling of a hole.

6.5.4. The Diffusion Coefficient from the Hole Model

Now that the viscous flow properties of an ionic liquid have been discussed, the next task is to derive an expression for the diffusion coefficient.

The elementary act of a transport process consists of hole formation followed by a particle jumping into the hole. The focus in this elementary act has hitherto been the center of the hole.

But what is the situation at the original site of the jumping ion? Alternatively stated: What has happened, as a consequence of the jump process, at the point where the ion was before it jumped? At this prejump site, there has been precisely that moving away of particles from a point which corresponds to hole formation (Fig. 6.41).

Thus, when a particle jumps, it leaves behind a hole. So then, instead of saying that a transport process occurs by particles hopping along, one could equally well say that the transport processes occur by holes moving.

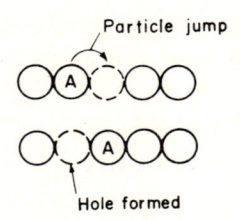

Fig. 6.41. When a particle jumps into a hole, there is the formation of a hole at the prejump site of the particle.

The concept is commonplace in semiconductor theory, where the movement of electrons in the conduction band is parallel by a movement of so-called "holes" in the valence band. It has already been assumed at the commencement of the viscosity treatment (Section 6.5.1) that the viscous flow of fused salts can be discussed in terms of the momentum transferred between liquid layers by moving holes.

The upshot of this analysis is that, when diffusion of particles occurs, there is a corresponding diffusion of holes. Hence, instead of treating ionic diffusion, one can consider hole diffusion and write Stoke's law [Eq. (4.176)] for the diffusion coefficient of holes

$$D = \frac{kT}{6\pi \langle r_h \rangle \eta} \tag{6.93}$$

But the hole-theory expression for viscosity is known. It is Eq. (6.92). Let this be introduced into Stoke's law [Eq. (6.93)]. The result is, using Eq. (6.29)

$$D = \frac{kT}{4\pi \langle r_h \rangle^2} \frac{(2\pi mkT)^{-1/2}}{n_h} e^{-A/RT}$$

$$= \frac{1}{4\pi (0.51)^2} \frac{\gamma}{n_h} (2\pi mkT)^{-1/2} e^{-A/RT}$$

$$= 0.31 \frac{\gamma}{n_h} (2\pi mkT)^{-\frac{1}{2}} e^{-A/kT} \tag{6.94}$$

The number of holes per unit volume can be expressed in terms of the known volume expansion of the liquid at a temperature T over that of the solid at the same T, ΔV_T, divided by the mean hole volume, i.e. $\Delta V_T / \langle v_h \rangle$, and reduced to the number per cubic centimeter, by dividing $\Delta V_T / v_h$ by the molar volume of the liquid at T, V_T. Hence,

$$D = 0.31 \gamma \frac{v_h V_T}{\Delta V_T} (2\pi mkT)^{-\frac{1}{2}} e^{-A/RT} \tag{6.95}$$

TABLE 6.28

Energies of Activation of Cation and Anion for Self-Diffusion in Simple Molten Electrolytes

Molten salt	E_D, kcal mole^{-1}	
	Cation	Anion
NaCl	7.1 ± 0.3	7.4 ± 0.8
KCl	6.9 ± 0.5	7.1 ± 0.5
CaCl$_2$	6.1 ± 0.7	8.9 ± 0.9
SrCl$_2$	5.4 ± 0.9	6.9 ± 0.7
BaCl$_2$	9.0 ± 1.2	9.5 ± 1.0
CdCl$_2$	6.8 ± 0.6	6.8 ± 1.0

Utilizing the value of the hole volume derived by substituting for $\langle r_h \rangle$ from Eq. (6.29) in $v_h = \frac{4}{3}\pi \langle r_h \rangle^3$:

$$D = \frac{0.17 kTV_T}{(2\pi m)^{\frac{1}{2}} \Delta V_T \gamma^{\frac{1}{2}}} e^{-A/RT} \quad (6.96)$$

The identity of the rate of hole diffusion and ionic diffusion is recalled. Thus, the final expression for the diffusion coefficient of ions in a fused salt is the same as that for holes, i.e., Eq. (6.96).

The first point to note about this expression, apart from the fact that it is of the form of the experimentally observed variation of D with temperature, i.e., $D = D_0 e^{-E_D/RT}$, is that the energy of activation for self-diffusion of cations and anions should be the same [cf. Eq. (6.33)]. This is essentially what is observed (Table 6.28).

Before going on to discuss the term A in terms of the structure of the salt, one must note what is meant by the term "Arrhenius activation energy." This term arose from the work of the early gas kineticists, who wrote equations of the type

$$\text{Rate} = Ze^{-E/RT}$$

and considered Z to have a negligible temperature dependence. Such an assumption would, of course, give

$$E = -R \frac{\partial \ln D}{\partial (1/T)} \quad (6.97)$$

and often it is simply this coefficient which is identified with the "energy of activation." Therefore, if one wants to calculate what a theory gives for this, one has to take into account whatever temperature dependence is possessed by the pre-exponential factor in the theory. Thus, if one knew (as one hopes to know later on in this chapter) a theoretical expression for A, of (6.96), one would have to calculate

$$-R\frac{\partial \ln D}{\partial (1/T)}$$

from (6.96) (including the effect of the temperature dependence of the pre-exponential) and compare its theoretical value with the experimental value of E calculated from Eq. (6.97). From (6.96),

$$-R\frac{\partial \ln D}{\partial (1/T)} = A + RT - RT^2\left(\frac{1}{\Delta V_T}\frac{d\Delta V_T}{dT} + \frac{1}{2\gamma}\frac{d\gamma}{dT} - \frac{1}{V_T}\frac{dV_T}{dT}\right) \tag{6.98}$$

In the following sections, a value of A will be calculated. Utilized in the right-hand side of (6.98), it gives there the theoretical prediction of the experimental energy of activation, i.e. the left-hand side of Eq. (6.98) (*cf.* 6.97).

6.5.5. A Critical Test of a Model for Ionic Liquids is a Rationalization of the Heat of Activation of $3.7RT_m$ for Transport Processes

It has been pointed out (Sections 6.4.2b and 6.4.3) that the heats of activation for viscous flow, and for self-diffusion, are given by the empirical generalization $E_\eta = E_D = 3.7RT_m$, where T_m is the melting point in degrees Kelvin. Some of the data which support this statement are plotted in Figs. 6.23 and 6.25. The empirical law seems to be applicable for all nonassociated liquids, i.e., not only ionic liquids. An empirical generalization which encompasses the rare gases, organic liquids, the molten salts, and the molten metals is a challenge for theories of the liquid state, and hence for fundamental electrochemists interested in pure liquid electrolytes.

Several *approaches* to the theory of liquids can be distinguished (*cf.* Section 6.2.4). One may begin by expressing the properties of the liquid in terms of a distribution function (an expression which indicates the probability of finding particles at a distance r from a central reference ion). Ideally, one should be able to calculate the distribution function itself from a knowledge of the intermolecular forces between the particles. However, this is in fact usually too difficult a task, and one falls back upon an experimental determination of this function (X rays), or, to making educated guesses about the scenarios inside the liquid, i.e., one agrees to assume temporarily some model of the structure, and then goes to develop the mathematical consequences of the assumption. The results of predictions from such developments for alternative models (*cf.* Sections 6.2.2 and 6.2.3) are compared with experimental data and then one may decide upon the most consistent model and use it as a working hypothesis which gives a rough idea of the structure of the liquid until absolute calculations of this are possible (*cf.* Section 6.2.4).

Tests of models have been referred to in the above-named sections. In general, the present comparative status of the various viewpoints is that some of them do give rise to reasonably close calculations of equilibrium properties, e.g., compressibility and surface tension. What it has not been possible to do up to recent times, however, is validly to rationalize the data of irreversible phenomena, e.g., conductance, viscous flow, and self-diffusion. For this reason, the discovery in 1963–1965 of the $3.7RT_m$ laws (Nanis, Richards) is of particular significance: it puts a simple, clear, and very challenging target for testing models of liquids.

6.5.6. An Attempt to Rationalize $E_D = E_\eta = 3.7RT_m$

In order to attempt to find what the hole theory predicts for the heat of activation in viscous flow, it is necessary to attempt to utilize this theory to calculate the term A in, e.g., Eq. (6.96). The meaning of A has already been defined in Section 6.5.2. It is the work done in transferring a mole of particles from the surroundings of a hole into its interior.

An *assumption* will now be added to the model. It is that, near or at the melting point, a hole is annihilated by the "evaporation" into it of *one* particle, i.e., one particle just fills it. There is no violation of physical sense in this assertion, for use of (6.29) shows that the size of the holes predicted by the hole theory is near to that of the ions which are supposed, in the model, to jump into them. Correspondingly, the work done to annihilate a hole is numerically equal to the work done in forming a surface of radius r_T, namely $4\pi \langle r_T \rangle^2 \gamma$, where γ is the surface tension. Hence, if n_T particles must jump into a hole to annihilate it at temperature T

$$\frac{n_T A_T}{N_A} = 4\pi \langle r_T \rangle^2 \gamma_T \qquad (6.99)$$

where A_T is the term A for the temperature T. Hence, with the assumption made ($n_{m.p.} = 1$),

$$A_{m.p.} = N_A 4\pi \langle r_{m.p.} \rangle^2 \gamma_{m.p.} \qquad (6.100)$$

What of n_T at T's other than the melting point, at which temperature it has been assumed to be unity? From (6.29), the hole volume (and hence its surface area) should *increase* as T increases (and γ decreases, as it does with increase of T). Of course, ions surround the hole, and it seems reasonable to suppose that, as the hole volume increases, the number of ions which surround the hole increases, and the number which is needed to fill it (thus causing the work $4\pi\langle r_T\rangle^2 \gamma_T$) increases.

Let it be assumed that the difference (ΔV_T) between the volume of the liquid salt and that of the corresponding solid is due only to holes. Then, the number of holes, per mole of salt, is, at temperature T,

$$\frac{\Delta V_T}{\langle v_{h,T}\rangle} \tag{6.101}$$

where $\langle v_{h,T}\rangle$ is obtained from (6.29).

Thus, the number of ions per hole at T is

$$\frac{2N_A}{\Delta V_T/\langle v_{h,T}\rangle} \tag{6.102}$$

Hence,

$$\frac{\text{Ion per hole at } T}{\text{Ions per hole at m.p.}} = \frac{\Delta V_{\text{m.p.}}}{\Delta V_T} \frac{\langle v_{h,T}\rangle}{\langle v_{h,\text{m.p}}\rangle} \tag{6.103}$$

One assumes that the number of ions which will be needed to fill the hole would be proportional to the number of ions per hole. Thus,

$$\frac{n_T}{n_{\text{m.p.}}} = \frac{\Delta V_{\text{m.p.}} v_{h,T}}{\Delta V_T v_{h,\text{m.p.}}} \tag{6.104}$$

But

$$n_T A_T = 4\pi N_A \langle r_T\rangle^2 \gamma_T \tag{6.105}$$

$$A_T = \frac{4\pi N_A \langle r_T\rangle^2 \gamma_T \Delta V_T v_{h,\text{m.p.}}}{\Delta V_{\text{m.p.}} v_{h,T}} \tag{6.106}$$

with $n_{\text{m.p.}} = 1$.

Thus [and with (6.29)]

$$\frac{A_T}{A_{\text{m.p.}}} = \frac{T_{\text{m.p.}}^{\frac{1}{2}}}{T^{\frac{1}{2}}} \frac{\Delta V_T}{\Delta V_{\text{m.p.}}} \frac{\gamma_T^{\frac{3}{2}}}{\gamma_{\text{m.p.}}^{\frac{3}{2}}} \tag{6.107}$$

One can obtain numerical values for the term on the right. It has been calculated from experimental data for some fourteen simple molten salts, and, if one restricts the range of experimental data used to about

IONIC LIQUIDS 583

200°C above the melting point, it is found that

$$\frac{T_{\text{m.p.}}^{\frac{1}{2}}}{T_T^{\frac{1}{2}}} \left(\frac{\gamma_T}{\gamma_{\text{m.p}}}\right)^{\frac{3}{2}} \frac{\Delta V_T}{\Delta V_{\text{m.p.}}} \simeq 1 \qquad (6.108)$$

Under such circumstances from (6.107) and (6.108)

$$A_T = A_{\text{m.p.}} \qquad (6.109)$$

But

$$A_{\text{m.p.}} = 4\pi \langle r_{\text{m.p.}} \rangle^2 \gamma_{\text{m.p.}} \qquad (6.110)$$

From (6.29)

$$A_{\text{m.p.}} = 4\pi (0.51)^2 k T_{\text{m.p.}}$$

$$= 3.3 k T_{\text{m.p.}} \text{ per ion} \qquad (6.111)$$

$$= 3.3 R T_{\text{m.p.}} \text{ per gram ion} \qquad (6.111a)$$

But, from (6.109),

$$A_T = A_{\text{m.p.}} = 3.3 R T_{\text{m.p.}} \qquad (6.112)$$

This result may be compared with the empirical heat of activation by substituting it in (6.98). Thus, the terms on the left of (6.98), the experi-

TABLE 6.29

Comparison of Abilities of Three Models of Ionic Liquids to Predict Data

Quantity	Hole	Vacancy	Liquid free volume
Molar vol. change on fusion	Good	Not yet calc.	Good
Molar entropy change on fusion	Fair	Very good	Theory not yet formulated
Compressibility	Very good	Not yet calc.	Not yet calc.
Expansivity	Very good		
D_0, E_D	D_0 good, $E_D = 3.74 RT_m \times \left[1 + \frac{T_m}{\gamma_m}\left(\frac{\partial \gamma}{\partial T}\right)\right]$ excellent	D_0 good, E_D qualitative	Satisfactory, but predicts curvature of log D vs $\frac{1}{T}$
E_η	Excellent	Qualitative	

mental values, are found to be about $3.7RT_{m.p.}$. Using $A = 3.3RT_{m.p.}$, for which the derivation (Emi) has been given here, one obtains agreement between observed and calculated values of the heat of activation to within about 5–10%. (The correction terms in RT and RT^2 of (6.98) affect the situation little because they work in opposite directions.)

The theory of holes is thus able to give some account of the heat of activation in transport in ionic liquids. Nor is the major assumption of the derivation (that only one ion suffices to fill a hole at $T = T_{m.p.}$, but a larger number at higher T) a difficulty, for (so long as it is near to one) it is the *change* of this number with temperature (rather than its absolute value) which is of importance.

A brief comparison of three different important models of ionic liquids is given in Table 6.29.

6.5.7. The Hole Model, the Most Consistent Present Model for Liquid Electrolytes

The Swiss-cheese model approach is consistent with experimental parameters and structural data (e.g., X-ray experiments, which show that the distance apart of ions remains constant or decreases on melting) which is better than that of other models for liquid electrolytes at this time.

The ability to reproduce experimental data numerically is well shown in Table 6.30, which gives a comparison of experimental compressibilities

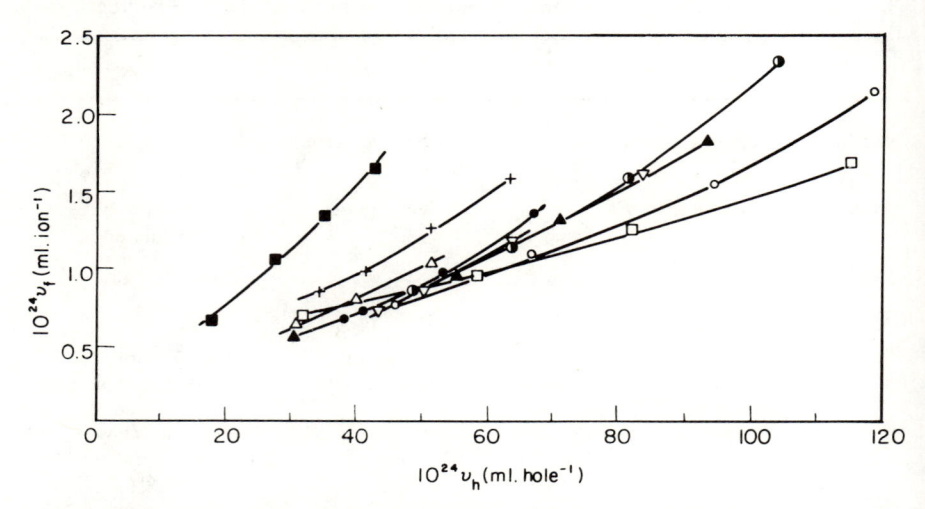

Fig. 6.42. Plot of the free volume per ion against the average hole volume; (■) LiCl, (△) NaCl, (●) KCl, (▲) CsCl, (+) NaBr, (▽) KBr, (◐) CsBr, (□) NaI, (○) KI.

TABLE 6.30
Comparison of Calculated Isothermal Compressibilities and Expansivities with the Experimental Values

Salt	Temp. °C	$10^{12}\beta$, calc, cm² dyn⁻¹	$10^{12}\beta$, obs, cm² dyn⁻¹	$10^4\alpha$, calc, °C⁻¹	$10^4\alpha$, obs, °C⁻¹
LiCl	614	20.8	19.4		
	800	30.5	24.7	4.0	3.0
	900	39.1	28.6		
	1000	47,9	33.0	4.0	3.3
NaCl	800	27.8	28.7	3.3	3.6
	900	36,9	22.8		
	1000	49,6	40.0	3.6	3.9
KCl	772	27.6	36.2		
	800	30.2	38.4	2.8	3.9
	900	42.3	45.7		
	1000	56.7	54.7	3.2	4.2
CsCl	642	20.2	38.0		
	800	39.0	51.8	4.2	4.1
	900	57.0	62.6		
	1000	72.5	76.3	4.4	4.5
CdCl₂	600	31.0	29.8	2.7	2.4
	700	37.3	33.1	2.7	2.4
	800	46.9	36.9	2.9	2.5
NaBr	747	32.8	31.6		
	800	40.1	33.6	3.2	3.1
	900	47.4	38.6		
	1000	59.5	44.9	4.0	3.4
KBr	735	28.0	39.8	2.5	3.7
	800	34.5	43.8	2.6	3.8
	900	51.3	52.1		
	1000	69.9	62.1	3.0	4.1
CsBr	636	56.3	49.1	4.0	4.4
	800	76.0	67.1	4.2	4.6
	900	97.5	82.7		
	1000	130.5	103.1	4.4	5.1

with the values which the hole theory yields. An interesting aspect of the evidence for this model is the relation of the free volume (*cf.* Fig. 6.13) to the volume of the expansion of melting. The free volume, in the sense referred to here, is the space which is free to each atom on the average. This relation is shown in Fig. 6.42. The continuous increase of the free volume with the hole volume would be consistent with a model in which, for approximately every fifth atom, there is a hole, so that, when a vibrating atom comes into contact with this space, its free volume is increased. The free volume is thus related to the hole volume.

One must not give the reader the impression that all is calm in the field of molten electrolytes. Firstly, the model which gives the greatest degree of agreement with experiment is a very crude one indeed, and suffers a fairly large objection because it is, in a sense, not a proper model and attempts to deal in a curious—if perhaps ingenious—way by analogy with the bubble in a near-boiling liquid. One might at first not take it seriously, but one's opinion may change when one sees the ability of the model to predict experimental data, without the calibration of any constants, and in such superiority (with respect to transport phenomena) to other present models, which do not yet yield an attempted theoretical value of the heat of activation for transport, or of the volume change on melting.

Further consideration shows that there is a reasonable likelihood that a liquid molten salt is *really* somewhat as the hole theory indicates. The ions tend to cling together in clusters, and between these are many gaps and cavities of varying sizes, in rapid change. It is difficult to avoid such a model if one is to attain consistency with the considerable increase of volume on melting *and* the fact that the internuclear distance decreases or stays the same. No *type* of model is consistent with these facts except a hole model, and, after that, it is largely a matter of how to describe it in terms of physical chemistry. Of course, a much better mathematical treatment than that above must be given, but there are difficulties. In some respects, the model bears resemblance to that containing molecular clusters in gases.

In this chapter, an attempt has been made to give some account of the structure of liquid electrolytes. One does not imply, necessarily, that the hole theory may be invoked as a helpful description of all nonassociated liquids. In ionic liquids, the lattice cement is stickier and hence there is more tendency to cluster than, say, in liquid argon where the free space demanded by expansion may be made up by increasing the gap between the atoms.

Further Reading

1. N. E. Richards, *Proc. Roy. Soc. (London)*, **A241**: 44 (1957).
2. H. Eyring, *J. Chem. Phys.*, **4**: 249 (1937); *Proc. Nat. Acad. Sci.*, **44**: 683 (1958); **46**: 333 (1960).
3. H. Reiss, H. L. Frisch, E. Helfand, and J. L. Lebowitz, *J. Chem. Phys.*, **32**: 119 (1960).
4. G. W. Hooper, *Discussions Faraday Soc.*, **32**: 318 (1962).
5. L. Nanis, *J. Phys. Chem.*, **67**: 2865 (1963).
6. F. H. Stillinger, Chapter 1 in: M. Blander, ed., *Molten Salt Chemistry*, Interscience Publishers, Inc., 1964.
7. H. Bloom, Chapter 1 in: B. R. Sundheim, ed., *Fused Salts*, McGraw-Hill Book Company, New York, 1964.
8. S. R. Richards, *J. Phys. Chem.*, **69**: 1627 (1965).
9. J. D. Bernal, *Discussions Faraday Soc.*, **43**: 60 (1967).
10. K. D. Luks and H. T. David, *Ind. Eng. Chem. Fundamentals*, **6**: 194 (1967).
11. R. Vilcu and C. Misdolea, *J. Chem. Phys.*, **46**: 906 (1967).
12. H. T. Davis and J. McDonald, *J. Chem. Phys.*, **48**: 1644 (1968).

6.6. MIXTURES OF SIMPLE IONIC LIQUIDS—COMPLEX FORMATION

6.6.1. Mixtures of Simple Ionic Liquids May Not Behave Ideally

A measure of understanding has been gained of the structure and transport properties of simple ionic liquids. In practice, however, mixtures of simple liquid electrolytes are more important. One reason for their importance is that mixtures have lower melting points. But what happens when two ionic liquids, for example, $CdCl_2$ and KCl, are mixed together?

Consider, for instance, the electrical conductance of fused $CdCl_2$ and KCl mixtures. If the equivalent conductivity of the mixtures (at a fixed temperature) were given by a simple additivity relation, then a linear variation of equivalent conductivity with mole fraction of KCl should be observed (dotted line in Fig. 6.43). The straight line should run from the equivalent conductivity of pure liquid $CdCl_2$ at a particular temperature to that of pure liquid KCl at the same temperature. Some binary mixtures of single ionic liquids do indeed exhibit the simple additivity implied by the dotted line of Fig. 6.43.

There are, however, other systems in which deviations occur from a simple additive law for conductance. The system of $CdCl_2$ and KCl is a case in point (full line in Fig. 6.43). A minimum in the conductivity curve is observed. What is the significance of this minimum?

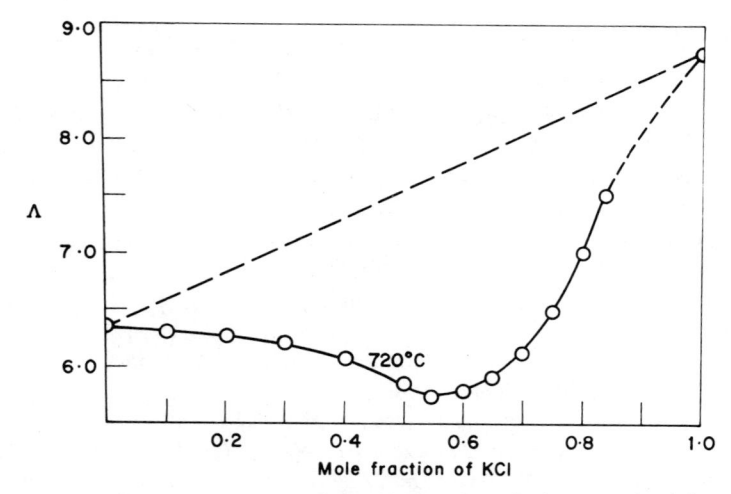

Fig. 6.43. The variation of the observed equivalent conductivity of CdCl$_2$–KCl mixtures as a function of the mole fraction of KCl. The dotted line corresponds to the variation which would be given by the additivity behavior.

6.6.2. Interactions Lead to Nonideal Behavior

The situation is reminiscent of some happenings in aqueous solutions. At infinite dilution of, say, a solution of KCl, the properties (e.g., equivalent conductivity) due to the K$^+$ and Cl$^-$ ions are additive (see Section 4.3.10). This is ideal behavior (see Section 3.4.1). With increasing concentration, however, there is a departure from ideality: the equivalent conductivities are not simply additive.

The nonideality in aqueous solutions was ascribed to *interactions* between K$^+$ and Cl$^-$ ions, and the ion–ion interaction theories were evolved for aqueous solutions. Thus, a whole gamut of interactions was considered. A pair of ions could experience long-distance coulombic forces. There could be short-range forces between them. The electrostatic attraction between a pair of oppositely charged ions could overwhelm thermal jostling and result in the formation of ion pairs. And, finally, the ions could undergo chemical binding and be joined in complex ion formation.

One can resort to similar explanations of departures from ideality in mixtures of simple ionic liquids. But there are specific differences between the situation in fused salts and that in aqueous solutions. In pure liquid KCl, there is no concentration variable, and therefore, fused KCl has a single value of equivalent conductivity (at a particular temperature). The mean distance between the K$^+$ and Cl$^-$ ions cannot be altered, as it can be in aqueous solutions, by interposing varying amounts of solvent because there is no solvent. Hence, the equivalent conductivity of pure liquid KCl embodies the effects of all the possible interactions for the temperature concerned. The thermodynamics of mixtures of molten salts has been intensively studied by Kleppa and by Bloom.

The interactions which one proposes to account for the deviations from ideality in mixtures of ionic liquids are interactions between the ions of one component of the mixture considered as a solvent and the ions of the other component which is added. In the case of KCl added to pure $CdCl_2$, one can consider, for example, the interactions between Cd^{++} ions and Cl^- ions.

6.6.3. Can One Meaningfully Refer to Complex Ions in Fused Salts?

It is intended here to discuss nonideality arising from complex ion formation. In the case of mixtures of pure liquid electrolytes, however, the idea of complex ion formation raises some conceptual problems.

Consider complex ion formation in the $CdCl_2$-KCl system, and let it be assumed for the moment that a $CdCl_3^-$ complex ion is formed. If such complex ions were formed in an aqueous solution of $CdCl_2$ and KCl, they would exist as little islands separated from other ions by large expanses of water. In fused salts, there are no oceans of solvent separating the ions. Thus, a Cd^{++} ion would constantly be coming into contact, on all sides, with chloride ions, and yet one singles out three of these Cl^- ions and says that they are part of (or belong to) a $CdCl_3^-$ complex ion (Fig. 6.44). It appears that, in the absence of the separateness possible in aqueous solutions, the concept of complex ions in molten salts is suspect. As will be argued below, however, what is dubious turns out to be not the concept but the

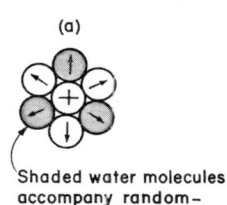

(a) Shaded water molecules accompany random-walking ion

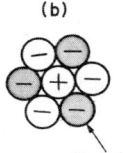

(b) Shaded ligands accompany random-walking ion

TABLE 6.31

Lifetime of Complex Ion in Molten Salts

Complex ion	Molten solvent	$\tau_{\text{complex ion}}$, sec
CdBr$^+$	KNO$_3$–NaNO$_3$	0.3, at 263 °C

comparison of complex formation in fused salts with complex formation in aqueous solutions.

It is more fruitful to compare complex formation in ionic liquids with the phenomenon of hydration of ions in aqueous solution (Chapter 2). It will be recalled that, though an ion was seen as constantly nudged by the water molecules of the surrounding medium, some of the water molecules were considered part of the primary hydration sheath of the ion. The criterion by which the two kinds of water were distinguished was that all those water molecules which surrendered their translational degrees of freedom and participated in the random walk of the ion constituted its solvation sheath. Thus, a solvated ion is an entity in which a certain number (the primary solvation number) of solvent molecules participate in the random walk of the ion.

Similarly, a complexed ion is an entity in which a certain number of *ligand* ions (e.g., three chloride ions in a CdCl$_3^-$ complex) participate in the random walk of the ion (i.e., the Cd^{++} ion in the CdCl$_3^-$ complex). The other Cl$^-$ ions only undergo promiscuous contacts with the Cd^{++} ion of the complex, not long-term affairs. The implication is that a complex ion (i.e., the ion and its ligands) is an entity with a lifetime that is at least several orders of magnitude longer than the time required for a single vibration. Experiments on the lifetimes of complex ions confirm this view (Table 6.31).

From the standpoint of this comparison (Fig. 6.44), it is seen that the concept of a complex ion in a molten salt is at least as tenable as that of a hydrated ion in aqueous solutions. But what experimental evidence exists for complex ions in fused salt mixtures? To answer this question, one must discuss some results of investigating the structure of mixtures of simple ionic liquids.

6.6.4. Raman Spectra, and Other Means of Detecting Complex Ions

Some of the techniques for examining the constitution of electrolytes have been described in Section 3.8. In particular, the usefulness of Raman spectroscopy for the detection of complex ions was stressed. The unique advantage of this spectroscopic method is that analysis of its data may give the formula of the complex entity present, together with detailed information on its structure, for example, the angles between the constituent bonds in a complex ion.

A simple comparative procedure in utilizing the Raman method is to compare the spectra obtained in *fused* salts with those obtained from the salts in other states of existence (aqueous solutions, solid state, or gas phase), in which the spectra have been shown by independent methods to be due to certain identified species.

An example of identification by this comparative method is given by the study of liquid mercuric halides $HgCl_2$ and $HgBr_2$ performed by Janz. Pure liquid mercuric halides yield Raman spectra which show Raman frequencies identical to those from the solid. In the crystalline state, however, it is known that the chloride and the bromide have distinctly molecular lattices with linear X—Hg—X (X = Cl or Br) triatomic molecules. It can therefore be concluded that the essential elements of the solid structure are retained in the liquid state, i.e., the mercuric halides from "molecular" melts, not ionic liquids.

Upon the addition of KCl to a melt of $HgCl_2$, the frequencies corresponding to the linear triatomic molecule persist but with diminishing intensity. On the other hand, new frequencies appear that indicate the formation of new species, i.e., complex ions (*cf.* Section 3.8). These new frequencies can be attributed to tetraatomic and pentatomic complex ions, $HgCl_3^-$ (planar) and $HgCl_4^{2-}$ (tetrahedral), respectively (Fig. 6.45).

The case of the $CdCl_2$–KCl system can be quoted as an example of the absolute method of using Raman spectroscopy for the identification of complex ions in fused salt mixtures (*cf.* Chapter 3). The Raman spectra of the $CdCl_2$–KCl system in the liquid state give four Raman peaks for the 50% KCl system in the liquid state (Fig. 6.46). It was concluded from these frequencies that a pyramidal $CdCl_3^-$ is the predominant complex ion

Planar
$HgCl_3^-$

Tetrahedral
$HgCl_4^-$

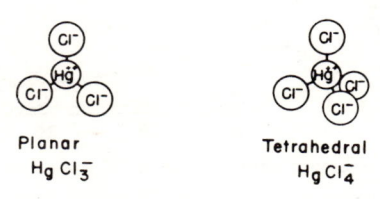

Fig. 6.45. Planar $HgCl_2$ and tetrahedral complex ions.

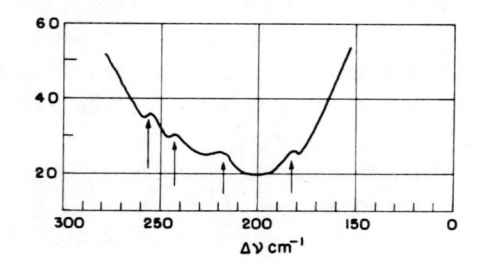

Fig. 6.46. The Raman spectrum of a CdCl$_2$–KCl system (with 50% KCl) showing four absorption peaks.

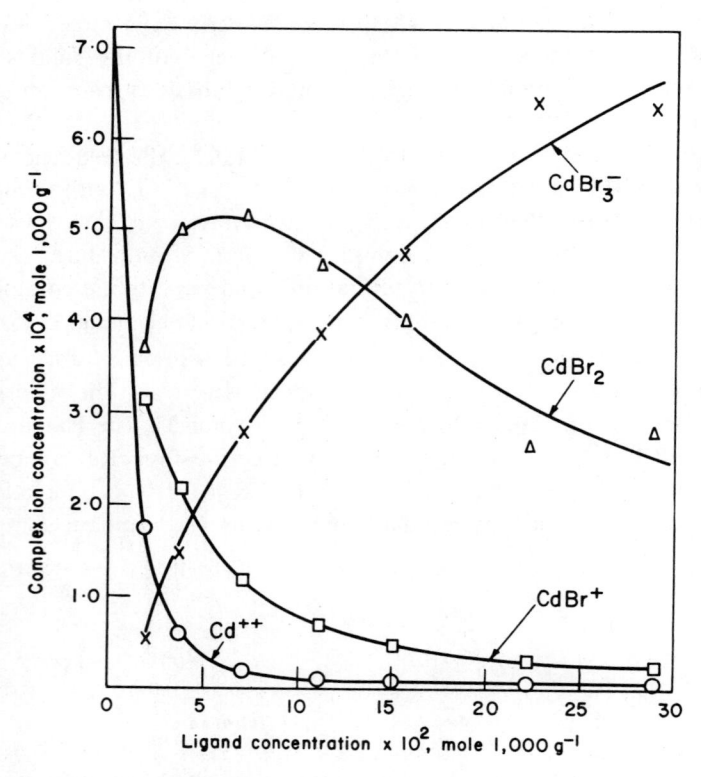

Fig. 6.47. Dependence of the concentration of various complex ions Cd(Br)$_x$ upon the addition of KBr.

in molten $CdCl_2$–KCl over the composition range examined, i.e., 33.3 to 66.6%.

Other methods of investigation (e.g., studies of the self-diffusion coefficient of Cd in the mixtures) had suggested that the behavior of the $CdCl_2^-$–KCl system could be interpreted in terms of the formation of complex ions. However, such methods only give a qualitative indication of the presence of a complex ion and indicate its variation with composition. Raman spectroscopy, on the other hand, has two distinct advantages. Firstly, the spectra reveal in an unambiguous way the presence of complex ions in molten salts; new frequencies (indicated by peaks in the intensity *versus* wavelength curves) arising from the addition of the ligand must be attributed to the formation of new species. Secondly, the number, height, and frequencies of the peaks can be compared with the corresponding quantities which are indicated by the theory of Raman spectra for various assumed ions. In this way, one can get comparatively detailed knowledge regarding the nature of the complex ions present. For example, in $CdCl_2$–KCl mixtures, it is possible to distinguish between the presence of tetrahedral and planar forms of assumed complexes with the same molecular formula.

A misapprehension sometimes exists that a molten mixture of two salts can give rise only to a single species of complex ion. This is not usually the case, as may be seen by the results of an examination of the complex ions formed by adding halide ligands to liquid cadmium nitrate dissolved in an $LiNO_3$–$NaNO_3$ solution. Which of several possible ions is formed depends largely on the amount of halide ions present per cadmium atom (Fig. 6.47).

Further Reading

1. W. Bues, *Z. Anorg. Allgem. Chem.*, **279**: 104 (1955).
2. D. Inman, *J. Electroanal. Chem.*, **5**: 476 (1963).
3. S. C. Wait, Jr., and G. J. Janz, *Quart. Rev. (London)*, **17**: 225 (1963); Chapter 5 in: H. A. Szymanski, ed., *Raman Spectroscopy*, Plenum Press, New York, 1967.
4. G. E. Walrafen and D. E. Irish, *J. Chem. Phys.*, **40**: 911 (1964).
5. K. Balasubramanyam, *Electrochim. Acta*, **8**: 621 (1963); *J. Chem. Phys.*, **40**: 2657 (1964); **42**: 676 (1965).
6. J. Lumsden, *The Thermodynamics of Molten Salt Mixtures*, Academic Press, Inc., New York, 1966.
7. H. Bloom, *The Chemistry of Molten Salts*, W. A. Benjamin, Inc., New York, 1967.
8. J. H. R. Clark and C. Solomons, *J. Chem. Phys.*, **47**: 1823 (1967).
9. C. Solomons, J. H. R. Clark, and J. O'M. Bockris, *J. Chem. Phys.*, **49**: 445 (1968).

6.7. MIXTURES OF LIQUID OXIDE ELECTROLYTES

6.7.1. The Liquid Oxides

Fused *salts* (and mixtures of fused salts) are not the only type of liquid electrolytes. Mention has already been made of *fused oxides* and, in particular, mixtures of fused oxides.

A typical fused oxide system is the result of mixing intimately a nonmetallic oxide (SiO_2, GeO_2, B_2O_3, P_2O_5, etc.) and a metallic oxide (Li_2O, Na_2O, K_2O, MgO, CaO, SrO, BaO, Al_2O_3, etc.) and then melting the mixture. The system can be represented by the general formula: M_xO_y–R_pO_q, where M is the metallic element and R the nonmetallic element.

There is obviously a case for *mentioning* these fused oxide systems in this brief presentation of pure liquid electrolytes, but why give them a special consideration? Are not the concepts developed for the understanding of molten *salts* adequate for the understanding of molten oxides? The essential features of fused salts emerge from models of the liquid state and, in particular, from the hole model based on density fluctuations. Is the hole theory adequate as a basis for a rationalization of the behavior of the fused oxides?

6.7.2. Pure Fused Nonmetallic Oxides Form Network Structures Like Liquid Water

Some of the special features of molten oxides must now be described for it is these features which do not permit the hole model of ionic liquids to be applied to fused oxides in the way it is to molten salts—namely, that, as far as transport properties are concerned, hole formation is the principal feature of the model that is consistent with experiment.

The first interesting feature of molten silica is that its conductivity is more like that of water (i.e., molten ice) than that of fused NaCl (Table 6.32).

TABLE 6.32

Specific Conductivity of Water and Molten SiO_2 and NaCl near the Melting Point

Substance	$\varkappa$, ohm^{-1} cm^{-1}	Temp., °C
SiO_2	7.7×10^{-4}	1800
NaCl	3.6	801
H_2O	4×10^{-8}	18

IONIC LIQUIDS 595

The dissimilarity in the conductivities of liquid NaCl, on the one hand, and liquid water and liquid silica on the other, is of fundamental import. When NaCl is fused, the ionic lattice (the three-dimensional periodic arrangement of ions) is broken down (see Section 6.1.2) and one obtains an *ionic liquid*. On the other hand, when ice is melted, the tetrahedrally directed hydrogen bonding involved in the crystal structure of ice (Fig. 2.24) is partially retained. Thus, water is not a collection of *separate* water molecules but an association (based on hydrogen bonding) of water molecules in a three-dimensional network. The network, however, does not extend indefinitely. There is a periodicity and only short-range order implying a certain degree of bond breaking. It is this network structure that is responsible for the *small* mole fraction of free ions (H^+ and OH^-) in water, in contrast to the almost total absence of any ion association (into pairs, complexes, etc.) in liquid NaCl. This great difference in the concentration of charge carriers is responsible for the several-orders-of-magnitude difference in the specific conductivities of liquid NaCl and liquid water.

Now, the specific conductivities of water and of fused silica are quite similar. This suggests that the structures of *crystalline* water [Fig. 6.48(a)] and *crystalline* silica [Fig. 6.48(b)] have much in common. Each oxygen atom in ice is surrounded tetrahedrally by four other oxygens, the oxygen–

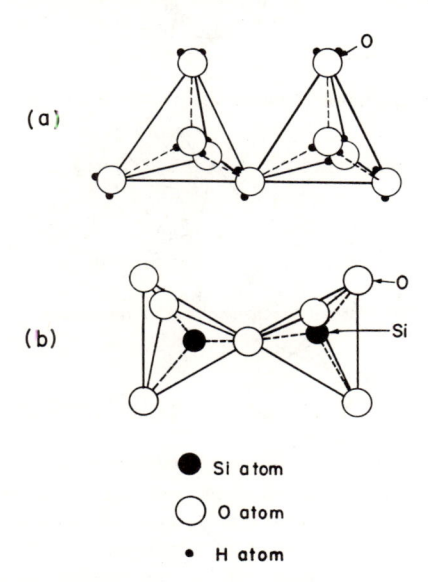

Fig. 6.48. The similarity between the basic building blocks of the ice and crystalline silica structures.

oxygen bonding occurring by a hydrogen bridge (the hydrogen bond). In crystalline silica, there are SiO$_4$ tetrahedra occurring through an oxygen bridge. The different forms of ice and the different forms of silica (Fig. 6.49) correspond to different arrangements of the tetrahedra in space.

It is reasonable, therefore, to consider that fused silica resembles liquid water. Just as liquid water retains, from the parent structure (ice), the three-dimensional network but not the long-range periodicity of the network, one would expect that liquid silica also retains the continuity of the tetrahedra, i.e., the space-network, but loses much of the periodicity and long-range order which are the essence of the crystalline state. This model of

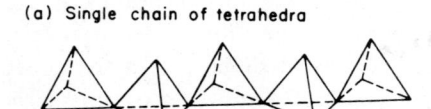

(a) Single chain of tetrahedra

(b) Double chain of tetrahedra

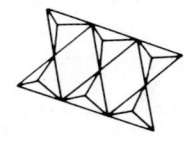

(c) Sheet of tetrahedra

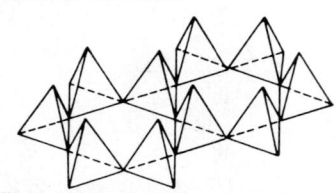

(d) Three-dimensional net of tetrahedra

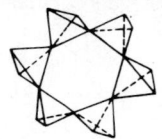

Fig. 6.49. Forms of silicates resulting from different ways of linking up SiO$_4$ tetrahedra: (a) single chain of tetrahedra; (b) double chain of tetrahedra, as in asbestos; (c) sheets of tetrahedra, as in clay, mica, and talc; and (d) networks of tetrahedra, as in ultramarine.

fused silica, based on keeping the extension of the network but losing the translational symmetry of crystalline silica, implies a low concentration of charge carriers in *pure* liquid silica and, therefore, the low conductivity in comparison with a molten salt (*cf.* Table 6.32).

6.7.3. Why Does Fused Silica Have a Much Higher Viscosity Than Do Liquid Water and the Fused Salts?

It has just been argued that the conductivities of simple ionic liquids, on the one hand, and liquid silica and water, on the other hand, are vastly different because a fused salt is an unassociated liquid whereas both molten silica and water are associated liquids with network structures. What is the situation with regard to the viscosities of fused salts, water, and fused silica? Experiments indicate that, whereas water and fused NaCl have similar viscosities, fused silica is a highly viscous liquid (Table 6.33). Here then is an interesting problem.

The theories of transport processes in liquids are based on elementary acts, each act consisting of two steps: (1) holes are formed; (2) then particles jump into these holes (*cf.* Section 6.5.4). For fused salts and other non-associated liquids, this theory based on holes was quite successful in explaining the movements and drift of particles. The *mean* volume of a hole is determined by the surface tension as follows [Eq. (6.29)]

$$\langle v_h \rangle = 0.55 \left(\frac{kT}{\gamma} \right)^{\frac{3}{2}}$$

from which it turns out (Table 6.11) that, in *fused salts*, the size of the holes is roughly equal to the size of the ions. Hence, holes can receive the jumping ions of the fused salt. Further, in simple ionic liquids, the free energy of activation for the jumping of ions into the holes is about an order of magni-

TABLE 6.33
Viscosities of Fused Silica, Water, and Fused NaCl

Substance	Temp., °C	Viscosity, poise
Liquid SiO$_2$	1720	3×10^6
Water	25	9×10^{-3}
Fused NaCl	850	13×10^{-3}

TABLE 6.34

Surface Tension of Molten SiO₂ and NaCl near the Melting Point

Molten salt	γ, dyne cm^{-1}	Temp., °C
SiO₂	250	1570
NaCl	114	801

tude smaller than the free energy for forming the holes. Once the hole is formed in a fused salt, the jump is easy.

The surface tension of fused silica is only about three times that of fused sodium chloride (Table 6.34). Hence [see Eq. (6.31)], in fused silica too, the range of sizes of holes present would be of atomic dimensions ($\sim$ 1 to 10 Å), as for fused ionic liquids.

In simple ionic liquids (e.g., NaCl), the holes are atomic sized; *so are the jumping ions.* That is why jumping is no problem. But what particles can jump into the atomic-sized holes of fused silica? Obviously, whole macro networks cannot jump into atomic-sized holes. How then can jump-dependent transport processes occur? There is a way. Small segments (one to a few atoms in size) can break off from the network, and these pieces (segments) can do the jumping (Fig. 6.50).

The comparison between transport processes in simple fused salts and in fused SiO₂ is interesting. In molten salts, the jump was relatively easy, but holes had to form. The rate of the whole process was controlled predominantly by this rate of formation. But now, with molten silica, the

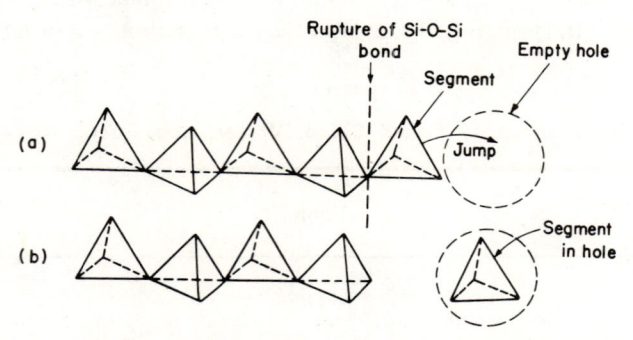

Fig. 6.50. Schematic diagram to show a segment of an SiO₄ chain breaking off and jumping into an empty hole; (a) before jump; (b) after jump.

TABLE 6.35

Comparison between Transport Processes in Simple Fused Salts and in Liquid Silica

	System	
	Fused salt	Liquid silica
Essence of situation	Particles waiting to jump into holes	Holes waiting for small enough segments of networks
Rate-determining step	Hole formation	Rupture of bonds between segments and network
Energy of activation for transport process (approx.)	$\Delta H_{\text{hole formation}}$	$\Delta H_{\text{bond rupture}}$

balance of influences is different. Holes are as easily formed as in the ionic liquids because the rate of hole formation is controlled by the vibrations of the atoms relative to each other, but it is difficult to produce the jumping particles by rupturing strong Si—O—Si bonds holding the network together (Table 6.35). Therefore, in the silicates, the rate-determining process is the bond-rupture step. Thus, in simple ionic liquids, the experimental activation energy[†] for the transport process, such as a viscous flow, is determined by the enthalpy of hole formation; in associated liquids with network structures, it is determined by the energy required to break the bonds of the network.

But, then, why is it that, even though both fused silica and water have network structures, their viscosities are quite different; in fact, the viscosity of water is similar to that of molten NaCl, which has no network? The difference in the viscosity behavior of water and of fused silica lies in the ease with which segments can be broken off the two networks. The Si—O—Si chemical bonds are much more difficult to rupture than the O—H—O hydrogen bonds (cf. Table 6.36). Thus, the small segments—probably H_2O molecules—are so easily produced in water that the holes do not have much

---

[†] It will be recalled that it was decided that the quantity obtained from the slope of the $\log \eta$ versus $1/T$ curve would be termed an *energy* of activation irrespective of whether it pertains to constant pressure or constant volume conditions, though, under the former, it is an enthalpy and, under the latter, it is an energy.

TABLE 6.36

Heat of Dissociation of Si—O and O—H Bonds

Bond	Heat of dissociation, kcal mole^{-1}
Si—O—Si	104
O—H—O, hydrogen bond	5

of a wait; an ease of flow, high fluidity or low viscosity, results. This is not the case with fused silica because of the much higher bond-breaking energy, and a high viscosity results.

Some support for the idea that the viscous-flow properties of associated liquids such as liquid silica and water are determined by the step of bond breaking rather than that of hole formation comes from the experimental plots of $\log \eta$ versus $1/T$. These plots suggest a slight trend away from linearity (Fig. 6.51), which is not the case for fused salts (Fig. 6.21). For

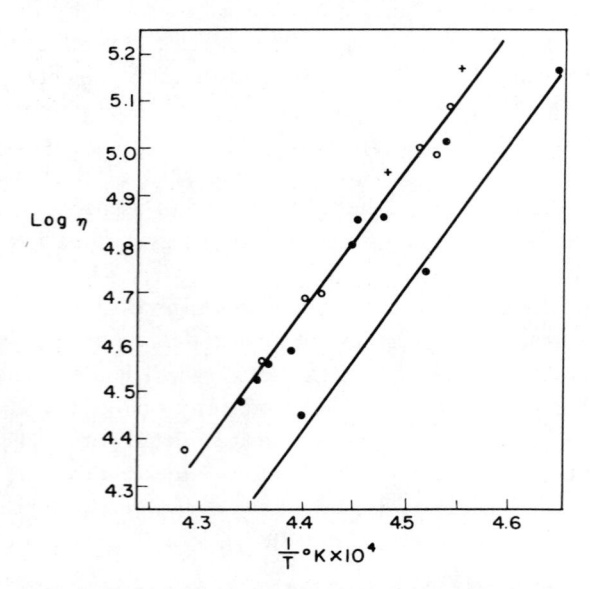

Fig. 6.51. Two independent sets of data for the difficultly determinable viscosity η of liquid SiO_2. Both suggest a slight tendency to curve in the sense that the energy of activation becomes higher at the lower temperatures.

water, too, the plot is curved slightly with the experimental energy of activation for viscous flow, $E_{\eta,p}$, decreasing with increasing temperature. The explanation for this phenomenon is as follows. Because the energy of activation for viscous flow depends upon the breaking of bonds and because, according to the Boltzmann distribution, the fraction of broken bonds increases with temperature, the fraction required to be broken by the shearing force in viscous flow decreases with an increase in temperature and, hence, there is a decreasing energy of activation with increasing temperature.

To summarize: Unlike fused salts, mixture of fused oxides are *associated* liquids. In fused oxides, hole formation is necessary but not the vital step which determines the rate of transport processes. It is the rate of production of individual small jumping units which controls them. This conclusion makes it essential to know what (possibly different) entities are present in fused oxides and what are the *kinetic* entities. In fused salts, the jumping particles are already present; the principal problem is the structure of the empty space or free volume or holes and the properties of these holes. In molten oxides, the main problem is to understand the structure of the macrolattices or particle assemblies.

6.7.4. The Solvent Properties of Fused Nonmetallic Oxides

If fused silica is a three-dimensional, aperiodic network, all the atoms are to some extent joined together, i.e., the liquid is a giant molecule. Can ions dissolve in such a structure? Water, too, is a network structure like SiO_2, and ions dissolve in water. Hence, liquid SiO_2 may well be expected to have solvent properties leading to the production of ionic solutions.

Water, it may be recalled (Chapters 2 and 5), has two modes of solvent action depending on the nature of the added electrolyte. The water can contact an ionic crystal (e.g., NaCl), detach the ions from the lattice through the operation of ion–dipole (or ion–quadrupole) forces, and convert them to hydrated ions (Chapter 2).

The water can also interact *chemically* with a potential electrolyte (e.g., CH_3COOH). The hydrogen atoms forming part of the hydroxyl group of the organic acid do not differentiate the oxygen atoms of the water network and those of the OH group. The hydrogen of the OH group detaches itself from the organic acid. Two ions are thus formed, a hydrogen bonded to a water molecule from the solvent, and an organic anion. This mode of solvent action is a proton-transfer or acid–base reaction.

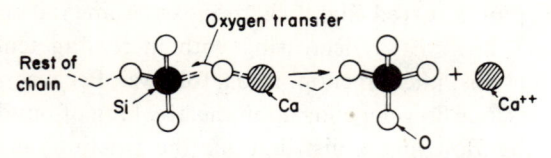

Fig. 6.52. The interaction between a metal oxide and a silicon atom of the silica network.

The type of solvent action which fused nonmetallic oxides have on metallic oxides may be likened to the second type of dissolution process, i.e., proton-transfer reactions. The process may be pictured as follows. The oxygens cannot discriminate between the metal ions M^+ (of the metallic oxide), with which they have been associated in the lattice of a metal oxide before dissolution, and the oxygen atoms of the SiO_4 tetrahedra contained in the solvent fused silica. The oxygens sometimes, therefore, leave the metal ions and associate with those of the tetrahedra. Dissolution has occurred with a type of oxygen-transfer reaction (see Fig. 6.52).

There is a further analogy between the solvent actions exercised by water and by a fused nonmetallic oxide. Just as water dissolves an electrolyte at the price of having its structure disturbed, so also the reaction resulting from the addition of a metallic oxide to a fused nonmetallic oxide like silica is equivalent to a bond rupture between the SiO_4 tetrahedrons (Fig. 6.53).

Solvent action occurs therefore in fused oxide systems along with a certain breakdown of the network structures present in the pure liquid solvent (e.g., in pure liquid silica).

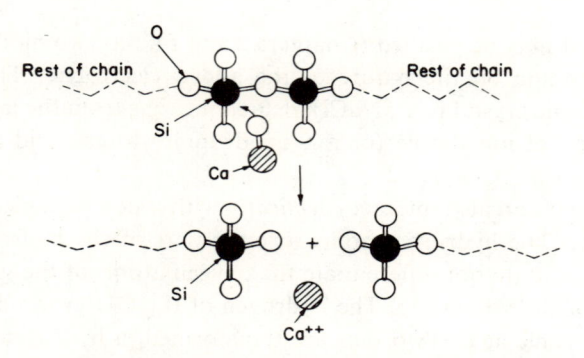

Fig. 6.53. The oxygen-transfer reaction leads to a rupture between the SiO_4 tetrahedra.

6.7.5. Ionic Additions to the Liquid-Silica Network: Glasses

An interesting aspect of molten oxide electrolytes may be mentioned at this point. Some liquids can appear to be solids, i.e., some solids are really liquids of such high viscosity that no significant flow occurs over tens or hundreds of years. These solidlike liquids are called *glasses*. The structure breaking which has just been described is an aspect of the basic mechanism behind the formation of glasses, which might be regarded as "cold molten silicates."

When ions with a relatively large radius and therefore small peripheral field are added to liquid silica, they produce structure breaking in the network. This is shown in a one-dimensional way in Figs. 6.52 and 6.53. With an increase in the number of ruptures in the network there is an increase in the number of "free" or "dangling" ends of the ruptured network. The network becomes, therefore, increasingly *distorted* with increase in the mole fraction of the metallic oxide present.

If, now, the "broken-down network" is at a sufficiently high temperature, it is known as a *liquid silicate*. Such a system results, for example, from adding an alkali oxide (e.g., Na_2O) in low concentration to SiO_2. The system can be, at sufficient temperature, distinctly a *liquid*, and the viscosity near the melting point may be, for example, 427 poises (at 1550 °C). When the temperature is dropped, the liquid silicate attempts to "freeze," but, to do this, the long-range order of the crystalline silicate has to be reestablished. The establishment of order, however, requires rearrangement of the structure, i.e., movements of the kinetic units of the broken-down network to get into the sites corresponding to order. Were these particles, or kinetic units, simple, they would be agile, i.e., their movements would be easy, the viscosity would be low, and the restoration of crystalline order would be accomplished almost immediately. A crystalline solid with sharp melting point would result.

But this reorganization is precisely what is not quickly accomplished by the entities in the broken-down networks in liquid silicates. The entities[†] resulting from the rupturing of three-dimensioned networks when metal oxides are added are big and sluggish. They cannot get into line in time; the viscosity is too high. The loss of thermal energy during cooling catches them still out of position as far as the regular three-dimensional arrangement of the crystalline silicate is concerned. Then it is too late for, as the

---

[†] The nature of these big anionic entities which exist in glass-forming silicates is discussed in Section 6.7.8.

temperature drops still further, they are still less likely to be able to get back into line. Thus, the loss of thermal energy freezes in the structure of the *liquid* silicate—a glass is formed. It is a "frozen liquid," i.e., a liquid which has been supercooled to such a high viscosity that it seems to have the essential requirement of a solid, absence of flow. A beam of X rays, however, would reveal an essential characteristic of the liquid state, namely, the absence of long-range periodicity.

If, however, sufficient time is allowed (e.g., a few hundred years), a sufficient number of the units of the broken-down network will get back into line. Long-range order will be partly reestablished; the glass *devitrifies* or "deglassifies."

How is the liquid-silicate network affected by the addition of various types of ions in the production of the peculiar and complicated kind of pure electrolyte, the glass? It is the answer to this structural question which provides the basis for the understanding of the *glassy* state.

6.7.6. The Extent of Structure Breaking of Three-Dimensional Network Lattices and Its Dependence on the Concentration of Metal Ions

The extent of breakdown of the network structures present in the pure liquid solvent can be treated in the following way. The SiO_4 tetrahedron is accepted as the basic structural unit in a mixture of a metallic oxide and fused silica. But are the tetrahedrons linked together at all, and, if so, how are they linked together? What is the number of links per silicon? Thus, water molecules are the basic structural unit in an aqueous solution, but extensive linkage and intermolecular bonding occurs in pure water. How is this affected by the presence of ions in solution (Section 6.1)?

In the fused-oxide system, the metal oxide is the structure breaker; in aqueous solutions, the electrolyte is the structure breaker. Does the extent of structural breakdown of the continuous Si—O network present in pure silica before addition of MO depend on the concentration of MO? The extent of breakdown depends on the concentration of the structure breaker; then one would expect that properties which depend on the size and nature of the structures present would also be concentration dependent.

In fused-oxide systems, a simple way of expressing the concentration of metal oxide M_xO_y in the fused nonmetallic oxide R_pO_q is sometimes used. This involves the so-called "O/R ratio." The O/R ratio is simply related to the mole fraction of the metallic oxide. For example, an O/Si ratio of 4 in a system of Li_2O and SiO_2 is obtained when the Li_2O has a mole fraction of 66% (i.e., $2Li_2O + SiO_2$ has four O's to one Si).

One way of probing the sizes of structures present in fused-oxide systems and their variation with the mole fraction of MO added to the nonmetallic oxide is through the variation of the ease of flow with composition. The viscosity of the system must be keenly sensitive to the size and nature of the kinetic entities present because it is these entities which must make the jumps from site to site involved in viscous flow.

The experimental results on the variation of the energy of activation for viscous flow, E_η, as a function of the mole per cent of the metal oxide, are shown in Fig. 6.54.

The basic feature of the results appears to be a very high ($\sim$ 150 kcal mole^{-1}) energy of activation for viscous flow of the pure nonmetallic oxide and then a rapid fall with addition of the metallic oxide, whereupon there is a leveling off to a value of about 40 kcal mole^{-1}, which remains relatively unchanged between about 10 and 50 mole % of Na_2O in SiO_2. This behavior can be used (Section 6.7.7) as a touchstone in deciding between alternate models for the structural changes accompanying changes in metal-ion concentration.

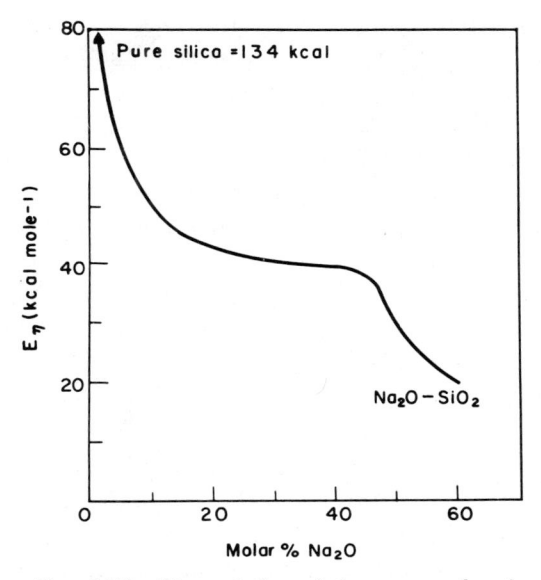

Fig. 6.54. The variation of the energy of activation for viscous flow in a $Na_2O + SiO_2$ melt as a function of the mole per cent of Na_2O.

The structural theories presented will be in terms of the liquid *silicates*—for most research in molten oxides has been done with them—but one can extend the basic structural ideas to fused-oxide systems involving metal oxides dissolved in B_2O_3 and P_2O_5 and probably to most liquid electrolytes in which there are largely continuous network structures at very low concentrations of M_xO_y.

6.7.7. The Molecular and Network Models of Liquid Silicate Structure

A naïve view of the happenings consequent to the addition of metallic oxides to molten silica is to think of different *uncharged* molecular species, the species changing with the mole fraction of the metallic oxide. This view—which was popular among metallurgists as late as 1950—has to be given up with alacrity because conductance studies of *mixtures* of M_xO_y and SiO_2 show that these systems are highly conducting and therefore rich in charge carriers (Table 6.37). One has to suggest ionic, rather than molecular, structures.

A second attempt at the interpretation of the structure of liquid silicates starts with a consideration of the curve of E_η versus mole % M_xO_y (Fig. 6.54). It is in terms of the gradual breakdown of the three-dimensional network of fused silica. Just as there is *thermal* bond breaking on going from crystalline to fused silica, one can consider that, with the addition of, say, Na_2O, the additional O atoms cause Si—O—Si bonds in the originally continuous network of SiO_2 to break. This gives structure breaking to various composition-dependent degrees.

A mole fraction of 66% of M_2O implies, as already stated, an O/Si ratio of 4 and must, therefore, be considered a composition at which simple SiO_4^{4-} ions exist.[†]

What model can be suggested for the corresponding structural changes "inside" the fused-oxide system for the M_2O mole-fraction range from 0 to 66%?

From the fact that there is such a sharp fall in the energy of activation for viscous flow between zero and about 10 mole % of M_2O (Fig. 6.54), one must think of a radical change (over this composition range) in the difficulty of causing a (possibly changing) kinetic entity to jump from site

---

[†] Correspondingly, for $M_2O > 66$ molar %, there are oxygen atoms in excess of the ability of the Si's present to coordinate them (O/Si ratio > 4). Hence, liquids with such compositions probably contain SiO_4^{4-} and O^{--} entities in addition to the cations M^+ present.

TABLE 6.37

Electric Conductance in the Liquid Silicates

Cation	Composition $XM_xO-YSiO_2$, $X:Y$	$\varkappa_{1750°C}$, ohm^{-1} cm^{-1}	$\Lambda_{1750°C}$	ΔH^*, kcal g^{-1} ion^{-1}
K$^+$	1:2	1.5	71.8	8.2
	1:1	2.4	82.7	8.0
Na$^+$	1:2	2.1	83.3	12.0
	1:1	4.8	126.0	13.5
Li$^+$	1:2	2.5	77.8	11.6
	1:1	5.5	109.0	10.6
	2:1	23.2	332.0	9.6
Ba^{++}	1:2	0.18	6.4	33.2
	1:1	0.60	16.2	17.5
	2:1	1.32	29.9	9.0
Sr^{++}	1:2	0.21	7.7	36.0
	1:1	0.63	15.7	26.7
	2:1	1.4	26.8	17.0
Ca^{++}	1:2	0.31	11.4	30.0
	1:1	0.83	18.4	20.0
	2:1	1.15	18.8	20.0
Mn^{++}	1:2	0.55	18.2	24.0
	1:1	1.8	35.1	16.0
	2:1	6.3	85.5	12.0
Fe^{++}	1:1	1.82	44.0	15.0
Mg^{++}	1:2	0.23	6.5	34.0
Mg^{++}	1:1	0.72	12.2	24.0
	2:1	2.15	24.7	17.0
Al^{+++}	10 wt %	3.10^{-3}	0.20	22.0
Ti^{++++}	10 wt %	6.10^{-4}	0.05	35.7

to site in the random walk which is the basis of diffusion (Section 6.5.4). The model must of course contain the explanation of the generation of more free ions to account for an increase of conductivity with an increasing amount of MO. The anions postulated as predominant for a given O/R ratio must meet some exacting requirements. Thus, (1) they must have formulas consistent with the overall O/R ratio; (2) they must have a total charge which compensates the total charge of the cations and thereby ensures overall electroneutrality; (3) they must be shaped in a way consistent with the bond angles, particularly the R—O—R angle, shown from X-ray data in the corresponding solids.

The broad approach by Zachariasen on the *network* model of liquid silicates is to break down the network present in the pure fused nonmetallic oxide thus

Three-dimensional network with some thermal bond breaking (e.g., SiO$_2$) $\xrightarrow{M_2O \text{ added}}$ Two-dimensional sheets

$\xrightarrow{\text{More } M_2O \text{ added}}$ One-dimensional chains $\xrightarrow{\text{Still More } M_2O}$ Simple SiO$_4^{4-}$ monomers

The details of the network theory of liquid-silicate structures, which was first suggested to rationalize the glassy state, are presented in Table 6.38. The chief defect of this model is that it argues for very large changes in the heat of activation for viscous flow in the composition range of 33 to 66% (but contrast the indication in Fig. 6.54). This is because, in the network model, the size and shape of the kinetic unit—the jumping entity—is supposed to undergo a radical change in the composition range of 33 to 66%. Thus (*cf.* Table 6.38), sheets are being broken up into chains. The kinetic unit of flow would therefore be expected, on this model, to change

TABLE 6.38

The Network Theory of Liquid Silicate Structure

Range of composition, mole % M$_2$O	Silicate entities present
0	Continuous three-dimensional (3-D) network of SiO$_4$ tetrahedrons with small degree of thermal bond breaking
0–33	Essential 3-D network of SiO$_4$ tetrahedrons with number of broken bonds equal to number of added O atoms from M$_2$O; end of 3-D boundary at 33%
33%	"Infinite" 2-D sheets of SiO$_4$ tetrahedrons; M$^+$ ions and O$^-$ ions between sheets
33–50%	Region of sheets and some chains of tetrahedrons
50%	Chains of infinite length
50–60%	Chains of decreasing length
66%	SiO$_4^{4-}$

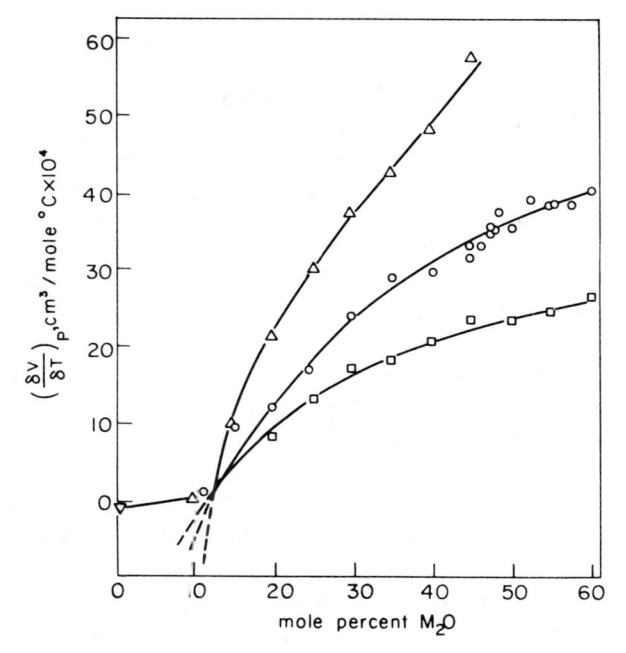

Fig. 6.55. The sharp change in the expansivity of M_2O–SiO_2 melts around the 10 mole % of M_2O composition; ($\triangle$) K_2O + SiO_2, ($\circ$) Na_2O + SiO_2, ($\square$) Li_2O + SiO_2, ($\triangledown$) SiO_2.

radically in size over *this* composition range, and this diminution in size of the flowing unit would be expected to make the heat of activation for viscous flow strongly dependent on composition in *this* composition range (33 to 66% M_2O).

In reality, however, the E_η changes by only 25% over the composition range of 10 to 50% M_xO_y (Fig. 6.54), whereas, from 0 to 10%, the change in energy of activation is about 200% (Fig. 6.54).

Another defect of the network model concerns the implications which it has for phenomena in the composition region around 10% M_xO_y. This is an important composition region. Experimentally, whether one measures the composition dependence of the heat of activation for viscous flow, of expansivity, of compressibility, or of other properties, in all cases, there is always a sharp and significant change (Fig. 6.55) around the 10% M_xO_y composition, which indicates a radical structural change at this point. But this composition has no special significance at all, according to the network model. Thus, one must—although the network model served well in

610 CHAPTER 6

an earlier stage of the development of the theory of glasses—reconsider the situation and develop a model which corresponds more, in its predictions, to the experimental facts.

6.7.8. Liquid Silicates Contain Large Discrete Polyanions

Consider the situation as one *decreases* the O/R ratio, i.e., *decreases* the mole per cent of the metallic oxide M_xO_y. Between 100 and 66%, there is little need for special modelistic thinking because the quadrivalency of silicon and the requirements of stoichiometry demand that the ionic species present are monomers of SiO_4^{4-} tetrahedrons.

It is in the composition range of 66 to 10% M_xO_y that the network model stumbles (*cf.* Section 6.7.7) in the face of facts.

What are the requirements of a satisfactory structural model for ionic liquids in this composition range? First, since transport number determinations show that the conduction is essentially cationic, the anions must be large in size compared with the cations (see Sections 4.4.8 and 4.5). Secondly, the marked change in properties (e.g., expansivity) occurring at 10% M_xO_y indicates radical structural changes in the liquid in the region of this composition. From the sharp rise of the heat of activation for the flow process to such high values (toward 100 kcal mole^{-1} at compositions below 10 mole % M_xO_y) with decreasing mole per cent of M_xO_y in this composition region, one suspects that the structural change which is the origin of all the sudden changes near 10 mole % M_2O must involve sudden aggregation of the Si-containing structural units into networks. The difficulty of bond breaking to get a flowing entity out of the network and into another site gives a rationalization to the very sluggish character of the liquid at 10 or less mole % of M_xO_y. Finally, from 66 to 10% *there must be no major changes in the type of structure*, except some increase in the *size* of the entities, because there is only a small increase in the heat of activation for viscous flow over this composition region. This relative constancy of the heat of activation for flow over this composition region means that the

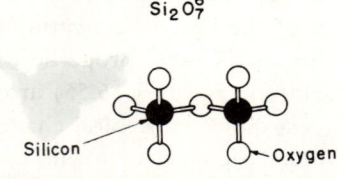

Fig. 6.56. The dimer ion $Si_2O_7^{6-}$.

various structural units present can become the kinetic entities of flow over this region without great change of the energy involved, i.e., the anions present must have similar structures.

The construction of a model can therefore start from the SiO_4^{4-} ions present at the *orthosilicate* composition (66% M_xO_y). With a decrease in the molar fraction of M_xO_y, the *size* of the anions must increase to maintain the stoichiometry. One can consider that there is a joining-up, or polymerization, of the tetrahedral SiO_4^{4-} monomers. For example, the dimer $Si_2O_7^{6-}$ (Fig. 6.56) could be obtained thus

$$SiO_4^{4-} + SiO_4^{4-} \xrightarrow{-O} {}^{3-}O_3Si—O—SiO_3^{3-}$$

This polymerization into larger polymerized anions (or polyanions) is the essential concept of the *discrete-polyanion* model suggested by Lowe, Mackenzie, *et al.* for the structure of mixtures of liquid oxides corresponding to composition greater than 10% M_xO_y and less than 66%.

At the outset, it seems no easy task to derive the structure of the polyanions predominant at each composition of the liquid oxides. However, several negative criteria limit the choice of possible polyanions. Thus, electroneutrality must be maintained at all compositions, i.e., the total charge on the polyanion group per mole must equal the total cationic charge per mole for a given composition. Since the cationic charge per mole must decrease with decreasing M_xO_y mole per cent, the negative charge on the polyanions per mole equivalent of silica must also decrease. Thus, the size of the polymerized anions must increase as the molar fraction of M_xO_y decreases.

After a dimer, i.e., $S_2O_7^{6-}$, is formed, the next likely anionic entity to appear as the M_xO_y/SiO_2 ratio falls might be expected to be the trimer

$$
{}^{3-}O_3Si—O—SiO_3^{3-} + SiO_4^{4-} \rightarrow {}^{3-}O_3Si—O—\underset{\underset{O^-}{|}}{\overset{\overset{O^-}{|}}{Si}}—O—SiO_3^{3-}
$$

Following the trimer, a polymeric anion with four units may be invoked to satisfy the requirements of electroneutrality and stoichiometry, etc., the general formula being $Si_nO_{3n+1}^{(2n+2)-}$. On this basis, however, when a composition of 50% M_xO_y is reached, i.e., when the mole fraction of M_xO_y is equal to that of SiO_2, the Si/O ratio is 1:3. However, from the general formula for the chain anion, i.e., $Si_nO_{3n+1}^{(2n+2)-}$, it is clear that O/Si = 3 when

$(3n + 1)/n = 3$, i.e., when $n \to \infty$. Hence, near to 50% M_xO_y, an attempt to satisfy the electroneutrality and stoichiometry by assuming that linear chain anions, extensions of the dimers and trimers above discussed, would imply a large increase in the energy of activation for viscous flow in the composition range (say 55 to 50 molar % of M_xO_y) because the linear polymer would here rapidly approach a great length. But no such sharp increase in the heat of activation for viscous flow is observed experimentally in the range of 55 to 50 molar % of M_xO_y. Hence, the composition range in the liquid oxides in which linear anions can be made consistent with the flow data is relatively small, between 66 and somewhat greater than 50 molar % of M_xO_y. The linear anion must be given up as a predominant anionic constituent before the metal oxide composition has dropped to 50 molar %.

An alternative anionic structure near the 50% composition which satisfies stoichiometric and electroneutrality considerations is provided by *ring formation*. If, in the composition range of 55 to 50% M_xO_y, the linear anionic chains (assumed to exist at compositions between some 50 and 60 mole per cent M_xO_y) link up their ends to form rings such as $Si_3O_9^{6-}$ or $Si_4O_{12}^{8-}$ (Fig. 6.57), then such ring anions satisfy the criteria of the O/Si ratio, electroneutrality, and also the Si—O—Si valence angle which X-ray data make expectable. Thus, the $Si_3O_9^{6-}$ anion corresponds to an O/Si ratio of 3, and, if one is considering a 50% CaO system, the charge per ring anion is 6−, and the charge on the three calcium ions required to give 3CaO/3SiO$_2$ is 6+. Further, the $Si_3O_9^{6-}$ anion is not very much larger than the $Si_2O_7^{6-}$ dimer, and, hence, there would not be any large increase in the heat of activation for viscous flow. With regard to the Si—O—Si bond angle, in the $Si_3O_9^{6-}$ and $Si_4O_{12}^{8-}$ ions, it is near that observed for the corresponding

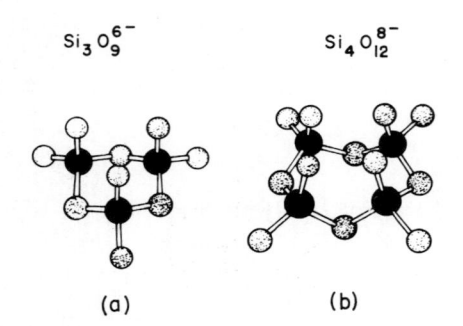

(a)   (b)

Fig. 6.57. The proposed ring anions: (a) $Si_3O_9^{6-}$ and (b) $Si_4O_{12}^{8-}$.

IONIC LIQUIDS 613

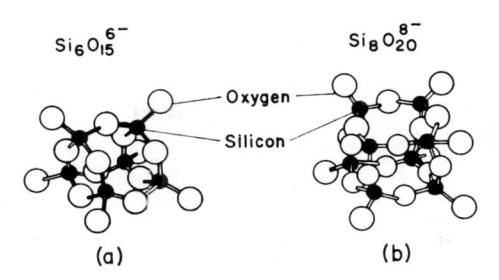

$Si_6O_{15}^{6-}$ $Si_8O_{20}^{8-}$

Fig. 6.58. The proposed large ring anions:
(a) $Si_6O_{15}^{6-}$ and (b) $Si_8O_{20}^{8-}$.

solids, i.e., the minerals wollastonite and poryphyrite, respectively, which are known to contain $Si_6O_9^{6-}$ and $Si_4O_{12}^{8-}$.

Further structural changes between 50 and 30% M_xO_y are based on the $Si_3O_9^{6-}$ and $Si_4O_{12}^{8-}$ ring systems. At the 33% compositions, polymers $(Si_6O_{15})^{6-}$ and $(Si_8O_{20})^{8-}$ (Fig. 6.58) can be postulated as arising from dimerization of the ring anions $Si_3O_9^{6-}$ and $Si_4O_{12}^{8-}$ (Fig. 6.57). As the M_xO_y concentration is continuously reduced, further polymerization of the rings can be speculatively assumed. For example, at $M_xO_y/3SiO_2$ when M_xO_y is 25%, the six-membered ring would have the formula $Si_9O_{21}^{6-}$ and would consist of three rings polymerized together (Fig. 6.59).

Ring stability might be expected to lessen with increase of size with increasing proportion of SiO_2 because the silicate polyanions which correspond to compositions approaching 10 molar % of M_xO_y would be very long. The critical 10% composition at which there is a radical change in many properties may be rationalized as that in the region of which a *discrete* polymerized anion type of structure becomes unstable (because of size) and rearrangement to the random three-dimensional network of silica occurs. That is, a changeover in structural type occurs to what is fundamentally the SiO_2 network with some bond rupture due to the metal cations. The very large increase in the heat of activation for flow which takes place at

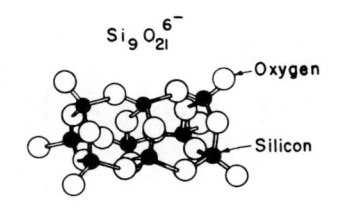

$Si_9O_{21}^{6-}$

Fig. 6.59. The proposed six-membered ring anion $Si_9O_{21}^{6-}$.

this composition (Fig. 6.54) would be consistent with this suggestion, as would also the sudden fall in expansivity.

These ideas of the discrete-polyanion model for liquid-silicate structure are summarized in Table 6.39. The description is, as in most models, highly idealized. Thus, all the silicate anions may not have the Si—O—Si angle of the crystalline state; only the *mean* angle may have the crystalline value. Further, the discrete anion suggested for a particular composition is only intended to be that which is there *dominant*, not exclusive. Mixtures of polymerized anions may be present at a given composition, the proportions of which vary with composition.

The discrete-polyanion model is a speculative one because no direct proof of the existence of, e.g., its ring-polymeric anions is available. It provides a much more consistent qualitative account of facts concerning the behavior of liquid silicates than does the network model. Thus, it predicts the observed marked changes in properties near 10% M_2O (Fig. 6.55), the relatively small variation in E_n over the concentration range of 10 to 50% (Fig. 6.54), etc. The structural suggestions for the nature of the anions receive some indirect support from solid-state structural analyses of certain mineral silicates.

TABLE 6.39

The Discrete Polyanion Model of Liquid Silicates

Range of composition, mole % M_2O	Type of silicate entities present
0	Continuous 3-dimensional networks of SiO_4 tetrahedrons with some thermal bond breaking and a fraction of SiO_2 molecules
0–10	Essentially SiO_4 network with number of broken bonds approximately equal to number of added O atoms from M_2O, having fraction of SiO_2 molecules and radicals containing M^+
10–33	Discrete silicate polyanions based upon a six-membered ring $(Si_6O_{15}^{6-})$
33–55	Mixture of discrete polyanions based on $Si_3O_9^{6-}$ and $Si_6O_{15}^{6-}$ or $Si_4O_{12}^{8-}$ and $Si_8O_{20}^{8-}$
~55–66	Chains of general form $Si_nO_{3n+1}^{(2n+2)-}$, e.g., $Si_2O_7^{6-}$
66–100	$SiO_4^{4-} + O^{2-}$ ions

6.7.9. The "Iceberg" Model

There are some facts, however, which cannot easily be rationalized in terms of the discrete-ion model.

First, the partial molar volume of SiO_2, related to the size of the SiO_2-containing entities, is relatively constant from 0 to 33 mole % M_2O (Fig. 6.60). On the basis of the discrete-ion model, the critical change at 10% M_xO_y involving the breakdown of three-dimensional networks and the formation of discrete-polyanions would require a decrease in molar volume of SiO_2 at about 10% for the following reason. The SiO_2 structure is a particularly open one and hence has a large molar volume; in contrast, a structure with discrete polyanions would involve a closing-up of some of the open SiO_2 volume and thus a decrease in partial molar volume compared with the SiO_2 networks. Secondly, it is a fact that, in certain ranges of composition, e.g., 12 to 33% M_2O, M_2O and SiO_2 are not completely *miscible*. The two liquids consist of an SiO_2 phase and a metal-rich phase. The discrete-anion model cannot accommodate this phenomenon.

It is suggested by White *et al.*, therefore, that, in the composition range of 12 to 33% M_2O, *two* structures are present, one similar to that which exists at 33% in the discrete-anion model. The other structure is a structure corresponding to glassy, or vitreous, silica, i.e., fused silica with the randomized three-dimensional networks frozen in. The vitreous silica is in the form

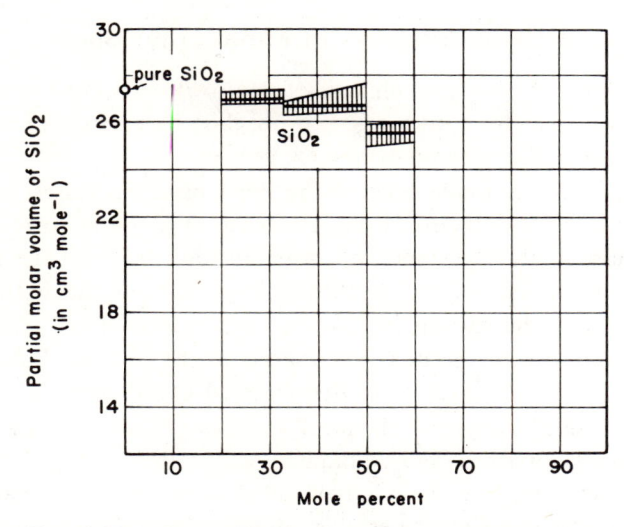

Fig. 6.60. The negligible change in the partial molar volume of an Li_2O–SiO_2 melt over the range of 0 to 33 mole % of Li_2O.

of "islets" or "icebergs." Hence, the name, the *iceberg model* of liquid-silicate structure. These icebergs are similar to the clusters that occur in liquid water (Section 6.7.2). The submicroscopic networks may be pictured as continually breaking down and reforming. Microphase regions of $M_xO_y \cdot 2SiO_2$ (the 33% structure) occur in the form of thin films separating the SiO_2-rich icebergs—hence, the possibility of a separation of the liquid into two phases, one rich in SiO_2 and the other in M_xO_y. Since most of the SiO_2 is present in the icebergs, the almost constant partial molar volume of SiO_2 in the icebergs has the same Si—O—Si angle as vitreous silica.

An estimate of the order of magnitude of the iceberg size can be made. For 12% M_2O, the radius of an (assumed spherical) iceberg is about 19 Å, and, at 33% M_2O, the iceberg of the iceberg model becomes essentially identical with the discrete ion of the discrete ion model, which has a radius of about 6 Å.

In the iceberg model, the structure of the medium on a microscale is heterogeneous. Flow processes would involve slip between the icebergs and the film. No Si—O—Si bonds need be broken.

At present, it seems that both the discrete-polyanion model *and* the iceberg model probably contribute to the structure of liquid silicates. In a sense, the iceberg model is the most complete model for it involves the discrete polyanions as well as the SiO_2 entities called icebergs.

6.7.10. Fused-Oxide Systems in Metallurgy: Slags

Knowledge of what goes on inside fused-oxide systems is of importance not only as a basis for future advances in glass technology (Section 6.6.4) but also to metallurgical processes.

Consider, for example, one of the most basic processes in industry, the manufacture of iron in the blast furnace (Fig. 6.61). Iron ore, coke, and flux (essentially limestone and dolomite) are fed into the top of the furnace. Compressed air fed in through openings in the bottom of the furnace converts the carbon in coke to carbon monoxide, which reduces the iron oxide to iron. Molten iron collects at the bottom. But, on top of the molten metal is a layer of *molten* material called *slag*.

What is slag? A typical chemical analysis (Table 6.40) shows that it consists mainly of silica SiO_2, alumina Al_2O_3, lime CaO, and magnesia MgO—in fact, precisely the kind of substances, i.e., nonmetallic oxides (e.g., SiO_2) and metallic oxides (e.g., CaO), the structure of which was under discussion here. (Some slags, in fact, might be regarded as molten glasses.) The constituents of the slags are present in the ores and in coke.

IONIC LIQUIDS 617

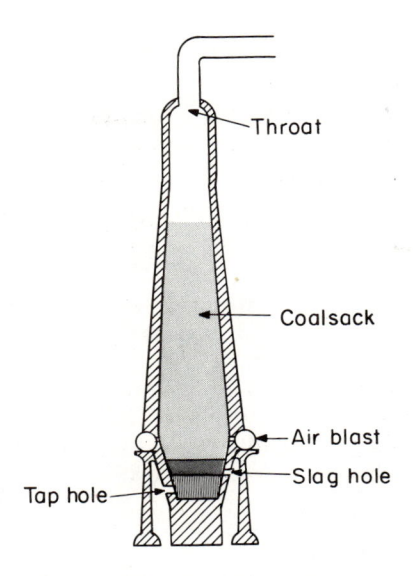

Fig. 6.61. Schematic of a blast furnace.

Successful operation of the furnace and production of the iron with the desired composition (and, hence, metallurgical properties) depends so much on making the *slag* with the right composition by controlling the raw materials fed into the furnace that it is sometimes said: "You don't make iron in the blast furnace, you make slag!"

Can one rationalize this importance of the slag? Measurements of conductance as a function of temperature and of transport number indicate that the slag is an *ionic* conductor (liquid electrolyte). But in the metal–slag interface, one has the classical situation (Fig. 6.62) of a metal (i.e., iron) in contact with an electrolyte (i.e., the molten oxide electrolyte, slag), with all the attendant possibilities of corrosion of the metal. Corrosion of metals is usually a wasteful process, but, here, the sequence of current-balancing partial electrodic reactions which make up a corrosion situation

TABLE 6.40

Analysis of a Typical Slag

Constituents	Fe	Mn	SiO_2	Al_2O_3	CaO	MgO	S
Percentage	0.34	1.18	34.67	14.58	44.78	3.21	1.36

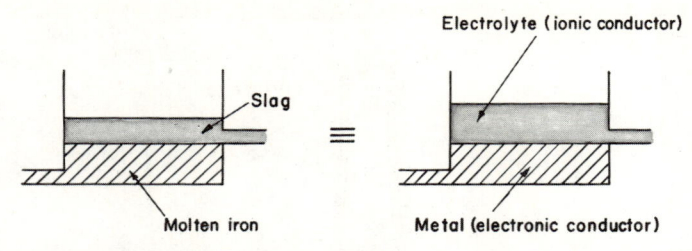

Fig. 6.62. Molten metal in contact with slag is electrochemically equivalent to a metal in contact with an electrolyte.

in fact are the factors which control the equilibrium of various components (e.g., S^{--}) between slag and metal and, hence, the properties of the metal, which depend so much on its trace impurities. For example,

$$S[\text{in metal}] + 2e \to S^{--}[\text{in slag}]$$
$$Fe \to Fe^{++} + 2e$$

The quality of the metal in a blast furnace is thus determined largely by electrochemical reactions at the slag–metal interface.

Further Reading

1. W. Zachariasen, *Z. Krist.*, **74**: 139 (1930); *J. Am. Chem. Soc.*, **54**: 3841 (1932).
2. K. Endell and J. Hellbrügge, *Naturwiss.*, **30**: 421 (1942).
3. A. E. Davies, *J. Chem. Phys.*, **19**, 225 (1951).
4. J. O'M. Bockris and D. C. Lowe, *Proc. Roy. Soc. (London)*, **A226**: 423 (1954).
5. J. D. Mackenzie, *Trans. Faraday Soc.*, **51**: 1734 (1955).
6. J. D. Mackenzie and J. A. Kitchener, *Chem. Rev.*, **56**: 455 (1956).
7. J. W. Tomlinson and J. L. White, *Trans. Faraday Soc.*, **52**, 299 (1956).
8. J. W. Tomlinson and M. S. R. Heynes, *Trans. Faraday Soc.*, **54**, 1822 (1958).
9. H. Bloom, Chapter 3 in: J. O'M. Bockris, ed., *Modern Aspects of Electrochemistry*, Vol. 2, Butterworths' Publications, Ltd., London, 1959.
10. E. Kojonen, *J. Am. Chem. Soc.*, **82**: 4493 (1960).
11. J. D. Mackenzie, Chapter 8 in: J. D. Mackenzie, ed., *Modern Aspects of the Vitreous State*, Vol. 1, Butterworths' Publications, Ltd., Washington, D.C., 1960.
12. H. Bloom, Chapter 1 in: B. R. Sundheim, ed., *Fused Salts*, McGraw-Hill Book Company, New York, 1964.
13. J. L. Barton, *Compt. Rend.*, **264**: 1139 (1967).
14. L. I. Sperry and J. D. Mackenzie, *Phys. Chem. Glasses*, **9**: 91 (1968).

Appendix 6.1. The Effective Mass of a Hole

The pressure gradient in a fluid in the direction x, as a result of an instantaneous velocity u in that direction, can be expressed as

$$\frac{dp}{dx} = \varrho X - \varrho \frac{du}{dt} - \left(u \frac{du}{dx} + v \frac{du}{dy} + w \frac{du}{dz}\right) + \eta \left(\frac{d^2u}{dx^2} + \frac{d^2u}{dy^2} + \frac{d^2u}{dz^2}\right)$$
$$+ \frac{\eta}{3} \frac{d}{dx} \left(\frac{du}{dx} + \frac{dv}{dy} + \frac{dw}{dz}\right) \quad (A6.1.1)$$

Similar equations exist for the pressure gradients along the other two mutually perpendicular axes y and z. In these equations, p is pressure; ϱ is density of fluid, and η is its viscosity; u, v, and w are the instantaneous fluid velocities at the points x, y, and z in the directions of the three coordinate axes; and X is the component of the accelerating force in the x direction.

Stokes has shown that, in cases where the motion is small, the fluid is incompressible and homogeneous, etc., these equations can be simplified to a set of three equations of the form

$$\frac{dp}{dx} = \eta \left(\frac{d^2u}{dx^2} + \frac{d^2u}{dy^2} + \frac{d^2u}{dz^2}\right) - \varrho \frac{du}{dt} \quad (A6.1.2)$$

plus an equation of continuity

$$\frac{du}{dx} + \frac{dv}{dy} + \frac{dw}{dz} = 0 \quad (A6.1.3)$$

Solution of Eqs. (A6.1.2) and (A6.1.3) in spherical coordinates leads to a general expression of the form

$$\text{Force} = 2\pi a \int_0^\pi (-P_r \cos\theta + T_\theta \sin\theta)_a \sin\theta \times d\theta \quad (A6.1.4)$$

for the force of the fluid acting on the surface ($r = a$) of a hollow sphere oscillating in it with a velocity V given by $ce^{i\omega t}$, where P_r and T_θ are the instantaneous normal and tagential pressures at the points r and θ, c is the velocity at time $t = 0$, and ω is the frequency of oscillation. The term in T_θ can be ignored for present purposes since it corresponds to a viscous force acting on the sphere owing to its motion through the liquid.

Inserting the appropriate expressions for P_r, ignoring the terms arising from the viscosity of the liquid since there can be no viscous slip between a liquid and a hole, and proceeding through a number of algebraic stages yields

$$\text{Force} = -(\tfrac{2}{3}\pi a^3 \varrho) c\omega i e^{i\omega t} \quad (A6.1.5)$$

where ϱ is the density of the fluid. Writing now

$$m' = \tfrac{2}{3}\pi a^3 \varrho \quad (A6.1.6)$$

for the mass of fluid displaced by the sphere gives

$$\text{Force} = -\left(\frac{m'}{2}\right)c\omega i e^{i\omega t} \quad (A6.1.7)$$

But, from the equation

$$\frac{dv}{dt} = \ddot{x} = ci\omega e^{i\omega t} \quad (A6.1.8)$$

and on remembering that, by Newton's law of motion, action and reaction are equal and opposite, it follows that

$$\text{Effective force due to spherical hole} = \left(\frac{m'}{2}\right)\ddot{x} \quad (A6.1.9)$$

This force thus corresponds to the force $(m'/2)\ddot{x}$ which would be produced by a solid body of mass $m'/2$ operating under conditions where the fluid is absent; it is thus produced by a hole of *effective* mass $m'/2$ or $\tfrac{2}{3}\pi a^3 \varrho_{\text{liq}}$.

Appendix 6.2. Some Properties of the Gamma Function

The gamma function, $\Gamma(n)$ is defined thus

$$\Gamma(n) = \int_0^\infty e^{-t} t^{n-1}\, dt$$

Some of its properties are as follows:

1. When $n = 1$,

$$\Gamma(1) = \int_0^\infty e^{-t}\, dt = 1$$

2. When $n = \tfrac{1}{2}$,

$$\Gamma(\tfrac{1}{2}) = \int_0^\infty e^{-t} t^{-\tfrac{1}{2}}\, dt$$

Put $t = x^2$, in which case

$$dt = 2x\, dx$$
$$t^{-\tfrac{1}{2}}\, dt = 2\, dx$$

and

$$\Gamma(\tfrac{1}{2}) = 2 \int_0^\infty e^{-x^2}\, dx$$

Using Eq. (6.11), i.e.,

$$\int_0^\infty e^{-ax^2}\, dx = \frac{1}{2}\sqrt{\frac{\pi}{a}}$$

one has

$$\Gamma(\tfrac{1}{2}) = \sqrt{\pi}$$

3. $$\Gamma(n+1) = \int_0^\infty e^{-t} t^n\, dt$$

Integrating by parts,

$$\int_0^\infty \underset{u}{t^n} \underset{dv}{e^{-t}\,dt} = -[t^n e^{-t}]_0^\infty + n \int_0^\infty \underset{v}{e^{-t}} \underset{du}{t^{n-1}\,dt}$$

$$= n\Gamma(n)$$

Hence,

$$\Gamma(n+1) = n\Gamma(n)$$

Appendix 6.3. The Kinetic Theory Expression for the Viscosity of a Fluid

Consider three parallel layers of fluid, T, M, B (Fig. 6.63), moving with velocities $v + (\partial v/\partial z)\lambda$, v, and $v - (\partial v/\partial z)\lambda$, respectively, where z is the direction normal to the planes and λ is the mean free path of the particles populating the layers, i.e., the mean distance traveled by the particles without undergoing collisions. In the direction of motion of the moving layers, the momenta of the particles traveling in the T, M, and B layers is $m[v + (\partial v/\partial z)\lambda]$, mv, and $m[v - (\partial v/\partial z)\lambda]$, respectively.

When a particle jumps from the T to M layers, the net momentum gained by the M layer is $mv - m(v + (\partial v/\partial z)\lambda) = -m(\partial v/\partial z)\lambda$. If $\langle\omega\rangle$ is the mean velocity of particles in the direction normal to the layers, then, in 1 sec, all particles within a distance $\langle\omega\rangle$ will reach the M plane. If one considers that there are n particles per cubic centimeter of the fluid and the area of the M layer is A cm², then $n\langle\omega\rangle A$ particles make the $T \rightarrow M$ crossing per second, transporting a momentum per second of $-[n\langle\omega\rangle Am(\partial v/\partial z)\lambda]$ in the downward direction.

When a particle jumps from the B to M layers, the net momentum gained by the M layer is $mv - m[v - (\partial v/\partial z)\lambda] = +m(\partial v/\partial z)\lambda$, i.e., the momentum transported per particle in the *downward* direction is $-m(\partial v/\partial z)\lambda$. Hence, the

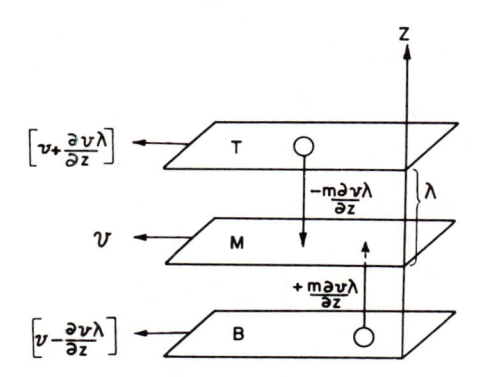

Fig. 6.63. Viscous forces arise from the transfer of momentum between adjacent layers in fluid.

momentum transferred per second in the downward direction owing to $B \to M$ jumps is $-[n\langle\omega\rangle Am(\partial v/\partial z)\lambda]$.

Adding the momentum transferred owing to $T \to M$ and $B \to M$ jumps, it is clear that the momentum transferred per second in the downward direction, i.e., the rate of change of momentum, is $-2n\langle\omega\rangle m\lambda[A(\partial v/\partial z)]$. This rate of change of momentum is equal to a force (Newton's second law of motion). Thus, the viscous force F_η is given by

$$F_\eta = -(2n\langle\omega\rangle m\lambda)A\frac{\partial v}{\partial z} \qquad (A6.3.1)$$

But, according to Newton's law of viscosity, the viscous force is proportional to the area of the layers and to the velocity gradient, and the proportionality constant is the viscosity, i.e.,

$$F_\eta = -\eta A\frac{\partial v}{\partial z} \qquad (A6.3.2)$$

From Eqs. (A6.3.1) and (A6.3.2), it is clear that

$$\eta = 2nm\langle\omega\rangle\lambda \qquad (A6.3.3)$$

INDEX

absolute mobility, and Einstein relation, 374
absolute potential, impossibility of
    measurement, 645
absolute potential difference
  attempts to measure, 644
  structured, 660
absolute strength of acid or base, 494
absorption spectroscopy, in study of solutions, 275
acceptor particles, need for in tunneling, 977
acid(s)
  Bronsted's view, 489
  dissociation constant of, 496
  relative strength of, 495
acid–base strength(s), 501, 509, 511
acid strength, 506, 510
acid strength theory, 505
action potential, and nerve impulse, 940
activation, of electrocatalysis, 1170
activation barrier, and rate-determining step, 1003
activation energy, 457
  for diffusion, related to melting point, 547
  electrical contribution to, 872
  for multistep reaction, 1002
  for viscous flow in fused salts, related to melting point, 550
activation-transport control, of electrode reaction, 1054
activity coefficient(s)
  concept, 202
  cube-root law, 269
  Debye–Hückel model, 223
  Debye–Hückel, parameters of, 211, 212
  further reading on, 246
  at high concentration, 245
  and hydration number, 240
  and ideal solutions, 205
  and ion pairs, 260

activity coefficient(s) *(cont.)*
  and ionic strength, 215
  mean, 207, 269
  model for effect of solvent molecule, 240
  of single ionic species, 206
  and solvation, 241
  theory and experiment, 209, 213, 216, 228
  theory of solvent effects on, 242
adatoms, 1177
additives, organic, 1222
adiabatic change, 974, 975
adion(s)
  and adatoms, 1177
  charge on, 1177
  concentration, determination of, 1197
  concentration profile of, 1200
  concentration as function of time, 1189
  and electric field, 1181
  existence of, 1177
  formation of hydrated ions from, 1187
  function in electrodeposition, 1177
  lattices from, and spiral growth, 1203
  surface, deposition from, 1187
  tabulated, 1198
adsorbed intermediates, 1016
adsorption
  of atomic hydrogen, and coverage, 1246
  of cations and anions, at interfaces, 680
  contact (or specific), 748
  definition of, 682
  of desolvated ions, 742
  free energy change in, 742
  variation with potential, 638
  in electrochemistry and chemistry, 683
  organic, 792, 793
    and free energy of flip-flop water molecules, 799
  superequivalent, 638, 748
  and surface excess, distinction, 684

adsorption energy, effect on tunneling, 965
adsorption intermediates, further reading on, 1036
adsorption isotherm, for ions, theory and experiment, 773, 774
adsorption step, in hydrogen evolution, 1153, 1233
aeration, differential, 1302
aggregation of colloidal particles, theory, 838
aims, of this book, vi
air
  containing CO_2, effect on metals, 1268
  electrochemically burned, 1382
air electrode, in storage, 1427
alkali metals, and storage, 1423
alloy(s), 1224, 1225
American convention, 1118
American space program, contributions to fuel cells, 1388
analogy(ies)
  mass transfer and heat transfer, 1041
  semiconductors and electrolytic solutions, 812
angel, Gibbs', 679
anode(s), 1311, 1312
anodic protection, 1318
artificial organs, and electrochemistry, 43
association, of ions, temporary and permanent, 273
atom(s)
  of hydrogen, recombination of, in void, 1334
  labeled, and surface coverage, 1036
atom–atom step, in hydrogen evolution, 1234
atomic energy
  and electricity storage, 1397
  and electrochemical power sources, 1395
  and electrochemistry, 43
atomic power, coming era of, and electrochemistry, 43
autodissociation, proton transfer reactions in, 492
autodissociation constants, 499
automobiles
  necessary electric energy density for driving, 1420
  exhaust products from, 1357
auxiliary electrode, and cathodic protection, 1312
auxiliary electrode, use of in determining overpotential, 890
averaging process, Debye, 141
axon
  as a cell, 937
  potential across, 938
  of squid, 940
Balkans, map of, 1430
bands, bending of, near surface of semiconductor, 816
band picture, 804, 807, 810, 820
barrier(s)
  for consecutive steps, 1002
  electron leak through, 945, 952, 962
  transfer of charge across, further reading on, 946

barrier(s) *(cont.)*
  width of, and probability of tunneling, 954
base, 494, 495
batteries
  in ancient times, 1265
  classical, 1413
  and dendrites, 1221
  mission time in which useful, 1432
benzene, 1098, 1099
biological cells, and charge transfer processes, 841, 1266
biological membranes, 937, 981
biological processes, 4, 840
biological reactions, quantum nature of, 21
biological situation, 941
biology, and electrochemistry, 29, 42, 43
Bjerrum approach, to ion pairs, 253, 257, 260
Bjerrum's integral, 256
blister, bursting, 1334
blood, clotting of, and electrochemistry, 840
bond, stretching of, and electronation reaction, 971
bond strength, effect on electrocatalysis, 1157
Born charging process, 84
Born cycle, 50, 86
Born equation, 57, 69
  and strength of acids, 510
Born model, 49, 50, 70, 71
Bridgman technique, for single crystals, 1218
Bronsted's view of acids, 489
bronzes, 1169
Brownian motion, and movement of piston, 300
bulk properties, and interphase properties, 641
bunching, 1209
butanol
  adsorption of, on mercury, 795, 798
Butler–Volmer equation, 862, 984, 1054
  and biological situation, 941
  and catalysis, 1141
  and contact adsorption, 911
  deduction, 880
  effect of water coverage?, 1015
  and effective potential in catalysis, 1142
  and electric car, 928
  final form for multistep reactions, 1000
  further details, 910
  further reading on, 929
  in galvanostatic transients, 1190
  high-field approximation, 885, 888, 889, 1001
  for iron dissolution and deposition, 1084
  low-field approximation, 882, 892
  for multistep reaction, 998, 1090
  in terms of stoichiometric number, 1006
  and order of reaction, 1009
  in quantum mechanics, 980
  and rate-determining step, 1139
  and structure of interface, 911
  a summary, 928
  and surface coverage, 1014, 1015
  and zeta potential, deduction, 913
cadmium–nickel battery, 1401
calomel electrode, 654

capacitance
  constant, equation for, 760
  and contact adsorption, 751
  dipole, 798
  of diffuse layer, 731, 732
  electrical, determination at interface, 703
  and electrocapillary curves, 721
  and Gouy–Chapman theory, 731
  hump, 754, 761
  at interface, as function of dipole orientation, 788
  of semiconductor-solution interface, 817
  of whole electrode, 790
capacitance–potential curve, lateral adsorption model, 778
capacitance hump, equation for, 761
capacitor(s)
  and dielectric, 134
  and dipoles, 135
  in series, 735
capacity, constant region of, 753
capillary electrometer, 689
cars
  and batteries, 1419
  and fuel cells, 1158
  and pollution, 1419
carbon dioxide
  in atmosphere, as function of time, 1355
  and possible rise in sea level, 1356
carbon monoxide–air cells, 1393
Carnot limitation, 1358, 1360
  and electrochemistry, 1364
  and internal combustion, 1358
  physical interpretation, 1367
catalysis
  Butler–Volmer equations in, 1142
  and electrocatalysis, 987
  for oxygen on doped tungsten bronzes, 1422
  tabulated, 1377
catalyst(s)
  charge transfer, 10
  chemical and electrocatalysts, 1141
  distribution of, and porous electrode, 1384
  distribution of, in porous electrodes, 1172
catalytic activity
  for the oxidation of ethylene, 1161
  oxide-free and oxide surfaces, 1258
cathodic protection, 1309, 1312, 1313
cavity, 150, 1335
cell
  driven, 1128
  electric, potentials in, 649
  electrochemical, discussion of dependence on current, 1128, 1129
  entire, and Nernst equation, 904, 1114
  local, in corrosion, 1270
  and metal–metal potentials in electrocatalysis, 1148
  for the observation of transients, 1184
  relations in, further reading on, 1137
  short-circuited, and stability of metals, 1269
cell model, for liquid electrolytes, 529
cell potential
  and current, in electrochemical energy conversion, 1371

cell potential *(cont.)*
  maximum, 1137
  minimum, 1137
  in self-driving cell, 1131
cellulose, oxidation of, 1169
central ion, excess charge density as function of, 194
charge
  at boundary, 627
  of individual ions, in diffuse layer, 747
  storage of, 703
charge density
  in double layer, components of, 712
  on electrode, determination of, 702
  excess, and ionic atmosphere, 636
charge transfer
  across barriers, further reading on, 946
  and blockage of electrode surface, 1014
  chemical and electrical implications, 846
  and Fermi level, 977
  and formation of intermediates, 1027
  and instability of surfaces, 1268
  and metal–slag equilibrium, 618
  in perspective, 974
  quantification of quantum-mechanical picture, 977
  and rate-determining step, 1185
charge transfer catalysts, 10
charge transfer theory, summary, 893
charged sphere, and Born model, 50, 52
charging process, effect of double layer on, 1190
cheap heat, and electrochemistry, 43
chemical desorption step, in hydrogen evolution, 1234
chemical and electrochemical reactions, further reading on, 989
chemical energy, conversion to electrical, and symmetry factor, 1138
chemical potential
  change arising from ionic cloud, 201
  and computation at interfaces, 672
  and flux of ions, 397
  standard, 203
  and thought experiment, 694
chemical reaction
  and electrical energy, 15
  and electrode reactions, 896
chemistry, and electrochemistry, 28, 38, 869
chi potential, 667
Christmas trees, mini, 1221
chronopotentiometry, 1051
circuitry, and electrochemistry, 43
circuits, 33, 1316
circuits in the body, and electrochemistry, 43
civilization, and surfaces, 1267
cliff, and symmetry factor, 937
closest approach, 225, 741
clotting, of blood, and zeta potential, 841
cloud, near central ion, 193
clusters of ions, 82, 265
coagulation, 839
codeposition, of hydrogen, with metal, 1227
cold combustion, 1369
cold emission, of electrons, 944

colloids, 835, 838
colloid chemistry, further readings on, 841
collodial particle(s)
  energy of interaction for coagulation, 840
  and potential distance relation, 729
  space charge near, 217
colloidal nature of biological processes, 4
combustion
  cold, 1369
  products of, other than carbon dioxide, 1353
competition, between adsorbed water and hydrogen, 1015
complex formation, and mixtures of ionic liquids, 587
complex ion(s)
  concentration as a function of added ligands, 592
  lifetime, 590
  in molten salts, further reading on, 594
  tests for, 589, 593, 594,
  and Raman spectra, in molten salts, 590
compressibility
  calculated from hole model, for molten salts, 585
  and hydration number, 127
concentrated solutions, skepticism on theory of, 281
concentration gradient, 1059
  and chemical potential, 291
  and diffusion flux, 295, 1040
  ion diffusion in, 288
  linear, and Planck–Henderson equation, 419
  in tracer diffusion, 543
concentration overpotential, 1052, 1053
concentration perturbation, Laplace transform of, 329
condenser
  parallel-plate, and double layer, 634
  and storage of charge, 703
conductance
  in migration, further reading on, 367
  in nonaqueous solutions, further readings on, 452
  theory, further reading on, 439
  theory of, in terms of Debye–Hückel–Onsager equations, 435
conductance of true electrolyte, in nonaqueous solution, 452
conduction, 345, 351
conduction band, 809
conductivity
  equivalent, 358
  and ion association, 448
  molar, 357
  of molten salts, 517
  of pure liquid electrolytes, 553
  specific, and current density, 354
conductor(s), 806
  electronic, and passivation, 1320
  ionic, and effect on corrosion, 1273
configuration, of water molecules around proton, 470

conservation of momentum, and radiationless transition, 950
constant-flux diffusion problem, and solution of other problems, 323
constitution of proton, further reading on, 470
contact adsorption, 748
  and Butler-Volmer equation, 911
  and capacitance, 751
  and capacitance of interface, 749
  definition of, 682
  of desolvated ions, 742
  determination of, 743
  free energy change in, 742
  as function of charge, 748, 763
  and image charge, 767
  and ionic radius, 744
  lateral repulsion model, 766
  measurement of, 745
  of negative ions, 637
  and stability of colloids, 840
  and surface state, 821
  tests for isotherm, 769
contact adsorption model, tests for, 775
contact potential difference, 647
convection, 1051–1057
convention
  American, 1118
  international, 1118
conversion, of chemical energy to electricity, 14, 42, 1266, 1358
converters, photogalvanic, and electrochemistry, 43
converter-storers, 1430
coordination number, of proton, 468
copper, deposition of, summary of mechanism, 1202
correspondence principle, 20, 224
corrosion, 1266, 1267
  as affected by solid phases, 1283
  and agitation of solution, 1300
  basic kinetic conditions for, 1274, 1286
  bird's eye view of, 1347
  common examples, 1301
  cost of, 1346
  effect of equilibrium potential, 1294
  effect of transport difficulties, 1294
  effect of purity, 1273
  effect of Tafel slope, 1294
  and electrodics, 1272, 1275, 1285
  embrittlement in, 1347
  electrochemical mechanism, 1268
  and flip-flop water dipole molecule, 790
  and future of fuel cells, 1388
  and hydrogen evolution, 1232
  inhibition of, and electrodics, 1306
  local cell theory of, 1270, 1273
  Nernst equation and potential–pH diagrams in, 1279
  and oxygen electronation, 1252
  potential of, 1285
  rate of, 1276, 1284, 1285
  rate-determining step, 1296
  and reversible potential, 1274
  in sand, 1304
  spontaneous energetics, 1278

corrosion *(cont.)*
  summary of mechanisms, 1345
  thought experiments in, 1272
  through oxygen starvation, 1303
  through paint, 1302
  of ultrapure metals, 1273
  and undevice, 860
  and wastage, 861
  at water line, 1303
  yield-assisted, 1337
corrosion inhibition, and film-forming inhibitors, 1309
corrosion and passivity, further reading on, 1349
costs, reduction of, in fuel cells, 1385
coupled reactions, 1235
coverage
  determination of, 1029
  of electrode, 1235, 1245
    with inhibitors, tabulated, 1309
    and mechanism determination, 1097
  with hydrogen, determination of, 1245
  of organic molecules, on electrodes, 796
  of surface with hydrogen, variation with potential, 1246
crack(s), 1335–1341
crack initiation, testing of, 1343
cracking
  of hydrocarbons, 1158
  stress corrosion, 1335, 1338
crevices, corrosion associated with, 1302
cross coefficient, 828
cryogenics, and storage of hydrogen, 1421
crystal, growth of, and fast-growing face, 1216
crystal faces, rates of deposition on, theory of, 1216
crystal facets, 1212, 1213, 1214
crystal growth, faster at projection under electric field, 1217
crystal growth, morphology of, for copper, 1222
crystal lattice, 65, 1205
crystal lattice plane, kink site on, 1178
crystal plane, 1179, 1213, 1216
crystallization, 1174, 1202, 1204, 1218
crystalline solids, band theory of, 804
cube root law, for activity coefficients, 269
current-centric view, 16
current–distance relation, along meniscus in single pore, 1385
current–potential curve, and internal resistance of fuel cells, 1373
current–potential diagram, for passivation, characteristic, 1317
current–potential laws, 930, 1113
current–potential relation
  and alloy composition, in oxygen reduction, 1259
  and alloy deposition, 1259
  in cells, 1133
  characteristic shape in passivation, 1317
  for Gemini fuel cells, 1389
  and mass transfer limitation, 1373
  at n–p junction, 936

current–potential relation *(cont.)*
  in terms of cell potentials, 1135
current efficiency, 1229, 1232
current transients, and Fick's law, 316
cybernetic organism, 1267
cyborg, 1267
cytochrome
  quantum mechanical tunneling to, 981
  tunneling to, and enzymes as electrodes, 1253
Daniell cell, 858
  sign of voltage of, 1125
d-band character of metals, and oxygen adsorption on, 1163
de Broglie wavelength, 20, 21, 948, 949
Debye charging process, 248, 249
Debye equation, for dielectric constant, 142
Debye theory of dielectric constant of gas, 140
Debye–Hückel activity coefficient, parameters of, 211, 212
Debye–Hückel constant, 190, 197
Debye–Hückel length
  and diffiuse charge, 730
  in semiconductors, 816
Debye–Hückel model
  approximations of, 219
  breakdown of, 266
Debye–Hückel radius, of ionic atmosphere, 197
Debye–Hückel theory, 180, 189
  an assessment, 230
  basis of, further reading on, 212
  comparison with experiment, 212
  further reading on, 238
  parentage, 237
  postulates, 236
  summary of derivation, 233
  triumphs and limitations, 212
Debye–Hückel thickness, thickness of ionic cloud in, 220
Debye–Hückel–Onsager equations, 434, 438
  comparison with experiment, 436
  for nonaqueous solutions, 442
decay, 4, 861
decoration, surface, 1266
de-electronation, 352, 847
  definition of, 853
  and desorption of hydrogen, 1246
  effect of field, 875
  and quantum mechanics, 973
dehydration, and electrodeposition, 1176, 1177, 1180
delay time, and electronics, 33
demon, Maxwell's, 679
dendrites, 1220–1221
deposition
  of alloys, 1223
  of alloys, equations for, 1224
  and crystallization, 1202
  metal, advance of growth step in, 1204
  metal, and screw dislocation, 1206
  metal, steps in, 1203
  metal, and nucleation, 1204
  random walk process in, 1186
  on single crystal, 1218

deposition overpotential, and exchange current density, 1215
desorption step, 1153, 1233, 1234
deuterons, mobility of protons and, 472
device
 electrochemical systems as, 851
 electronic, 324
diagnostic coefficients, and propane oxidation, 1107
diagnostic criteria, for de-electronation of ethylene on platinum, 1162
dielectric constant, 55, 152
 of aqueous solutions, 157
 in bulk near ion, 71
 Debye equation for, 142
 and deformation polarization, 145
 and dipoles, 139
 of electrolytic solutions, 157
 gas, Debye theory of, 140
 and ionic solutions, 132
 and ionic solvation, 155
 and oriented dipole layer, 136
 and polarizability, 138
 and relative strength of acids, 507
 and solvation sheath, 156
 of solvent and solution, further reading on, 158
 theory of, for water, 146
 in water, alignment of group, 146
 of water, calculation for, at various temperatures, 154
 of water, in double layer, 756
 of water, theory of, 153
diesel oil, burned electrochemically, 1391
differential equation, for diffusion, integrated, 417
diffuse charge, and Debye–Hückel length, 730
diffuse layer
 charge of individual ions in, 747
 effect on Tafel relation, 915
 and streaming current, 830
diffusion, 27
 of adion to kink site, 1182
 at constant current, quantitative, 1044
 and convection, 1051
 driving force for, 290
 further reading on, 345
 in fused salts, 542
 of hydrogen by interstitial and intergranular paths, 1330
 of hydrogen into metals, 1328
 of hydrogen, into regions of stress, 1330
 linear, independent of time, 1077
 in molten salts, at constant temperature, 548
 of neutral ion pairs, 383
 nonsteady state, gross view, 307
 an overall view, 342
 to peak of crystal, 1219
 as pseudoforce, 306
 spherical, 1002, 1220
 in solution, and electrocrystallization forms, 1219
 on surface, contribution to overpotential, 1199

diffusion *(cont.)*
 surface, deduction of equations for, 1188
 surface, and lattice formation, 1195
 time in, by Einstein–Smoluchowski equation, 333
diffusion coefficient, 296
 and critical concentration for hydrogen, 1328
 and Einstein relation, 374
 and Einstein–Smoluchowski relation, 339
 and holes, in molten salts, 577
 and mobility, 374, 376
 and molecular quantities, 338
 rate process expression for, 342
 and structural properties, 344
 of substances near melting point, 546
 tracer, 545
diffusion control, 1046
diffusion equation, and Nernst-Planck flux equation, 411
diffusion flux
 and Kirchhoff's laws, 1039
 produced by sinusoidal variation of current, 328
diffusion layer, 1055–1058
diffusion layer thickness, and microrough surface, 1219
diffusion potential
 as a function of transport number, 418
 further reading on, 420
 and transport number, 406
diffusion process, boundary conditions, 313
digestion, biochemical, 40
dimerization, in liquid silicates, 610, 612
dipole
 difference of contribution to potential of flip and flop positions, 787
 and electric field, interaction, 784
 interaction with electrode, 784
 orientation of, and capacitor, 135, 143
 water, flip-flop model, 790
 of water, in interior of electrolyte, 624
dipole-covered phase, potential difference through, 667
dipole–dipole interaction potential, at electrodes, 786
dipole orientation, net, at interphase region, 627
dipole potential
 and electrocatalysis, 1145
 at interface, 667
 at electrodes, 783
direct energy conversion, 1358, 1359
 and Carnot limitation, 1360
 by electrochemical means, 1360, 1361
 further reading on, 1400
 summary, 1398
discharge of ions, and dependence on lattice site, 1178
discrete polyanions, 610, 614
dislocation, 1201, 1205, 1206
dispersion forces, 166, 167
dissociation constant, of acid, 496, 497, 498, 500

dissolution
  field-assisted, 1337
  of iron, intermediates in, 1087
  of iron, mechanism for, 1085
  of iron, prediction of various mechanisms, and experiment, 1092
  of kinky surface at bottom of crack, 1337
  of metals under stress, and Miller index, 1335
dissolution–precipitation mechanism, of passivation, 1321, 1324
distribution
  of electrons, among energy levels in metal, 956
  time average, spatial, 182
distribution function, for holes in liquid electrolyte, 537
distribution law
  in Einstein–Smolchowski diffusion, 335
  for size of hole, 534
Doddario Committee, and John Malone's baleful prediction, 1357
donor, 499
doping agent, 819
doping of bronzes, and electrocatalysis of oxygen, 1169
double layer
  charge density components, 712
  charging of, 1027
  concentration of reactants in, and work function, 1150
  constant capacitance of, model for, 758
  "constant" value for capacitance, 753
  effect upon electrocatalysis, 1012
  electrical, becomes trouble layer, 750
  and Gauss's law, 727
  interaction of, and stability of colloid, 837
  ionic cloud at, 722
  isotherm for ions in, 764
  and parallel plate condenser, 634
  and Poisson equation, 724
  potential difference at, 635
  special position of mercury in studies of, 687
  structure of, further reading on, 790
  thickness of, and colloids, 835
double-layer charging, in galvanostatic transients, 1190
double-layer structure, further reading on, 717
double-layer studies at solids, further reading on, 803
double-layer theories, review of, 752
double-layer treatments, history of, 724
drift, ionic, to interface, 391, 845
drift velocity
  calculation of relaxation components, 427
  electrophoretic components of, 427
  and interacting ions, 425
  relaxation component of, 430
driven cell, 1128, 1131
driving force, 343, 412
droplets, corrosion under, 1305
economics, and social importance of overpotential, 16

edge vacancy, and electrodeposition, 1179
effective mass, of hole, 619
efficiency, 1358–1371
e–i junction, law for, 936
Einstein relation, and absolute mobility, 374
Einstein–Smoluchowski equation, 333, 334
  how many ions diffuse?, 333
electric car
  and Butler–Volmer equation, 928
  and electrocatalysis, 1155
electric conduction, and liquid silicates, 606
electric energy storage, needed for automobiles, 442
electric field
  effect upon electrocatalysis, 1168
  effect on rate, 869
  influence on random walk, 350
  local, in polar dielectric, 147
electric reactions, and electrochemistry, 14
electrical energy
  and free energy, 15
  production by thermal and electrochemical means, 40
electricity
  from chemical energy, by electrochemical means, 1362
  conversion of energy to, and electrochemistry, 42
  from heat, 1360
electricity storage
  in alkali metals, 1423
  and atomic energy, 1397
  in hydrogen, 1420
  important quantities in, 1404
electricity storage density, 1404, 1406
electricity storer(s), 1402, 1412, 1413
  future ones, 1420
electrification, 6, 7, 623
electrified interface(s)
  absolute potential difference at, 675
  further reading on, 717
  importance of, in practical situations, 642
  mobile, electrokinetic properties, 826
  retrospect and prospect, 715
  structure of, 718
  thermodynamics of, 688, 698
electrocapillary curves
  basic equation for, 698
  and capacitance, 721
  differentiation of, 704
  facts on, 690
  as perfect parabola, 705
  and surface excess, 710
electrocapillary equation, final general form, 701
electrocapillary maximum, 691
electrocapillary thermodynamics, 713, 714
electrocatalysis, 1141
  activation in, 1170
  and cancellation of thermionic work function, 1147
  and cars, 1155
  and chemical catalysis, 987
  of copper deposition, on various surfaces, 1202

electrocatalysis *(cont.)*
  dependence on electronic properties, 1147
  difference from chemical catalysis, 1143
  and dipole potential, 1145
  effect of double layer on, 1012
  effect of metal–metal potentials, 1148, 1155
  and exchange current density, 1146
  and Galvani potential difference at nonpolarizable electrodes, 1149
  heat of activation in, of nonbonding reactions, 1149
  of hydrocarbons, 1156, 1158
  and hydrocarbon oxidation, 1391
  of the hydrogen evolution reaction, 1155, 1232
  irrelevance to, of experiments with porous electrodes, 1376
  lack of effect of work function on, 1148
  of oxygen, and doping of bronzes, 1169
  potential of, comparison, 1145
  and potential of zero charge, 1142
  rate equations for, compared with those for catalysis, 1143
  in reactions involving adsorbed species, 1153
  reactivity at low temperatures in, 1169
  of redox reactions, 1149
  reference potential for, 1143, 1144
  secondary effects due to double layer, 1151
  secondary effect of work function on, 1151
  and simple redox reactions, 1146
  in space vehicles, 1158
  special position of platinum in, theory of, 1166
  tunneling condition for hydrogen evolution, 1154
  volcano relations in, 1165
electrocatalyst, determination of adsorbed entities at, 1030
electrochemical cells, 1114, 1132
electrochemical converter(s)
  Carnot efficiency limitation avoided in, 1364
  efficiency, 1364, 1366
electrochemical desorption, in hydrogen evolution, 1234
electrochemical device(s)
  as energy producer, 855
  as substance producer, 851
electrochemical electricity storers, 1412
electrochemical energy conversion
  and atomic energy, 1395
  its central problem, 1369
  dominating role of electrocatalysis, 1372
  and Tafel relation, 1370
electrochemical energy converters
  cost of, and porous electrodes, 1385
  deduction of real efficiency, 1366
  and power-rate relation, 1378
electrochemical energy producer, and power density, 859
electrochemical energy storage
  feasible goals, 1431
  summary of, 1430
electrochemical engine, 1157

electrochemical era, 43
electrochemical generators, 1386
  examples of, 1385
electrochemical methods, for surface coverage, 1035
electrochemical model, for slag–metal equilibrium, 618
electrochemical potential(s)
  digression on, 693
  equality of, in different phases, rationalized, 864
  gradient of, 395
  and work of bringing charge particles into material phase, 695
electrochemical producer, 1361
electrochemical reaction(s), 7
  and chemical reactions, 8, 987
  always quantal, 985
electrochemical reactor(s), 9, 1405, 1406
electrochemical system(s)
  as devices, 851
  and metal–metal potential, 1113
  series of potential drops in, 1112
electrochemical vista, 43
electrochemist(s)
  frustrations of, 1042
  training of, 44
electrochemistry
  advances expected, 42
  and artificial limbs, 43
  and atomic reactors, 43
  awakening, 24
  as basic science for advances in postindustrial era, 42
  and biology, 29
  brilliant beginning, 14
  and cheap heat, 43
  and chemistry, 28, 869
  and circuits in the body, 43
  and coming era of atomic power, 43
  and conversion of energy to electricity, 42
  conversion of, to charge transfer orientation, 17
  delay in development of, 18
  and development of circuitry, 43
  and developments in molecular biology, 42, 43
  and direct energy conversion, 1361
  disciplines in, 27
  and electric reactions, 14
  and electronics, 18, 27, 33
  and electron transfer, 25
  future role, 41
  and geology, 29
  an interdisciplinary area, 1, 25, 29, 31
  and interfaces, 22, 23
  interfacial, degree of ubiquity, 39
  involvement of, in many sciences, 28
  and machining, 42, 43
  and medical developments, 43
  need for books, vi
  and new towns, 43
  and other fields, 25
  perspective from afar, 23
  perspective from a medium distance, 24

electrochemistry *(cont.)*
  perspective in time, 39
  and photogalvanic converters, 43
  place in science, 26, 41
  and polluted liquids, 43
  and possible fuel cell heart, 43
  and powering of ships, 43
  and powering of vehicles, 43
  quantum nature of reactions in, 20, 21
  as separate discipline, 38
  and sewage disposal, 43
  sign convention in, 1115
  and stabilization of materials, 42
  and storage of energy, 42
  and time, 38
  and tools, 43
  and transportation, 43
  and urban living, 42
  and water purification, 42
  wider significance?, 38
electrocrystallization, 1129, 1173, 1174, 1219
electrode(s)
  as catalyst, 34, 1139
  de Broglie wavelength at, 20
  porous, 1171
    activity near tip of meniscus, 1384
    diffusion of reactant in, 1384
    and distribution of catalyst, 1384
    vital importance in fuel cells, 1172
  sick, 1231
electrode kinetics
  and double-layer effects, 916
  and organic reactants, 916
  transfer coefficient as center of, 918
  and zeta potential, 912
electrode processes, quantum-mechanical approach to, 947
electrode reactions
  and heterogeneous reactions, 989
  history of quantum-mechanical developments, 983
  and tunneling, 955
electrode surfaces, 36
electrode–electrolyte potential difference, analysis of, 659
electrodeposits, organic, 1222
electrodeposition
  consecutive stages in, 1180
  and dehydration, 1180
  electronation in, 1176
  function of adions in, 1177
  and hole vacancy, 1179
  influence of potential of zero charge, 1180
  of metals, and tunneling, 1177
  rotation of a spiral in, 1205
  stepwise dehydration in, 1177
  and surface adions in random walk, 1181
  and surfaces which change with time, 1182
  in terms of consecutive reactions, schematic, 1183
elecrodeposition rate, as a function of crystal plane, 1216
electrodics, 19, 846
  and corrosion, 1285
  and electronics, 30, 31, 32

electrodics *(cont.)*
  elementary, further reading in, 909
  elementary, summary, 908, 983
  and inhibition of corrosion, 1306
  and quantum mechanics, 30
  transient techniques in, 34
  transport aspects of, summary, 1076
  in the west before 1950, 1380
electrodissolution, burst of, and Laplace transformation, 330
electrogrowth, 1215
  basic aspects of, 1173
  and kink sites, 1203
  topographical features, 1184
electrokinetic properties, 826
  further reading on, 835
electrolyte(s)
  forces at boundary of, 623
  glasses as, 603
  potential and true, conductance in different media, 179
  pure liquid, 513
  true and potential, 176
electrolytic solutions
  electromagnetic radiation in study of, 274
  and infrared spectroscopy, 278
  and nuclear magnetic resonance spectroscopy, 278
  and Raman spectra, 277
  and semiconductors, 811
electromagnetic methods for investigating solutions, further reading in, 279
electromagnetic radiation, in study of electrolytic solutions, 274
electromagnetic theory of light, 35
electron(s)
  cold emission of, 944
  collision with impurity atom, 956
  distribution among energy levels in metal, 956
  in holes, and the double layer, 825
  image energy of, as function of distance from metal, 945
  leak through barrier, 945
  their mechanics, 947
  near interfaces, potential of, 943
  number of which strike surface of metal, 990
  penetration into forbidden region, 950
  in space region, and wave function, 669
  and tunneling to solution, 959
  in vacuum, and probability of passage through, 952
  which are free to tunnel, 959
electron overlap potential difference, 670
electron sink, 853
  electrode, 1126
electron source electrode 853, 1126
electron transfer
  and electrochemistry, 25
  probability for tunneling, expression for, 978
  type of, in electrochemistry, 20

electron transfer reactions, 494
  and electroneutrality difficulties in
    conduction, 351
  the 1950's, 17
electron transfer theory, beginnings, 18
electron tunneling, 946
  condition for, 972
electronation, 352, 847
  in corrosion, 1275
  of oxygen, 1251
    and enzymes, 1253
electronation reaction
  and bond stretching, 971
  effect of field, 874
  of hydrogen on platinum, energy terms in, 968
electroneutrality
  conflict with conduction, 351
  and coupling between ionic species, 410
  in fused salts, 543, 566
  principle of, 414
electronics, 18, 27, 30, 31, 32, 33, 43
electro-osmosis, 826
  theory of, 827
electro-osmotic motion, of phases relative to each other, 831
electrophoresis, 832, 834
electrophoretic effect, 424, 425
electrostatic potential
  and charged sphere, 52
  and field strength, 347
  near ion, as function of distance, 193
  variation of, near interface, 664
electrostriction, 126
ellipsometric spectroscopy, 37
ellipsometry, 37, 1319
embrittlement
  in corrosion, 1347
  by hydrogen, 1314, 1338, 1339, 1344
emission
  cold, 944, 952, 955
  hot, of electrons, 941
  thermionic, 953
empty space, in fused salts, 523
energetics, of certain corrosion reactions, 1278
energy(ies)
  of activation, for self-diffusion, 579
  of crack, 1341
  electrical, 11
  free and electrical, 15
  as function of repulsion between water molecules, 970
  of interaction between colloid particles, 838
  of strain, 1340
energy barrier, and rate theory, 341
energy consumption, electrochemical, 1350
energy conversion, 1266, 1358, 1361
  to electricity, and electrochemistry, 42
  as an interdisciplinary field, 31
  and storage, terminology of, 1402
energy conversion efficiency, maximum of, 1376
energy converter(s)
  electrochemical, efficiency of, 1374

energy converter(s) *(cont.)*
  power and efficiency in, 1379
  thermionic, 1359
  thermoelectric, 1359
energy density, 1407, 1408
  idealized maxima, tabulated, 1408
  and rate of working of cell, 1410
  of stores, feasible values, 1418
  versus power density, 1411
energy gap, 810, 931
energy levels, 805, 807, 956
energy producer, 855
energy-producing device, current and potential in, 1131
energy sources, 1350
  distribution of, 1351
energy states
  discrete, 806
  those accounting for free electrons, 959
energy storage, 1266
  electrochemical, summary, 1412, 1430
  terminology, 1402
energy storage density
  and non-aqueous electrolyte, 1427
  for some realized cells, 1409
energy storers, 1420, 1426, 1428
energy waster, 859
enthalpy
  by Born, and ionic radii, 60, 69
  and electrochemical energy conversion, 1368
  and entropy, of ion–solvent interaction, 59
entropy, and electrochemical energy conversion, 1364, 1368
enzymes, 1253
equilibrium
  difficulty of observing rates near to, 1263
  and electrochemical potential, 696
  at interface, 876
  and Nernst equation, 898
  and steady state, 1018
equilibrium cell potentials, useful ?, 1124
equilibrium potential, 876
  and activity in solution, 905
  limitation in usefulness, 876
equivalent circuit(s)
  for galvanostatic transients, 1026
  for interface involving ideally polarizable electrode, 654
  and pseudocapacitance, 1029
equivalent conductivity, 358
  and concentration, 360
  and ionic mobility, 372
  significance of, 360
era, electrochemical, 43
error function, 321, 1065
Esaki tunnel diode, and tunneling, 956
ethylene
  adsorption of, on platinum, 1160
  rate-determining step in the oxidation of, 1159, 1160
ethylene oxidation
  and catalytic activity, 1161

ethylene oxidation *(cont.)*
 diagnostic criteria for de-electronation on platinum, 1162
 negative pressure effect in, 1160
 radiotracer measurements of coverage, 1160
 rate-determining step, 1160, 1161, 1164
 volcano relation in, 1164
European convention, 1118
evolution, of gases, 1102, 1104
excess charge density, as function of distance from central ion, 194
exchange current density, 876, 877
 catalytic effects due to double layer properties, 1151
 and deposition overpotential, 1215
 determination of, for hydrogen evolution, 1238
 and electrocatalysis, 1146
 and heat of activation, 1150
 for hydrogen evolution, tabulated, 1238
 for metal deposition, 1202
 on noble metals, and platinum–rhodium alloys, 1260
 and polarizability, 895
 and rate-determining step, 997
 and reaction order, 1012
 schematic diagram of, 878
 small, difficulty of measurements with, 1260
 for various crystal faces, 1215
 and work function, 1149
exclusion principle, 963
expansivity
 calculation from hole model, 585
 of liquid silicate models, 609
faces, fast growing, 1217
facets, 1213, 1214, 1217
faradaic rectification, 885
faradaic resistance, 996
fatigue of metals, 1267
Fermi–Dirac distribution, deduced, 958
Fermi energy, 959, 962
Fermi level, 957, 964, 977
Fick's first law, 293, 315, 316, 343, 1056, 1065
Fick's second law, 308, 344, 1040
field
 current produced, 883
 in double layer, 630
 excess, 882
 induced reorientation, 484
 at interphase, 630
 nonlinear, 348
 producing current, 882
 in semiconductors, 814
field strength
 effect on current, 881
 and electrostatic potential, 347
 and flux, 350
film(s)
 electronic conductivity of, and passivation, 1320

film(s) *(cont.)*
 passive, formation of, at the bottom of pits, 1335
 precursor, 1319
film-covered surfaces, determination of properties by ellipsometry, 37
Flade potential, 1316
flash photolysis, 34
flip dipoles, at electrodes, lateral interaction between, 785
flip-flop model, 790, 797
flip-flop water on dipoles, 779
flocculation, 839
flux
 and forces, 894
 as a function of field strength, 350
 of ions, and chemical potential, 397
 of ions, and electrostatic potential, 397
 at low field gradient, 295
 sinusoidally varying, near electrode–solution interface, 327
 time derivitive of, in step function, 330
flux equality condition, 1038
flux equations, and Onsager equations, 411
forbidden region, electron penetration into, 950
forces
 anisotropic, 626
 at boundary of electrolyte, 623
 and fluxes, 894
 in organic adsorption, at electrodes, 792
fossil fuels
 available for centuries if used mainly for food and textiles, 1350
 converted to food and textiles, 1358
 lack of rationality of burning up, 1352
Fourier's law, 894
fraction, of ions, produced in pulse, near electrode, 332
free electrons, energy states for, 959
free energy
 change of, when ion goes from OHP to IHP, 763
 determination by chemical and electrochemical means, 40
 and electrical energy, 15
 of ion–ion interactions, 181
 of organic adsorption, at electrodes, 792, 796
free energy of activation, as function of potential, 390
free energy change
 in contact adsorption, 742
 in ion–solvent interactions, 49, 50
free energy level, of proton, 500
frog, electrical movement of its nerve, 11
Frumkin, effect of leadership in Russia on electrochemistry throughout world, 18
fuel cell
 catalyst, 1383
 first one, 1380
 future of, and corrosion, 1388
 history of, 1386, 1387

fuel cell *(cont.)*
 immediate uses of, 1396
 mission time in which useful, 1432
 practical applications of, 1396
 and production of water, 1420
 as source of drinking water, 1389
 vital importance of porous electrodes in, 1172
fuel cell heart, possible electrochemical development, 43
fuel cell research, funding of, and pollution, 1386
fuels, and electrocatalysis, 1387
Fuoss approach, to ion pairs, 261
Furth approach to work of hole formation, 536
fused oxides, as slags, 616
fused salts (*see also* molten salts)
 and activation energy for diffusion, 547
 and activation energy for viscous flow, 550
 atomistic theory of transport, 574
 diffusion in, 542
 empty space in, 523
 holes and diffusion coefficients in, 577
 heat of activation for viscous flow, 551
 internuclear distances in, 524
 radiotracer method for transport number determination, 571
 and self-diffusion, 542
 Stokes–Einstein relation in, 552
 transport, and holes, 574
 transport numbers, Stokes' law approach to, 572
 viscous forces and momentum in, 575
Galvani potential, thought experiment synthesis of, 672
Galvani potential difference, 670
 at nonpolarizable electrodes and electrocatalysis, 1149
galvanostatis rise time, 1193
galvanostatic transient, 1021, 1185
 and Butler–Volmer equation, 1190
 and double-layer charging, 1190
 equations for, 1024
galvanostatic transient method, for surface coverage, 1030
Garrett–Brattain space charge region, 812
Gaussian box, 728
Gaussian surface, 137, 151
gels, 839
Gemini fuel cells, used in space, 1388
generators, electrochemical, 1385, 1386
geology, and electrochemistry, 29
Gibbs, his angel, 679
Gibbs' surface excess, 680, 683
glasses, 603
 electrolytic structures of, 603
 as ionic liquids, 603
 liquid silicate as, 603
goals, in storers, 1420
Gouy–Chapman and Helmholtz–Perrin, relative contributions, 737, 822

Gouy-Chapman model, and potential dependence of capacitance, 732
Gouy–Chapman region, and colloidal stability, 839
Gouy–Chapman theory, and capacitance, 731
grains, 1218
greenhouse effect, 1356
Griffith crack, 1340
group dipole, 147
growth step, 1210, 1211
Guntelberg charging process, 248
happenings, thermal and electrochemical, alternative versions, 40
heart, artificial, possible electrochemical development, 43
heat of activation
 change of, with electrode potential, 924
 in electrocatalysis, of nonbonding reactions, 1149
 and exchange current density, 1150
 for flow in silicate melt, 605
 for proton transport, in aqueous solutions, 473
 and temperature, in proton mobility, 486
 for viscous flow, and melting point, 551
heat to electricity conversion, 1360
heat engine, essence of working of, 1352
heat of hydration
 of hydrogen ion, 105, 467, 468
 by quadrupole model, 101
 relative, as a function of radius, 97
 of transition metal ions, as a function of atomic number, 112
 of transition metal ions, and water stabilization energy, 113
heat of solution, 65, 67
heat of solvation, 63, 66, 88, 96
Helmholtz and Gouy capacities, in series, 736
Helmholtz–Perrin model, 718
Helmholtz–Perrin and Gouy–Chapman, relative contributions, 737
Henderson–Planck equation, solution for, 459
heterogeneity, surface, on solid electrodes, 803
Hiatus, the Great Nernstian, 16
history of double-layer treatments, 724
hole
 average size of, 539
 concept of formation of, 528
 and diffusion coefficients, in fused salts, 577
 effective mass of, 619
 formation of, in valency band, 810
 lifetime of, in fused salts, 576
 in liquid electrolyte, size of, 537
 and transport in fused salts, 574
 viscosity in terms of, 577
hole current, and electron current, 933
hole mobility, values of, 931
hole model, 527
 and compressibility, 584
 and expansivity, 585

hole model *(cont.)*
  for liquid electrolytes, further reading on, 541
  most consistent model at present, 584
  normalizing conditions, 536
  probability of finding hole of radius r, 535
  and rationalization of relation of heat of activation to melting point, 582
hole motion, and electron motion, 811
hole vacany, and electrodeposition, 1179
Hooke's law, 1331, 1334
hot emission of electrons, 941
hump
  of capacitance, 754
  experimental, in capacity–charge curve, 762
hydrated ions
  distance of closet approach to electrodes, 741
  formation from adions, 1187
hydration, 80
  calculations involving quadrupoles, 104
  effect of ligands, 109
  and orbitals, 110
  of transition metal ions, 106
  of transition metal ions, and stabilization of field, 111
hydration number(s)
  activity coefficient as function of, 240
  of alkali and halide ions, by independent methods, 118
  and compressibility, 127
  primary, table of, 131
  and radius, 130
  by various methods, 118
hydration of proton, heat of, 467
hydrazine–oxygen cells, performance tabulated, 1390
hydrazine–oxygen fuel cells, 1389
hydrocarbons
  catalysis of and cars, 1158
  cracking of hydrocarbons, reforming of hydrocarbons, 1158
  electrocatalysis of, 1158
  reformed by steam, 1390
  reforming of, 1391
  saturated, 1391
    mechanism determination of, 1107
    mechanism of oxidation of, 1107
    rate-determining step in the de-electronation of, 1110, 1158
  unsaturated, electrochemical data concerning, 1392
hydrocarbon–air cells, 1391
hydrocarbon fuels, how used?, 1351
hydrocarbon oxidation, 1158–1159
hydrodynamic flow, and convection, 1050
hydrogen
  accumulated, in cracks inside metal, 1339
  accumulation of, at regions of stress, 1333
  adsorbed, and stability of metals, 1328
  adsorption of, and change the path of crystallization, 1129
  and change of mechanical properties, 1129
  diffusion into
    metals, 1328
    regions of stress, 1330
  diffusion by interstitial and intergranular paths, 1330
  effects, and passivation, 1349
  electricity storage in, 1420
  initiation of cracks by, 1335
  and instability of metals, 1338
  kinetics of discharge of, favorable effects on storage of hydrogen, 1421
  partial molar volume of, 1331, 1332
  penetration into bulk of metal, 1329
  pressure of in metals, 1333
  storage of, cryogenics, 1421
hydrogen–air battery, 1421
hydrogen–oxygen cell, 1388, 1390
hydrogen adsorption, atomic, on electrodes, and coverage, 1246
hydrogen bond(s)
  and proton mobility, 468
  in solvation process, 90
  and clusters of water molecules, 88
hydrogen coverage, 1245
  of surface, variation with potential, 1245, 1246
  and various mechanisms, 1247
hydrogen codeposition, 1227, 1229
hydrogen desorption, and de-electronation, 1246
hydrogen embrittlement, 1314, 1338
hydrogen evolution
  adsorption step in, 1153, 1233
  atom–atom step in, 1234
  chemical desorption step in, 1234
  and corrosion, 1232
  and current efficiency, 1232
  deduction of values for transfer coefficient, 1241
  desorption in, 1153, 1234
  determination of path and rate-determining, step, 1237
  determination of transfer coefficient by various means, 1238
  and electrocatalysis, 1232
  equations for various mechanisms, 1240
  exchange current density for, tabulated, 1238
  further readings on, 1250
  general, 1231
  history of, 1231
  ion–atom recombination, 1234
  mechanisms, 1235
  on metals, equations for, 1228
  paths, 1233
  reaction paths, 1235
  recombination mechanism in, 1234
  and separation factor, values for, 1249
  tabulated summary of probable mechanisms, 1250
  and transfer coefficient, 1102

hydrogen evolution *(cont.)*
  tunneling conditions for, and electro-
    catalysis, 1154
hydrogen ion, trigonal pyramid structure of,
    466
hydrogen overpotential, experimental
    characteristics, 1232
hydronium ion, existence of, 487
  proton mobility in, 486
  and water, structure of, 76
iceberg model, for liquid silicates, 615
Ilkovic's equation, 1068
image charge, and contact adsorption, 767
image energy
  of electron, as function of distance from
    metal, 945
    its place in charge transfer kinetics,
      960
  and tunneling, 961
image forces, 661–662
  and quantum mechanical tunneling, 960
image interactions, with charged electrodes,
    660
impedance, 896
  high, and measurement of overpotential,
    897
  and measuring circuit, 896
impurity atoms, collision of electron with,
    956
impurity conduction, in silicon and
    germanium, 930
inclusions, microscopic, of copper, effect on
    reaction rate, 1272
indifferent electrolyte, 1060, 1069
induction time, in passivation, 1322
infrared spectroscopy, and electrolytic
    solutions, 278
inhibition, of corrosion, and electrodics,
    1306
inhibitor(s)
  coverage, tabulated, 1309
  adsorption, 1310
  practical examples of, tabulated, 1311
initiation, of cracks, 1335
inner potential, 673
inner potential difference
  between dissimilar phases, 677
  between two identical phases,
    measurability of, 679
instability
  of metals, 1338, 1343
  of surface, and internal decay, 1335
instantaneous pulse, 329
instruments, high impedance, need for, 896
instrumentation, for potentiostatic
    transients, 1033
interaction(s)
  ion–ion compared with ion–electrode, 723
  metal–water, 740
  minimal, between dipoles at electrodes, 785
interaction energy, and orientation of
    dipole near ion, 81
interatomic spacing, 807, 808

interface(s)
  accumulation of substances at, 679
  adsorption of ions at, 738
  affected by image forces, 662
  bird's eye view of structure, 824
  creation by thought experiment, 5
  current–potential laws at, 930
  dipole potential at, 667
  electrified
    importance of, 642
    retrospect and prospect, 715
    review, 716
    structure of, 718, 632
    thermodynamics of, 688
  electron potential near, 943
  examination in terms of transients, 1026
  exchange of electrons through moisture
    films, 6
  ionic drift to, 845
  at metals, other than mercury, 801
  metal–solution, and Volta potential, 665
  nonpolarizable, 697, 701, 894
  polarizable, 653, 700, 894
  potential differences at
    further readings on, 687
    and surface tension, 688
  profile of concentration at, in adsorption,
    681
  semiconductor–electrolyte, 803
  structure of, and Butler–Volmer equation,
    911
  thermodynamic deduction of surface
    excess equations, 700
  two-way electron traffic across, 873
  under transient conditions, 1017
  water molecules in oriented layer at, 633
interfacial tension
  and applied potential, 701
  measurement of, 688
intermediates
  adsorbed, 1016, 1029
  determination of, in benzene oxidation,
    1099
  in dissolution of iron, 1087
  and potential–time transients, 1026
  formed by charge transfer, 1027
  and propane oxidation, 1110
internal combustion engine, 1358
  wasting fuel as heat, 1352
internal resistance of fuel cells, effect upon
    current–potential curve, 1373
internuclear distance, in solid and liquid
    fused salts, 524
interphase, 2, 630
  surface excess at, 683
interphase properties, and bulk properties,
    641
interphase region, 626, 627
ion(s)
  adsorbed, on kink sites, and passivity, 1325
  association of, temporary and permanent,
    273
  clusters of, 265

hydrated, distance of closest approach to
    electrodes, 741
  in sheath of water molecules, 77
ion–atom combination, in hydrogen evolution,
    1234
ion-cloud theory, 180
ion–dipole interaction(s), 83
  equations deduced, 169
ion–dipole model, of solvent interaction, 80
ion–dipole theory, of solvation, evaluation,
    93
ion–electrode interactions, 723
ion–electrode and ion–ion interactions, 723
ion–ion interaction(s), 175
  and activity coefficients, 202
  and Debye–Hückel theory, 725
  free energy of, 181
  parentage of theory of, 273
  in perspective, 279
ion-quadrupole theory, of solvation, 103
ion-size parameter, 224, 225, 230
ion–solvent interaction, 49
  and cavities, 81
  effect on activity coefficient calculation,
    238
  and effect of quadrupole theory, 115
  equations for, 59
  experiment and theory, 94
  free energy change in, 49, 50, 57, 58
  further reading on, 116
  heat of, and thermodynamic cycle, 66
  improvement, by quadrupole model, 100
  of individual ions, 114
  quadrupole model of, 99
  summarizing remarks, 113
  thought experiments in, 80
ion–solvent–nonelectrolyte interactions, 158
ion–water interactions, quadrupole theory,
    171
ion association, 447
  and conductivity, 448
ion association constant, 257, 259
ion migration, as function of electrostatic
    potential gradient, 288, 289
ion pairs
  and activity coefficients, 260
  Bjerrum approach, 260, 263, 264, 265
  Fuoss approach, 261
  further reading on, 266
  and triple ions, 265
ion pair formation, 251, 255
ion pair fraction, 258
ion size parameter, skepticism, 280
ionic association, 251
ionic atmosphere, 428, 636
  and ionic migration, 420
  radius of, Debye–Hückel, 197, 199
  and variation with potential, 636
ionic cloud
  asymmetric, 422
  catching up with moving ions, 422
  chemical potential change arising from,
    201

ionic cloud *(cont.)*
  and Debye–Hückel constant, 200
  further reading on, 202
  relaxation of, 423
  shape of, 431
ionic drift
  interdependence of, 399
  to interface, 845
ionic fluxes, and Onsager equations, 411
ionic groups, in silicates, 596
ionic liquids, 513
  further reading on general aspects of, 522
  and glasses, 603
  lattice-oriented models for, 522
  and liquid silicates, 603
  mixture of, and complex formation, 587
  and Nernst–Einstein relation, 555
  slags, 617
ionic migration, 367, 387, 420
ionic mobility, 401
ionic movement, as function of random walk,
    299
ionic product, 499
  and semiconductors, 819
ionic radius, and contact adsorption, 744
ionic solutions, and dielectric constant, 132,
    155
ionic solvation, 155
ionic species, single
  activity coefficient of, 206
  ambiguity of measurement of properties,
    64
ionic strength, definition, 210
ionic transport, and electrochemical potential,
    395
ionics, 19, 846
  analogy to behavior of electrons in holes,
    32
  and charge transfer, 1036
  definition of, 3
  rise and fall, 24
ionization potential, and energy released
    from electronation of proton, 967
iridium, 1253, 1255
iron
  mechanisms of dissolution of, tabulated,
    1091
  potential of, as affected by solid phase,
    in corrosion, 1283
  potential–pH diagram for, 1280
irradiation, ultrasonic, and electrocatalysis,
    1170
isotherm
  for adsorption of oxygen on platinum, 1257
  contact adsorption, tests of the value of,
    770
  deduction of, for contact adsorption, 768
  for ions in the double layer, 764
  lateral repulsion model, discussion, 777
  and organic adsorption, 797
  tests for, 771, 772
isotopes
  radioactive, self-diffusion, 542

INDEX

isotopes *(cont.)*
  substitute, dependence of reaction rate on, 1106
IUPAC sign convention, in detail, 1119
journals, international, in electrochemistry, v
kilowatt hours per kilogram, 1407, 1410
kinetics, of hydrogen discharge, and storage, 1421
kinetic theory, for viscosity, 621
kink sites
  adsorbed ions on, and passivity, 1325
  diffusion of adion to, 1182
  and electrogrowth, 1203
Kirchhoff's law, and diffusion flux, 1039
Laplace transform
  for concentration perturbation, 329
  of constants, 454
  in diffusion problems, 1041
  and Fick's second law, 344
  use in polarography, 1063
  of pulse of flux, 331
Laplace transformation
  and burst of electrodissolution, 330
  definition of, 310
  explanation of, 309
  and Fick's law, 315
  initial and boundary conditions for diffusion process stimulated by constant current, 313
  use in Fick's second law, 312
  and transients, 1041
lattice approach, to concentrated solutions, 266
lattice dislocations, and spiral, 1206
lattice energy, and heat of solution, 67
lattice formation, after surface diffusion, 1195
layer growth, 1208, 1222
leveling, 1222
life, and oxygen electronation, 1252
lifetime
  of complex ions, in molten salts, 590
  of proton in solution, 485
ligands, effects on hydration heat, 109
light
  absorption of, and passivity, 1319
  polarized, application to electrode surfaces, 36
  reflected, and passivation, 1320
  theory of, electromagnetic, 35
  visible, application to electrode surfaces, 36
limiting current
  and hydrogen evolution, 1229
  practical importance of, 1059
  typical experimental values of, 1060
limiting current density, and electric migration, 1075
limiting law, breakdown of, 267
linear diffusion, time-independent, 1077
Lippman equation, 702
liquids, comparison of properties of various kinds, 519

liquid electrolytes
  cell model for, 529
  distribution function for holes in, 537
  further reading on, 533
  gas-oriented models for, 529
  hole model for, further reading, 541
  hole size in, 537
  liquid free volume model for, 530
  summary of models for, 532
  transport numbers in, 566
liquid free volume model
  for pure liquid electrolytes, 530, 531
liquid oxides, 594
liquid silicates
  dimers in, 610, 612
  and electric conduction, 606, 609
  as glasses, 603
  model of discrete anions, 610
  model of icebergs, 615
  network model of, and table, 606
  and polymerization, 611
  structure breaking in, and concentration of metal ions, 604
  ring anions in, 613
  ring formation in, 612
  as slags, 616
local cell, diagrammatic, 1271
local field, 147
  calculation of, 148
logistics, 1036
Los Angeles, and the smoggy state, 1354
Luggin capillary, 891
machining, and electrochemistry, 42, 43, 1267
macrostep(s), 1207, 1208
  and bunching, 1209
  and irregular edge, 1209
magnetohydrodynamic conversion, 1360
map of the Balkans, 1430
mass of hole, effective, 619
materials
  decay of, 4
  dependence of properties on surface, 3
materials science, 3, 5
  definition of, 5
  as an interdisciplinary field, 31
maximum, electrocapillary, 691
maximum cell potential, 1137
maximum efficiency, of electrochemical converter, 1364
maximum energy density, for electrode couples, idealized, 1408
Maxwell, his demon, 679
mean jump distance, 340
mean square distance, time, 453
measurement
  of overpotential, and high impedance, 897
  of Volta potential difference, 841, 842
mechanical properties
  change of, due to hydrogen, 1129
  and hydrogen codeposition, 1229
mechanisms
  biological, 1266

INDEX   xlix

mechanisms *(cont.)*
  of ethylene oxidation, on various metals, 1160
  of hydrogen evolution, 1235
    affected by coverage of hydrogen, 1235
    and coverage, 1247
  of the oxidation of saturated hydrocarbons, 1107
  of oxygen electronation, evaluation, 1253
  of passivation, 1319
  of porous electrode, 1384
  probable, of hydrogen evolution, 1250
  of reaction
    and Faraday's law, 1096
    and surface coverage, 1096
  on surface which change with time, 1182
mechanism determination, 1080
  and coverage of electrodes, 1097
  for ethylene, 1094
  further reading on, 1093, 1110
  and mixed potentials, 1105
  and separation factor, 1106
  on surfaces which change with time, 1182
  techniques for, 1099
mechanistic studies, summarizing remarks, 1090
medical electronics, and electrochemistry, 43
membrane
  effect on gravitational flow, 569
  Tafel relation at, 942
membrane potential, 410, 941
meniscus, 1383, 1384, 1388
mercury
  and double layer studies, 687
  and hydrogen evolution, 1232
metal(s)
  band picture of, with interatomic spacing, 808
  break-up when strained, 4
  of great strength, and electrochemistry, 43
  stretched, dissolution of, 1337
  sudden failure of, 4
metal deposition
  consecutive steps in, 1183
  further reading on, 1230
  with hydrogen, 1227
  and rise time, 1185
  transients in, 1183
metal–metal potential difference, and reaction rates, 1147
metal oxide and silicon atom, interaction in silica network, 602
metal–slag equilibrium, and charge transfer, 618
metal–slag system, as electrode–solution system, 618
metal–solution interface, and Volta potential, 665
metal–solution junction, and semiconductor, 850
metal–water interactions, 740
metallurgical extraction, and electrochemistry, 43

metallurgy, 27, 1266
  and electrochemistry, 28
methanol–water mixtures, abnormal conductivity in, 473
microrough surface, and diffusion layer thickness, 1219
microscopy, 37, 38
microspiral, 1207
microsteps, 1207, 1208, 1209
migration
  conductance in, further reading on, 367
  and diffusion, further reading on, 399
  of ions, as function of electrostatic potential gradient, 289
  ionic, 367
  and radiotracers, 570
Miller index, of metal surface, and cracking under stress, 1335
mixed potential, 1286
  determination of, 1105
  as a function of concentration, 1104
  and mechanism determination, 1105
mobile electrified interfaces, 826
mobility
  absolute, 370
  of charge carriers, in liquids and solids, 471
  conventional, 371
  and diffusion coefficient, 374, 376
  of ions, 369
  of proton, 471, 472, 476
  solvent effect on, at infinite dilutions, 443
mobility method
  in determining solvation number, 125
  for solvation number, 130
  for ionic liquids
    comparison of ability to predict, 583
    facts, 522
    further reading in, 587
    gas–oriented, 529
      lattice–oriented, 522
      summary, 532
    comparison of predictions, 485
model-oriented approach, 1
molecular biology, developments in, and electrochemistry, 42, 43
molten oxides, properties near melting point, 518
molten salts
  average size of hole in, 539
  compressibility calculated from hole model, 585
  detection of ions by Raman spectra, 590
  deviations from Nernst–Einstein relation, 557
  diffusion in, at constant temperature, 548
  distribution law for hole size, 534
  electroneutrality in, near electrode, 566
  future of model, 586
  lifetime of complex ions in, 590
  liquid free volume model, diagrammatic, 531
  models in, further reading, 587

INDEX

molten salts *(cont.)*
  nonideal behavior due to interactions in, 588
  quasi-lattice model, 527
  Schottky defects in, 527
  tests for complex ions in, 594
  vacancy model for, 526
  volume change on fusion, 525
Morse curve
  and acid–base strength, 501
  and symmetry factor, 922, 926
Morse equation
  a gross approximation, 981
  in proton conductance calculations, 477
moving ion, asymmetric cloud around, 429
multistep reactions, 991
  activation energy for, 1002
  and Butler–Volmer equation in final form, 1000
  further reading on, 1017
  involving stoichiometric number, 1005
n–p junction(s), 930
  diffusion of holes and electrons across, 932
  exponential laws for, 936
  potential difference at, 932
n–p product, 819
n-type semiconductors, 818
  band picture, 820
natural convection, 1050
natural gas, electrochemical burning, 1393 1394
negative potential, superposition of, and stabilization,1314
Nernst diffusion layer, 1056
Nernst–Einstein equation, limitations, 382
Nernst–Einstein relation, 377, 381
  and diffusion of ion pairs, 583, 586
  gross view of deviations in molten salts, 557
  and heats of activation, 556
  and ionic liquids, 555
  irreversible thermodynamic view of deviations, 558
  and paired vacancy theory of diffusion, 562
Nernstian Hiatus, 16, 23, 24
  and smog, 16
  and pollution, 17
Nernst–Planck equation, and electrolytic transport, 405
Nernst–Planck flux equation, 398
  and diffusion equation, 411
Nernst equation
  dilemmas associated with, 907
  discussion of deduction, 899
  kinetically deduced, 898
  physical significance associated with, 908
  and potential–pH diagrams, in corrosion, 1279
  sphere of relevance, 906
  as zero-current special case, 898
Nernst's relation, and alloys, 1223
nerve cell, diagram of, 937
network, and associated water, 77

network model
  for liquid silicates, 606
  defects of, 609
  table describing, 606
neutral ion pairs, diffusion of, 383
new towns, and electrochemistry, 43
Nobel Prize, given to electrochemist, 1060
nonaqueous solutions, 400
  and electricity storers, 1424
  standard electrochemical potential in, 1425
nuclear magnetic resonance spectroscopy, and electrolytic solutions, 278
nuclear power, direct conversion to electricity, 1396
nucleation
  conditions for, 1204
  and metal deposition, 1204
ohmic resistance, 997
Onsager equations, and ionic fluxes, 411, 828
optical method, of examining passivation, 1319, 1326
optimization, for energy conversion, 1374
orbitals, and hydration, 110
order, of electrodic reaction, 1008
order of reaction in electrodic reactions, 1009
  compared with order of chemical reactions, 1011
organics, effect on smoothing, 1222
organic adsorption, 791
  and electrode charge, 795
  flip-flop model for, 797
  forces in, 792
  further reading on, 800
  maximum, and potential, 798, 799
organic electrolytes, use in electrochemical electricity storers, 1413
organic reactants, and electrode kinetics, 916
organic substances, and morphology, 1222
outer Helmholtz plane, 635
  and Butler–Volmer equation, 911
  locus or reaction, 917
  position of, 757
outer potential, 673
  definition, 663
  diagrammed, 665
  and double-layer studies, 665
overall rate, and rate-determining step, 1002
overpotential, 883
  activation, 1053
  attitude toward, in pre-electrodic days, 1131
  classical picture, 1231
  concentration, 1052
  consequences of lack of understanding, 1381
  definition of, 880
  deposition, and exchange current density, 1215
  and electron queue, 993
  and electronation reaction, 1133
  and exchange current density, 896

overpotential *(cont.)*
  measurement of, and high impedance, 897
  and multistep reaction, a near-equilibrium relation, 994
  and pollution, 1420
  and self-driving cell, 1136
  social importance of knowledge of, 16
  for various steps in multistep reaction, 996
overvoltage of hydrogen, 1231
oxidation of ethylene, dependence of rate upon substrate, 1161
oxide film
  and electrocatalysis, 1258
  electronic conduction of, and passivation, 1320
  and oxygen catalysis, 1258
oxide path, in oxygen evolution, 1104
oxygen
  adsorption on noble metals, related to d-band character, 1163
  catalysis, and doped tungsten bronzes, 1169, 1422
  electronation
    and corrosion, 1252
    evaluation of mechanism, 1253
    and life, 1252
  evolution, oxide path in, 1104
oxygen pressure, and corrosion rate, 1276
oxygen reactions, further reading on, 1263
oxygen reduction
  chemical step in, 1255
  rate-determining step in, 1255
  and stoichiometric number, 1254
oxygen reduction reaction
  catalysis of, 1256
  observation of equilibrium potential, 1263
oxygen starvation, and corrosion, 1303
ozone, connected with pollution, 1353
pH, and current efficiency, 1229
pH–potential diagrams, 1120, 1278
  for iron, 1280
  for lead, 1284
  and Nernst equation, 1279
  and solution concentration, 1280
  uses and abuses, 1281, 1284
paint, corrosion through, 1302
paired vacancies
  diffusion of, 1124
  and Nernst–Einstein relation, 562
parabola
  in electrocapillary curves, 705
  in surface tension–potential relation, 691
parallel plate condenser model, surface, tension–potential relation for, 719
partial molar volume
  of hydrogen, in metal, 1331
  of hydrogen in metals, and distortion of lattices, 1334
particles and formation of holes, 578
passivation, 1315
  and absorption of light, 1319
  characteristic current–potential course, 1317

passivation *(cont.)*
  competition in models, 1324
  dissolution–precipitation, mechanism of, 1321
  electrochemical model, 1325
  and electronic conductivity of films, 1320
  ellipsometry in, 1319
  induction time in, 1322
  mechanism of, 1319
  monolayer model, 1325
  optical method of examining, 1319
  and protection from corrosion, 1348
  and reflected light, 1320
  solid state model, 1325
passivation potential, 1316, 1317
passive films, formation at the bottom of pits, 1335
passivity
  and adsorbed ions on kink sites, 1325
  competing theories of, 1325
  electrodic model of, 1326
  criteria for distinction between models, 1326
  electrodic model of, 1326
  and oxide formation, 1327
  and potential–pH diagram, 1327
  and thickness of oxide in critical region, 1325
passivity and corrosion, further reading on, 1349
paths of hydrogen evolution, 1233
Pauling equation, use in electrocatalytic theory, 1164
penetration of hydrogen, into bulk of metal, 1329, 1334
Pennsylvania, Electrochemistry Laboratory in [University of], vii
permeation
  affected by embrittlement, 1344
  as function of stress, 1331
permeation currents, and evidence for cracking, 1343
perturbation methods, 1019
phases, moving, and the double layer, 826
phase boundary, 624
  double layer at, 630
photochemical reactions, and electrical, 11
photoelectric effect, 674
photogalvanic converters, and electrochemistry, 43
photolysis, flash, 34
pipes, corrosion of, 1304, 1345
Planck's relation, 948
Planck–Henderson equation, 417, 419
platinum
  hydrogen electronation reaction on, 968
  special position in electrocatalysis, reasons for, 1166
platinum electrocatalyst, difficulties, 1158
platinum–gold, catalysis on, 1259
poisons, 1262
Poisson equation
  applied to diffuse charge region inside

Poisson equation *(cont.)*
  semiconductor, 812
  deduction of, 282
  solution of, for double-layer situation, 726
Poisson–Boltzmann equation
  linearized form of, 190, 234
  so-called rigorous solution, 247
  tests for validity of solution, further reading on, 250
polarizability, 895
  degree of, at various electrode surfaces, 802
  effect in salting out calculations, 161
  and exchange current density, 895
polarizable interface
  and interfacial thermodynamics, 700
  and lack of equilibrium, 697
polarographic wave, 1066
polarography, 1060, 1062, 1065, 1069
pollution
  of atmosphere, with products of internal combustion, 1352
  and cars, 1419
  and electrochemistry, 43
  and funding of fuel cell research, 1386
  and Nernstian Hiatus, 17
  and overpotential, 1420
  prediction of, 1357
polyanion, discrete, 610, 614
pore
  thin, high limiting current at, 1383
  wetted, and contact angle, 1383
porous electrode(s)
  activity near tip of meniscus, 1384
  and cost of electrochemical energy converters, 1385
  distribution of catalyst in, 1172
  and three-phase boundary, 1383
  use of, 1171
  vital importance in fuel cells, 1172
possible recharge cycles, as function of depth discharge, 1417
potential(s)
  absolute, impossibility of measurement, 645
  anode, 1126
  of average force, 183
  between metals, in cell, 652
  changes during increase of current, as function of type of cell, 1131
  of charged sphere, 52
  chemical, 672
    thought experiment in, 694
  of comparison, in electrocatalysis, 1145
  of corrosion, 1285
  in diffuse layer, as function of distance, 729
  electrochemical, 693, 694
  of electrochemical energy converter, its regions, 1372
  at electrodes, due to dipoles, 783
  electrostatic, variation of near interface, 664
  Flade, 1316
  at membrane, 410

potential(s) *(cont.)*
  of metals, in equilibrium with $1M$ solution of ions, 1282
  metal–solution, measurement of changes in, 650
  mixed, as a function of concentration, 1104
  of oxide formation, 1282
  of passivation, tabulated, 1317
  produced by electrochemical energy converter, and current, 1370
  reversible, of oxygen, determined, 1261
  standard, of certain electrode reactions, tabulated, 1116
  and surface tension at interface, 689
  variation of film thickness with, and passivity, 1325
  variation of, and ionic atmosphere, 636
  Volta, and potential of zero charge, 707
  of zero charge, 691
potential–pH diagram, 1278
potential–distance relation, for colloid particles, 729
potential difference
  absolute
    attempts to measure, 644
    at electrode–electrolyte interface, 670
    across single interface, 648
    structured, 660
  across barrier, 871
  across electrochemical system, 1112
  across interphase, generality, of 631
  between two electrodes, 346
  contact, 647
  of dipole, at electrode–electrolyte interface, 670
  due to electron overlap at interface, 670
  in electro-endosmotic motion, 831
  Galvani, 670
  inside semiconductors, 804
  at interface, 688, 912
  at interfaces, further reading on, 687
  measurability of outer and inner, tabulated, 677
  and n–p junctions, 933
  periodically varying, 1020
  Volta, measurement of, 841, 842
potential drop, in diffuse layer, 728
potential electrolytes, 176, 488
  dissolution of, 491
  and proton transfer reactions, 491
  and true electrolytes, difference between, 178
potential energy
  and Morse curves, 920
  of water system, during rotation, 477
potential energy–distance profile, 866
  for successive motions of ions, 868
potential energy–distance relations, and tunneling through barrier, 962
potential energy–distance theory, and Morse equation, 919
potential energy barrier, for proton transfer, 479

potential energy barrier *(cont.)*
  in acid strength theory, 505
potential energy curves, 921, 925, 971
  and effect of increased M–H bond
    strength, 1156
  in proton transfer, 969
  and stretching of bonds, 971
  vertical shift under potential change, 924
potential pulse, 1020
potential step, 1020
potential sweep method, 1033
potential of zero charge, 706
  and electrocatalysis, 1142
  and electrodeposition, 1180
  and heat of activation, for metal
    deposition, 1181
  and organic adsorption, 798
  and potential of maximum adsorption,
    tabulated, 800
  and surface potential, 707
  tabulated, 864
  and Tafel slope, 915
  in terms of Volta potential, 707
potential–time transients
  effect of intermediates on, 1026
  under diffusion control, 1046
potentiodynamic method, 1033, 1034
potentiostatic circuit, for examination of
  passivation, 1316
potentiostatic transient(s), 1032
  and adsorption of benzene on platinum,
    1033
  instrumentation for, 1033
Pourbaix diagrams, 1121, 1123, 1281
  and spreading of cracks, 1343
power, 1410
  and efficiency, in energy converters, 1379
  and energy density, 1429
power density
  feasible goals for, 1431
  as function of energy density, for
    silver–zinc cell, 1411
  and lithium–chlorine cell, 1423
power density *versus* energy density, 1411
power output, 1171, 1376
power stations, and fuel cells, 1396
powering of vehicles, and electrochemistry, 43
precipitation, of materials, near electrode,
  1322
precursor film, 1319
pre-electrolysis, purification by, 1261
prepassive film, 1319
pressure
  critical, for spreading of crack, 1341
  of hydrogen in metals, 1333
    high, 1333
primary hydration number
  of proton, 469
  table of, 131
primary solvation, 79, 160
probability
  of electron transfer, 978
  of finding one ion near another, 253

probability *(cont.)*
  that hole has radius r, 538
  of tunneling, on both sides of the barrier,
    952
probability amplitude, and electron passage,
  947
projection, 1266
  and concentrated electric field at, 1217
propane oxidation
  diagnostic coefficients, 1107
  and intermediates, 1110
  possible rate determining steps, 1109
protection
  anodic, diagrammatic, 1318
  cathodic, 1309, 1312
    theory of, 1309
  surface, 1266
protein, quantum mechanical tunneling to,
  981
proton(s), 461
  affinity, 466
  conductance and Morse equation, 477
  chain mechanism, 474
  constitution of, further reading on, 470
  free energy level, 500, 503
  heat of hydration of, absolute, 105
  hydrated, and electron tunneling to, 962
  jumping, from water, 475
  lifetime in solution, 485
  mobility of
    abnormally high, 472, 485
    in aqueous solutions, model, 484
    further reading on, 488
    in ice, 486
    rate-determining step in, 485
    and water rotation, detailed scheme, 483
  as nonconforming ion, 461
  in solution, existence of, 465
  solvation, 462
proton jumps
  and classical laws?, 478
  quantum-mechanical, 480
  water re-orientation necessary for, 481
proton transfer
  in autodissociation, 492
  conditions for, 968
  between hydroxonium ions and favorably
    oriented water molecules, 478
  and potential electrolytes, 491
  potential energy barriers for, 505, 479
  and tunneling conditions, 966
proton transport, 470, 473
pseudocapacitance, equivalent circuit for,
  1028, 1029
pseudo-equilibrium, 1236
pseudoforce, and diffusion, 306
pulse
  double, and staircase pulsing, 1019
  square wave, 1019
  with step reversal, 1019
pulse generator, 329
pulsing, and double pulse, 1019
pure liquid electrolytes, 513, 518

liv   INDEX

purification, 1260, 1261, 1263
purity, effect on corrosion, 1273
quadrupole model, 99, 100, 101, 104
quantum electrochemistry, in 1960's, 19
quantum-mechanical charge transfer,
    relation to proton transfer, 974
quantum-mechanical proton jumps, 480
quantum-mechanical theory, or radiationless
    tunneling, 960
quantum-mechanical transfer, basic condition
    for, 945
quantum mechanics, 27, 41
  deduction of Butler–Volmer equation in,
    980
  desirable refinements in electrochemical
    application to, 981
  of electrode processes, 947
  of electrode processes, further reading
    on, 985
  of electrode reactions, history of
    developments, 983
  and electrodics, 30
  symmetry factor theory in, 974
  tunneling conditions for, 972
  of tunneling to proteins, 981
quantum nature of biological reactions, 21
quantum nature of electrochemical reactions,
    21
quasi-lattice theory, 271, 272
queues, 992
radicals
  examination of, 1034
  intermediate detection of, 34
  on surface, nature of, and mechanism, 1105
radiotracer method
  and electrochemical method, 1035
  for surface concentration, 1030
radiotracers
  and determination of transport numbers
    in fused salts, 570, 571
  and electrochemical methods, comple-
    mentary nature of, 1098
Raman scattering, 277
Raman spectra
  and detection of complex ions in molten
    salts, 590, 591
  and electrolytic solutions, 277
ramp, various types, 1019
random walk
  in deposition, 1186
  distances moved and time, 306
  influence of electric field on, 350
  ionic movement as a function of, 299
  sum of distance from origin, squared, 303
  and surface adions, in electrodeposition,
    1181
rate(s)
  of deposition, on different planes, theory
    of, 1215
  as a function of substrate, for oxidation
    of ethylene, 1161
rate-determining step, 1138
  and activation barrier, 1003

rate-determining step *(cont.)*
  and charge transfer, 1185
  with chemical step, 1002
  concept of, 997
  in context, 1138
  in corrosion processes, 1296
  deduction of in terms of barrier height,
    1003
  and energy barrier for multistep reaction,
    1002
  in ethylene oxidation, 1160
  and exchange current density of partial
    reactions, 997
  and highest standard free energy, 1003
  and hydrogen evolution, 1237
  in hydrocarbon oxidation, 1159
  and iron deposition and dissolution, 1084
  in some metal depositions, 1195
  in oxygen reduction, 1255
  and path, 1139
  in propane oxidation, 1109
  in proton motion, 485
rate constant
  for adion diffusion, as function of
    dislocation density, 1201
  determined from transient measurements,
    for surface diffusion, 1196
  of non-rate-determining reaction, first
    determination of, 1195
rate equations, for electrocatalysis, compared
    with catalysis, 1143
rate processes, 340, 341, 342, 387, 391,
    455, 867
ratio of currents, in alloy deposition, 1226
rationality, of burning fossil fuels, lack
    of, 1352
Rayleigh scattering, 277
reaction(s)
  chemical and electrochemical, 8, 40
  coupled, 1235
  on electrodes, several simultaneous, 1274
  electron transfer, in 1950's, 17
  multistep, 991
  photo- and electro-, 11
  thermal, and electrochemical, 14
reaction coordinate, 456
reaction mechanism
  determination of, 1080
  elucidation of its stages, 1095
  and solution entities, 1095
  and surface coverage, 1096
reaction order, 1012, 1013, 1099, 1241
reaction paths, for oxygen electronation,
    1255
reaction rate
  its dependence on isotope substitution, 1105
  electrochemical, and relation to
    metal–metal p.d., 1147
reactivity, at low temperatures, in
    electrocatalysis, 1169
reciprocity relation, 413
recovery of materials, and electrochemistry,
    43

rectification, faradaic, 885
redox reactions, 982, 1046, 1146
reference potential, for electrocatalysis, 1143
reforming, of hydrocarbons, 1158, 1391
relations in cells, further reading on, 1137
relative strength of acid, 495
relative strength of base, 495
relative potential differences, 655, 656
relaxation, of ionic cloud, 423
relaxation methods, and various types of stimuli, 1019
resistance, 996, 997
response
  in concentration, to switch-on of current flux, 317
  of dielectric medium to field, 145
  of system to outside stimulus, 325
reversibility, microscopic, 874
ring anions, in liquid silicates, 613
rise time, 1185–1194
rotating disk electrode, 1058
  development of, 1070
rotation
  potential energy of water system during, 477
  of screw dislocation, schematic, 1206
  of a spiral, in electrodeposition, 1205
  of water and proton mobility: detailed scheme, 483
sacrificial anode, 1312
salting in, 159
  anomalous theory, 164
  normal, 163
salting, out, 158
  further reading on, 168
saturated hydrocarbons, 1107, 1391
scattering, Rayleigh and Raman, 277
scavenger electrolysis, theory of, 1262
Schottky defects, 526
Schrödinger equation, 947
science, advances in, and electrochemistry, 39, 41
screw-thread, and spiral, analogy, 1207
screw dislocations, 1203, 1206, 1210, 1211, 1212
secondary solvation, 79
  effect on solubility, 161
self-diffusion
  coefficients, of various substances, 545
  energies of activation for, 579
  experiments, results of, 544
  experiments with radioactive isotopes, 542
self-driven cell, 1128
self-driving cell, 1131, 1136
semiconductors, 806
  analogy to electrolytes in solution, 32
  conduction of, 809
  and Debye–Hückel length, 816
  doped, 819
  and electrolytic solutions, 811
    analogies tabulated, 812
  impurity additions to, 818

semiconductors *(cont.)*
  intrinsic, 910
  and metal–solution junction, 850
  and tunneling, 956
  values of mobilities in, 931
semiconductor–electrolyte interfaces, 803
semiconductor electrochemistry
  further reading on, 823
  and nonmetals, 823
semiconductor junctions, 930
separation, of isotopes, and hydrogen evolution, 1246
separation factor, 1106, 1231, 1248, 1249
separators, and dendrites, 1221
servicing center, 992
sewage disposal and electrochemistry, 43
SHE and IUPAC convention, 1117
signs, of electrode potentials, 1120
sign convention
  in electrochemistry, 1115
  IUPAC, in detail, 1119
silica, fused, and liquid water, 596
silicates
  liquid, and transport processes in simple fused salts, 609
  structure of ions of, 596
  liquid network model of, 606
silicate melt, heat of activation for flow in, 605
silver–zinc cell, 1411, 1416, 1417
single crystals
  Bridgman technique for, 1218
  deposition on, 1218
single ions, 61, 62
sink, for electrons, 853
sinusoidal stimulus, rectified, 1019
slags, 616–618
slow discharge, 1153, 1236
smog
  its formation, 1354
  and Nernstian Hiatus, 16
smoothing, 1222
sodium–sulfur cell, 1417, 1418
sols, 839
solid metals, and mercury, in double-layer studies, 802
solubility, 160, 163
solution(s)
  agitation of, and corrosion, 1300
  entities in, and reaction mechanism, 1095
  ideal, and activity coefficient, 205
  ionic, quasi-lattice approach to, 266
  ionic, standard state in, 204
  visible and ultraviolet absorption spectroscopy in study of, 275
solvated ions, at interface, 633
solvation, 80
  effect on activity coefficient, calculation of 241
  evaluation of ion-dipole of, 93
  as function of orientation time of water molecule, 122
  heats of, for pairs of ions, 63

solvation *(cont.)*
  ion-quadrupole theory of, 103
  primary, 79, 160
  of protons, 462, 469, 470
  secondary, 79
  as time-dependent phenomenon, 121
  of transition metal ions, 106
solvation effects, on activity coefficients, consistency of theory?, 243
solvation number,
  and compressibility, 125, 126
  dynamic model, 124
  further reading on, 132
  and hopping, 122
  and mobility method of determination, 125
  primary, explanation of, 119
  and theory of activity, 244
  ultrasonic method of determination of, 25
  usefulness of concept, 124
solvation theory, electrostatic, further reading on, 72
solvent, its effect upon the strength of an acid, 504
solvent effect(s)
  on activity coefficient, theory of, 242
  on mobility, at infinite dilutions, 443
solvent levels, and strength of acids, 506
solvent properties, fused nonmetallic oxides, 602
solvent sheath, structure of water in, 78
source, for electrons, 853
space charge, 813
  capacitance associated with, 816
  inside semiconductor, 815
  in ionic solution, 217
species, in solution, and mechanism determination
specific adsorption, 748
  and constant adsorption, 638
specific conductivity
  of aqueous and nonaqueous solutions, 446
  and ionic mobility, 372
spectra, Raman, and electrolytic solutions, 277
spectroscopy, 27
  ellipsometric, 37
  infrared, and electrolytic solutions, 278
  nuclear magnetic resonance, and electrolytic solutions, 278
  Raman, and study of solutions, 276
  in study of electrolytic solutions, 272
  visible and ultraviolet absorption, in study of solutions, 275
sphere, 50, 52, 55
spherical cavity
  field in, 148
  with reference dipole, 148
spherical diffusion, 1062
  and growth at surface, 1219
spiral, formation of, from screw dislocation, 1206
spiral growth, 1203
spiral tip, concentration at, 1220

square wave pulse, 1019
stability
  electrodic principles of, 1348
  of interior of metal, and surface hydrogen, 1328
  of metals, 1268
    and adsorbed hydrogen, 1328
  electrodic approach to increase, 1306
  of surface and charge transfer, 1268
stabilization of materials, and electrochemistry, 42, 1314, 1318, 1323
standard electrochemical potentials, in nonaqueous solutions, 1425
standard electrode potentials, table of, problems, 1115
standard free energy of activation, for multistep reactions, 1002
standard free energy of intermediate states, and water coverage, 1016
standard hydrogen electrode, 1011
steady state
  and concentration of intermediates, 1038
  and convection, 1052
  equality of velocity for all steps in, 1038
  and equilibrium, 1018
  and Fick's first law, 343
  and galvanostatic transients, 1025
  how long to establish?, 1018
steam hydrocarbon process, 1390, 1391
steel, and cathodic protection, 1312
step function flux, as function on time, 331
step site, surface diffusion to, 1182
Stern model, 733, 734, 735
stimulation, of interface, to show variation, 1019
stimuli, 1019, 1020, 1021
stoichiometric number, 1004
  and Butler–Volmer equation, 1006
  determination of, 1101
  and hydrogen evolution, 1241
  and multistep reactions, 1005
  and oxygen reduction, 1254
  and symmetry factor, 1007
  tabulated, for hydrogen evolution, 1245
  and transfer coefficient, 1007
Stokes–Einstein relation, 377, 379, 380, 551, 552
Stokes' law
  and proton mobility, 471
  and transport numbers in fused salts, tests, 573
Stokes' law approach, to transport numbers in fused salts, 572
Stokes mobility, 381
storage
  of charge, 703
  of electricity, 40, 1266, 1421
strain, in metals, caused by hydrogen, 1333
streaming current
  and diffuse layer, 830
  relative motion of phases in, 827, 829
streaming current density, equation relating to zeta potential, 831

streaming potential, and specific conductivity, 828
strength
    of acid, absolute, 494
    of acid, and solvent effect, 504
    of base, 494
stress, 1330, 1331
stress corrosion, examples, tabulated, 1338
stress corrosion cracking, 1335, 1336, 1337, 1349
stretching, of bond, in ions at electrodes, 920
structure breaking in liquid silicates, and concentration of metal ions, 604
structure of charged interfaces, birds'-eye view, 824
substance producer, 854, 1127
    as electrochemical device, 851
summarizing remarks, on mechanistic studies, 1090
summary
    charge transfer theory, 893
    of corrosion mechanisms, 1345
    of criteria in good electrochemical energy storage, 1412
    of direct energy conversion, 1398
    of electrochemical electricity storers, 1412
    of electrochemical energy storage, 1430
superequivalent adsorption, 638, 748
superimposition of negative potential, and stabilization, 1314
superposition, of potential of ion and cloud, 199
surface(s)
    and civilization, 1267
    decoration of, 1266
    electrification of, consequences, 7
    electrode, state of, 36
    of metals
        and cracks by hydrogen, 1333
        and instability, 1328
        protection of, 1266
        stability of, and charge transfer, 1268
    and time change, in respect to electrodeposition, 1182
surface adion(s)
    deposition from, 1187
    in random walk, 1181
surface coverage
    during benzene oxidation, 1098
    and Butler–Volmer equation, 1014
    and charge transfer reaction, 1014
    by galvanostatic transients, 1030
    and mechanism of reaction, 1096
surface diffusion
    of adions, in electrodeposition, 1177
    contribution to overpotential, in steady state, 1199
    dependence upon time, 1194
    equations for, 1188
    to kink site, 1182
    and lattice formation, 1195
    to a step site, 1182
    and stepwise dehydration, 1177

surface diffusion parameter, in metal deposition, 1202
surface excess
    and amount adsorbed, 684
    determination, 710
    and distribution of species in the interface region, 683
    electrocapillary equations for, 710
    of individual species, and surface tension, 711
    as macroscopic concept, 684
    and radiotracer measurements, 686
surface potential, 667, 673
    across interface, 677
    between two wires in cell, 675
    measurement of change of, 674
    origin of, 669
    and potential of zero charge, 707
surface state
    and contact adsorption, 821
    and Gouy–Chapman diffuse charge, 822
    in semiconductors, and contact–adsorbed ions, 822
surface tension
    and capacitance, 705
    at interface, and potential, 689
    and potential difference at interface, 688
    and surface excess, 711
    and variation with electrolyte concentration, 693
surface tension–potential relation, from parallel plate model, 719
sweep, triangular, 1019
Swiss cheese, and hole model, 527
symmetry, dependence on hill-shaped potential barrier, 936
symmetry factor, 923
    basic condition for the presence of, 977
    and biological situations, 941
    deduction in terms of potential energy curves, 925
    dependence on potential, 926
    elementary theory, 923
    first model, 871
    and flow of chemical to electrical energy, 1138
    and Morse curves, 922, 923, 926
    potential dependence as a function of of exchange current density, 927
    in quantum mechanics, third model, 674, 977
    second model, 922
    and stoichiometric number, 1007
    and transfer coefficient, 1007
synthesis, 40
Tafel's law, 15
Tafel lines
    for alloy deposition, 1225
    for oxygen reduction on doped bronzes, 1422
Tafel relation
    effect of diffuse layer on, 915

Tafel relation *(cont.)*
  and electrochemical energy conversion, 1370
  at membranes, 942
Tafel slope
  effect on corrosion, 1294
  and potential of zero charge, 915
  and reaction order, 1090
tanks, stabilization of, 1318
teflon, use in porous electrodes, 1391
temperature
  of atmosphere, as function of time, 1356
  and dielectric constant, 60
terminology, of electrodes, historical, 1126
terminology, in energy conversion and storage, 1402
test charge, and definition of Galvani potential, 672
textbooks, electrochemical in English, absence of, v
thermal combustion, and waste of energy, 1357
thermionic emission, 953
thermionic energy converter, 1359
thermionic work function, and electrocatalysis, 1147, 1150
thermoelectric energy converter, 1359
thickness of atmosphere, variation of, with concentration, 198
Thompson's hypothesis, 416
thought experiments
  at charged interface, and surface potential, 668
  and definition of chemical potential, 694
  with image forces, 661
three-electrode system, 891
three-phase boundary, at porous electrode, 1383
  importance of, 1384
transients, 32, 1020
  basic equations for, 1027
  cell for measurement, 1184
  and equivalent circuits, 1026
  galvanostatic transients, 1021, 1185
    arrangement for, 1022
    and Butler–Volmer equation, 1190
    and determination for surface radical concentration, 1197
    and double-layer charging, 1190
    solution of equations for, 1191
    and steady state, 1025
  at interfaces, 1017
  and Laplace transformation, 1041
  in metal deposition, 1183
  potential–time
    under diffusion control, 1046
    effect of intermediates on, 1026
  potentiostatic, 1032
    instrumentation for, 1033
transient behavior, and equivalent circuit, 1026
transient methods, 1019
transient state, and Fick's second law, 344

transient techniques, and prospects for electrodic research, 34
thrombosis, electrochemical mechanism of, 840
transfer
  of hydrated ion, to hole site, 1180
  mass transfer and heat transfer, analogies between, tabulated, 1041
  representation of, for ion to kink or edge vacancy, 1180
transfer coefficient
  and Butler–Volmer equation, 1007
  for cathodic and anodic reactions, 1007
  as center of electrode kinetics, 918
  determination of, 1100
  and hydrogen evolution mechanisms, tabulated, 1235
  and stoichiometric number, 1007
  and symmetry factor, 1007
  tabulated summary of values, 1241
  and Tafel plots, 1103
  for various mechanisms, 1007
transition(s)
  electronic, adiabadic, 974, 977
  radiationless, and conservation of momentum, 950
transition metal ions
  heat of hydration and atomic number, 108
  hydration of, and stabilization of field, 106, 111
  interaction with water, 106
transition time, 1043
  and concentration, 1048
  concept of, 1047
  and convection, 1052
  relation to potential, 1048
  and charge transfer, 1036
  to electrode, effect on kinetics, 916
  control, further reading on, 1079
  forces in phenomenological treatment of, 294
  in fused salts, atomistic theory, 574
  in fused salts, further reading on, 573
  of proton, 470
  in solution, work done in, 292
transport aspects of electrodics, summary, 1076
transport numbers, 400, 401, 402, 403, 406, 415, 418, 565, 566, 568, 571, 572
transport phenomena, 541
transport theory, and history, 1042
transport, and electrochemistry, 43
triangular sweep, 1019
triple ions, 265
  and nonaqueous solutions, 449
triple layer, 750
tritium
  use in adsorption measurements, 1030
  use in mechanism determination, 1106
tungsten bronzes, doped, and oxygen catalysis, 1422
tunneling, 952

INDEX lix

tunneling *(cont.)*
  conditions for, effect of adsorption
    energy on, 965
  condition for, in hydrogen evolution, 1154
  to cytochromes, and enzymes as electrodes,
    1253
  and electrode reactions, 955
  and electrodeposition of metals, 1177
  of electron, 946
    conditions for, 972
    and de-electronation reaction, 973
    through barrier, probability of, 952
  and Esaki tunnel diode, 956
  and Fermi level, 964
  forbidden conditions for, 972
  to hydrated proton, 962
  and image energy, 961
  and Morse equation, 981
  need for acceptor particles in, 977
  probability of, and width of barrier, 954
  of protons, 1155
    in ice, 487
    in water, 487
  quantum-mechanical, and biological
    membranes, 981
  radiationless, 961
  and semiconductors, 956
  simultaneous, of two electrons,
    unlikelihood of, 1082
  work function in, 963
turbulence, 1055
ultrasonics, and hydration number, 128
ultrasonic irradiation, and electrocatalysis,
  1170
ultraviolet and visible adsorption
    spectroscopy, in study of solutions, 275
undevice, 859, 860
unsaturated hydrocarbons, electrochemical
  data concerning, 1392
urban living, and electrochemistry, 42
vacancies, types of, 1179
vacancy model for fused salts, 526
vacuum, work function in, 963
virial, definition of, 237
viscosity
  kinetic theory expression for, 621
  of molten salts, 547
visible and ultraviolet absorption
    spectroscopy, in study of solutions,
    275
volcano relation, in electro-organic chemistry,
    1164
volcano relation, interpretation of, in
    electrocatalysis, 1165
Volta potential
  measurability of, 666
  and metal–solution interface, 665
Volta potential difference, theory of
    measurement, 841
volume change on fusion, molten salts, 525
Wagner–Traud mechanism for corrosion,
  1274

waiting lines, 992
water
  adsorption of, and Butler–Volmer
    equation, 1015
  bound near to ion, and deviations from
    ideality, 238
  constitution of, in vapor, 464
  de-electronation reaction, and ethylene
    oxidation, 1160
  de-ionization of, thermal and
    electrochemical, 40
  as dielectric, 132
  dielectric constant, and association of
    dipoles, 154
  dielectric unit in, 146
  on electrodes, flip-flop, 779
  immobilized near ion, 77
  interaction, with transition metal ions, 106
  a new form, 76
  orbitals around, 73
  potable, from fuel cells, 1389
  primary region of solvation, 79
  production of, and fuel cells, 1420
  as quadrupole, 98
  in secondary region of solvation, 79
  and silica, 595
  as solvent, 440
  structure of, 72, 80
  theory of its dielectric constant, 1461
water coverage, possible effect on
    intermediate concentration, 1016
water dipoles
  at electrodes, and capacitance, 788
  flip-flop model, 741, 780
water molecules, adsorption on electrodes,
  739
  condensation of, in solvation calculation,
    89
  orientation of, at electrodes, 741
water purification, and electrochemistry, 42
water reorientation, 482
  and proton mobility, 482
wave form, sinusoidal, 1052
wave function, of electron, near interface,
  670
wavelength, de Broglie, 949
work
  chemical and electrical, separation of, 695
  of transfer, of ion from vacuum to
    solvent, 50
work done
  in transport in solution, 292
  in transport of unit charge from infinity
    to interior of phase, 694
work function
  cancellation of, in electrocatalysis, 1147
  and electrocatalysis, 1146
  and exchange current density, 1149
  influence of concentration of reactants
    in double layer, 1150
  influence upon rate of nonbonding
    electrochemical reaction, 1149
  of metals, tabulated, 944

work function *(cont.)*
   in tunneling, 963
   and zero charge potential, 1151
water discharge, in ethylene oxidation, 1160
water line, corrosion at, 1303
X-rays, and constitution of protons in solution, 462
Young's modulus, 1340
zero charge, potential of, 691, 706
   and electrocatalysis, 1142
zero charge potential, relation to work function on, 1151
zero charge situation, 863
zero current, and Nernst's law, 897
zeta potential
   and clotting of blood, 841
   and dependence on concentration, 913
   and electrode kinetics, 912
   relation to concentration, 914
   and streaming current, 831
zinc, and air electrode, storage, 1427
zinc–air cell, and storage, 1428
zinc-containing cells, and dendritic growth, 1428